T0202779

Springer Textbooks in Earth Sciences, Geography and Environment

The Springer Textbooks series publishes a broad portfolio of textbooks on Earth Sciences, Geography and Environmental Science. Springer textbooks provide comprehensive introductions as well as in-depth knowledge for advanced studies. A clear, reader-friendly layout and features such as end-of-chapter summaries, work examples, exercises, and glossaries help the reader to access the subject. Springer textbooks are essential for students, researchers and applied scientists.

More information about this series at http://www.springer.com/series/15201

Jibamitra Ganguly

Thermodynamics in Earth and Planetary Sciences

Second Edition

 Springer

Jibamitra Ganguly
Department of Geosciences
University of Arizona
Tucson, AZ, USA

ISSN 2510-1307   ISSN 2510-1315 (electronic)
Springer Textbooks in Earth Sciences, Geography and Environment
ISBN 978-3-030-20881-3   ISBN 978-3-030-20879-0 (eBook)
https://doi.org/10.1007/978-3-030-20879-0

This Springer imprint is published by the registered company Springer Nature Switzerland AG
The registered company address is: Gewerbestrasse 11, 6330 Cham, Switzerland

A theory is the more impressive the greater the simplicity of its premises, the more different kind of things it relates, and the more extended its area of applicability. Therefore the deep impression that classical thermodynamics made upon me. It is the only physical theory of universal content which I am convinced will never be overthrown, within the framework of applicability of its basic concepts.

Albert Einstein

*Dedicated to the Pioneers who led the way
and Students, Colleagues and Mentors
who helped me along the way*

Preface to the Second Edition

The first edition of the book was written when I had to carry a full load of teaching, maintain well-funded research programs and carry out many other activities that are typically expected of someone holding an academic position in USA. There was hardly any free time to write a book of this nature, and consequently writing of the book became a stressful undertaking. This, however, is a common situation with active scientists who write books. After my retirement a few years ago that allowed me considerable freedom of how I use my time, I felt that I should work on the book again for a second edition that would give me the opportunity to correct many typographical errors that I have spotted in the meanwhile (these were corrected in the Chinese translation that was published in 2015), improve the clarity of presentation in several places and add new materials. In the last category, I have added a new chapter on Statistical Thermodynamics and also significant amount of new materials in many of the existing chapters; most of these additions are in Chaps. 4 – 8, 10, 12, 13 and Appendices A and C. Additionally, there is a new Appendix (D) containing solutions of selected problems that are marked by asterisks in the chapters. Answers and hints for solutions have been provided for the problems for which solutions are not included in this Appendix. As in the first edition, I have inserted the problems into appropriate places within the text that the problems relate to instead of following the usual practice of collecting them at the end of each chapter.

I am thankful to Dr. Manga Venkateswara Rao for helpful discussions and reviews of selected chapters.

In the Introduction of their ground breaking work on paleothermometry using Statistical Thermodynamics in 1951, Harold Urey and co-workers remarked:

> *Geologists have drawn many conclusions from the purely qualitative evidence of geological studies in regard to the past climatic conditions on the earth. These deductions are based upon a great variety of evidence, and the ability of the geologists to deduce as much as they have in regard to these conditions excites the wonder and admiration of all the uninitiated who examine their work even casually.*

Although this statement is about the geologists' early contribution to paleoclimate studies, it also applies, at least in my judgment, to geologists', and in a broader sense Earth scientists' contributions to a variety of large-scale problems. However, as in the paleoclimate studies, one of the most important developments in Geological and related aspects of Planetary sciences has been the integration of quantitative analysis using thermodynamics with the "qualitative evidence" that the Geologists are used to dealing with and their ability to extract major new insights from that to address large-scale natural processes.

Hopefully, the extent and scope of application of thermodynamics to the quantitative analysis of complex natural processes would continue to grow.

Tucson, AZ, USA   Jibamitra Ganguly
March 2019

Preface to the First Edition

When the knowledge is weak and the situation is complicated, thermodynamic relations are really the most powerful

Richard Feynman

Thermodynamics has played a major role in improving our understanding of natural processes and would continue to do so for the foreseeable future. In fact, a course in thermodynamics has now become a part of Geosciences curriculum in many Institutions despite the fact that a formal thermodynamics course is taught in every other department of physical sciences, and also in departments of Chemical Engineering, Materials Sciences, and Biological Sciences. The reason thermodynamics is taught in a variety of departments, probably more so than any other subject, is that its principles have wide ranging applications but the teaching of thermodynamics also needs special focus depending on the problems in a particular field.

There are numerous books in thermodynamics that have usually been written with particular focus to the problems in the traditional fields of Chemistry, Physics, and Engineering. In recent years, several books have also been written that emphasized applications to Geological problems. Thus, one may wonder why there is yet another book in thermodynamics. The primary focus of the books that have been written with Geosciences audience in mind has been chemical thermodynamics or Geochemical thermodynamics. Along with expositions of fundamental principles of thermodynamics, I have tried to address a wide range of problems relating to geochemistry, petrology, mineralogy, geophysics, and planetary sciences. It is not a fully comprehensive effort but is a major attempt to develop a core material that should be of interest to people with different specialties in the Earth and Planetary Sciences.

The conditions of the systems in the Earth and Planetary Sciences to which thermodynamics have been applied cover a very large range in pressure-temperature space. For example, the P-T conditions for the processes at the Earth's surface are 1 bar, 25 °C, whereas those for the processes in the deep interior of the Earth are at pressures of the order of 10^6 bars and temperatures of the order of 10^3 °C. The pressures for processes in the solar nebula are 10^{-3}–10^{-4} bars. The extreme

range of conditions encompassed by natural processes requires variety of manipulations and approximations that are not readily available in the standard textbooks on
thermodynamics. Earth scientists have made significant contributions in these areas
that have been overlooked in the standard texts since the expected audience of these
texts rarely deal with the conditions that Earth scientists have to. I have tried to
highlight the contributions of Earth scientists that have made possible meaningful
applications of thermodynamics to natural problems.

In order to develop a proper appreciation of thermodynamic laws and thermodynamic properties of matter, it is useful to look into their physical picture by
relating them to the microscopic descriptions. Furthermore, in geological problems,
it is often necessary to extrapolate thermodynamic properties of matter way beyond
the conditions at which these have been measured, and also to be able to estimate
thermodynamic properties because of lack of adequate data to address a specific
problem at hand. These efforts require an understanding of the physical or microscopic basis of thermodynamic properties. Thus, I have occasionally digressed to
the discussion of thermodynamics from microscopic viewpoints, although the
formal aspects of the subject of thermodynamics can be completely developed
without appealing to the microscopic picture. On the other hand, I have not spent
too much effort to discuss how the thermodynamic laws were developed, as there
are many excellent books dealing with these topics, but rather focused on exploring
the implications of these laws after discussing their essential contents. In several
cases, however, I have chosen to provide the derivations of equations in considerable detail in order to convey a feeling of how thermodynamic relations are
manipulated to derive practically useful relations.

This book has been an outgrowth of a course on thermodynamics that I have
been teaching to graduate students of Earth and Planetary Sciences at the University
of Arizona for over a decade. In this course, I have meshed the development of the
fundamental principles with applications, mostly to natural problems. This may not
be the most logical way of presenting the subject, but I have found it to be an
effective way to keep the interest of the students alive, and answer "why am I doing
this?" In addition, I have put problems within the text in appropriate places, and in
many cases posed the derivation of some standard equations as problems, with hints
wherever I felt necessary based on the questions that I have received from my
students when they were given these problems to solve.

I have tried to write this book in a self-contained way, as much as possible. Thus,
the introductory chapter contains concepts from mechanics and quantum chemistry
that were used later to develop concepts of thermodynamics and an understanding
of some of their microscopic basis. The Appendix B contains a summary of some
of the mathematical concepts and tools that are commonly used in classical
thermodynamics.

Selected sections of the book have been reviewed by a number of colleagues:
Sumit Chakraborty, Weji Cheng, Jamie Connolly, Mike Drake, Charles Geiger,
Ralph Kretz, Luigi Marini, Denis Norton, Giulio Ottonello, Kevin Righter,
Surendra Saxena, Rishi Narain Singh, Max Tirone, and Krishna Vemulapalli.

I gratefully acknowledge their help but take full responsibility for the errors that might still be present. In addition, feedbacks from the graduate students, who took my thermodynamics course, have played an important role in improving the clarity of presentation, and catching errors, not all of which were typographical. I will be grateful if the readers draw my attention to errors, typographical or otherwise, that might have still persisted. All errors will be posted on my web page that can be accessed using the link http://www.geo.arizona.edu/web/Ganguly/JG_page.html.

I started writing the book seriously while I was in the Bayerisches Geoinstitüt, Bayreuth, and University of Bochum, both in Germany, during my sabbatical leave in 2002–2003 that was generously supported by the Alexander von Humboldt Foundation through a research prize (forschungspreis). I gratefully acknowledge the support of the AvH foundation, and the hospitality of the two institutions, especially those of the hosts, Profs. Dave Rubie and Sumit Chakraborty. Research grants from the NASA Cosmochemistry program to investigate thermodynamic and kinetic problems in the planetary systems provided significant incentives to explore planetary problems, and also made my continued involvement in thermodynamics through the period of writing this book easier from a practical standpoint. I am also very grateful for these supports.

I hope that this book would be at least partly successful in accomplishing its goal of presenting the subject of thermodynamics in a way that shows its power in the development of quantitative understanding of a wide variety of geological and planetary processes.

And finally, as remarked by the noted thermodynamicist, Kenneth Denbigh (1955)

Thermodynamics is a subject which needs to be studied not once but several times over at advancing levels

October 2007   Jibamitra Ganguly
Tucson, Arizona, USA

Contents

Commonly Used Symbols

(Usual meanings, unless specified otherwise)

a_i^α or $a(i)^\alpha$	Activity of a component i in a phase α
C_p & C_P'	Isobaric-molar and isobaric-specific heat capacity, respectively
C_v	Heat capacity at constant volume
$D_i^{\alpha/\beta}$	Partition coefficient of i between the phases α and β $D_i^{\alpha/\beta} = (X_i^\alpha / X_i^\beta)$
dZ	An exact differential or total derivative of Z
δY	An inexact differential
∂X	A partial derivative of X
f_i^α, $f_i(\alpha)$	Fugacity of a component i in a phase α
f	Reduced partition function ratio (RPFR)
F	Helmholtz free energy
F'	Faraday constant
G, G_m	Total and molar Gibbs free energy
G_i^*, G_i^o	Gibbs free energy of i in the standard (*) and pure (^{o}) state, respectively
ΔG^{mix}, ΔG^{xs}	Gibbs free energy of mixing and excess (xs) Gibbs free energy mixing of a solution, respectively
$\Delta_r G$, $\Delta_r G^*$	Gibbs energy change and standard state (*) Gibbs energy change of a reaction
$\Delta G_{f,e}$, $\Delta G_{f,o}$	Gibbs free energy of formation of a compound from the constituent elements and oxides, respectively
G_f^x	Gibbs free energy of formation of a solute ion in a hypothetical "solute standard state" at unit molality in an electrolyte solution (Chap. 12.4.1)
g_i	Partial molar Gibbs free energy of the component i in a solution
g	Acceleration due to gravity
H, H_m	Total and molar enthalpy of a solution, respectively
H_i^*, H_i^o	Enthalpy of i in the standard (*) and pure (^{o}) state, respectively
ΔH^{mix} (also ΔH^{xs})	Enthalpy of mixing of a solution

$\Delta H_{f,e}$ & $\Delta H_{f,o}$	Heat of formation of a compound from elements and oxides, respectively
$\Delta_r H$	Enthalpy change of a reaction
h_i, h_i'	Partial and specific (') molar enthalpy of the component i in a solution
h	Planck constant, unless specified as height
K	Equilibrium constant of a reaction
k_T	Isothermal bulk modulus
k_S	Adiabatic bulk modulus
K_H, K_H'	Henry's law constant in fugacity-composition and activity-composition relations, respectively, of a dilute component
k_B	Boltzmann constant (1.38065×10^{-23} J/K)
$K_D(i\text{-}j)$	Distribution coefficient of the components i and j between a pair of phases
K_T, k_s	Isothermal and adiabatic bulk modulus, respectively
L	Avogadro's number (6.02217×10^{23} mol^{-1})
m_i	Molality of a component in a solution
N	Total number of moles
n_i	Number of moles of the component i
Q_{eq}	Equilibrium activity ratio of a reaction ($=K$)
Q	Canonical partition function in Chap. 14
q	Molecular partition function
R	Gas constant (8.314 J/mol-K)
S, S_m	Total, molar entropy of a solution, respectively
S_i^*, S_i^o	Entropy of the component i in the standard and pure state, respectively
ΔS^{mix}, ΔS^{xs}	Entropy of mixing and excess (^{xs}) entropy of mixing of a solution
$\Delta_r S$	Entropy change of a reaction
s_i	Partial molar entropy of the component i in a solution
T_c	Temperature of a critical point
$T_{c(sol)}$	Critical temperature of mixing of a solution
U, u'	Internal energy and specific internal energy ('), respectively
V, V_m	Total and molar volume of a solution, respectively
v_i or $\overline{V}_i$	Partial molar volume of the component in a solution
ΔV^{mix} (also ΔV^{xs})	Volume of mixing of a solution
$\Delta_r V$	Volume change of a reaction
W^+, δw^+	Total and infinitesimal work done **by** a system, respectively
W^-, δw^-	Total and infinitesimal work done **on** a system, respectively
$\delta\omega^+$, $\delta\omega^-$	Infinitesimal **non**-PV work done **by** and **on** a system, respectively
X_i or $X(i)$	Atomic or mole fraction of the component i in a solution
x_i^p	Atomic fraction of i in the sublattice p of a solution

y_i	Partial molar property of a component i in a solution with a total property Y (e.g., v_i: partial molar volume of the component i in a solution with a total volume V)
α	Coefficient of thermal expansion
β_T	Isothermal bulk modulus
β_S	Adiabatic bulk modulus
γ_i^α	Activity coefficient of a component i in a phase α
γ	Dihedral angle in Chap. 13
Γ_i	Concentration of component i per unit surface area of an interface
λ	Lagrangian multiplier
μ_i^α, μ_i^* & μ_i^o	Chemical potential of i in a phase α, in the standard state, and in the pure state, respectively
μ_{JT}	Joule-Thompson coefficient
$\bar{\mu}$	Reduced mass of a diatomic molecule
π'	Shear modulus
Γ_{th}	Thermodynamic Grüneisen parameter
Φ, ϕ	Fugacity and Osmotic coefficient, respectively
σ	Surface tension; also rate of entropy production
σ'	Symmetry number in rotational partition function

Physical and Chemical Constants

Avogadro's number	$L = 6.022(10^{23})\ \text{mol}^{-1}$
Boltzmann constant	$k_B = 1.38065(10^{-23})\ \text{J K}^{-1}$
Faraday constant	$F' = 9.6485(10^4)\ \text{C mol}^{-1}$
	$(\text{C: Coulomb} = \text{J/V})$
Gas constant	$R = 8.314\ \text{Jmol}^{-1}\text{K}^{-1}$
	$= 1.9872\ \text{cal mol}^{-1}\ \text{K}^{-1}$
	$= 83.14\ \text{bar cm}^3\text{mol}^{-1}\text{K}^{-1}$
Planck constant	$h = 6.62607(10^{-34})\ \text{J s}$
Acceleration of gravity at the Earth's surface	$g = 9.80665\ \text{m s}^{-2}$

Some Commonly Used Physical Quantities: SI Units and Conversions

Quantity	SI unit	Some conversions
Force	Newton (N): kg m s^{-2}	$1 \text{ N} = 10^5 \text{ dyne}$
Pressure	Pascal (Pa): N m^{-2} $= \text{kg m}^{-1}\text{s}^{-2}$	$1 \text{ bar} = 10^5 \text{ Pa}$ $1 \text{ atmosphere} = 1.01325 \text{ bar} = 760 \text{ mm of Hg}$ $1 \text{ GPa (Gigapascal)} = 10 \text{ kbar (kilobars)} = 10^4 \text{ bar}$
Energy	Joule (J): Nm $= \text{kg m}^2 \text{ s}^{-2}$	$1 \text{ cal} = 4.184 \text{ J}$ $1 \text{ J} = 10^7 \text{ ergs} = 10 \text{ cm}^3 \text{ bar} = 1 \text{ Pa m}^3$ $1 \text{ eV (electron volt)/atom} = 96.475 \text{ kJ/mol} = 23.058 \text{ kcal/mol}$ $1 \text{ eV} = 1.602(10^{-19}) \text{ J}$
watt (W)	J s^{-1}	
Length	meter (m)	$1 \text{ cm} = 10^4 \text{ μm (micron)}$ $1 \text{ nm (nanometer)} = 10 \text{ Å (angstrom)}$ $1 \text{ inch} = 2.54 \text{ cm}$

About the Author

Jibamitra Ganguly was educated in India and the University of Chicago, where he received his Ph.D. degree in Geophysical Sciences. This was followed by post-doctoral research at the Yale University and the University of California, Los Angeles, and appointment to a faculty position at the University of Arizona, where he is currently a Professor Emeritus of Geosciences. The author has made major contributions in a wide range of areas in the Earth and Planetary sciences relating to phase equilibria, thermodynamics, and diffusion kinetics that reflect an effective blend of experimental and theoretical studies with observational data. Besides the first edition of this book, the author has written a book entitled Mixtures and Mineral Reactions (co-author: S. K. Saxena) and edited a volume entitled Diffusion, Atomic Ordering and Mass Transport, all published by Springer-Verlag. In addition Prof. Ganguly is also the author of a book on the life and work of the Indian astrophysicist, Meghnad Saha (Meghnad Saha: His Science and Persona Through Letters and Writings), published by the Indian National Science Academy. In addition Prof. Ganguly is also the author of a book on the life and work of the Indian astrophysicist, Meghnad Saha (Meghnad Saha: His Science and Persona Through Letters and Writings), published by the Indian National Science Academy. He is a Fellow of the Mineralogical Society of America and the American Geophysical Union, and a recipient of the Alexander von Humboldt research prize.

Introduction

1

It must be admitted, I think, that the laws of thermodynamics have a different feel from most of the other laws of the physicist.

P. W. Bridgman

In this introductory chapter, I discuss the nature of thermodynamics and the type of problems that may be treated by the subject. I also collect together several introductory concepts regarding the nature of processes that are addressed by thermodynamics, concept of work from mechanics that lie at the foundation of thermodynamics, and several atomistic concepts that are important for developing insights into the thermal and energetic properties of matter, which are treated by thermodynamics at a macroscopic level. Finally, this chapter is concluded with a brief discussion of units and conversion factors.

1.1 Nature and Scope of Thermodynamics

Thermodynamics deals with the problem of conversion of one form of energy to another. **Classical** thermodynamics emerged primarily during the nineteenth century. Thus, the development of fundamental concepts of classical thermodynamics, like those of Mechanics and Electricity and Magnetism, precedes the development of modern concepts of the atomic or microscopic states of matter. There is also a non-classical arm of thermodynamics, known as **irreversible** thermodynamics, which is primarily a modern development. The laws of classical thermodynamics were formulated by deduction from experimental observations on **macroscopic scales**. Consequently, the thermodynamic laws are empirical in nature, and a thermodynamic system in which the laws are supposed to hold consists of a large number of atoms or molecules, of the order of Avogadro's number (10^{23}). We, of

© Springer Nature Switzerland AG 2020
J. Ganguly, *Thermodynamics in Earth and Planetary Sciences*,
Springer Textbooks in Earth Sciences, Geography and Environment,
https://doi.org/10.1007/978-3-030-20879-0_1

course, know now that all macroscopic properties of a system (such as pressure, temperature, volume etc.) have their origin in the motions and interactions of the atoms or molecules comprising the system. Thermodynamics, by itself, does not provide any fundamental insight as to the origin of thermodynamic laws and thermodynamic properties of matter.

The treatment of macroscopic properties in terms of statistical average of the appropriate properties of a large number of microscopic entities (atoms or molecules) constitutes the subject of **classical Statistical Mechanics**. While it provides analytical relationships between macroscopic properties and microscopic motions in a system, actual calculation of macroscopic properties from such relationships is a very difficult task. This is because of our lack of precise knowledge of the energetic properties of the microscopic entities, and computational difficulties. However, considerable progress has been made in both directions in recent years leading to what has become known as the **Molecular Dynamics** or **MD** simulations. These simulations represent a merger of statistical and classical mechanics, and hold great potential in predicting the thermodynamic and other macroscopic properties through considerations of microscopic interactions, and in refining our knowledge of the energetic properties in the atomic scale through comparison of the predicted and observed macroscopic properties. In addition, because of the enormous improvements in computational abilities, significant progress has also been made in the calculation of thermodynamic properties using **quantum chemical** approaches that have been briefly discussed in the Appendix C.

The fundamental concepts of classical thermodynamics have followed primarily from considerations of the problem of **conversion of heat into mechanical work and vice versa**, which inspired the great 'Industrial Revolution'. These have led to formal relationships among the macroscopic variables, and to descriptions of the equilibrium state of a macroscopic system under various sets of imposed conditions. (When a system achieves equilibrium consistent with the imposed conditions, all properties in the macroscopic scale not only remain unchanged, but also do not have any tendency to change with time as long as these conditions are not disturbed). Thermodynamics tells us that the macroscopic equilibrium state of a system depends only on the externally imposed conditions, such as pressure, temperature, volume, and is totally independent of the initial condition or the history of the system. Historically, this represented a major point of departure from the viewpoint of Newtonian mechanics that seeks to predict the evolutionary course of a system on the basis of its initial conditions.

Classical thermodynamics is a subject of great power and generality, and has influenced the development of important concepts in physical, chemical, biological and geological sciences, as well as in practical aspects of engineering. But it demands a moderate mathematical knowledge that is within the easy reach of a serious (or even not so serious) student of science or engineering. At the same time, thermodynamics has a rigorously logical structure that is often quite subtle. These aspects make the subject of thermodynamics apparently easy to learn, but yet difficult to completely appreciate in terms of its implications.

There are three laws at the foundation of thermodynamics, which are known as the first law, second law and third law, but most of the subject has been built on the first two laws. The second law of thermodynamics represents a supreme example of logical deduction of a revolutionary physical principle from systematic analysis of simple experimental observations. Because the basic concepts of thermodynamics are independent of any microscopic models, they have been unaffected by the developments in the microscopic description of matter—the validity of the laws were not threatened by discovery of errors in the microscopic models, nor the developments in thermodynamics took a quantum jump with exciting new discoveries in the microscopic domain.

1.2   Irreversible and Reversible Processes

Consider a gas inside a rigid cylinder fitted with a movable piston. Let P_{int} be the internal pressure of the gas and P_{ext} the pressure exerted on the gas from outside through the piston. If $P_{int} > P_{ext}$ then the gas will expand, and vice versa. Suppose now that the gas is allowed to expand rapidly to a particular volume, V_f. During this rapid of expansion the gas will be in chaotic motion, which will be visible even by macroscopic observation. Now let the gas be rapidly compressed back to its initial volume, V_i. After a while, the state of the gas will be the same as what it was at the beginning of the cyclic process, but the intermediate states during compression will be different from those during expansion. This is an example of an **Irreversible Process**.

Now imagine that the expansion of the gas from V_i to V_f is carried out in small incremental steps, as illustrated in Fig. 1.1, and that at each step the gas is held for a sufficiently long time to allow it to achieve equilibrium with the external pressure. If the process is reversed following the same procedure, then the state of the gas at a given position of the piston, say P_3, will be the same during both expansion and contraction, but not during the stage between two specific steps, say P_3 and P_4. However, the size of the steps can be made arbitrarily small, at least conceptually, so that the state of the gas during expansion is recovered during compression at any arbitrary position of the piston. This is an example of a **Reversible or Quasi-static Process**. Thus, reversible process is a process that is carried out at a *sufficiently slow*

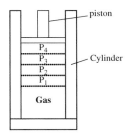

Fig. 1.1 Illustration of stepwise compression and expansion of gas within a rigid cylinder

rate such that the properties of the system at any state during the process differ by infinitesimal from those of its equilibrium state. The process is called reversible since a very small (i.e. infinitesimal) change in the external condition causes the system to reverse its direction of change.

All natural processes are irreversible, but a natural process may take place sufficiently slowly to approximate a reversible process. By this we mean that the time (Δt) over which a significant change of state of the system takes place is large compared to the time the system takes to achieve equilibrium, which is often referred to as the relaxation time, τ. The latter has a wide range of values, depending on the nature of the system and the perturbation produced in the system by the changing state conditions. As an example, for the problem of expansion of gas considered above, it can be shown that $\tau \sim V^{1/3}/C$, where V is the volume of the cylinder and C is the velocity of sound in the gas (Callen 1985), whereas for mineralogical reactions in geological or planetary processes, τ is often as high as millions of years.

1.3 Thermodynamic Systems, Walls and Variables

Any arbitrary but well defined part of the universe, subject to thermodynamic analysis, constitutes a thermodynamic system. The rest of the universe is called the surrounding. A system is separated from the surrounding by a **wall**. We can recognize the following types of systems.

Open System: A system which can exchange both energy and matter with the surrounding across its boundaries or walls.

Closed System: A system which can exchange energy with the surrounding, but not matter.

Isolated Systems: A system which can exchange neither energy nor matter with the surrounding.[1]

In order to make the existence of different systems possible, thermodynamics had also to device different types of wall, which are as follows.

Diathermal or non-adiabatic Wall: A wall that is impermeable to mass transfer, but permits transfer of heat through conduction. A closed system, in the sense defined above, is surrounded by diathermal wall.

Adiabatic Wall: A wall that does not permit either mass or heat transfer across it. Ignoring the effects due to fields (e.g. gravitational field), a system surrounded by an adiabatic wall can be affected from outside only through expansion or compression by moving the wall. The type of internally evacuated double wall used to make dewars for liquid nitrogen or helium is an example of an almost adiabatic wall. If we ignore the effects due to the fields, a system surrounded by a rigid adiabatic wall constitutes an isolated system.

---

[1]Some authors (e.g. Callen: Thermodynamics) use the term 'closed system' in the same sense as an 'isolated system' as defined here.

Semi-permeable Wall: This type of wall permits selective transfer of matter, and are also called semi-permeable membrane. For example, platinum and palladium are well known to be permeable to hydrogen, but not to oxygen or water (this property of the metals are made use of in some clever designs in experimental petrology to control oxygen partial pressure, e.g. Eugster and Wones 1962).

As we would see later, the thermodynamic walls play very important roles in the derivation of conditions that determine the evolution of a system towards the equilibrium state (see Lavenda 1978 for an insightful discussion). The thermodynamic potentials are defined only for the equilibrium states. Thus, one is faced with the paradoxical situation of determining the behavior of the potentials as a system evolves toward an equilibrium state, since the potentials are not defined for the non-equilibrium states. To resolve this problem, Constantin Carathéodory (1873–1950), a German mathematician of Greek origin, introduced the concept of **composite systems**, in which the subsystems are separated from one another by specific types of walls. Each subsystem is at equilibrium consistent with the restriction imposed by the internal and external walls, and thus has defined values of thermodynamic potentials. The internal walls separating the subsystems are then replaced by different types of walls and the system is now allowed to come to a new equilibrium state that is consistent with the new restrictions. This procedure reduces the problem of evolution of a system to one of a succession of equilibrium states. We would see several examples of the application of the concept of "composite system" later.

The thermodynamic variables are broadly classified into two groups, **extensive** and **intensive**. The values of the extensive variables depend on the extent or size of the system. They are additive, i.e. the value of an extensive variable E for an entire system is the sum of its values, E_s, for each subsystem ($E = \Sigma E_s$). Volume, heat, mass are familiar examples of extensive variables. The value of an intensive variable for a system, on the other hand, is independent of the size of the system. Familiar examples are pressure, temperature, density etc. These properties are not additive, and if the system is at equilibrium, then the value of an intensive variable at any point of the system is the same as in any other point.

For every extensive variable (E), it is possible to find a *conjugate* intensive variable (I) such that the product of the two variables has the dimension of energy. For example, for E = volume (V), conjugate I = pressure (P); for E = Area (A), conjugate I = surface tension (σ); for E = length (L), conjugate I = Force (F) etc.

1.4 Work

As defined in Mechanics, the mechanical work is the result of displacement of an object by the application of a force. If the applied force, F, is in a direction that is different from the direction of displacement, then one needs to consider the component of the applied force in the direction of displacement to calculate the work. If F is constant through a displacement ΔX along x, then the work (W) done by the force is simply given by the product of F_x and ΔX, where F_x is the component of F along x. In other words,

$$W = F.\Delta X = F\Delta X \cos \theta, \qquad (1.4.1)$$

where θ is the angle between the directions of applied force and displacement (Fig. 1.2). If the force is variable, then the work performed by a force on an object in displacing it from x_1 to x_2 along x is given by

$$W = \int_{x_1}^{x_2} F_x dx \qquad (1.4.2)$$

(If the displacement is along a curved path, then the work is given by the integral along the curved path; such integrals are known as line integrals.) In order to integrate F_x dx, F_x must be known as a function of x. It should be noted that if an applied force does not displace an object, it does not perform any work. Thus, a person pushing against a strong rigid wall does not perform any work by pushing against it for a long time; he or she simply gets tired. On the other hand, if there is no external force resisting the displacement, then there is no applied force either, and thus no work is performed.

If the angle between the direction of the applied force and displacement of an object is greater than 90° and less than 270°, then the force performs a negative work on the object, since $F(\cos\theta) < 0$ for θ values within this range. An example of negative work by a force that we would encounter later in this book is that performed by the gravitational force, mg, when an object is lifted upwards, where m is the mass of the object and g is the acceleration of gravity (force is mass times acceleration). Since the gravitational force is directed downwards, the angle between the directions of force and displacement is 180° (Fig. 1.1). Thus, the work performed by the force of gravity is mg(cos(180°))Δh = −mgΔh (Δh > 0).

In thermodynamics we speak of system and surrounding. A system can perform work on the surrounding or the surrounding can perform work on a system. We would use the symbols W^+ and W^- to indicate the works performed **by** and **on** a chosen system, respectively. Obviously, in a given process, $W^+ = -W^-$.

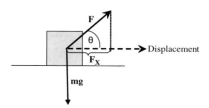

Fig. 1.2 Illustration of work done by a force (F) on an object when it is displaced along a horizontal direction. The gravitational force, **mg**, is directed downwards, and performs a negative work on an objected when it is displaced upwards

Of particular interest in thermodynamics is the work related to the change in volume of a system. For example, consider a gas contained in a cylinder that is fitted with a movable piston (Fig. 1.1). Now, if P is the pressure exerted by the gas on the cylinder walls, then the force exerted by the gas on the piston is P times the area of the piston, A (i.e. F = PA). Now if this pressure exceeds the external pressure, P_{ex}, on the piston, then the gas would expand. If the expansion is very rapid, then the gas will be in turbulence, and thus its pressure would be non-uniform, in which case we can no longer calculate the work done by the gas as a result of expansion. However, if the gas expands sufficiently slowly so that it has a uniform pressure throughout, and the piston is displaced against the external pressure from a position x_1 to x_2, then the work performed by the gas is given by

$$W^+ = \int_{x_1}^{x_2} (P_g A)dx \qquad (1.4.3)$$

But (Adx) is the infinitesimal change of gas volume, dV. Thus, for the displacement of the piston through the slow expansion of gas,

$$W^+ = \int_{v_1}^{v_2} P_g dV \qquad (1.4.4)$$

In a differential form, $\delta w^+ = P_g dV$, where the symbol δ denotes an *imperfect* differential (see Appendix B), and thus the work done on the gas, $\delta w^- = -P_g dV$. The value of the integral of an imperfect differential not only depends on the initial and final states of integration, but also on the path connecting these states. Thus, in general, the amount of work depends on the path followed to achieve a specific change of state. This concept is schematically illustrated in Fig. 1.3. The work performed **by** the gas on expansion from A to B along the solid line is given by the

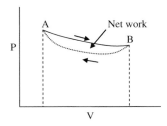

Fig. 1.3 Illustration of the P-V work done by a gas on expansion and contraction along specified paths. In expanding from A to B along the solid line, the gas performs a work that is given by the area under the solid line AB between the vertical dashed lines. When the gas returns from B to A along the dotted line, the work done **on** the gas is given by the area under the dotted curve bounded by the vertical dashed lines

line integral of PdV carried out along the solid line, that is by the area under the solid line bounded by the two vertical lines, whereas the work performed **on** the gas when it returns to A from B along the dotted line is given by the line integral of PdV carried out along the dotted line. Thus the net work performed **by** the gas in the cyclic process is given by the area bounded by the solid and dotted lines.

Equation (1.4.4) is valid regardless of the shape of the container (the interested reader is referred to Fermi 1956, for a proof). It is also valid, as emphasized by Zemansky and Dittman (1981), whether or not there is **(a)** any friction between the piston and the cylinder wall and **(b)** any non-mechanical irreversible process in the system, as long as pressure within the gas is uniform. Friction constitutes a part of the external force resisting the expansion of the gas. Now when the gas is compressed, the force exerted on the piston from outside has to overcome the resistance due to P_g and the friction of the piston. In this case, the infinitesimal work done on the gas (which we have chosen to be the system) is given by $\delta w^- = -P_{ex}dV$. However, if we want to use only P_g to calculate the work done on gas both during expansion and compression, then P_g and P_{ex} must be effectively equal, which requires an effectively frictionless condition.

Provided that there is negligible frictional resistance, the infinitesimal work done **on** a system due to a change of its volume, whether it is expansion or compression, can be expressed in terms of the (uniform) pressure P within the system, according to

$$\boxed{\delta w^- = -PdV} \tag{1.4.5}$$

When the gas expands, $dV > 0$, and therefore $\delta w^- < 0$, that is negative work is performed on the system (or positive work is performed on the surrounding), whereas when the gas is on compression, $dV < 0$, and thus, $\delta w^- > 0$, that is positive work is performed on the system.

In addition to work done by the expansion of a substance, which is commonly referred to as the PV work, there are other kinds of work resulting from other types of displacements against appropriate conjugate forces. For example, electrical work is performed by a charge as it moves through a potential difference, which may be utilized to drive a motor, and gravitational work is performed on a body as it is lifted against the force of gravity. Similarly, one can speak of work of magnetization, work against surface tension etc. All forms of work are important in thermodynamics, and the main problem is in the correct identification of the conjugate displacements and forces. However, the PV work has played a far greater role in the development of the fundamental concepts in thermodynamics. We would, thus, collectively denote the non-PV work by the symbol ω, using the plus and minus symbols in the same sense as in the PV work.

*__Problem 1.1__ Consider a mole of an ideal gas which has an equation of state $PV = RT$, where R is the gas constant (8.314 J mol^{-1}K^{-1} = 1.987 cal mol^{-1}K^{-1}) and T is the absolute temperature. Now express in terms of P and T, the reversible

---

*Solution to a Problem marked by an asterisk is given in the Appendix D

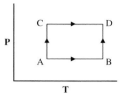

Fig. 1.4 Schematic illustration of the change of the state of a gas from A to D along two different paths in the P-T space, A → B → D and A → C → D

work done by the volume change associated with the change in the state of gas between $A(P_1, T_1)$ and $D(P_2, T_2)$ along two different paths, ABD and ACD (Fig. 1.4). You would get different answers for the work computed along these two different paths, even though the terminal states of integration are the same.

*Problem 1.2 Consider that the object in Fig. 1.2 is displaced horizontally on a rough surface. Is the work done by the force of friction, f_s, positive or negative? Write an expression for this work.

1.5 Stable and Metastable Equilibrium

Classical thermodynamics deals exclusively with equilibrium states of systems consistent with the imposed conditions. But what is an equilibrium state? We would discuss later formal thermodynamic criteria for describing the equilibrium states for different types of imposed conditions, but here we give a general description of stable and metastable equilibrium using familiar physical examples which are easy to appreciate.

Consider an example of a ball rolling down a mountain slope (Fig. 1.5). A ball rolls down the slope because it seeks progressively lower potential energy levels. However, the ball may get caught behind a small undulation on the slope (position a), or it may roll all the way down to the bottom (position b). When the ball is caught in position (a), it is said to be in a state of **metastable equilibrium**. It is a stable state of the ball not for all times, but for as long as the barrier remains or the position of the ball is not subjected to sufficient perturbation that could move it past the barrier. If the barrier is removed (say by erosion), the ball will eventually roll down the slope until it

Fig. 1.5 Illustration of **a** metastable, **b** unstable and **c** stable or steady state positions of a ball

reaches the bottom or gets caught in another barrier, but it will never move back on its own to its original position at (a) from a lower height.

The position of the ball at (b) on the hill slope represents an **unstable state**. The ball is said to be in **stable equilibrium** only when it has reached the lowest potential energy state among all the states that are accessible to it, if it is provided with enough energy to overcome barriers between different states. In our example, the state of the ball on a flat surface at the bottom of the hill could be viewed as of stable equilibrium. A **steady state** is a condition that does not represent the lowest energy state, but which does not change with time either.

1.6   Lattice Vibrations

The thermodynamic properties of molecules and crystals are related to the vibrational properties of the atoms around the equilibrium lattice sites. Here we discuss some elementary concepts of molecular and lattice vibrations that would be found useful in the discussion of thermodynamic properties of crystalline materials in the later sections.

A molecule is, in general, subject to translational, rotational and vibrational motions, each of which contributes to the total energy of a molecule. Almost all atomic mass is concentrated in a tiny nucleus, the mass of the electrons being negligible. In atomic mass units (amu) the mass of an electron is 0.000549, whereas those of proton and neutron are 1.0073 and 1.0087, respectively. The radius of the nucleus is of the order of 10^{-13} cm, whereas the overall dimension of a molecule is of the order of 10^{-8} cm. Consequently, one may consider that the atomic masses of the molecules are concentrated at individual points. Thus, we talk about **mass point** of a molecule (or of mass points of a system consisting of many molecules). In order to locate the instantaneous position of a mass point in space, we need three coordinates. The number of coordinates required to locate all mass points of a system is known as the **number of degrees of freedom**. Thus in a system consisting of N atoms, there are 3 N degrees of freedom.

The vibrational and rotational motions of a molecule constitute its internal motions. It is now well-known from quantum mechanics that the energies associated with the translational and internal motions of a molecule do not change continuously, but change in discontinuous steps. Thus, the energy spectrum of a molecule consists of a set of **quantized energy levels**. The separation D of the neighboring quantized energy levels of a molecule follows the order D(vibrational) > D(rotational) > D(translational). The separation of translational energy levels is, however, so close that for many purposes the translational energy can be thought to be continuous. The rotational energy is kinetic in nature, whereas the vibrational energy consists of both kinetic and potential components. The potential part arises from the relative positions of the atoms in a molecule during vibration, whereas the kinetic part arises from the velocity of atomic motion during the same action.

The simplest model of a vibrating diatomic molecule is that of a **harmonic oscillator**, in which the restoring force, F, is proportional to the displacement, x, from the equilibrium position according to $F = -Kx$, which is known as the **Hooke's law**, and where K is a force constant. Since force equals the negative gradient of potential energy, φ (i.e. $F = -d\varphi/dx$) the harmonic oscillator model leads to the following parabolic expression of potential energy as function of x,

$$\varphi(x) = 1/2Kx^2, \tag{1.6.1}$$

relative to that at the equilibrium position (x = 0) of the atoms. It follows from quantum mechanics that the vibrational energy levels, E_v, of a diatomic molecule behaving as a harmonic oscillator obeys the relation

$$E_v(n) = (n + 1/2)h\nu \tag{1.6.2}$$

where n denotes successive integers (quantum numbers), h is the Planck's constant (h = 6.626×10^{-34} J s) and v is the vibrational frequency. The quantity 1/2 hv is called the **zero-point energy**, because it represents the energy when n = 0, and is a consequence of the 'uncertainty principle' in quantum mechanics (see Sect. 14.3 for further discussion). The vibrational frequency of a specific oscillator is determined by the force constant and the masses of the vibrating atoms, and it typically has a value of 10^{12}–10^{14} per second. Thus, according to the above expression, the vibrational energy levels of a harmonic oscillator are equally spaced above the zero point level. The harmonic oscillator model of potential energy and vibrational energy levels of a hypothetical diatomic molecule is illustrated in Fig. 1.6.

The harmonic oscillator model is, however, not a generally satisfactory model for atomic vibrations in a molecule or a crystal. In reality, the vibration is **anharmonic** that leads to an asymmetry of the potential energy curve and decrease of the spacing interval between vibrational energy levels with increasing quantum number. As an example, we show in Fig. 1.7 the potential energy curve for hydrogen molecule along with the vibrational energy levels. Because of the anharmonicity effect, the

Fig. 1.6 Potential energy curve and vibrational energy levels of a diatomic molecule behaving as harmonic oscillator

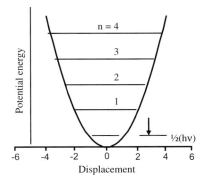

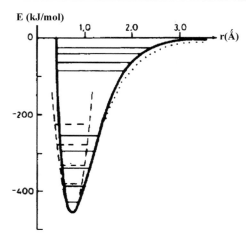

Fig. 1.7 Experimental (solid line) and calculated (dotted and dashed lines) potential energy curve of H_2 molecule, with the dashed line representing harmonic approximation. The horizontal lines represent the quantized energy levels. The harmonic approximations of vibrational energy levels are compared with the ones for the first five quantum numbers. From McMillan (1985). With permission from Mineralogical Society of America

restoring force becomes very weak and eventually becomes zero at large amplitude of vibration, leading to the dissociation of a molecule. If it were not for anharmonicity, there would be no dissociation. The thermal expansion of matter, of which dissociation is the extreme case, takes place by the displacements of the mean positions of the vibrating atoms in a crystal. If the potential energy changes along a parabolic curve according to the harmonic oscillator model, then the mean position will remain the same, preventing any thermal expansion. Similarly, diffusion of an atom within a solid would be impossible, except by quantum mechanical tunneling, if the potential energy well remains parabolic.

The effect of anharmonicity of vibration on the spacing of the energy levels is accounted for by adding additional terms to the right hand side of Eq. (1.6.2). In spite of its limitations, harmonic oscillator model has been frequently used, as we shall see later, to develop atomistic model of thermodynamic properties. The model gives reasonably good results at low temperatures where the potential energy curve approximately follows a parabolic form. This is known as **quasi-harmonic** approximation, as illustrated in Fig. 1.6.

The quantum of vibrational energy, $h\nu$, is called a **phonon**. Thermodynamic properties of a crystal can, in principle, be calculated from averages of vibrational energies, and for this purpose one needs to know a distribution function, $g(\nu)$, which gives the number of oscillators at a specific frequency of vibration (Fig. 1.8). The function $g(\nu)$ is known as the **phonon density of states**. The number of oscillators within a frequency range of ν_1 to ν_2 is given by the integral of $g(\nu)d\nu$ between the frequency limits, or the product $g(\nu)\Delta\nu$ when the frequency interval is small.

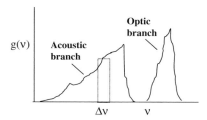

Fig. 1.8 Schematic illustration of phonon density of states of a crystal. The quantity $g(\nu)\Delta\nu$ approximately equals the number of oscillators within a small frequency interval $\Delta\nu$

The individual vibrations in a crystalline lattice are correlated. This correlation leads to a collective motion that produces travelling waves through a crystal. These travelling waves are called lattice modes, and have two branches that are known as **optic modes** and **acoustic modes** (Fig. 1.8). The lattice vibrations of all crystals have 3 acoustic and $3N - 3$ optic branches, where N is the number of atoms in the primitive unit cell. The optic modes interact with light waves whereas the acoustic modes interact with sound waves. The optic modes are of high frequencies and are, thus, excited mostly at high temperatures, whereas the acoustic modes are of relatively lower frequency and are, thus, excited mostly at low temperatures.

The earliest theory of lattice vibration is due to Einstein (1907), who introduced the fundamental idea of phonon or quantization of the energy of elastic waves in a solid. He assumed that atoms in a crystal vibrate around their individual equilibrium positions with the same frequency, and independently of one another. This frequency or the so-called Einstein frequency, ν_E, lies between the optic and acoustic frequencies of a crystal. Debye (1912) advanced Einstein's theory by considering that the density of states, $g(\nu)$, increases smoothly as a function of vibrational frequency, but up to an upper cut-off limit, which is now called the Debye frequency, ν_D. This is roughly the behavior of the acoustic branch of the phonon density of states. Further and more formal exposition of Einstein and Debye theories has been given in Sect. 14.4.

In the spirit of Einstein and Debye theories, one often speaks of **Einstein and Debye temperatures**, Θ_E and Θ_D, respectively. These are dimensionless quantities, and are defined as the product of $(h/2\pi k_B)$ and the respective frequencies, where h and k_B are the Plank constant and Boltzmann constant, respectively, that is,

$$\Theta_E = \frac{h\nu_E}{2\pi k_B} \tag{1.6.3}$$

and

$$\Theta_D = \frac{h\nu_D}{2\pi k_B} \tag{1.6.4}$$

1.7 Electronic Configurations and Crystal Field Effects

1.7.1 Electronic Shells, Subshells and Orbitals

According to the quantum theory, electrons in an atom revolve around a nucleus (consisting of protons and neutrons) in quantized or discrete energy levels. The energy levels are grouped together in shells and subshells according to quantum mechanical principles, as briefly reviewed below.

(a) A shell for electron energy levels is characterized by its principal quantum number, n, as **K** shell for $n = 1$, **M** shell for $n = 2$, **L** shell for $n = 3$, and **N** shell for $n = 4$.

(b) Within each shell, there are subshells that are characterized by the l or the azimuthal quantum numbers, which have integral values. For a given value of n, there are subshells with integral numbers of 0 to (n − 1). Thus, in the K shell, there is only one l subshell with value of 0, in the L shell, there are two l subshells with values of $l = 0$ and $l = 1$, in the M shell (n = 3) there are three l subshells with values of 0, 1 and 2, and so on (Note that the number of l subshells within a shell is the same as the value of the principal quantum number n that characterizes the shell).

(c) There is one electronic orbital in the s subshell ($l = 0$), and it is spherically symmetric, whereas the p subshell ($l = 1$) has three, d($l = 2$), has five and f ($l = 3$) has seven electronic orbitals. The p, d and f electronic orbitals have directional properties. The directional properties of the d-orbitals are illustrated in Fig. 1.9. The electronic orbitals within a subshell are characterized by the magnetic quantum number, m_l. For a given value of l, the magnetic quantum numbers have values of 0, ±1, ±2, … ± l. For example, for the p subshell,

Fig. 1.9 The spatial orientations of the five d-orbitals that are energetically degenerate in a spherically symmetric environment; the form of the d_{z^2} orbital can be generated by rotation about the z axis. From Fyfe (1964)

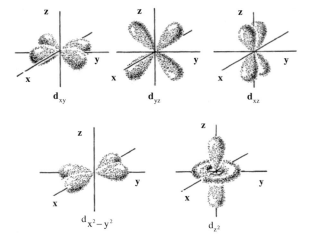

which has a value of 1 for the l quantum number, the m_l values are 0, +1 and −1, thus giving rise to three p orbitals. It can be easily seen that the total number of electronic orbitals in a subshell equals $2l + 1$.

(d) An electron has a spin quantum number, m_s, of +½ or −½. (One may imagine an electron to be spinning on its own axis and at the same time rotating about a nucleus in a manner analogous to the rotation of the Earth or a planet on its own axis and around the sun; the two types of spins are conventionally indicated as upward and downward pointing arrows, ↑ and ↓). According to the **Pauli exclusion principle**, no two electrons can have exactly the same quantum state, i.e. the same values of the n, l, m_l and m_s quantum numbers. Consequently, an electronic orbital cannot be occupied by more than two electrons. When there are two electrons in an orbital (and thus have the same values of n, l and m_l quantum numbers), they must have opposite spins, +½ and −½. Since a subshell has $2l + 1$ electronic orbitals, the total number of electrons in a subshell is $\leq 2(2l + 1)$

(e) When the electronic orbitals are degenerate, as in a free atom or ion, then the electrons are distributed among the orbitals such that there is a maximum number of unpaired spins. This is known as the **Hund's rule**. For example, in a free Fe^{2+} ion, there are 6 electrons in the 3d subshell, all of which have exactly the energy. According to Hund's rule, there must two electrons with opposite spin in one d orbital (resulting in a net spin of 0), whereas each of the four other d orbitals will have only one electron, all with the same spin. The electronic configuration of an atom or ion is reported as $n(subshell)^m$ for all n and subshells, where m is the number of electrons in the subshell. Thus, for example, the electronic configuration of Fe is $1s^2 2s^2 2p^6 3s^2 3p^6 3d^8$.

The quantum mechanical classification of shell, subshell and orbitals are summarized below.

Shell	Subshell	No. of orbitals
	$l = 0$ to $(n − 1)$	$2l + 1$
K (n = 1)	1s ($l = 0$)	1
L (n = 2)	2s ($l = 0$)	1
	2p ($l = 1$)	3
M (n = 3)	3s ($l = 0$)	1
	3p ($l = 1$)	3
	3d ($l = 2$)	5
N (n = 4)	4s ($l = 0$)	1
	4p ($l = 1$)	3
	4d ($l = 2$)	5
	4f ($l = 3$)	7

1.7.2   Crystal or Ligand Field Effects

The degeneracy of the electronic orbitals could be removed by interaction with the surrounding negatively charged ions or dipoles within a crystal. For transition metal ions with unfilled d orbitals, this removal of degeneracy of the d orbitals in a subshell leads to phenomena that have interesting thermodynamic consequences. The simplest analysis of the effect of the surrounding polyhedron on the d-orbitals of a central cation, which is adequate for our purpose, is provided by what is known as the **crystal field theory**, which was developed by Bethe (1929) and Van Vleck (1935). In this theory, the cation has orbitals with directional properties (Fig. 1.9), but the surrounding ions or ligands are considered to be orbital-less point charges or point dipoles. The nature of splitting of d-orbitals of a cation is dictated by the symmetry of the surrounding polyhedron, as schematically illustrated in Fig. 1.10 for *regular* octahedron, tetrahedron, dodecahedron and cube. The d-orbitals are divided into two groups, t_{2g}, which consists of the d_{xy}, d_{yz} and d_{zx} orbitals, and e_g, which consists of the $d_{x^2-y^2}$ and d_{z^2}. When the polyhedra are distorted, then there is further removal of d-orbital degeneracies depending on the nature of distortion (Fig. 1.11). Additionally, the degeneracy of the orbitals is removed if there are

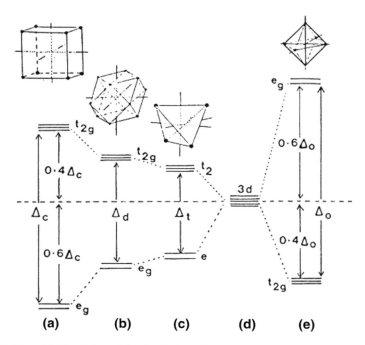

Fig. 1.10 Crystal field splitting of the d-orbitals (with the same principal quantum number) of a central atom in different types of *regular* coordinating polyhedra: (**a**) cubic (8); (**b**) dodecahedral (12); (**c**): tetrahedral (4); (**d**): spherical; (**e**) octahedral (6), where the parenthetical numbers indicate coordination numbers. From Burns (1993)

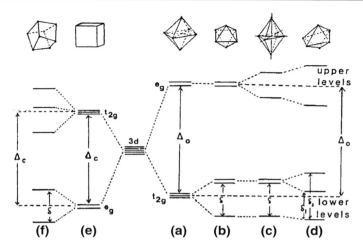

Fig. 1.11 Removal of the degeneracies of the e_g and t_{2g} orbitals by the distortion of regular octahedral and cubic sites. (**a**): regular octahedron; (**b**): trigonally distorted octahedron; (**c**): tetrahedrally elongated octahedron; (**d**): distorted 6-coordinated site as in the M1 and M2 sites of pyroxene and olivine; (**e**): regular cube; (**f**): distorted cube (e.g. dodecahedral site of garnet). From Burns (1993)

more orbitals than the number of electrons in a symmetrical non-linear molecule. This is known as the **Jahn-Teller effect**. In all cases the splitting is such that if all d-orbitals have the same number of electrons, then there is no net change of energy. For example, in a regular octahedron there are three d-orbitals in a lower energy state, which is $2/5\Delta$ below the initial energy level, and two d-orbitals in a higher energy state, which is $3/5\Delta$ above the initial energy level, where Δ represents the magnitude of splitting between the e_g and t_{2g} orbitals. Thus, if each d-orbital has one electron, then the net change of energy is $3(-2/5\Delta) + 2(3/5\Delta) = 0$. The magnitude of crystal-field splitting of d-orbitals can be determined by optical absorption spectroscopy.

More sophisticated theories are the **ligand field theory** in which the ligands are also considered to have orbitals, and the **molecular orbital theory**, in which the effect is evaluated through construction of molecular orbitals as linear combination of the atomic orbitals. Relatively simple but authoritative discussion of these theories can be found in Orgel (1966). All theories make use of the symmetry properties of the polyhedron surrounding a cation.

When the crystal- or ligand-field becomes strong, the magnitude of the d-orbital splitting increases. Thus, if a mineral is compressed, then the value of Δ is expected to increase. In the pure crystal field analysis of the problem, $\Delta \propto R^{-5}$, where R is the distance between a transition metal ion and the surrounding anion. When the magnitude of Δ exceeds a threshold value, the lowering of energy (stabilization effect) achieved by having electrons with unpaired spins is overcompensated by the increase of energy (destabilization effect) due to electron occupancies of the higher energy orbitals, after each lower energy orbitals have acquired an unpaired electron. Under this condition, there would be spin pairing in the lower energy orbitals. This type of transition from the high spin to low spin state of transition metal ion is expected to take place in the high pressures of the Earth's interior. Linn et al. (2007) found Fe^{2+} in ferropericlase, $(Fe,Mg)O$, to undergo a gradual transition from the high-spin to low-spin state over a P-T range corresponding to conditions in the Earth's lower mantle (1000 km or 38.6 GPa, 1900 K to 2200 km or 97.3 GPa, 2300 K).

Problem 1.3 (a) Common natural garnets can be described by the chemical formula $X_3Y_2Si_3O_{12}$ where X and Y represent, respectively, a dodecahedral site with eight oxygen coordination and an octahedral site with six oxygen coordination. The common X-site cations are Fe^{2+}, Mg, Ca and Mn^{2+} whereas Y-site cations are Al, Fe^{3+} and Cr^{3+}. What explains the absence of Mn^{3+} in the Y site in spite of the fact that Fe^{3+}, which is the neighboring metal cation in the Periodic Table, substitutes in the Y site in an oxidized environment? (b) From the point of view of ionic radius, both Fe^{3+} and Cr^{3+} are almost equally favorable in the octahedral site of garnet (radius ratio with respect to 4-coordinated oxygen is 0.467 for Fe^{3+} and 0.445 for Cr^{3+}, whereas the cation/oxygen radius ratio should be between 0.414 and 0.732 for a 6-coordinated cation, according to Pauling's rules). But considering other factors, which one would you expect to be more favorable in this site (favorable means lowering of energy)?
Hint: Consider the electronic configurations of the ions. (b) **Answer**: Cr^{3+} (explain why).

1.8   Some Useful Physical Quantities and Units

There are different types of units for a given physical or chemical quantity that have been used in the thermodynamics, and other branches of physical science. The units recommended by IUPAC (International Union of Pure and Applied Chemistry) are those of the International System of Units (Système International d'unités) or SI units, which originated in France. It is founded on seven mutually independent **base units** of seven base quantities, such as length (meter: m), mass (kilogram: kg), time (seconds: s) etc., and **derived units** of quantities related to the base quantities. We briefly review here some of the quantities and their units in the SI system that are useful in thermodynamics, along with the conversions to other units that have also been used from time to time.

Momentum (M): Since it is a product of mass and velocity, in the SI system the momentum has the unit of (kg)(m/s) or kg m s^{-1}.

Force (F): Force is defined as the rate of change of momentum, d(M)/dt. Thus, the unit of force in the SI unit is (kg m s^{-1})(s^{-1}) or kg m s^{-2}, which is known as a Newton (N). In the non-relativistic domain, mass (m) is constant so that F = m(dv/dt) = ma, where v is the velocity and a is acceleration.

Pressure: Pressure is force per unit area (A). Thus, the unit of pressure in the SI system is N/m^2 or $kg\text{-}m^{-1}s^{-2}$, which is known as the Pascal (Pa), and is related to another commonly used pressure unit, namely bar, as $1\ Pa = 10^{-5}$ bar (1 atmosphere = 1.01325 bars). The gigapascal (GPa) that is now commonly used as a pressure unit in the Earth Science literature is related to kilobar (kb), which is the most frequently used pressure unit in the earlier literature, as $1\ GPa = 10$ kb (Giga: 10^9; kilo: 10^3).

Energy: Energy is given by the product of force and displacement. Thus, the SI unit of energy is Nm ($kg\ m^2\ s^{-2}$), which is the energy required to displace a mass of 1 kg through a distance of 1 m. A Nm is known as Joule (J), which is related to other energy units used in thermodynamics as follows.

$$1\ \text{thermochemical cal (cal)} = 4.184\ J$$

$$1\ J = 10\ cm^3 bar$$

$$1\ eV = 1.602 \times 10^{-}19\ J\ (\text{eV: electron volt}).$$

Note that the product of $cm^3 bar$ represents energy. This can be easily understood by rewriting $cm^3 bar$ as cm^3(Force/A), since bar is a measure of pressure. Thus, $cm^3 bar$ has the dimension of force times the displacement (cm^3/A), which is the dimension of energy.

References

Bethe H (1929) Termauf spalltung in Kristallen. Ann Physik 3:133–208

Burns RG (1993) Mineralogical applications of crystal field theory. Cambridge, p 551

Callen HB (1985) Thermodynamics and an introduction to thermostatics. Wiley, New York Chichester Brisbane Toronto Singapore, p 493

Debye P (1912) Zur Theorie der spezifischen wärmen. Ann der Physik 39:789–839

Einstein A (1907) Die Plancksche Theorie der Strahlung und de Theorie der spezifischen Warmen. Annal der Physik 22:180–190

Eugster HP, Wones DR (1962) Stability relations of the ferruginous biotite, annite. J Petrol 3: 82–125

Fermi E (1956) Thermodynamics, Dover, p 160

Fyfe WS (1964) Geochemistry of Solids – An Introduction, McGraw Hill

Lavenda BH (1978) Thermodynamics of irreversible processes. Dover, New York, p 182

Linn J-F, Vanko G, Jacobsen SD, Iota V, Struzhkin VV, Prakapenka VB, Kuzentsov A, Yoo C-S (2007) Spin transition zone in the Earth's mantle. Science 317:1740–1743

McMillan P (1985) Vibrational spectroscopy in mineral sciences, In: Kieffer SW and Navrotsky A (eds) Reviews in Mineralogy, vol 14, Mineralogical Society of America, Washington DC, pp 9–63

Orgel LE (1966) An introduction to transition-metal chemistry – Ligand-field theory. Methuen, p 184

Van Vleck JH (1935) Valence strength and magnetism of complex salts. J Chem Phys 3:807–813

Zemansky MW, Dittman RH (1981) Heat and thermodynamics. McGraw-Hill, New York, p 543

First and Second Laws

2

> *Die Energie der Welt ist konstant (The energy of the Universe is constant) Die Entropie der Welt strebt einem maximum zu (The entropy of the Universe increases to a maximum)*
>
> Clausius (ca. 1867)

The above statements by Rudolf Clausius (1822–1888) are cosmological expressions of the first and second laws of thermodynamics that resulted from the failure of engineering efforts to develop machines that could (a) create energy and (b) also convert energy to work **without any limitation**.

The first law of thermodynamics, which is based on the works of James Prescott Joule (1818–1889), Julius Robert von Mayer (1814–1878) and Herman von Helmholtz (1821–1894) during the period 1842–1848, addresses the question of energy change in a system interacting with the surrounding, but not subjected to mass flow. One can look upon the total energy of a system as the sum of its external and internal energies. The former results from the position and motion of the system as a whole (potential and kinetic energies, respectively), whereas the latter is an intrinsic property of the internal state of matter comprising the system. (In some treatments, only the kinetic energy of the system is considered to be the external energy, but in this section we will consider both kinetic and potential energies of the system as its external energies).

It was known in Mechanics, which preceded the development of Thermodynamics, that in a purely mechanical system (i.e. a system not subjected to heating or friction), the external energy is conserved. Thus, when an object is thrown upwards in a gravitational field the sum of its potential and kinetic energies is conserved at every stage of its motion. This conservation principle was first recognized by Leibnitz in 1963 (Leibnitz is also a co-founder of the subject of calculus, along with Sir Isaac

© Springer Nature Switzerland AG 2020
J. Ganguly, *Thermodynamics in Earth and Planetary Sciences*,
Springer Textbooks in Earth Sciences, Geography and Environment,
https://doi.org/10.1007/978-3-030-20879-0_2

Newton). Thermodynamics brings into this picture of conservation of external energy in mechanics, the principle of conservation of internal energy of an isolated system.

The first law of thermodynamics introduced the concept of the **conversion** of one from energy to the other, specifically of work to heat and vice versa, and ruled out the possibility of constructing a device, often referred to as the perpetual motion machine of the first kind, which could create energy (without involving nuclear reaction). However, it did not impose any restriction on the efficiency of trans- formation of one form of energy into another. **The second law of thermodynamics imposes definite limitation on the conversion of heat into work**. There is, however, no restriction on the conversion of work into heat. The second law was developed almost simultaneously with the first law on the basis of the works of William Thompson (also known as Lord Kelvin, 1824–1907), Rudolf Clausius (1822–1888) and Sadi Carnot (1796–1832). It led to the introduction of a new property, namely the **entropy**, which always increases with time in an isolated system as a consequence of processes taking place within the system. This idea was in apparent conflict with the dominant scientific idea of the time, the Newtonian or classical mechanics. In the latter, the equations of motion are symmetrical with respect to time, which implies that if a system evolves from one configuration S_1 return to another configuration S_2 with time, then it should also be possible for the system to S_1. The second law of thermodynamics was indeed one of the greatest scientific revolutions, and as noted by Feynman (1963), it was rather unique in the sense that it came through an engineering effort instead of a fundamental inquiry about the nature of physical laws.

2.1   The First Law

If two stationary systems, which are in mutual contact, can exchange energy, then the gain in the internal energy of one system must be compensated by the loss in the internal energy of the other. Thus, the net change of internal energies of the two systems, which together constitute an isolated system, is zero. If the boundary between the two systems is impermeable to mass transfer, then energy can be transferred between the two systems only by means of (a) work done by one on the other, and (b) heat transfer from one to the other.

With the above background, the first law of thermodynamics can be stated as follows. (a) The internal energy, U, of a system depends only on the state of the system, and (b) the change in the internal energy of a **closed** system is the sum of the energies <u>absorbed</u> by the system from its surrounding in the form of heat and work, i.e.

$$\Delta U = Q + W^- \qquad (2.1.1)$$

where ΔU is the difference between internal energies of the system in its terminal and initial states, Q is the heat absorbed by the system from the surrounding and W^- is the work done by the surrounding on the system, which is the same as the work absorbed by the system.[1] ΔU has a positive value when the system gains energy and a negative value when it loses energy. At the outset note that it is the **change** in internal energy rather than internal energy itself that is addressed by the first law.

The principle of conservation of energy is so deeply rooted in science that if it seemed to fail, a search is made to find a new form of energy. Indeed, Wolfgang Pauli (1900–1958) was led to propose the existence of an electrically neutral particle possessing little or no mass but a definite energy, in order to preserve the energy balance in the nuclear reaction that is known as β decay, which obeyed the mass balance, but not the energy balance. In this nuclear process, a neutron is transformed into a proton, or vice versa, by emission of a β-particle according to neutron → proton + β⁻-particle (electron) or proton → neutron + β⁺-particle (positron). Wolfgang Pauli in 1938 suggested that the "missing energy" that was required to preserve energy conservation between the initial nucleus and the transformation products was carried by a new particle that interacted extremely weakly with the other particles and thus escaped detection. This particle is called **neutrino**, a name given by the great Italian physicist Enrico Fermi (1901–1954), and which in Italian means very little neutral body. (Confirmation of Pauli's prediction of the existence of neutrino 25 years later led to the Nobel prize in Physics to Frederick Reines and Clyde Cowan. Both Fermi and Pauli also received Nobel prize in Physics for other contributions).

Since U depends only on the state of the system, Eq. (2.1.1) implies that even though the individual values of Q and W depend, in general, on the path along which the state of a system has changed, the sum of Q and W is **independent** of the path, and depends **only** on the initial and final states of the system. Thus,

$$\int_A^B dU = \int_A^B (\delta q + \delta W^-) \qquad (2.1.2a)$$

The integrals on both sides are line integrals along the same path connecting the two states, A and B, or since U is a state function (which implies dU is an exact differential, see Appendix B.2)

$$\oint dU = 0 = \oint (\delta q + \delta W^-) \qquad (2.1.2b)$$

where the symbol $\oint$ denotes line integral around a closed loop, beginning and ending in the same state. If a system is contained within an adiabatic enclosure, then $(\Delta U) = W^-$, in which case the amount of work done on (or performed by) the system is *independent* of the path. In other words, w becomes a state function, so that $\delta w(Q = 0)$ becomes an exact differential (i.e. $\delta w = dw$). Thus, the change in

---

[1]Since, according to Einstein's theory of relativity, internal energy can be created at the expense of mass through nuclear reaction (according to the famous relation $E = mc^2$, where c is the velocity of light and m is the rest mass), Eq. (2.1.1) must be restricted to systems that are not subjected to nuclear reaction. Alternatively, we can think of ΔU as the change of internal energy due to absorption of heat and work and the change of energy due to a change of the rest mass, Δmc^2.

the internal energy of a system between two states, which is a unique quantity, can be measured by knowing the adiabatic work performed on the system to achieve the same change of state. Without knowing or inquiring anything about the microscopic nature of internal energy, we can give a thermodynamic and measurable definition of the **change** of internal energy of a system when it moves from one state to another as the work done on the system when the same change of state is brought about by an *adiabatic* process.

Most often, we are concerned with P-V work (i.e. the work related to the change of volume). As discussed earlier (Sect. 1.4), there are also various other forms of work. Recalling that the differential form of all these 'other' forms of reversible work performed by the system have been denoted by $\delta\omega$, we have δw^- (i.e. the differential of the total work done on a system) = $-PdV + \delta\omega^-$, and consequently,

$$dU = \delta q - PdV + \delta\omega^- \qquad (2.1.3)$$

(When asked to state the first law, it is not very uncommon for a student to have some confusion if the sign before PdV should be positive or negative. To avoid the confusion, one should remember that when a system is compressed (that is dV is negative), it gains energy. This is possible only if there is a negative sign before PdV. Conversely, when a system expands (dV > 0), it performs work on the surrounding and thus loses energy. Again this condition is satisfied with a negative sign before PdV).

The first law also establishes the **mechanical equivalent of heat**. It is evident from Eq. (2.1.1) that the internal energy of a closed system can be changed, as reflected by its change of temperature, without the intervention of work by simply bringing it into contact with another system at a different temperature. Here the mode of energy transfer is purely heat conduction. The amount of work that must be performed on the same system **adiabatically** to achieve the same change of internal energy or temperature constitutes the mechanical equivalent of heat. The amount of heat necessary to raise the temperature of 1 g of water by 1 K is arbitrarily defined as a calorie. The same thermal effect on a gram of water can be achieved adiabatically by performing on it 4.184 J of work. Thus 1 cal = 4.184 J (Recall that joule is the MKS unit of work resulting from the application of a force of one Newton (kg m/s^2) through a displacement of one meter).

2.2 Second Law: The Classic Statements

The second law can be stated in a variety of ways, all of which are, of course, equivalent. The classic statements of the second law, which are due to William Thompson (1824–1907), who is also known as Lord Kelvin, and Rudolf Clausius (1822–1888), are as follows.

Kelvin Statement: **A transformation whose only final result is to transform into work heat extracted from a source which is at the same temperature throughout is impossible.**

Clausius Statement: **A transformation whose only final result is to transfer heat from a body at a given temperature to a body at a higher temperature is impossible.**

As such, these statements do not convey any impression of the revolutionary nature of the second law. The latter emerges only through a careful analysis of either statement that leads to the development of the concept of entropy and its property of unidirectional change with respect to time for natural processes in an isolated system.

Before going into entropy, let us first discuss the meaning of the Kelvin statement. Consider an ideal gas in contact with a heat bath at a uniform temperature. The resulting expansion of the gas would deliver work on the surrounding. Now an ideal gas has the unique property that its internal energy depends only on its temperature. Consequently, for the isothermal expansion, $\Delta U = 0$, so that $Q = -W^- = W^+$ (i.e. work done by the system), where $Q = \int \delta q$. Here heat absorbed from a source at a uniform temperature is completely converted into work. However, this is not the only final result. At the end of the process, the gas is left with a larger volume than at the beginning. The Kelvin statement would have been violated if it were possible to return the gas to its initial volume without the intervention of a heat sink. In order to return the gas to its initial volume, some heat must be withdrawn from the gas by bringing it into contact with a heat bath at a lower temperature. In this case, the final result would consist of conversion of heat into work plus a definite amount of dissipation or wastage of heat.

If it were not for the validity of the 'Kelvin statement', it would have been possible to construct a 'perpetual motion machine', often referred to as the perpetual motion machine of the second kind, which will perform work endlessly by withdrawing heat from its surrounding environment which is at uniform temperature and has virtually inexhaustible amount of energy. Failure of all efforts to construct a perpetual motion machine constitutes the experimental evidence in support of Kelvin's postulate.

The Clausius statement says that heat by itself cannot flow from a lower to a higher temperature. In a hot summer day, an air conditioner withdraws heat from a house and dissipates it into the atmosphere. But there is no way we can make the heat flow from a house to the surrounding atmosphere which is at a higher temperature without paying money to the power company. The validity of the Clausius statement also implies validity of the Kelvin statement. Thus, if in the above example of expansion of gas, the heat dissipated into the heat sink could have returned by itself to the heat source, we would have had complete conversion of the **net** heat withdrawn from the source into work. This would have been a violation of the Kelvin statement. The problem of thermal pollution of environment (which is treated as a heat sink) in the modern industrialized society is a direct consequence of the second law of thermodynamics.

2.3   Carnot Cycle: Entropy and Absolute Temperature Scale

Once it was realized that there must be two heat baths at two different temperatures, one acting as a heat source and the other as a heat sink, in order for a system to perform work using a cyclic process, the French engineer Sadi Carnot (1796–1832) set out to analyze the efficiency of conversion of heat to mechanical work in a cyclic process that is commonly referred to as the **Carnot cycle**. We would first see how the analysis of the properties of Carnot cycle led to the development of the concept of entropy, discuss some ramifications and mineralogical applications of this concept, and then discuss the limiting efficiency of conversion of heat to work that followed from Carnot's analysis.

The Carnot cycle consists of **two isothermal** and **two adiabatic steps**, as illustrated in Fig. 2.1, all four steps being sufficiently slow to be effectively reversible (it can be shown that the magnitude of P-V slope of an adiabatic step must be greater than that of an isothermal step involving the same body). The sequence of the steps is as follows.

(a)   An isothermal expansion of a gas from A to B by withdrawing an amount of heat Q_2 from a heat bath at an uniform temperature Θ_2 defined by an empirical temperature scale;

(b)   further expansion from B to C under adiabatic condition as a result of which the gas cools from Θ_2 to a temperature Θ_1 defined by the same empirical scale;

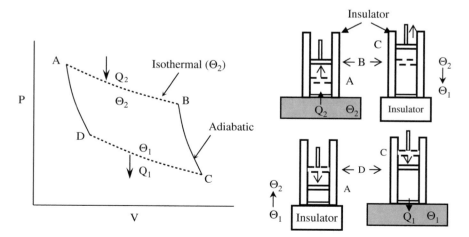

Fig. 2.1   Schematic illustration of the Carnot cycle in the P-V space. The dotted and solid lines indicate isothermal and adiabatic paths, respectively. The path A → B → C represents continuous reversible expansion whereas the path C → D → A represents continuous reversible compression. In the right panel, the shaded boxes represent heat baths. Heat is transferred into the material within the cylinder only through the bottom

(c)   isothermal compression from C to D by delivering an amount of heat Q_1 to a heat bath at temperature Θ_1, with which the gas is now in contact; and finally

(d)   further adiabatic compression until the gas reaches the initial temperature Θ_2.

From a systematic treatment of the postulate of Kelvin, it can be shown (e.g. Denbigh 1981, pp. 27–29) that for a reversible cycle operating between two heat baths, the ratio of the heat withdrawn to heat delivered by a body operating the reversible cycle depends only on the temperature of the two heat baths, i.e. $Q_2/Q_1 = f(\theta_2, \theta_1)$. Furthermore, it is possible to define a temperature scale such that the functional relation has the specific form $Q_2/Q_1 = \theta_2/\theta_1$ so that

$$\frac{Q_2}{\theta_2} = \frac{Q_1}{\theta_1}, \tag{2.3.1}$$

This temperature scale is called the **thermodynamic temperature scale**. Stated in words, the above equation says that it is possible to define a temperature scale such that the ratio of heat withdrawn to the thermodynamic temperature of the heat bath equals the ratio of heat delivered to the thermodynamic temperature of the heat sink. The ratio Q_2/Q_1 is independent of the property of the body that operates the Carnot cycle.

The temperature scale that is used in thermodynamics is known as the **Kelvin temperature** scale in which temperature is denoted by the symbol **T**. The name is in recognition of the contribution of Lord Kelvin towards its development. The temperature measured using the Kelvin scale not only satisfies Eq. (2.3.1) but has the same step size for a degree as in the already established Celsius or centigrade scale. Specifically, the Kelvin and centigrade scales are related according to T K = T °C + 273.15, where T °C is the temperature measured in the centigrade scale (A numerical value of temperature, say 400°, in the two scales is written as 400 K and 400 °C). The Kelvin scale was developed by assigning a temperature of 273.16 K to the triple point of water at which liquid water, ice and vapor in the pure H_2O system are in thermodynamic equilibrium. This specific value for the triple point was chosen to make the degree step size, which is the only arbitrary aspect of the thermodynamic temperature scale, exactly the same as in the centigrade scale (The latter objective could have been achieved by assigning an appropriate temperature to some other state of a substance, but the triple point of pure water was chosen because of the relative ease with which it can be reproduced in the laboratory).

Let us now evaluate the quantity $\oint \delta q/T$ for the Carnot cycle (cc), where $\oint \delta q$ is the heat absorbed in the (reversible) cyclic process. We have

$$\oint \left(\frac{\delta q}{T}\right)_{cc} = \int_A^B \frac{\delta q}{T_2} + \int_c^D \frac{\delta q}{T_1}$$

$$= \frac{Q_2}{T_2} - \frac{Q_1}{T_1} \qquad (2.3.2)$$

(Note that Q_1 is the heat given off by the system in the isothermal process $C \rightarrow D$, so that $-Q_1$ is the heat absorbed by the system in the same process). But, according to Eq. (2.3.1), the right hand quantity is zero, so that

$$\oint \left(\frac{\delta q}{T}\right)_{cc} = 0 \qquad (2.3.3)$$

It can be shown that the above result holds for **any reversible cyclic process** and not just for the special type of reversible cyclic process depicted by Carnot cycle, that is consisting of two isothermal and two adiabatic steps (see, for example, Denbigh 1981). Thus, in general

$$\oint \left(\frac{\delta q}{T}\right)_{rev} = 0 \qquad (2.3.4)$$

This equation should be contrasted with the fact that, in general, $\oint \delta q \neq 0$. Thus, we find that in a *reversible* process, the inexact differential δq can be transformed into an exact differential by multiplication with **an integrating factor** $1/T$. This property of $(\delta q/T)_{rev}$ was first discovered by Clausius, who designated it by the special symbol dS, where S was called the **entropy** of a system. Thus,

$$dS = \left(\frac{\delta q}{T}\right)_{rev} \qquad (2.3.5)$$

The above expression constitutes the **thermodynamic definition of entropy**. Expressed in words, *the entropy change associated with a change of state of a system is given by the heat absorbed by the system divided by its temperature when the same change of state is brought about reversibly*. To appreciate this statement, let us consider a schematic illustration, Fig. 2.2, which shows two states, A and B, of the same system in the P-V space. The states can be connected by an isothermal reversible path R, and multitude of irreversible paths, one of which is the path I. Let us say that the heat absorbed by the system on moving from A to B along the irreversible path I is Q_I, whereas that absorbed for the same change along the reversible path R is Q_R. Now suppose that the system has changed from state A and B along the path I. Equation (2.3.5) implies that the entropy change of the system in this transformation is **not** equal to Q_I/T, but is given by Q_R/T. Regardless of how the actual transformation is carried out from the state A to the state B, the entropy change associated with that transformation is always Q_R/T.

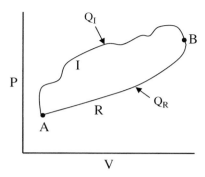

Fig. 2.2 Schematic illustration of two paths, R and I, joining two states of a system, A and B, in the P-V plane at isothermal condition. Q_R and Q_I are the heat absorbed by the system for the change of state along the paths R and I, respectively. The path R is assumed to be reversible, and I is assumed to be irreversible

An alternative approach to the development of the concept of entropy is that due to the Greek mathematician Constantin Carathéodory (1873–1950), who demonstrated mathematically, and without recourse to any hypothetical cyclic process, the existence of an integrating factor that can convert the inexact differential δq to an exact or perfect differential (For an exposition of Carathéodory's work in English language, see Margenau and Murphy 1955, and Chandrasekhar 1957).

2.4   Entropy: Direction of Natural Processes and Equilibrium

We now explore the role played by the state function entropy in (a) establishing formal criterion for the direction of spontaneous or natural processes, and (b) defining the state of thermodynamic equilibrium. For this purpose, we accept a well-known result that for a general cyclic process in a closed system

$$\oint \frac{\delta q}{T} \leq 0, \tag{2.4.1}$$

the equality holding only when the process is reversible (see Denbigh 1981 or Fermi 1956 for proof).

Referring to Fig. 2.2, let us now consider a cyclic process in which the state of a system has changed isothermally from A → B along the irreversible path I, and returned to the state A isothermally along the reversible path R. Now the overall cyclic process is irreversible since a part of it is irreversible. Consequently, we have, according to Eq. (2.4.1)

$$\oint \frac{\delta q}{T} = \int_A^B \left(\frac{\delta q}{T}\right)_I + \int_B^A \left(\frac{\delta q}{T}\right)_R < 0 \qquad (2.4.2)$$

By definition, the last integral equals $(S_A - S_B)$ (Eq. 2.3.5). Thus,

$$\int_A^B \left(\frac{\delta q}{T}\right)_I + (S_A - S_B) < 0 \qquad (2.4.3)$$

or

$$\int_A^B \left(\frac{\delta q}{T}\right)_I - (S_B - S_A) < 0, \qquad (2.4.4)$$

and, consequently,

$$S_B - S_A > \int_A^B \frac{\delta q}{T} \qquad (2.4.5)$$

Combining Eqs. (2.3.5) and (2.4.5), we can now write the following general expression for any arbitrary process in a **closed** macroscopic system:

$$\boxed{dS \geq \frac{\delta q}{T},} \qquad (2.4.6)$$

where the equality holds only when the process is reversible (i.e. when equilibrium is achieved). This is the **commonly used statement of the second law**, and constitutes one of the most revolutionary expressions in the history of science, as we would appreciate by exploring its physical implications.

If the closed system under consideration is in an adiabatic enclosure, then $\delta q = 0$. Thus, for a system that is closed with respect to both mass and heat transfer (that is an adiabatically closed system), $dS \geq 0$. Now, recall that an isolated system does not exchange either energy (in the form of work and heat) or mass with the surrounding. Thus, we can state that

$$(dS)_{\text{isolated}} \geq 0 \qquad (2.4.7)$$

or

$$\left(\frac{dS}{dt}\right)_{\text{isolated}} \geq 0 \qquad (2.4.8)$$

that is, *in an **isolated macroscopic** system, entropy can never decrease*; it either increases due to irreversible processes within the system, or stay the same when equilibrium is achieved. (However, note that for the entropy of a system to increase, it is only necessary for it to be adiabatically closed instead of being isolated, which is more restrictive, since the transfer of energy in the form of reversible work does not affect the entropy of a system).

The entropy of a system can decrease if heat is withdrawn from it and delivered to another system. But the entropy of the two systems together, which constitute an isolated system, must either increase or remain the same. Extending this approach to its extreme, we can view the Universe as the ultimate isolated system. **Thus the entropy of the Universe must always increase due to the spontaneous processes taking place inside it**. The famous astrophysicist, Sir Arthur Eddington (1882–1944), thus, called entropy as the **arrow of time**: the future is the direction of increasing entropy of the Universe. Given two snapshots of a macroscopic system, entropy provides us with the only non-subjective criterion by which one can tell which of the two represents the later stage in the time evolution of the system.

Equation (2.4.6) may be recast as

$$
\begin{aligned}
dS &= \frac{\delta q}{T} + (dS)_{int} \\
&= (dS)_{ext} + (dS)_{int} \\
&= (dS)_{ext} + \sigma, \quad (dS)_{int} = \sigma \geq 0
\end{aligned}
\tag{2.4.9}
$$

where the first and second terms on the right indicate, respectively, the entropy change of the system due to heat exchange with the surrounding (δq: heat *absorbed* from the surrounding; $(dS)_{ext}$: entropy absorbed from the external environment) and the entropy created by irreversible processes within the system. Examples of irreversible processes contributing to the internal entropy production, $(dS)_{int}$, are chemical reactions, heat and chemical diffusions, viscous dissipation. The ramifications of entropy production in irreversible processes constitute the field of **Irreversible Thermodynamics**, some aspects of which have been discussed in the Appendix A.

Problem 2.1 Show that Eq (2.4.7) is equivalent to the statement $(dS)_{U,V} \geq 0$ for a closed system subjected only to P-V work.

2.5 Microscopic Interpretation of Entropy: Boltzmann Relation

The second law of thermodynamics introduced a new description of natural processes. That is, (a) successive states attained by spontaneous processes in an **isolated** and **macroscopic** system are characterized by progressively increasing values of a quantity known as entropy, and (b) regardless of its initial state, the final goal of an isolated system is unique, being defined by its state of maximum entropy. This **independence** of the final state of a macroscopic system on its initial condition, and the idea of existence of a system property that cannot be reversed or repeated during a natural process in an isolated system were completely foreign to the spirit of the contemporary science. A microscopic or fundamental picture was, therefore, needed for the property of entropy to understand what it is physically that must always increase during a spontaneous process in an isolated system.

In the late 19th century, the Austrian Physicist, Ludwig Boltzmann (1844–1906) took this giant step, and developed the relationship between the entropy of a macroscopic system and its microscopic states (and, thus, in effect, anticipated the existence of atoms). His work led to a proper appreciation of the physical nature and the domain of validity of the second law of thermodynamics. Boltzmann showed that every macroscopic state, χ, of a system is associated with a certain number of microscopic states, $\Omega(\chi)$, and that the entropy of the given macroscopic state, $S(\chi)$, is proportional to $\ln\Omega(\chi)$. Subsequently, Planck (1858–1947; Nobel Prize: 1918), who was one of the founders of quantum mechanics, modified the Boltzmann relation to the following form

$$S(\chi) = k_B \ln \Omega(\chi) \qquad (2.5.1)$$

where k_B is known as the Boltzmann constant ($k_B = 1.381 \times 10^{-23}$ J K^{-1}). Thus the above equation is sometimes referred to as the Boltzmann-Planck relation, but more commonly simply as the Boltzmann relation. This equation is engraved in Boltzmann's tombstone in Vienna, Austria (Box 2.7.1).

To understand the meaning of the above equation, let us consider two boxes, one consisting of six black balls and the other of six white balls, each sitting in a cavity within a box (Fig. 2.3). All balls and cavities are of equal size and balls of a given color are indistinguishable among themselves. The boxes are now brought into physical contact, placed on a vibrator, and the barrier between the boxes is removed. Let us now imagine that the boxes are covered so that the balls are not visible to the naked eye, but we have some way of knowing if a ball has moved into the 'wrong' box as a result of the vibration. That is we can say, if there is one black ball to the right (hence, one white ball to the left), or two black balls to the right (hence, two white balls to the left), and so on, but we have no way of knowing

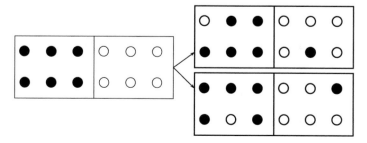

Fig. 2.3 Illustration of microscopic states associated with a given macroscopic state. The two right panels show two of the 36 possible configurations (microscopic states) that are possible for the single macroscopic state of one ball in the "wrong" position

exactly which of the available cavities in a given box are occupied by the white or black balls. The state of a box with a certain number of "wrong balls" is its macroscopic state, whereas a state with a specific arrangement of the balls within the cavities constitutes a microscopic state. If all balls can move and occupy the cavities with equal ease (which means that all microscopic states are equally accessible), then for the state of one black ball to the right, and correspondingly, one white ball to the left, there are 36 possible positions of the balls within the two boxes. In other words, for the macroscopic state of one white ball and one black ball in the 'wrong' positions, there are 36 microscopic configurations. According to Eq. (2.5.1), the entropy of this macroscopic state is S(one wrong pair) = $k_B \ln(36)$. If there are two white balls to the left and two black balls to the right, then the number of microscopic states in each box is 15, and hence the total number of microscopic states is $15 \times 15 = 225$, so that S(two wrong pairs) = $k_B \ln(225)$.

A simple way to calculate the number of distinguishable configurations within a given box, when all configurations corresponding to a given macroscopic state, χ, are equally probable is given by the combinatorial formula

$$\Omega_1(\chi) = \frac{N!}{n!(N-n)!},$$ (2.5.2)

where N is the total number of sites and n is the number of wrong balls within the particular box (note that rearrangement of balls of the same type does not lead to configurations that are distinguishable). The total number of distinguishable con-figurations, $\Omega(\chi)$, for the two boxes is given by $\Omega(\chi) = \Omega_1(\chi)\Omega_2(\chi)$. The above relation yields the number of configurations for **random** distribution over the microscopic states since all configurations are assumed to be equally probable.

We can say that the larger the number of microscopic configurations accessible to a macroscopic state of a system, the larger is the degree of microscopic disorder of that particular state of the system. Thus, the Boltzmann interpretation of entropy has led to the popular statement that the **entropy of a state of a system is a measure of the degree of disorder of that state**. Higher the disorder, the larger is the entropy. It should, however, be emphasized that this connection between entropy and order is strictly valid for an isolated system. A system that is not isolated may develop ordered structures without necessarily decreasing its entropy. For further discussion of this interesting topic, the reader is referred to Nicolis and Prigogine (1989).

Using Eq. (2.5.2), one finds that there are 924 total microscopic configurations corresponding to all possible macroscopic states (1(W)-l: 36 states; 2(W)-l: 225 states, and so on, where 1(W)-l means 1 wrong ball in the left box, and thus 1 wrong ball in the right box, and so on). If all microscopic states are equally accessible, then the probability of finding a random distribution of the balls (three white and three black balls in each box, $\Omega = 400$) after the boxes have been vibrated for a sufficiently long time, is $400/924 = 0.43$, that of finding of 2 (or 4) balls in the wrong position within each box is $225/924 = 0.24$, that of finding 1 (or 5) ball(s) in the wrong position in each box is $36/924 = 0.04$, and that of finding 6 balls in the wrong (or right) positions in each box is $1/924 = 0.001$. Thus, we are much more likely to find the system of two boxes of balls to be in a state of random distribution than in any other state after the balls have been allowed to move around for a sufficiently long period of time. After the random distribution is achieved, the system will time to time reverse back to the states of lower entropies in proportion to the probabilities of the states. Thus, the statement of second of law that the entropy of an isolated system is maximum when equilibrium is achieved is valid in a statistical sense. Also the progression of entropy towards the maximum value (corresponding to the random distribution) is not completely monotonic. **The entropy fluctuates with time, and it is the entropy of the system averaged over a certain period of time that increases with time**, as illustrated in Fig. 2.4. Larger the size of the system, smaller is the time scale over which we need to consider the average entropy value that must increase with time, and lower is the probability of the system of returning to its initial state. Thus, for a **macroscopic system under macroscopic scale** of observation, the entropy increases with time if it is **isolated** from the surrounding.

Figure 2.5, which is modified from Reif (1967), shows a computer simulation of entropy fluctuations as a function of time. There are 40 particles in a box, and at the beginning of the simulations, there were 21 particles in the left-half and 19 in the right-half, as illustrated in the left panel. Each particle was given an initial position and velocity, and it is assumed that a collision between any pair of particles does not lead to any loss of kinetic energy and momentum (this is known as **elastic collision**). The particle positions were tracked at periodic intervals so that the elapsed time is given by $t = \tau j$, where j is a frame index of the sequential snap shots (1, 2, 3 …) and τ is the time between two successive frames. The right panel shows the number of particles in the left-half as function of j, which is a proxy for time. The average number of particles in the left half of the box for 30 frames is 20, which is the expected value for maximum entropy corresponding to the most disordered state, but there are obviously fluctuations around this value in several frames.

The important relations according to, and related to, the first and second laws are summarized in the box below, in which the expressions of dU and dS apply to systems with fixed masses of all chemical components:

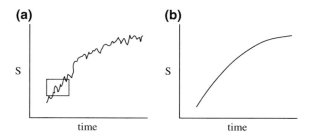

Fig. 2.4 Change of entropy of an isolated system with time: (**a**) fluctuations of entropy at the microscopic level; (**b**) progressive increase of the average value over a certain period, such as shown by the box in (**a**), towards a maximum value

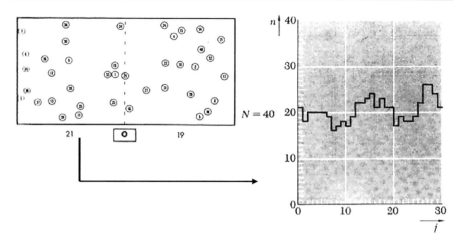

Fig. 2.5 Computer simulation of particle distribution as a function of time. The initial distribution is shown in the left, with 21 particles in the left side and 19 particles in the right side of a box. All collisions were assumed to be elastic, and the initial positions and velocities of the particles were specified. The time evolution of the number of particles in the left half of the box is shown in the right panel. The elapsed time $t = \tau j$, where j is a frame index and τ is the time between two successive frames. Modified from Reif (1967)

$$dU = \delta q + \delta w^- \quad (\delta w^- : \text{work done on the system})$$

$$dS = \frac{\delta q_{rev}}{T} \quad (\delta q: \text{heat absorbed by the system})$$

$$dS = \frac{\delta q}{T} + \sigma; \quad \sigma(\text{Internal entropy production}) \geq 0$$

$$S(\chi) = k_B \ln \Omega(\chi); \quad \Omega(\chi) : \text{number of microscopic states}$$
$$\text{related to the macroscopic state } \chi$$

Box (2.5.1)

2.6 Black Hole and Generalized Second Law of Thermodynamics

In discussing the first law of thermodynamics as a law of conservation of energy (Sect. 2.1), it was pointed out that the classic statement of the law requires a modification to account for the fact that, according to the special theory of relativity, energy can be created at the expense of mass. Coming now to the Clausius statement of the second law that the entropy of the universe always increases to a maximum, we ask what happens to the entropy of the universe when an ordinary mass falls into a black hole? This question is said to have been posed by a young physics student and later a distinguished physicist, Jacob David Bekenstein (1947–2015), to his famous physics professor John Archibald Wheeler (1911–2008).

To an observer, the entropy of the visible universe decreases as entropy-bearing objects from the latter falls and disappears into a black hole that is not amenable to observation from outside. Thus, in order for the second law to be valid, the entropy of the black hole must increase by a magnitude that more than compensates for the decrease of entropy in the visible universe. Bekenstein (1973) showed that the entropy of a black hole is proportional to its surface area and the latter increases as a result of annihilation of matter by a black hole from the exterior region. Also the increase of surface area of the black hole is such that it more than compensates for the loss of entropy of the visible universe. This led to the Bekenstein's **Generalized Second Law** of thermodynamics:

The sum of black hole entropy, S_{BH}, and the ordinary entropy of matter (and radiation field) in the visible universe, S_{VU}, cannot decrease, or

$$\Delta(S_{BH} + S_{VU}) \geq 0 \qquad (2.6.1)$$

In a landmark paper (Hawkins 1974) almost immediately following Bekenstein's work, the legendary physicist Stephen Hawking (1942–2018) showed that a black hole emits thermal radiation (now referred to as **Hawking Radiation**), leading to a reduction of its surface area as a result of decrease of mass. However, it has been found that the entropy increase associated with the Hawking radiation more than compensates for the entropy decrease due to the reduction of the surface area of a black hole, thus upholding the validity of the GSL (Bekenstein 1975; Hawkins 1976).

2.7 Entropy and Disorder: Mineralogical Applications

There are various sources of microscopic disorder in a substance that contribute to its entropy. In a crystalline substance, the two most important sources of microscopic disorders are those due to the existence of multitude of configurational and vibrational states that are consistent, respectively, with the composition and energy of the crystal, as discussed below. We would first discuss the configurational disorder, which is easier to understand, and the associated entropy, which is commonly referred to as configurational entropy.

2.7.1   Configurational Entropy

2.7.1.1   Random Atomic Distribution: Complete Disorder

Consider, for example, a solid of solution of olivine $(Mg_X, Fe_{1-X})_2SiO_4$, in which Mg and Fe occupy the octahedral lattice sites. There are two types of geometrically distinct octahedral sites or sublattices in olivine, known as M1 and M2 sites, which are present in equal number in a mole of a crystal. However, at temperatures of common geological interests, these sites are not distinguished by Mg^{2+} and Fe^{2+} ions in that there is equal population of either type of atoms in both sites (i.e. $x_{Fe}(M1) = x_{Fe}(M2)$ and $x_{Mg}(M1) = x_{Mg}(M2)$, where x_i stands for the atomic fraction of i in the specified site). Thus, we can say, at least for the purpose of calculation of configurational entropy of olivine, that effectively there is just one type of site over which Fe^{2+} and Mg^{2+} ions are distributed in olivine as long as the above condition is valid. Now, for a given composition of olivine, the Mg and Fe^{2+} ions (we will henceforth drop the superscript 2^+) can be distributed in a variety of ways over these octahedral sites without changing their fractional amounts (just as the in the example of multiple arrangements of black and white balls within a box for any specification of the content of a box).

In order to calculate the configurational entropy of a ferromagnesian olivine of specified composition, X (e.g. atomic fraction of Mg, or $X_{Mg} \equiv X = 0.2$), we need to evaluate the number of microscopic configurations associated with this composition, and then apply the Boltzmann relation, Eq. (2.5.1). Assuming, for simplicity, that the distribution of Fe and Mg are random over the available octahedral sites, we can apply Eq. (2.5.2) to calculate $\Omega(X)$. However, since we are dealing with very large number atoms (of the order of 10^{23} that are present in molar quantities), the factorial terms in Eq. (2.5.2) become very cumbersome in their standard forms, but can be easily evaluated by using what is known as **Sterling's approximation** for the factorial of large numbers (see Appendix B.7),

$$\ln N! = N\ln N - N \tag{2.7.1}$$

Now let n_{Mg} stand for the number of Mg atoms and n_{Fe} for the number of Fe atoms and N for the total number of atoms so that $N = n_{Mg} + n_{Fe}$. We can then write from Eq. (2.5.2)

$$\begin{aligned}\ln \Omega(X)_{conf(r)} &= \ln N! - \ln(n_{Mg})! - \ln(n_{Fe})! \\ &= [N\ln N - N] - [n_{Mg}\ln(n_{Mg}) - n_{Mg}] - [n_{Fe}\ln(n_{Fe}) - n_{Fe}]\end{aligned} \tag{2.7.2}$$

where $\Omega(X)_{conf(r)}$ means the number of (distinguishable) geometric configurations for random distribution of atoms corresponding to a specific bulk composition, X. Substituting N for $(n_{Fe} + n_{Mg})$, we obtain

$$\ln \Omega(X)_{conf(r)} = (n_{Fe} + n_{Mg}) \ln N - n_{Mg} \ln n_{Mg} - n_{Fe} \ln n_{Fe}$$

$$= -n_{Fe} \ln\left(\frac{n_{Fe}}{N}\right) - n_{Mg} \ln\left(\frac{n_{Mg}}{N}\right) \qquad (2.7.3)$$

$$= -n_{Fe} \ln X_{Fe} - n_{Mg} \ln X_{Mg}$$

Expressing n_i in the above equation as $N(n_i/N) = NX_i$, we now have

$$\ln \Omega(X)_{conf(r)} = -N\left(X_{Fe} \ln X_{Fe} + X_{Mg} \ln X_{Mg}\right) \qquad (2.7.4)$$

Now for a mole of olivine crystal, $(Mg_X Fe_{1-X})_2 SiO_4$, the number (N) of Fe plus Mg is 2L, where L is the Avogadro's number. Thus, substituting Eq. (2.7.4) into the Boltzmann relation, Eq. (2.5.1), we have

$$S(X)_{contin} = -k_B(2L)\left(X_{Mg} \ln X_{Mg} + X_{Fe} \ln X_{re}\right)$$

Noting that $k_B L = R$, where R is the gas constant, we finally obtain

$$S(X)_{conf(r)} = -2R\left(X_{Mg} \ln X_{Mg} + X_{Fe} \ln X_{Fi}\right) \qquad (2.7.5)$$

per mole of olivine with formula unit written as $(Mg,Fe)_2 SiO_4$. For the general case of **random** mixing of several types of atoms or ions within a particular type of crystallographic site or sublattice, the above relation can be generalized as

$$\boxed{S(X)_{conf(r)} = -\nu R \sum_i X_i \ln X_i} \qquad (2.7.6)$$

where ν is the total number of moles of all mixing units that are distributed **randomly** over the available crystallographic sites. Note that the total number of moles of mixing units equals the number of moles of the available sites in a crystal since all sites are effectively filled. The qualification "effectively" is due to the fact that there are always some equilibrium vacancies within a crystallographic site. A specific type of vacancy may be treated as a component and included in the above equation. But since the mole fraction of vacancy is very small, of the order of 10^{-4} or less, the vacancy terms make negligible contribution to the configurational entropy, and are thus usually dropped.

2.7.1.2   Ordering with Random Atomic Distribution Within Each Sublattice

If the distributions of Fe and Mg are not uniform between the two sublattices in olivine, i.e. $x_i^{M1} \neq x_i^{M2}$ where x_i is the atomic fraction of i in the specified site, then we need to distinguish the two types of sites for the purpose of calculation of configurational entropy. In this case, since $\Omega_T = \Omega_{M1}\Omega_{M2}$, we have

$$S_{conf(r)} = k_B \ln \Omega_T = k_B (\ln \Omega_{M1} + \ln \Omega_{M2})$$
$$= S_{conf(r)}(M1) + S_{conf(r)}(M2) \tag{2.7.7}$$

where the last two terms stand for the configurational entropies in the specific sites. Note here the **additive property of entropy**. The total configurational entropy of the system is the sum of the configurational entropies of the subsystems. Combining the last two equations

$$S_{conf(r)} = -R\left(\nu_{M1} \sum x_i^{M1} \ln x_i^{M1}\right) - R\left(\nu_{M2} \sum x_i^{M2} \ln x_i^{M2}\right) \tag{2.7.8}$$

where ν_{M1} and ν_{M2} are the number of moles of M1 and M2 sites per mole of olivine formula, as written. Note that in a mole of olivine crystal written as $(Mg,Fe)_2SiO_4$, there are one mole of M1 and one mole of M2 sites, so that $\nu_{M1} = \nu_{M2} = 1$ per mole of $(Mg,Fe)_2SiO_4$, and also since M1 and M2 are present in equal proportion, $X_{Fe}(total) = \frac{1}{2}[x_{Fe}^{M2} + x_{Fe}^{M1}]$.

In general, then, we have, for random of mixing of atoms in a multisite solid solution

$$S_{con(r)} = -R\left[\nu_{s(1)} \sum x_i^{s(1)} \ln x_i^{s(1)} + \nu_{s(2)} \sum x_i^{s(2)} \ln x_i^{s(2)} + \ldots\right] \tag{2.7.9}$$

where $\nu_{s(i)}$ is the number of moles of s(i) sublattice per mole of a crystal over which the mixing units are distributed, it being assumed that the distribution *within* (not between) each type of site is random.

2.7.1.3 Solved Problem: Change of Configurational Entropy Due to Random Mixing of Gases

Consider that a mole of an inert gas such as Ar (atomic No. 18) is separated within a box from three moles of another inert gas such as Xe (atomic No. 54) by an impermeable partition, with Ar in the left compartment (L) and Xe in the right compartment (R). Now the partition is replaced by one that permits the transport of both gas atoms, which eventually leads to a random distribution of the two gases within the box, without changing the total number of moles of gas in each compartment. What is the change of entropy of the system?

According to Boltzmann relation (Eq. 2.5.1), S(initial) = 0 since $\Omega = 1$. The entropy of the system after the replacement of the partition and attainment of random distribution within the box may be calculated in two different but equivalent ways: (a) we can calculate the configurational entropies of each compartment separately, and then add these to obtain the total configurational entropy of the box, or (b) we can treat the box as not having any compartment since both compartments become equivalent after attainment of random distribution of the two gases within the box.

For the entire system, we have $X_{Ar} = 0.25$ and $X_{Xe} = 0.75$. Thus, after random distribution is achieved, $X_{Ar}(L) = X_{Ar}(R) = 0.25$ and $X_{Xe}(L) = X_{Xe}(R) = 0.75$.

Now using the approach (a):

$$
\begin{aligned}
S_{conf(r)} &= \left[S_{conf(r)}(L)\right] + \left[S_{conf(r)}(R)\right] \\
&= -R[0.25\ \ln(0.25) + 0.75\ \ln(0.75)] - 3R[0.25\ \ln(0.25) + 0.75\ \ln(0.75)] \\
&= -4R[0.25\ \ln(0.25) + 0.75\ \ln(0.75)] \\
&= 18.701\ \text{J/K}
\end{aligned}
$$

Alternatively, using the approach (b), we obtain directly:

$$
S(\text{random}) = -4R[0.25\ \ln(0.25) + 0.75\ \ln(0.75)]
$$

since there are four moles of gas in the system. Thus, the change of entropy, $\Delta S = S_{conf(r)} - S(\text{initial})$, is given by $\Delta S = 18.701 - 0 = 18.701$ J/K. The change of entropy due to mixing, as in the above example, is known as the **entropy of mixing**, and is commonly denoted as $\Delta S(\text{mix})$ (note that the mixing need not be random).

Instead of Ar and Xe, if the two gases were two isotopes of a gas, for example, ^{16}O and ^{18}O, the entropy of mixing is still the same. **The entropy of random mixing does not depend on the extent of difference in the properties of the mixing units**, as long as the mixing units are non-interacting and are not identical. For mixing of identical particles or chemical species, $\Delta S(\text{mix}) = 0$ since rearrangements of different atoms of the same type do not lead to different arrangements that are distinguishable.

Problem 2.2 The mineral albite, $NaAlSi_3O_8$, has four non-equivalent tetrahedral sites that are occupied by Al and Si. These sites are labelled as $T_1(O)$, $T_1(m)$, $T_2(O)$ and $T_2(m)$. At low temperature (<650 °C), albite has an ordered distribution of Al and Si in that Al substitutes preferentially in the $T_1(O)$ site, whereas Si substitutes preferentially in the other tetrahedral sites (this structural form of albite is called low-albite). On the other hand, at higher temperature, the distribution of Al and Si tend to become random. Assuming that the Al and Si distribution in 'low albite' is completely ordered and that in 'high albite' is completely random (i.e. equal occupancy of Al and Si in all T sites), calculate the $\Delta S_{conf(r)}$ when 'low albite' transforms to 'high albite. Then continue to Problem 6.3 when you have gone up to that level.

Answer: 18.70 J/mol-K.

2.7.1.4 Constrained Random Atomic Distribution Within a Sublattice

If the distribution within a sublattice is not random, then $\Omega_{S(i)}$ has to be calculated taking into account the restriction on the distribution of atoms. A problem of the latter type is encountered in connection with the distribution of Al and Si over the tetrahedral sites of a crystal because the distribution should avoid an energetically unfavorable configuration Al-O-Al (known as the 'aluminum avoidance principle') that will be sometimes encountered if the distribution of Al and Si are completely random (Loewenstein 1954). As an example, let us consider the configurational entropy of an orthopyroxene solid solution in the system $MgSiO_3$-Al_2O_3. Here the charge balanced substitution is $(MgSi)^{6+} \Longleftrightarrow (2Al)^{6+}$, with Mg occupying two non-equivalent crystallographic sites, M1 and M2, and Si occupying two non-equivalent tetrahedral sites, A and B. Al enters both M1 and M2 octahedral sites, but only the B tetrahedral site.

The distribution of Al and Si in each M site may be assumed to be random, in which case the configurational entropy arising from the mixing of Mg and Al within the M1 and M2 sites is given by Eq. (2.7.6). However, the calculation of S_{conf} due to the mixing of Al and Si in the tetrahedral B site needs to account for the effect of exclusion of Al-O-Al linkage. That is, we need to calculate S_{conf} in the tetrahedral B site for random mixing of Al and Si, but with the constraint that Al-O-Al linkages are avoided in the pyroxene structure which consists of a single tetrahedral chain. This problem was first addressed by Ganguly and Ghose (1979) who deduced that the total number of sites that are available to the interchange of Al and Si in the pyroxene structure, subject to the "aluminum avoidance principle", is $(N^B - Al^B + 1)$, where N^B is the total number of B sites and Al^B is the total number of Al within the B sites. Thus, according to Eq. (2.5.2)

$$\Omega_{conf}^B(Al - avoid) = \frac{(N^B - Al^B + 1)!}{Al^B!(N^B - 2Al^B + 1)!} \qquad (2.7.10)$$

Now let the total number of B sites be α times the number of the Al atoms in the site, so that $N = \alpha Al^B$. Using Sterling's approximation for the factorial of large numbers (Eq. 2.7.1), we then obtain

$$\begin{aligned} S_{conf}^B(Al - avoid) &= k_B \ln \Omega_{conf}^B(Al - avoid) \\ &= k_B Al^B \{(\alpha - 1)\ln[Al^B(\alpha - 1) + 1] - \ln[Al^B(\alpha - 2) + 1]\} \\ &\quad + k_B \{\ln[Al^B(\alpha - 1) + 1] - \ln[Al^B(\alpha - 2) + 1]\} \end{aligned}$$
$$(2.7.11)$$

Now the fraction of Al in the B site, x_{Al}^B, is Al^B/N^B. Thus, $k_B(Al^B) = k_B(N^B x_{Al}^B)$ so that, if N^B equals the Avogadro's number (i.e. if there is one mole of B site), then $k_B(Al^B) = Rx_{Al}^B$. Also since Al^B is very large, we can easily drop the term +1 within the square brackets of the above equation. Thus,

$$S_{conf}^B(Al - avoid) = Rx_{Al}^B[(\alpha - 1)\ln(\alpha - 1) - (\alpha - 2)\ln(\alpha - 2)] \qquad (2.7.12)$$

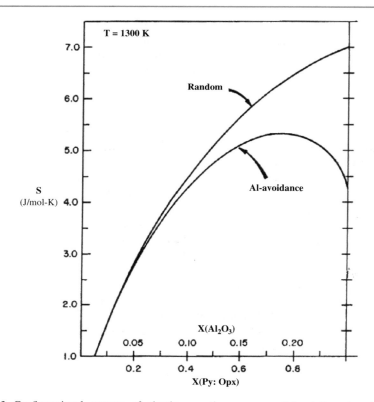

Fig. 2.6 Configurational entropy of aluminous orthopyroxene solid solution, in which Al substitutes in both octahedral and tetrahedral B sites according to $^{VI}(Mg)^{IV}(Si) \leftrightarrow {}^{VI}(Al)^{IV}(Al)$. The distribution of Mg and Al in the octahedral sites is assumed to be completely random whereas that of Si and Al in the tetrahedral B sites is assumed to be either completely random or random but subject to the Al-avoidance principle. X(Py:Opx) is the mole fraction of an aluminous end-member component, $Mg_3Al_2Si_3O_{12}$, in orthopyroxene that is treated as a solid solution of this component with $Mg_4Si_4O_{12}$. From Ganguly and Ghose (1979)

Note that by definition $\alpha = 1/x_{Al}^B$. The S_{conf} and molar entropy of orthopyroxene solid solution, as calculated by Ganguly and Ghose (1979) from the site occupancy data of Al, Mg and Si, are shown in Fig. 2.6, in which the results of calculations for completely random mixing in the M1, M2 and B sites are compared with those that consider *conditional* random mixing in the B site, as expressed by the above equation.

2.7.2 Vibrational Entropy

As discussed in Sect. 1.6, a crystal may be viewed as a collection of atomic oscillators with quantized vibrational energy levels. For a given energy of a crystal, the

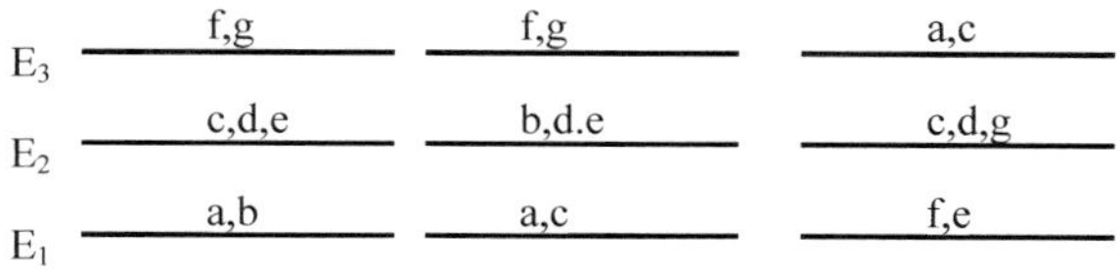

E$_3$	f,g	f,g	a,c
E$_2$	c,d,e	b,d.e	c,d,g
E$_1$	a,b	a,c	f,e

Fig. 2.7 Some examples of the distribution of seven oscillators over three energy levels, E$_1$, E$_2$, E$_3$, such that there are two oscillators in E$_1$, three in E$_2$ and two in E$_3$

oscillators can be distributed in a number of ways over the vibrational energy levels. As an example, let us consider a collection of 7 oscillators (a, b, c, d, e, f and g), each with a frequency of v, and 3 vibrational energy levels (E$_1$, E$_2$ and E$_3$) with the following **distribution** of the oscillators: 2 in level E$_1$, 3 in level E$_2$ and 2 in level E$_3$. Three examples of such distribution are illustrated in Fig. 2.7. The number of possible **arrangements** of the oscillators for a specific distribution is given by

$$\Omega_{vib} = \frac{N!}{n_1!n_2!n_3!}, \qquad (2.7.13)$$

where N is the total number of oscillators, and n$_i$ is the number of oscillators in the energy level E$_i$. Thus, for the above distribution of oscillators, there are 210 possible arrangements or configurations (7!/(2!3!2!)). However, the total number of possible arrangements of the oscillators that we have calculated above is without any restriction on the energy of the crystal. In practice, each arrangement of the oscillators must be such that the total energy of the crystal is conserved. The most probable distribution of oscillators over the quantized vibrational energy levels is the one that maximizes the function Ω_{vib} subject to the energy conservation constraint (recall that the energy of each oscillator is given by (n + ½)hv - see Eq. 1.6.2). Just as the configurational entropy is the result of multitude of possible distribution of atoms over the available lattice sites subject to the conservation of the bulk composition of a crystal, as discussed above, the vibrational entropy is the result of the multitude of arrangements of the atomic oscillators over the quantized vibrational energy levels subject to the constraint of energy conservation. The method of calculation of vibrational entropy is discussed in Chap. 14 dealing with statistical thermodynamics.

2.7.3   Configurational Versus Vibrational Entropy

The requirement that the entropy or the overall state of disorder of a solid must increase with increasing temperature commonly leads to progressive disordering or random-ization of the distribution of atoms among the non-equivalent crystallographic sites in a solid, such as noted in the Problem 2.2 for the temperature dependence of the distribution of Al and Si among the tetrahedral sites of albite, NaAlSi$_3$O$_8$. Another example is orthopyroxene, (Fe,Mg)SiO$_3$, in which Fe and Mg occupy two types of

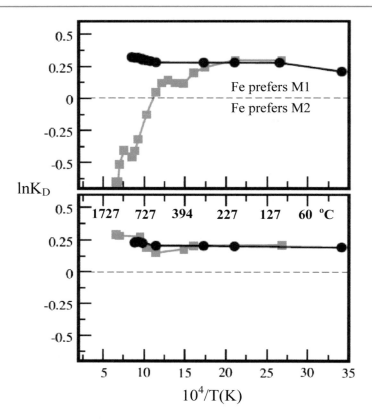

Fig. 2.8 Temperature dependence of the Fe-Mg distribution coefficient, K_D, between the M1 and M2 sites of olivine, $(Fe,Mg)^{M1}(Fe,Mg)^{M2}SiO_4$. K_D is defined as the $(Fe/Mg)^{M1}/(Fe/Mg)^{M2}$ ratio. Upper panel: Experimental data of Heinemann et al. (2006), circles, and Redfern et al. (2000), squares. Lower panel: theoretical calculations using structural data from Heinemann (circles) and Redfern (squares). Modified from Chatterjee et al. (2011)

octahedral or six coordinated sites, M1 and M2, with a preference of Fe for the M2 site and that of Mg for the M1 site. The site preferences decrease with increasing temperature, that is the distribution of Fe and Mg between the two types of sites progressively approach the state of random distribution, although complete random distribution is never achieved (the temperature needed to achieve the state of random distribution lies beyond the melting temperature of orthopyroxene).

An interesting exception to the common trend of progressive configurational disordering with increasing temperature is found in the temperature dependence of the state of Fe-Mg ordering between the M1 and M2 sites of olivine. The results of experimental study (Redfern et al. 2000; Heinemann et al. 2006) and theoretical calculations (Chatterjee et al. 2011) of the temperature dependence of intra-crystalline Fe-Mg distribution coefficient, K_D, are illustrated in Fig. 2.8. Here K_D is defined as

$$K_D = \frac{(Fe/Mg)^{M1}}{(Fe/Mg)^{M2}} \tag{2.7.14}$$

For completely random distribution of Fe and Mg between the M1 and M2 sites, $K_D = 1$ (ln $K_D = 0$).

The experimental data of Redfern et al. (2000) show that after achieving a state of complete disorder as the temperature increased to ~ 600 °C, Fe partitioned preferentially into the M2 site (and correspondingly Mg to M1 site) with continued heating. This leads to progressive configurational *ordering* and hence lowering of configurational entropy (S_{conf}) as temperature increases above 600 °C. The data of Heinemann et al. (2006), on the other hand, show K_D to be nearly independent of temperature within their investigated temperature range of ~ 15–900 °C and are similar to the subsequent theoretical predictions. Since the overall state of disorder must increase with increasing temperature, we then have $\partial S/\partial T \approx (\partial S/\partial T)_{vib} + (\partial S/\partial T)_{conf} > 0$ with $(\partial S/\partial T)_{conf} \leq 0$.

As another example of the role of vibrational entropy in increasing the overall disorder even when there is a decrease of configurational disorder, consider the phenomenon of spontaneous crystallization of supercooled water, as discussed by Denbigh (1981). A supercooled water is metastable, and would eventually crystallize to ice. Since ice has a more ordered arrangement of H_2O molecules than water, there is a decrease of configurational entropy. Thus, it may seem that a spontaneous natural process in an isolated system has led to a decrease of entropy, in contradiction to the second law (this is sometimes referred to as the **Bridgman paradox**, after the name of Percy Williams Bridgman (1882–1961; Nobel prize: 1946) who first posed this problem, presumably to test the understanding of the second law). However, it is not so. There is a more than compensating gain in the vibrational entropy in the transformation from water to ice as the latter has much larger number of vibrational energy levels over which the vibrational modes can be disordered without changing the total energy of the ice.

Box (2.7.1): Boltzmann's Struggle and (Posthumous) Triumph
The description of the equilibrium state of an isolated system as a state of maximum disorder of its microscopic entities was due to what is known as Boltzmann's **H-theorem**. According to this theorem, which Boltzmann proposed in 1872, the elementary entities of a gas will fill a confined space after it is released from one corner, and stay in that state indefinitely. The theorem faced challenge from leading contemporary scientists mainly for two reasons: (a) it appealed to atoms, the existence of which was not proved during Boltzmann's time (and which many thought to be merely the product of "overactive imagination" rather than real physical entities), and (b) it was in apparent conflict with Newtonian mechanics, which predicts time reversibility in the evolution of a system, that is if a state A evolves to a state B, then with the progress of time the reverse must also be possible. In

deriving the H-theorem, Boltzmann also made a questionable assumption that
the atoms comprising a gas are at all times distributed independently of one
another, which implies no interaction among them, and his approach was
statistical in nature describing the behavior of the average property of the gas
particles. Boltzmann's theorem faced strong challenge from another contemporary Austrian Physicist, Josef Loschmidt, but the strongest challenge
seems to have been posed by the work of the famous French mathematician,
Henri Poincaré (1854–1912). In a work that is known as the *recurrence
theorem*, Poincaré proved that a group of three particles moving in a confined
space and obeying Newton's laws of motion would repeatedly return very
close to their initial configuration. This initiated a bitter debate between
Boltzmann and the German mathematician Ernst Zermelo, who argued that
Poincaré's theorem is valid for any number of particles, and therefore the
H-theorem must be invalid. Boltzmann responded by showing that the
number of particles even in a moderate volume of gas is so large that the time
scale of recurrence would exceed the age of the universe, and therefore
recurrence would not be observed in a macroscopic system. However, this
bitter debate took a toll on Boltzmann's psyche, and depressed with the
feeling that he had failed to discover a law that is valid universally and that so
little of his works seemed to have been accepted by the leading scientists and
mathematicians of his time, Boltzmann took his own life in 1906 while on
vacation with his family in Duino, near Trieste, Italy (it is speculated that
Boltzmann might have suffered from bipolar disease). We now know that one
of the two greatest developments of modern physics, namely quantum
mechanics, is indeed statistical in nature, and that atoms are not products of
"overactive imagination", and also that physical laws, including Newtonian
mechanics, are often valid within certain domains, rather than universally.
There is now a revival of interest among mathematicians about the
H-theorem, to see if it is still valid for systems with large number of particles
without Boltzmann's assumption of their completely non-interacting behavior
at all times. But since no one has yet found a violation of Boltzmann's
theorem for a macroscopic system, and his ideas are so much in tune with
modern developments in physics, mathematicians no longer see the problem,
as the mathematician Shinbrot (1987) puts it, as "how to prove Boltzmann
wrong but how to prove him right". An excellent but quite non-technical
account of the modern developments can be found in the referred article by
Shinbrot.

Tombstone of Ludwig Boltzmann, Viena, Austria

2.8 First and Second Laws: Combined Statement

We can now combine the first and second laws as follows. According to the first law, we have for a **closed** system

$$dU = \delta q + \delta W^-$$
(2.8.1)

If the process is reversible, then according to the second law,

$$dS = \left(\frac{\delta q}{T}\right)_{rev}$$
(2.8.2)

Furthermore, for a reversible process,

$$\delta W^- = -PdV + (\delta\omega^-)_{rev},$$
(2.8.3)

where P is the pressure within the system itself, and $(\delta\omega^-)_{rev}$ is the reversible work other than P-V work absorbed by the system, i.e. performed by the surrounding on the system (recall that if the process is not reversible then P is the external pressure on the system). Thus, for a *reversible* process, we can write the following combined statement of the first and second laws for systems with fixed mass of all chemical species

$$\boxed{dU = TdS - PdV + (\delta\omega^-)_{rev}}$$ (2.8.4)

This equation embodies the definitions of absolute temperature and pressure as partial derivatives terms of internal energy of a closed system:

$$T = \left(\frac{\partial U}{\partial S}\right)_{V,\omega}; \quad P = -\left(\frac{\partial U}{\partial V}\right)_{S,\omega}$$ (2.8.5)

From Eq. (2.8.4), we may write for a system with fixed masses of all chemical species and restricted only to P-V work,

$$\boxed{U = U(S, V),}$$ (2.8.6)

This equation constitutes a **fundamental relation** of thermodynamics. As we would see later, a number of practically useful thermodynamic potentials can be derived from this fundamental relation.

Problem 2.3 Show that the reversible work performed by a system on the surrounding upon changing its state from condition A to condition B is greater than the irreversible work performed by it for the same change of state, i.e.

$$(\delta w^+)_{rev} > (\delta W^+)_{irrev}$$ (2.8.7)

Hint: Use first law for both processes, and the fact that dU is an exact differential. Then use the second law, i.e. $TdS > (\delta q)_{irrev}$.

*Problem 2.4 Seven moles of a perfect gas expand adiabatically against no external restraint. The initial P-T condition is 5 bars, 300 K, and the final pressure is 0.5 bars. Calculate ΔT, ΔV, ΔU, Q, and W^+ (i.e. the work done by the body), making use of the knowledge that the internal energy of an ideal gas depends only on temperature. (Hint: pay close attention to the language of the problem; also read Sect. 1.4 for calculating work)

2.9   Condition of Thermal Equilibrium: An Illustrative Application of the Second Law

As an illustration of how the second law of thermodynamics leads to the determination of the direction of spontaneous change and condition of equilibrium, let us consider a "composite system" (Fig. 2.9), which is isolated from the surrounding, and consists of two subsystems, 1 and 2 that are initially separated by an adiabatic wall so that no heat can flow from one subsystem to another. (The important role played by "composite systems" in the derivation of thermodynamic condition of equilibrium has been discussed in Sect. 1.3.) Each subsystem has a uniform temperature, but the temperature in one is different from that in the other. Thus, each subsystem is in internal equilibrium and therefore the overall system is in equilibrium consistent with the restrictions imposed by the walls. Consequently, each subsystem and the overall system have definable thermodynamic properties.

Now consider that the internal adiabatic wall is replaced by a thin but rigid diathermal wall that permits heat transfer. It is well known from common experience that heat will flow from higher to lower temperature until a uniform temperature is established throughout the system. We want see if this 'common knowledge' readily follows from the second law.

For the convenience of derivation, we want to maintain uniform temperature within each body, even if heat is flowing from one to the other. This can be accomplished by making the diathermal wall to be poorly conducting so that the heat transfer across the partition is much slower than that needed for the attainment of thermal equilibrium within each body or subsystem.

Let S be the total entropy of the system and S_1 and S_2 the entropies of the subsystems. We assume that the entropy of the wall is negligible compared to the total entropy of the system. In that case, since the composite system is isolated from the surrounding, and entropy is additive (Eq. 2.7.7), we have $S = S_1 + S_2$. Also since the composite system is isolated from the surrounding, the sum of the internal energies of the two subsystems, i.e. $U_1 + U_2$, must remain constant, so that $dU_1 = -dU_2$. Thus, for a constant value of U, the entropy change of the composite system with respect to a change in the internal energy of one of the subsystems, say U_1, is given by

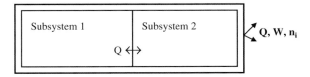

Fig. 2.9 Cross-sectional view of heat transfer across a rigid, impervious and weakly diathermal wall separating two subsystems (1 and 2). The composite system is isolated from the surrounding

$$\frac{\partial S}{\partial U_1} = \frac{\partial S_1 + \partial S_2}{\partial U_1} = \frac{\partial S_1}{\partial U_1} - \frac{\partial S_2}{\partial U_2} = \frac{1}{T_1} - \frac{1}{T_2} \qquad (2.9.1)$$

(We write partial derivatives in the above equation since the volume of each sub-system has been held constant.) Now, since each part of the system is considered to be rigid, we have $dU_1 = \delta q_1 - PdV_1 = \delta q_1$. Thus,

$$\partial S = \delta q_1 \left(\frac{1}{T_1} - \frac{1}{T_2} \right) \qquad (2.9.2)$$

But since according to the second law, $dS \geq 0$ for an isolated system, we have from the above relation

$$\delta q_1 \frac{(T_2 - T_1)}{T_1 T_2} \geq 0 \qquad (2.9.3)$$

Thus, if $T_2 > T_1$, then the heat absorbed by the system 1, δq_1, must be positive (i.e. heat must flow from 2 to 1), and vice versa. This process will continue as long as there is any finite temperature difference between the two bodies. At equilibrium $\partial S = 0$, and consequently $T_2 = T_1$.

Problem 2.5 Show that in the above problem, $dS_1 = \delta q_1/T$ even though the process is irreversible.

Hint: From the first law, and the fact that U is a state function, show that $(\delta q_1)_{rev} = (\delta q_1)_{irrev}$.

2.10   Limiting Efficiency of a Heat Engine and Heat Pump

2.10.1   Heat Engine

A heat engine is an engineering device that withdraws heat from a heat source and converts it into mechanical work. Thermodynamic restriction (second law) requires that some heat must be wasted (i.e. delivered to a heat sink) in the process of conversion of heat to work and continued operation of the heat engine (Fig. 2.10). The conversion factor or efficiency, η, of such a device is defined by the ratio of the mechanical work performed by it to the amount of heat that it absorbed, which has a price tag. Thus,

$$\eta = \frac{W^+}{Q_2} \qquad (2.10.1)$$

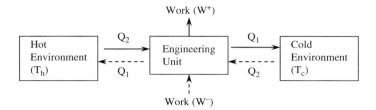

Fig. 2.10 Schematic illustration of the operation of a heat engine (solid lines) and heat pump (dashed lines) between two environments at two different temperatures. Q_2 and Q_1 stand for the heat extracted and heat delivered, respectively. W^+ and W^- are, respectively, the work done and work absorbed by the engineering unit

Thermodynamic considerations permit an evaluation of the maximum possible conversion of heat to work by considering that attainable in a reversible cyclic process (i.e. in a Carnot cycle), since the work performed by a system in a reversible process is greater than what it can perform in an irreversible (i.e. real) process (Eq. 2.8.7).

Application of the first law to a reversible cyclic process yields

$$\Delta U = 0 = (Q_2 - Q_1) + W^- = (Q_2 - Q_1) - W^+ , \qquad (2.10.2)$$

where $(Q_2 - Q_1)$ is the **net** heat absorbed by the system, Q_2 being the heat absorbed from a source and Q_1 being the heat delivered to a sink. Both Q_2 and Q_1 are treated as positive quantities (Note that while the relation $\Delta U = 0 = Q + W^-$ is true for any type of cyclic process, the relation $Q = Q_2 - Q_1$ is valid only when the cyclic process is extremely slow or reversible so that the isothermal expansion and isothermal compression are the only reasons for the change of heat content in the system. For example, if the process is not sufficiently slow to maintain a constant temperature in the body then heat could be lost by thermal diffusion). Thus,

$$\eta_{max} = \frac{(W^+)_{rev}}{Q_2} = \frac{Q_2 - Q_1}{Q_2} = 1 - \frac{Q_1}{Q_2} \qquad (2.10.3)$$

However, for a reversible process $Q_1/Q_2 = T_1/T_2$ (Eq. 2.3.1) so that

$$\boxed{\eta_{max} = 1 - \frac{T_1}{T_2} \equiv \frac{T_h - T_c}{T_h}} \qquad (2.10.4)$$

where the subscripts h and c stand for 'hot' (high temperature), and 'cold' (low temperature), respectively. Thus, the *limiting* efficiency or conversion factor of a heat engine operating between two heat baths, each maintained at a constant temperature, depends only on the temperatures of the heat baths.

In the design of a real engine, the property of interest is not necessarily the efficiency of conversion of energy to work, but the efficiency of **power output** that generates profit after paying for the cost of fabrication and maintenance. Here one considers an endoreversible engine in which the processes of heat transfer to and from the engine are the only irreversible processes. It can be shown (Callen 1985), using the above result on maximum efficiency of a Carnot engine, that the maximum efficiency of power output, ε_{erp} from an endoreversible engine is given by

$$\varepsilon_{erp} = 1 - \left(\frac{T_c}{T_h}\right)^{1/2} \qquad (2.10.5)$$

It is interesting to note that the above expression is independent of the conductivities of the materials through which heat is transferred into and out of the heat engine. The efficiencies of large power plants closely match the limiting efficiencies given by the above expression.

2.10.2 Heat Pump

The operation of a heat pump is just the opposite to that of a heat engine in that it withdraws heat from a cold source and delivers heat to a hot sink. Familiar examples are the devices that keep the interior of a refrigerator cool by withdrawing heat from inside it and delivering heat to the room (just feel the temperature at the back of refrigerator), or an air-conditioning unit that keeps the inside of a room/house cool by withdrawing heat from inside and delivering heat to the atmosphere, or a heating unit that keeps a house warm during winter by performing just the reverse operation. In order for a heat pump to perform the desired action of transferring heat from a cold to a hot environment, work needs to be performed *on* the heat pump (which is to be purchased from a power company), as illustrated in Fig. 2.10. An appropriate measure of the efficiency of the performance of a heat pump which is supposed to cool an environment would be the ratio of heat extracted from that environment to the work performed **on** the pump. We would call this refrigeration efficiency, ε_r. It is easy to see from Eq. (2.8.7) that the work absorbed by a system (i.e. performed *on* a system) in order for it to achieve a given change of state is less when the work is performed along a reversible path than when it is performed in an irreversible path (if we multiply both sides of Eq. (2.8.7) by -1, then it becomes $-(\delta w^+)_{rev} < -(\delta w^+)_{irrev}$ or $(\delta w^-)_{rev} < (\delta w^-)_{irrev}$). Thus, since $W^- = Q_1 - Q_2$, as shown above (Eq. 2.10.2), the maximum value of refrigeration efficiency is given by

$$(\varepsilon_r)_{max} = \frac{Q_2}{(W^-)_{rev}} = \frac{Q_2}{Q_1 - Q_2}, \qquad (2.10.6)$$

Dividing both the numerator and denominator by Q_1 and substituting the relation $Q_1/Q_2 = T_1/T_2$, we obtain

$$(\varepsilon_r)_{max} = \frac{T_c}{T_h - T_c} \qquad (2.10.7)$$

On the other hand, if the purpose of the heat pump is to extract heat from a cold environment and deliver heat inside a hotter environment (e.g. keeping a house warm in winter), as illustrated by the dashed lines in Fig. 2.10, the performance efficiency of the heating unit, ε_h, should be measured by the amount of heat delivered to the amount of work performed *on* it. In that case we have, following the above train of arguments,

$$(\varepsilon_h)_{max} = \frac{T_h}{T_h - T_c} \qquad (2.10.8)$$

It is obvious that the performance efficiency of a heat pump (whether it is supposed to cool or heat) decreases as the cold environment, from which heat is to be extracted, gets cooler, if the temperature of the hot environment remains the same.

There are four common sources of **irreversibility** in heat engines (Kittel and Kroemer 1980): (a) part of the heat withdrawn from the high temperature source may get conducted directly to the heat sink, for example through the cylinder wall containing the heat engine, (b) there may be thermal resistances to the flow of heat to and from the heat engine, (c) part of the work produced by it may get converted to heat by friction, and (d) gas may expand irreversibly within the pump.

Problem 2.6 A house is kept cool at a temperature of 75 °F by an air-conditioner on a hot summer day when the outside temperature is 110 °F. In order to do this, the air-conditioner needs to withdraw heat at the rate of 740 kJ/min. Calculate the minimum power that must be consumed by the air-conditioner. Note that unit of power is watt or J/s.

Hint: The minimum power in watt consumed by the air-conditioner is the same as the minimum rate of work in J/s done on it. The latter may be calculated from the ratio of the rate of heat withdrawn from outside and the maximum efficiency of the air-conditioner.

Answer: 807.3 W.

2.10.3 Heat Engines in Nature

At this point, it should be interesting to discuss two Carnot-type heat engines in nature that drive convection in the Earth's mantle and hurricane. Figure 2.11 schematically illustrates the nature of a convecting cycle. We note that unlike the Carnot cycle depicted in Fig. 2.1, there are no isothermal paths in the convecting cycle since lateral temperature gradients must be present in order to drive the convection. However, we can reduce the convecting cycle into one that involves two

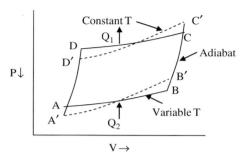

Fig. 2.11 Schematic illustration of the reduction of a cyclic process that involves two variable temperature paths (A → B, C → D) into a cyclic process involving two isothermal paths (dashed lines, A′ → B′, C′ → D′). The area enclosed by ABCD, which represents the work done by the variable temperature cyclic process, is the same as that enclosed by A′B′C′D′

isothermal paths at the source and sink, which are represented by A′B′ and C′D′, respectively (dashed lines), such that the P-V area defined by ABCD, which represents the work done by the system, is the same as that defined by A′B′ C′D′.

The earth's mantle below a thermal boundary layer or lithosphere is convecting efficiently and is generally believed to have nearly adiabatic or isentropic temperature gradient, as discussed in a later Sect. 7.3. The convection involves deformation of mantle material. As discussed by Stacey (1992), the mechanical power required for the deformation is derived from the convection itself. However, the mechanical power is dissipated into the convecting medium and gets returned to the heat source. Thus, using Eq. (2.10.1), the efficiency of the convecting heat engine (che) in the mantle (work done divided by the effective heat input) is given by

$$\eta_{che} = \frac{W^+}{Q_2 - W^+} = \frac{W^+/Q_2}{1 - W^+/Q_2} \tag{2.10.9}$$

From Eqs. (2.10.3) and (2.10.4), the maximum value of W^+/Q_2 is given by $(T_h - T_c)/T_h$. Substituting this relation in the last equation, we obtain the expression of maximum efficiency as (Stacey 1992)

$$\eta_{che}(\text{max}) = \frac{T_h - T_c}{T_c} \tag{2.10.10}$$

Note that the maximum efficiency expressed by the last equation differs from that in Eq. (2.10.4). The latter gives the efficiency of a reversible Carnot engine and has T_h instead of T_c in the denominator. Thus, in principle, η_{che} could be greater than 1. However, as pointed out by Stacey (1992), it should not be concluded that the efficiency of a convecting "mantle engine" actually exceeds the reversible thermodynamic engine, but that the dissipation of mechanical work into the source helps increase the mechanical power.

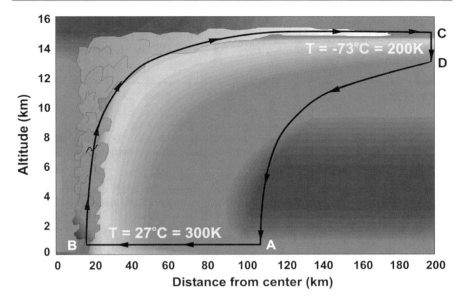

Fig. 2.12 Representation of hurricane as a Carnot type heat engine. AB and CD are nearly isothermal heat source (ocean) and heat sink (space), respectively. BC and CD are the adiabatic limbs, representing expansion and compression, respectively. The storm center is at the left edge. From Emanuel (2006). With permission from the American Institute of Physics

Figure 2.12 shows hurricane as a Carnot engine, whose storm center lies along the left edge (Emanuel 2006). The stage A to B indicates nearly isothermal expansion in contact with the surface of ocean which is effectively an infinite heat reservoir. At B, where the surface wind is strongest, the air turns abruptly upwards as a nearly adiabatic flow until it reaches the point C. Between C and D, heat is isothermally radiated to space as infrared radiation. Finally, adiabatic compression from D to A brings the cycle to completion. As shown by Emanuel (2003, 2006), the wind speed of the hurricane is given by

$$\upsilon^2 = \left(\frac{T_h - T_c}{T_c}\right)E, \qquad (2.10.11)$$

where E is a measure of the thermodynamic disequilibrium between ocean and atmosphere that allows convective heat transfer. Here again, note that T_c instead of T_h appears in the denominator. This is due to the added contribution from dissipative heating.

It is evident from the above expression that the thermodynamic efficiency term, that is the term within the parentheses, has small effect on the wind velocity of a hurricane; for example, an increase of ocean temperature by 5 K would raise the wind velocity by only $\sim 2\%$. The main effect on υ comes from the thermodynamic disequilibrium term E that can be shown to increase with increasing greenhouse gases and water vapor that increases with global warming. The increase of E saturates at ocean surface temperature of 35–38 °C (Emanuel, pers.com).

References

Bekenstein JD (1973) Black holes entropy. Phys Rev D 7:2333–2346
Bekenstein JD (1975) Statistical black-hole thermodynamics. Phys Rev D 12:3077–3085
Callen HB (1985) Thermodynamics and an introduction to thermostatics. Wiley, New York, Chichester, Brisbane, Toronto, Singapore, p 493
Chandrasekhar S (1957) An introduction to stellar structure. Dover
Chatterjee S, Bhattacharyya S, Sengupta S, Saha-Dasgupta T (2011) Crossover of cation partitioning in olivines: a combination of ab initio and Monte Carlo study. Phys Chem Min 38:259–265
Denbigh K (1981) The principles of chemical equilibrium. Dover
Emanuel K (2003) Tropical cyclones. Ann Rev Earth Planet Sci 31:75–104
Emanuel K (2006) Hurricane: tempests in a greenhouse. Phys Today 59:74–75
Fermi E (1956) Thermodynamics. Dover, p 160
Feynman R (1963) The Feynman lectures on physics, vol 1. Addison Wesley
Ganguly J, Ghosh SK (1979) Aluminous orthopyroxene: order-disorder, thermodynamic properties, and petrological implications. Contrib Mineral Petrol 69:375–385
Hawkins SW (1974) Black-hole explosions. Nature 248:30–31
Hawkins SW (1976) Black-holes and thermodynamics. Phys Rev D 13:191–197
Heinemann R, Kroll H, Kirfel A, Barbier B (2006) Order and anti-order in olivine I: Structural response to temperature. Eur J Mineral 18:673–689
Kittel C, Kroemer H (1980) Thermal physics. Freeman, San Francisco, p 473
Loewenstein W (1954) The distribution of aluminum in the tetrahedra of silicates and aluminates. Amer Min 39:92–96
Margenau H, Murphy GM (1955) The mathematics of physics and chemistry. Van Nostrand, Toronto, New York, London
Nicolis G, Prigogine I (1989) Exploring complexity. W.H. Freeman, New York, p 313
Redfern SAT, Artioli G, Rinaldi R, Henderson CMB, Knight KS, Wood BJ (2000) Octahedral cation ordering in olivine at high temperature. II: an in situ neutron powder diffraction study on synthetic MgFeSiO$_4$ (Fa50). Phys Chem Min 27:630–637
Reif F (1967) Statistical physics: Berkeley physics course, vol 5, pp 398
Shinbrot M (1987) Things fall apart: no one doubts the second law, but no one's proved it - yet. The Sciences:32–38
Stacey FD (1992) Physics of the earth. Brookfield Press, Brisbane, Australia, p 513

Thermodynamic Potentials and Derivative Properties

3

From an operational standpoint, it is convenient to have thermodynamic state functions that can be minimized to obtain the equilibrium state of a system under condition of fixed temperature and pressure or a combination of one these intensive variables and an extensive variable. These state functions are **Gibbs free energy (G), Helmholtz free energy (F) and Enthalpy (H)**. As we would see later, the equilibrium state of a system under a prescribed set of conditions that involve at least one intensive variable is obtained by **minimizing** one of these state functions, depending on which variables have been specified (such as constant P and T, or constant V and T or constant S and T). These state functions are, thus, often referred to as the **thermodynamic potentials**, by analogy with our common knowledge that a stable state of a system is obtained by minimizing an appropriate potential. These thermodynamic potentials are often introduced in an *ad hoc* fashion in terms of U and/or S (see Box 3.1.1). However, these can be derived in a systematic way from the fundamental relation $U = U(S, V)$, by a mathematical technique known as **Legendre transformation**, as shown below. In this section, we would always assume that the system under consideration has fixed masses of all chemical species.

3.1 Thermodynamic Potentials

Following Callen (1985), let us consider a simple geometric example to illustrate the principle of Legendre transformation. Figure 3.1 shows a curve in the Y–X plane, which can be represented by an equation $Y = Y(X)$ (e.g. $Y = X_1 + X_1^2 + X_1^3 + \ldots$). One can now draw a family of closely spaced tangents to the curve as shown in the figure. This specific family of tangents can be represented in terms of the intercept as a function of slope by an equation of the form $I = I(P)$, where I is the intercept on Y axis and P is the slope at a point (X, Y) on the curve. The original curve can be recovered from the equation $I = I(P)$; it is simply a curve that is tangent to the family of straight

© Springer Nature Switzerland AG 2020   57
J. Ganguly, *Thermodynamics in Earth and Planetary Sciences*,
Springer Textbooks in Earth Sciences, Geography and Environment,
https://doi.org/10.1007/978-3-030-20879-0_3

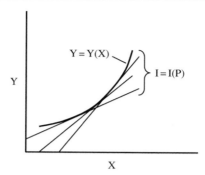

Fig. 3.1 Illustration of the concept of expressing a curved line, $Y = Y(X)$, in a two dimensional space in terms of a family of closely spaced tangents to the curve

lines represented by $I = I(P)$. The equations $Y = Y(X)$ and $I = I(P)$ are, thus, alternative descriptions of the same curve. It is simply a matter of convenience dictated by a specific purpose as to how one wants to represent the curve.

The mathematical technique by which an equation of the form $Y = Y(X)$ is transformed to $I = I(P)$ is known as *Legendre transformation*. The function I is called the Legendre transform of the function Y. It is easy to see from the equation of a straight line that

$$I = Y - \frac{dY}{dX} X \qquad (3.1.1)$$

If we have a function $Y = Y(X_1, X_2, \ldots X_n)$, then the *partial* Legendre transform of Y with respect to the variable X_i, keeping all other variables constant, is given by

$$I_{X_i} = Y - \left(\frac{\partial Y}{\partial X_i}\right)_{X_j \neq i} X_i \qquad (3.1.2)$$

where the subscript $X_{j \neq i}$ implies that all variables except X_i are kept constant in the partial differentiation. The variables can be chosen one at a time or in various combinations. A Legendre transform of Y with respect to the variables X_1 and X_2 is simply

$$I_{X_1, X_2} = Y - \left(\frac{\partial Y}{\partial X_1}\right)_{X_i \neq X_1} X_1 - \left(\frac{\partial Y}{\partial X_2}\right)_{X_i \neq X_2} X_2 \qquad (3.1.3)$$

For brevity, we will henceforth denote the partial derivative of a function Y with respect to a variable Z at constant x as $\left(Y'_z\right)_x$.

For the function $U = U(V, S)$, there are 3 possible Legendre transforms, which are as follows (the total number of Legendre transforms of the function Y equals $2^n - 1$, where n is the number of independent variables).

$$\left(I_v\right)_s = U - \left(U_v'\right)_s V = U + PV \qquad (3.1.4)$$

$$\left(I_s\right)_v = U - \left(U_s'\right)_v S = U - TS \qquad (3.1.5)$$

and

$$I_{v,s} = U - U_s'S - U_v'V = U - TS + PV \qquad (3.1.6)$$

where $(I_v)_s$ means the partial Legendre transform with respect to the variable V at constant S, and similarly for $(I_s)_v$. These new auxiliary functions $(I_v)_s$, $(I_s)_v$ and $I_{v,s}$ are called **Enthalpy** (H), **Helmholtz's free energy** (F) and **Gibbs free energy** (G), respectively. These are obviously state functions since they represent combinations of state functions. Summarizing, we now have three important, and as we would see later, practically useful state functions defined as

$$H = U + PV$$
$$F = U - TS$$
$$G = H - TS$$
$$= F + PV$$

Box (3.1.1)

It will be shown later that these relations also hold if the restriction of the system as one of fixed masses of all chemical species is removed.

Differentiation of the expression of H yields

$$dH = dU + PdV + VdP. \qquad (3.1.7)$$

But since dU = TdS − PdV for a reversible process in a closed system that is restricted only to P–V work, we have

$$dH = TdS + VdP. \qquad (3.1.8)$$

Thus, we have introduced the intensive variable P as an independent variable in the representation of the new function H. Similarly, by differentiating F and G, and substituting the expression of dU for a reversible process, we can derive

$$dF = -PdV - SdT, \qquad (3.1.9)$$

and

$$dG = VdP - SdT \qquad (3.1.10)$$

These important differentials of the auxiliary state functions for a closed system are summarized within a box along with the fundamental differential of U for a reversible process.

$$dU = TdS - PdV$$
$$dH = TdS + VdP$$
$$dF = -PdV - SdT$$
$$dG = VdP - SdT$$

Box (3.1.2)

Note that the above differential expressions of H, F and G are valid for a *reversible process in a closed system that is restricted to only P–V work*. If we are to include other types of work, then dU = TdS − PdV + $(\delta\omega^-)_{rev}$ (Eq. 2.7.4), and consequently the term $(d\omega^-)_{rev}$ must be added to the right hand side of the last three equations.

3.2 Equilibrium Conditions for Closed Systems: Formulations in Terms of the Potentials

In conjunction with the second law, the thermodynamic potentials developed above can be used to define the directions of spontaneous change and conditions of equilibrium under various types of conditions. As an illustration, let us seek the condition of equilibrium at constant P–T condition in a closed system that is restricted only to P–V work. Under these conditions, the appropriate function to deal with is Gibbs free energy, G, since it is a function only of P and T for the system under consideration. Using the relation G = U + PV − TS (Box 3.1.1), we have, at constant P–T

$$dG = dU + PdV - TdS \qquad (3.2.1)$$

Now, when restricted only to P–V work

$$dU = \delta q + \delta w^- = \delta q - PdV. \qquad (3.2.2)$$

(Recall from 1.4 that the relation $\delta w^- = -PdV$ does not necessarily imply a reversible process, but requires uniformity of pressure throughout the system. If the process is irreversible, then P may be either internal or external pressure, depending on whether the system is under expansion or compression, but it is the same pressure in the last two equations.) Combining these equations, we obtain

$(\partial G)_{P,T} = (\delta q - PdV) + PdV - TdS = \delta q - TdS$. According to the second law, $dS \geq \delta q/T$, the equality holding only at equilibrium. Consequently,

$$(\partial G)_{P,T} \leq 0, \tag{3.2.3}$$

In other words, in a system restricted only to P–V work and held at constant P and T, the direction of any spontaneous change is such that it reduces the Gibbs free energy. In general, $dG = 0$ is satisfied when G has reached either a maximum or a minimum value, but since G must decrease by a spontaneous change at constant P–T condition, G must be at **minimum** when equilibrium is achieved. If we now permit non-PV work, then we must add the term $\delta\omega^-$ to the right hand side of Eq. (3.2.2). Consequently, we have

$$(\partial G)_{P,T} \leq \delta\omega^- \tag{3.2.4}$$

or

$$-(\partial G)_{P,T} \geq -\delta\omega^- \geq \delta\omega^+$$

Pursuing the above analysis and using the second law relation $dS \geq \delta q/T$ or $(\delta q - TdS) \leq 0$, we can derive a further set of relations for systems (a) restricted only to P–V work and (b) exposed to other types of work. All these results are summarized in the Box (3.2.1), emphasizing the fact that all are consequences of the second law. The state functions U, H, F and G are also called **thermodynamic potentials**, since for a system restricted only to P–V work, the direction of spontaneous change of a system is dictated by the change of one of these functions, depending on the imposed conditions, toward a minimum. We are commonly concerned, especially in geological problems, with the equilibrium properties at constant P, T condition in systems restricted commonly only to PV work, and therefore seek to minimize the Gibbs free energy of the system. However, there are special situations in geological and planetary problems in which we hold different types of variables constant, other than the combination of variables discussed above. Determination of equilibrium conditions for those cases requires minimization of **new types of potentials**. We return in Sect. 10.13 to the derivation of those potentials through Legendre transformations.

Problem 3.1 Prove that at constant T and V, the direction spontaneous change is dictated by the condition

$$(\partial F)_{T,V} < 0 \tag{3.2.5}$$

until F reaches a minimum, in which case $(\partial F)_{T,V} = 0$.

Box 3.2.1 Change of the thermodynamic potentials of a closed system under various sets of imposed conditions, as dictated by the second law. The arrows indicate the fact that all the relations in terms of S, G, H, U and F on the two sides are consequences of the second law.

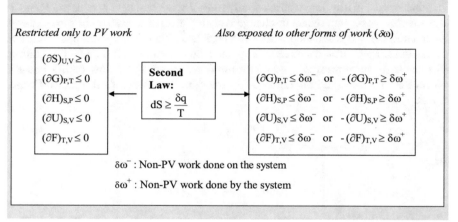

3.3   What Is Free in Free Energy?

For a finite change of state of a closed system from A to B at constant T and V, we have from the relations summarized in the preceding box

$$-\int_A^B (\partial F)_{T,V} = F(A) - F(B) \geq \int_A^B \delta\omega^+ \qquad (3.3.1)$$

Now, the last integral, which denotes the non-PV work performed **by** the system, must be positive in order to be useful. This requires that $F(A) > F(B)$. Thus, the useful non-PV work performed by the system by changing its state from A to B at constant T, V condition must be either less or equal to the decrease of the thermodynamic potential, F. Helmholtz (1821–1894), thus, coined the term **free energy** for the function F in order to emphasize that the energy released by decreasing F is the **maximum** amount of energy that is *free* or available to be transformed into useful work by a closed system under constant T, V condition. Subsequently, the terminology has been modified to call F as **Helmholtz free energy** in order to distinguish it from G, which came to be known as **Gibbs free energy** since, as should be obvious from the relation in the Box (3.2.1), that the decrease of G also represents the maximum energy that can be transformed into useful work, other than the P–V work, by a closed system at constant P–T condition. In some modern usage, the adjective *free* is often dropped, and H and G are simply referred to as Helmholtz energy and Gibbs energy, respectively.

3.4 Maxwell Relations

Since each of the differentials dU, dH, dF and dG are exact, one can derive the following relations for a closed system restricted only to PV work (i.e. $\delta\omega = 0$) from the equations summarized in the Box (3.1.2) through the application of Euler's reciprocity relation (Eq. B.3.3). The expressions are listed sequentially beginning with the exact differential property of dU.

Box 3.4.1

$$\left(\frac{\partial T}{\partial V}\right)_S = -\left(\frac{\partial P}{\partial S}\right)_V \tag{3.4.1}$$

$$\left(\frac{\partial T}{\partial P}\right)_S = \left(\frac{\partial V}{\partial S}\right)_P \tag{3.4.2}$$

$$\left(\frac{\partial P}{\partial T}\right)_V = \left(\frac{\partial S}{\partial V}\right)_T \tag{3.4.3}$$

$$\left(\frac{\partial V}{\partial T}\right)_P = -\left(\frac{\partial S}{\partial P}\right)_T \tag{3.4.4}$$

These relations are commonly known as **Maxwell relations**. There are also other Maxwell relations that follow from the total or exact differentials of U, H, F and G in open system, but the ones above are the most useful relations in classical thermodynamics. Note that the above Maxwell relations connect the derivatives of the extensive variables S and V and those of their conjugate intensive variables T and P in various combinations, but excluding the derivative of an extensive variable with respect to its *own* conjugate intensive variable. The importance of the Maxwell relations lies in the fact that a required derivative relation, which we want to calculate and for which data are not available, may be replaced by another relation for which data are available, or which may be combined with other parameters in an equation to reduce it to a tractable form. Furthermore, these relations can be used to cross-check the internal consistency of data. We would encounter many examples of this type of operations later.

3.5 Thermodynamic Square: A Mnemonic Tool

At this point, the large numbers of thermodynamic relations that have been presented so far are likely to seem exasperating to most students who are not gifted with extraordinary memory. To help alleviate this problem, Max Born (1882–1970; Nobel prize: 1954), developed a simple mnemonic diagram, which is sometimes referred to as the **thermodynamic square**, for recalling the exact differentials of U, F, G and H, the associated Maxwell relations, and conditions of equilibrium under various sets of imposed conditions (He presented this mnemonic method in a lecture in 1929. This lecture was attended by Professor Tisza of M. I. T., and the mnemonic method appeared in a well-known book called Thermodynamics (John Wiley) written by one of Tisza's students, Herbert B. Callen).

The construction and use of the square, which is illustrated in Fig. 3.2, can be summarized as follows. Write the thermodynamic potentials that are minimized to achieve equilibrium (i.e. F, G, H and U) in descending alphabetical order on the four sides of a square, beginning with the top side and proceeding in a clockwise fashion. Then insert the variables that affect G (i.e. P and T) at the corners on the right hand side of the square (that is the side containing G) and the variables that affect U (i.e. S and V) at the corners of the left hand side of the square (i.e. the side containing U), as shown, and write the +ve sign on the two top corners, and −ve sign on the two bottom corners of the square. Note that the conjugate intensive and extensive variables at the corners of the square are linked by the diagonals (S is linked with the conjugate intensive variable T, and V is linked with the conjugate intensive variable P). This arrangement can be ensured by writing the variables in the same alphabetical order, ascending from the bottom (S → V, P → T) to the top side. The square is complete, and we can state the operating principles to derive different thermodynamic equalities. However, it is best to discover the operating principles by recalling the often used relations involving G, namely $G = H - TS$, $dG = -SdT + VdP$ and $(\partial G)_{P,T} = 0$ at equilibrium (for a system restricted to P–V work), and finding out the schemes which recover these relations from the square.

Fig. 3.2 Illustration of thermodynamic square. The symbols have usual meanings

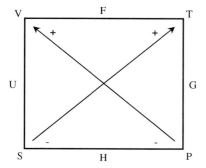

Now consider how the relation $G = H - TS$ can be recovered from the square by picking the potential moving clockwise from G, i.e. H, then going to the end of the side containing H and picking the symbol at the corner, i.e. $-S$, and multiplying it by the one at the other end of the associated diagonal, i.e. T. This scheme of operation can be applied to H to obtain $H = U + VP$; to U to obtain $U = F + TS$ and reorganized into the more familiar form $F = U - TS$; to F to obtain $F = G - PV$ or the more familiar equation $G = F + PV$.

Next, see how the relation $dG = -SdT + VdP$ is recovered from the diagram by moving to the other side of the square from G (i.e. the side with U), picking the variables at the two ends of this side ($-S$ and V), and multiplying each one by the differential of the conjugate variable connected by the diagonal lines ($-S \rightarrow dT$, $V \rightarrow dP$). Following similar steps, one obtains $dH = VdP + TdS$; $dU = TdS - PdV$ and $dF = -SdT - PdV$.

Next, we want to obtain the directions of spontaneous change and conditions of equilibrium of a closed system restricted only to PV work under various sets of imposed conditions by noting that for such a system $(\partial G)_{P,T} \leq 0$, i.e. the potential to minimize is the one labelled to the side containing the variables that are held constant. Thus, we have $(\partial H)_{S,P} \leq 0$, $(\partial U)_{S,V} \leq 0$, $(\partial F)_{V,T} \leq 0$.

In order to determine the scheme for recovering the Maxwell relations, first note that from the expression $dG = VdP - SdT$, we have (applying the reciprocity relation - see Appendix B.3) $(\partial V/\partial T)_P = -(\partial S/\partial P)_T$. The two sides of the equality sign represent two similar operations on opposite sides, in which only the sign of the variable at the numerator is important. Now apply the analogous operations to the other sides, and you will have the Maxwell relations given by Eqs. (3.4.1)– (3.4.4). For example, what is the equivalent of $(\partial V/\partial S)_P$? Applying similar operation on the opposite side of the square, we have $(\partial T/\partial P)_S$.

3.6   Vapor Pressure and Fugacity

At any temperature, every substance has a finite equilibrium vapor pressure, which we would refer to simply as the vapor pressure of the substance. When a condensed substance is introduced into a sufficiently evacuated container, it generates a quantity of vapor such that, when equilibrium is achieved, the pressure exerted by the vapor equals the vapor pressure of the condensed substance at the specific temperature. A volatile substance has a high vapor pressure whereas a non-volatile substance has a very low vapor pressure. At room temperature, alcohol has a high vapor pressure, a moth-ball has a moderate vapor pressure and a rock-forming mineral has extremely low vapor pressure. These are familiar concepts from elementary chemistry.

Consider now the phase diagram of pure H_2O (Fig. 3.3). The lines separating the fields of ice and liquid from vapor indicate, respectively, the vapor pressure of ice and liquid (water) as functions of temperature. At the triple point at 0.0061 bars, 0.01 °C, the vapor pressures of ice and water are the same. If we extrapolate the

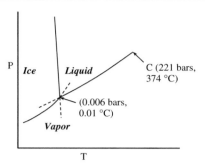

Fig. 3.3 Schematic phase diagram of H_2O showing the triple point (that is the point of equilibrium coexistence of ice, liquid and vapor) at 0.006 bars, 0.01 °C, and the critical end point, C, on the liquid-vapor coexistence line at 221 bars, 374 °C (see Chap. 5.1 for the discussion about critical phenomenon)

vapor pressure curves of ice and liquid beyond the triple point, we find that in the field of stability of liquid, the vapor pressure of liquid is lower than that of ice, and vice versa. At the triple point, where ice and liquid are in equilibrium, the vapor pressures of these two phases are exactly the same. Thus, instead of Gibbs free energy, we can describe the stability conditions of ice and liquid in terms of their vapor pressures. However, this alternative description of the relative stabilities and equilibrium conditions of phases in terms of their vapor pressures is correct only as long as the vapor behaves as a perfect gas. When it does not, one needs to make a correction to the vapor pressure to describe relative stability of phases. This "corrected" vapor pressure is known as **fugacity**. The concept of fugacity as an alternative measure to Gibbs free energy to describe stability or escaping tendency of phases was introduced by G. N. Lewis (1875–1946) in 1901.

Fugacity bears the same formal relation to G as the vapor pressure of a perfect gas. For one mole of a perfect gas, which obeys the relation V = RT/P, we have, at constant temperature, dG = VdP = RTdlnP. By analogy, fugacity, f, of a pure substance has been related to G at constant temperature according to

$$dG = RTdlnf = RTdln(P\Phi) \qquad (3.6.1)$$

where Φ is called the fugacity coefficient. Also, since at sufficiently low pressure all gases approach the perfect gas behavior, the fugacity must obey the limiting relation

$$P \rightarrow 0, \quad f = P \qquad (3.6.2)$$

or

$$P \rightarrow 0, \quad \Phi = 1 \qquad (3.6.3)$$

Equations (3.6.1) and (3.6.2) constitute the fundamental definition of the fugacity.

Consider now the line separating the fields of water and ice. At any point along this line, $G_{H_2O}(ice) = G_{H_2O}(liquid)$. But it is also correct to say that at any such point, $f_{H_2O}(ice) = f_{H_2O}(liquid)$, and if the gas behaved as a perfect gas, then $P_{H_2O}(ice) = P_{H_2O}(liquid)$.

An expression for the pressure dependence of fugacity can be easily derived by differentiating both sides of the first equality of Eq. 3.6.1 with respect to P, which yields

$$\left(\frac{\partial \ln f}{\partial P}\right)_T = \frac{1}{RT}\left(\frac{\partial G}{\partial P}\right)_T = \frac{V}{RT} \tag{3.6.4}$$

The fugacity of a gas at a given pressure and temperature can be measured from a knowledge of the difference between its volume, V, and that of an ideal gas, V_{ideal}, at the specified condition, as follows.

For a mole of gas, let $V_m = V_m(ideal) + \varphi$ so that we have from the last expression

$$(V_m(ideal) + \varphi)dP = RTd\ln f$$

Thus, at constant T, using the P–V–T relation for a mole of ideal gas (i.e. $V_m(ideal) = RT/P$)

$$\int_{P^*}^{P'} \left(\frac{RT}{P} + \varphi\right) dP = RT \int_{f(P^*)}^{f(P')} d\ln f \tag{3.6.5}$$

or

$$RT \ln P' - RT \ln P^* + \int_{P*}^{P'} \varphi \, dP = RT \ln f(P') - RT \ln f(P^*) \tag{3.6.6}$$

From Eq. (3.6.2), as $P^* \to 0$, $f(P^*) = P^*$, in which case the second term on the right becomes $RT\ln P^*$, and consequently we have for one mole of a gas

$$RT \ln P' - RT \ln P^* + \int_0^{P'} \varphi \, dP = RT \ln f(P') - RT \ln P^*$$

or

$$
\boxed{
\begin{aligned}
RT \ln f(P') &= RT \ln P' + \int_0^{P'} \varphi \, dP \\
&= RT \ln P' + \int_0^{P'} \left(V_m - \frac{RT}{P} \right) dP
\end{aligned}
}
$$

$$(3.6.7)$$

Tunell (1931) suggested that the last equation should be used as the definition of fugacity, since the desired properties of fugacity (Eqs. 3.6.1 and 3.6.2) that were sought by Lewis follow from this definition. However, this approach is rarely followed probably because it is a complex definition, and it does not convey the physical insights of Lewis' approach in developing the concept of fugacity.

3.7 Derivative Properties

We present here a set of useful functions, which are known as isobaric heat capacity, C_P, constant volume or isochoric (constant density) heat capacity, C_V, the isobaric expansivity (or the coefficient of thermal expansion), α, isothermal and isentropic compressibilities, β_T and β_s respectively. These functions represent partial derivatives of the thermodynamic state functions, as discussed below, and are experimentally measurable.

3.7.1 Thermal Expansion and Compressibility

The coefficient of thermal expansion (α), and the isothermal (β_T) and isentropic (β_S) compressibilities are defined by

$$
\alpha = \frac{1}{V} \left(\frac{\partial V}{\partial T} \right)_P
$$

$$(3.7.1)$$

$$
\beta_T = -\frac{1}{V} \left(\frac{\partial V}{\partial P} \right)_T
$$

$$(3.7.2)$$

$$
\beta_S = -\frac{1}{V} \left(\frac{\partial V}{\partial P} \right)_S
$$

$$(3.7.3)$$

The **bulk modulus** is the inverse of compressibility (in other words, bulk modulus is the incompressibility), and is commonly designated by the symbols k_T and k_S for

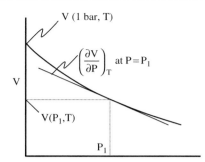

Fig. 3.4 Illustration of the volumetric parameters used to define compressibility at an arbitrary pressure P_1 according to Eq. (3.7.2). Both $V(P_1, T)$ and $V(1 \text{ bar}, T)$ are used as normalizing factors

the isothermal and isentropic conditions, respectively. The isentropic properties are usually referred to as **adiabatic** properties, since according to the Second Law, a reversible adiabatic process implies an isentropic process. Here the implicit assumption is that pressure has changed sufficiently slowly for the system to effectively maintain equilibrium.

The normalizing factors $1/V$ in the above equations represent the instantaneous volume, as illustrated in Fig. 3.4. A commonly used alternative form of defining α and β uses the normalizing factor $1/V$ (P, 298 K) and $1/V$ (1 bar, T), respectively, instead of the instantaneous volume V. However, the normalizing volumes are **not** V (1 bar, 298 K) in either case.

Note that since volume always decreases with increasing pressure, $(\partial V/\partial P)_T < 0$, and consequently there is a negative sign in front of this derivative in Eq. (3.7.3) so that the compressibility and bulk modulus are positive quantities. On the other hand, the volume of a substance does not always increase with increasing temperature. Anharmonicity of vibration is responsible for the usual phenomenon of expansion of volume with increasing temperature, as discussed in 1.6, but a substance can also exhibit zero and even negative thermal expansion. Design of ceramic materials that can withstand thermal shock requires extremely small values of α. A review of such materials, which are of great interest in the ceramic industry, may be found in Hummel (1984). Examples of materials that show negative thermal expansion are ZrW_2O_8 (Mary et al. 1996), β-quartz and several other minerals that have been discussed by Welche et al. (1998). Heine et al. (1999) have shown that a negative thermal expansion in a framework structure is a consequence of the geometric effect of rotation of the rigid octahedral and tetrahedral units such as SiO_4 and AlO_6 around the bridging oxygens at high temperature, resulting in a reduction of the unit cell dimension. This concept is illustrated in Fig. 3.5. In practical applications, a compound with negative α may be mixed with an appropriate amount of a compound with positive α to yield a composite material with effectively zero thermal expansion. Such materials can be heated in an oven without mechanical failure, except that caused by poor fabrication.

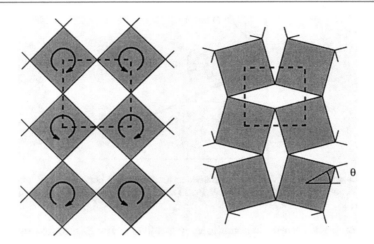

Fig. 3.5 Illustration of the origin of negative thermal expansion by the rotation of rigid polyhedral units around the points of linkage. The unit cell is shown by dashed lines. Note the reduction of the unit cell size as a result of rotation. From Welche et al. (1998)

3.7.2 Heat Capacities

The average heat capacity, C_{av}, is defined as the ratio of the heat energy absorbed by a system, Q, to the associated temperature rise, ΔT. The instantaneous heat capacity at temperature T is the limiting value of this ratio. Thus, $(C)_{av} = Q/\Delta T$ or $(C)_{@T} = \delta q/dT$. However, since δq is an inexact differential, the heat absorbed by a system for a specific change of temperature depends on the manner the temperature change is brought about. It is, therefore, customary to define two types of limiting heat capacities, C_p and C_v, the first relating to temperature change at constant pressure, and the second to temperature change at constant volume.

$$C_P = \left(\frac{\delta q}{\partial T}\right)_P \tag{3.7.4a}$$

$$C_v = \left(\frac{\delta q}{\partial T}\right)_v \tag{3.7.4b}$$

From Eqs. (3.1.7) and (3.1.8), for a reversible process at constant pressure, $dH = dU + PdV = TdS$. Also, from the first law, $dU + PdV = \delta q$. Thus, for a reversible process at constant pressure, $\delta q = dH = TdS$, so that

$$C_P = \left(\frac{\partial H}{\partial T}\right)_P = T\left(\frac{\partial S}{\partial T}\right)_P \tag{3.7.5}$$

Problem 3.2 Prove that in a system restricted to P–V work, the following relations hold for a reversible process

$$C_V = \left(\frac{\partial U}{\partial T}\right)_V = T\left(\frac{\partial S}{\partial T}\right)_V \qquad (3.7.6)$$

(Prove both equalities)

*Problem 3.3 Consider a mole of CO_2, which has an average heat capacity (C_p) of around 35 J/mol K between 300 and 200 K; the gas is allowed to expand reversibly against a frictionless piston from an initial state at 30 bars, 300 K to a final pressure of 1 bar within a thick metal cylinder that has a C_p of 85 J/mol-K and is thermally insulated from the surrounding. The effect of pressure on thermodynamic properties within the above pressure range is negligible. Assuming CO_2 to behave as an ideal gas and that its mass is negligible compared to that of the container, calculate (a) the final temperature of the gas, (b) the work done by the gas, given that the internal energy of an ideal gas depends only on its temperature, and (c) the entropy change of the system (gas + cylinder).

Hint: (a) First show that $(dH)_{system} = d(U + PV) = (VdP)_{gas}$ for the stated conditions. Then use the relation between H and C_p, and go from there to calculate T_{final}. (b) Calculate $W^+(gas)$ through its relationship with ΔU, according to the first law, under adiabatic condition.

Answers: (a) $T_f = 215$ K; (b) $W^+(gas) = 2276$ J/mol; (c) $(\Delta S)_{sys} = -26.8$ J/mol.

One can derive an important relation between C_p and C_v by starting with the relation S = S(T, V), which yields

$$dS = \left(\frac{\partial S}{\partial T}\right)_V dT + \left(\frac{\partial S}{\partial V}\right)_T dV \qquad (3.7.7)$$

Differentiating both sides of this expression with respect to T at constant P, we have

$$\left(\frac{\partial S}{\partial T}\right)_P = \left(\frac{\partial S}{\partial T}\right)_V + \left(\frac{\partial S}{\partial V}\right)_T \left(\frac{\partial V}{\partial T}\right)_P \qquad (3.7.8)$$

From Eqs. (3.7.5) and (3.7.6), the first two terms in the above equation equals C_p/T and C_V/T, respectively, whereas, using Maxwell relation (Box 3.4.1) and the property of implicit function (Appendix, Eq. B.4.3)

$$\left(\frac{\partial S}{\partial V}\right)_T = \left(\frac{\partial P}{\partial T}\right)_V = -\frac{\left(\frac{\partial P}{\partial V}\right)_T}{\left(\frac{\partial T}{\partial V}\right)_P} \qquad (3.7.9)$$

(here the first equality is a Maxwell relation and the second equality follows from the property of implicit function). Using the definitions of α and β_T (Eqs. 3.7.1 and 3.7.2, respectively), the above relation yields $(\partial S/\partial V)_T = \alpha/\beta$, so that Eq. (3.7.8) reduces to

$$
\begin{aligned}
C_P &= C_V + \frac{\alpha^2 VT}{\beta_T} \\
&= C_V + \alpha^2 VTk_T
\end{aligned}
\tag{3.7.10}
$$

It follows that for one mole of an ideal gas (which obeys the relation $PV = RT$),

$$
C_P - C_V = R \tag{3.7.11}
$$

The derivation of this relation is left to the reader as an exercise.

The heat capacity of a solid derived from atomistic lattice theories and measurements of the vibrational properties is C_V, as discussed in Sects. 4.2 and 14.4, whereas phase equilibrium calculations require C_P. Equation (3.7.10) permits conversion of C_V to C_P. The latter is also the quantity determined by calorimetric measurements. Saxena (1988) applied Eq. (3.7.10) to derive C_V, α and k_T from calorimetric C_P data of enstatite (Mg_2SiO_4) and forsterite (Mg_2SiO_4) as a function of temperature. He expressed these quantities in terms of polynomial functions of T as

$$
\begin{aligned}
C_V &= C_o + C_1 T^{-1} + C_2 T^{-2} + C_3 T^{-3} \\
\alpha &= \alpha_0 + \alpha_1 T + \alpha_2 T^{-1} + \alpha_3 T^{-2} \\
k_T &= k_0 + k_1 T + k_2 T^{-1} + k_3 \ln T
\end{aligned}
$$

The constants of these functions were treated as floating variables that were adjusted by a non-linear optimization program to yield C_p values in a wide range of temperature that have the best match with the measured C_P versus T data. The values of α and k_T calculated from the optimized values of the constants in the last two equations are in good agreement with their measured values. This approach was also utilized in the development of a thermodynamic data base (Saxena et al. 1993).

The isothermal and adiabatic compressibilities are related to the two types of heat capacities, C_P and C_V, according to

$$
\frac{\beta_S}{\beta_T} \left(= \frac{k_T}{k_S} \right) = \frac{C_V}{C_P} \tag{3.7.12}
$$

This result can be derived as follows. From the definitions of β_T and β_s (Eqs. 3.7.2 and 3.7.3, respectively),

$$\frac{\beta_S}{\beta_T} = \frac{\left(\frac{\partial V}{\partial P}\right)_S}{\left(\frac{\partial V}{\partial P}\right)_T} = -\frac{\left(\frac{\partial V}{\partial P}\right)_S}{V\beta_T} \tag{3.7.13}$$

(Note that in deriving the above relation, the normalizing volume must be the same for both β_S and β_T). Now, writing the total differential of V with respect to P and T, and differentiating both sides with respect to P at constant S, we have

$$\left(\frac{\partial V}{\partial P}\right)_S = \left(\frac{\partial V}{\partial P}\right)_T + \left(\frac{\partial V}{\partial T}\right)_P \left(\frac{\partial T}{\partial P}\right)_S \tag{3.7.14}$$

It is shown later (Eq. 7.2.3) that the last parenthetical term on the right equals $(VT\alpha)/C_P$. Thus, using the definitions of α and β (Eqs. 3.7.1, 3.7.2 and 3.7.3), the above expression reduces to

$$\left(\frac{\partial V}{\partial P}\right)_S = -V\beta_T + (\alpha V)\left(\frac{VT\alpha}{C_P}\right) \tag{3.7.15}$$

Consequently, from Eq. (3.7.13),

$$\frac{\beta_S}{\beta_T} = 1 - \frac{\alpha^2 VT}{\beta_T C_P}, \tag{3.7.16}$$

which, on combination with Eq. (3.7.10), leads to Eq. (3.7.12). It is easy to show from the above equation that

$$\frac{k_T}{k_S} = 1 - \frac{\alpha^2 k_T T}{\rho C_P'} \tag{3.7.17}$$

where C_P' is the isobaric specific heat capacity.

3.8 Grüneisen Parameter

From Eq. (3.7.10), one obtains

$$\frac{C_P}{C_V} = 1 + \alpha T\left(\frac{V\alpha k_T}{C_V}\right) \tag{3.8.1}$$

The quantity within the parentheses is a dimensionless parameter, and is known as the thermodynamic Grüneisen parameter or ratio, Γ_{th}, named after Grüneisen (1926) who first introduced it from consideration of vibrational properties, as discussed below. Using Eq. (3.7.12), we also find that

$$\Gamma_{th} = \frac{V\alpha k_T}{C_V} = \frac{V\alpha k_S}{C_P} \tag{3.8.2}$$

Now, since $C_P/C_V = k_S/k_T$, we have, combining the last two equations,

$$\frac{k_S}{k_T} = 1 + \Gamma_{th}\alpha T \tag{3.8.3}$$

The thermodynamic Grüneisen parameter relates thermal and elastic properties in a dimensionless form, and has a restricted range of values for solids, usually between 1 and 2, and between 1.1 and 1.5 for the entire Earth (Stacey and Davis 2004), even though the individual terms in its expression may vary considerably. Figure 3.6 shows the variation of Grüneisen parameter of some silicates and oxides that are of interest in the Earth sciences as a function of temperature. Anderson (1995) found the assumption of constancy of $\rho\Gamma_{th}$ for a solid to be a better approximation than that of the constancy of Γ_{th} itself. This property was utilized to approximate the value of Γ_{th} of materials in the Earth's interior, as for example, by Jeanloz (1979).

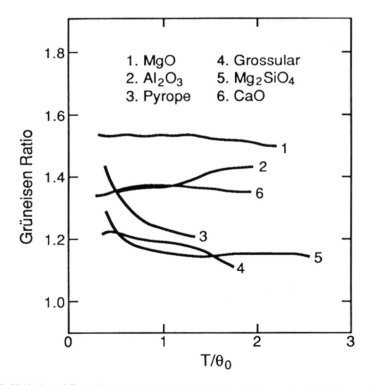

Fig. 3.6 Variation of Grüneisen parameter or ratio of some silicates and oxides that are of interest in the Earth sciences as a function of temperature normalized to the Debye temperature, θ_D. From Anderson (1995)

In the original development, Grüneisen assumed that the volume dependence of the frequency of the i th vibrational mode can be expressed as

$$\frac{\partial \ln \nu_i}{\partial \ln V} = -\gamma_i$$

where γ_i is a constant, and is commonly referred to as "mode gamma". It can be shown that if all vibrational modes have the same volume dependence, and this relation is independent of temperature, then the above equation leads to the expression of thermodynamic Grüneisen parameter, Eq. (3.8.2) (see, for example, Poirier 1991).

From Eq. (3.8.2) we can derive an expression that is of interest for the purpose of direct determination of the Grüneisen parameter in the laboratory and also for the calculation of adiabatic temperature gradient in the Earth's interior. Substituting the derivative expressions of the individual terms in this equation, we have

$$\Gamma_{th} = \frac{V\alpha k_T}{C_V} = -\frac{V\left(\frac{\partial V}{\partial T}\right)_P \left(\frac{\partial P}{\partial V}\right)_T}{T\left(\frac{\partial S}{\partial T}\right)_V} \tag{3.8.4}$$

Using now the property of implicit function (Eq. B.4.3),

$$\left(\frac{\partial V}{\partial T}\right)_P \left(\frac{\partial P}{\partial V}\right)_T = -\left(\frac{\partial P}{\partial T}\right)_V \tag{3.8.5}$$

Thus,

$$\Gamma_{th} = \frac{V}{T}\left(\frac{\partial P}{\partial T}\right)_V \left(\frac{\partial T}{\partial S}\right)_V = \frac{V}{T}\left(\frac{\partial P}{\partial S}\right)_V \tag{3.8.6}$$

Using Maxwell relation (Eq. 3.4.1) for the last derivative term, we have

$$\Gamma_{th} = -\frac{V}{T}\left(\frac{\partial T}{\partial V}\right)_S = -\left(\frac{\partial \ln T}{\partial \ln V}\right)_S \tag{3.8.7}$$

Replacing $(\partial \ln V)_S$ by $-(\partial P/k_S)_S$, which follows from the definition of k_s (Eq. 3.7.3 and the relation $\beta_S = 1/k_S$), we obtain

$$\Gamma_{th} = k_s \left(\frac{\partial \ln T}{\partial P}\right)_S \tag{3.8.8}$$

Boehler and Ramakrishna (1980) utilized the above expression to determine Γ_{th} by noting the temperature change of a sample due to *sudden* change of pressure in a piston-cylinder apparatus. It was assumed that there was no significant heat loss from the sample within the time scale of measurement of temperature change, and furthermore that there was no significant entropy production during the rapid change of pressure. Note that the effects of heat loss (which is equivalent to entropy loss) and internal entropy production would compensate one another, at least partly.

Using the hydrostatic relation $dP = \rho g dZ$, Eq. (3.8.8) can be recast as

$$\left(\frac{\partial \ln T}{\partial Z} \right)_S = \frac{\Gamma_{th} g}{(k_s/\rho)} \tag{3.8.9}$$

As we would see in the next section, the parenthetical quantity in the denominator is related to seismic velocities. Thus, this equation is of fundamental importance in relating seismic velocities, temperature gradient and Grüneisen parameter in the Earth's interior.

Problem 3.4 The Grüneisen parameter can be considered as the change of pressure of a crystal of constant volume with respect to a change of its internal energy density (i.e. density per unit volume). In other words,

$$\Gamma_{th} = \left(\frac{\partial P}{\partial (U/V)} \right)_V = V \left(\frac{\partial P}{\partial U} \right)_V \tag{3.8.10}$$

Derive Eq. (3.8.7) from this relation. This establishes the equivalence of the two definitions of Grüneisen parameter according to Eqs. (3.8.10) and (3.8.2). Note, incidentally, that the restricted range of values of Γ_{th} implies that pressure and internal energy density of a crystal change similarly.

3.9 P–T Dependencies of Coefficient of Thermal Expansion and Compressibility

In principle, the isobaric thermal expansion coefficient of a substance, α, depends on pressure and its isothermal compressibility, β_T, depends on temperature. The pressure dependence of α and the temperature dependence of β_T are, however, interrelated through the fact that dV is an exact differential. From $V = f(P, T)$, we have

$$dV = \left(\frac{\partial V}{\partial P} \right)_T dP + \left(\frac{\partial V}{\partial T} \right)_P dT \tag{3.9.1}$$

or, using Eqs. (3.7.1) and (3.7.2),

$$dV = -V\beta_T dP + \alpha V dT \tag{3.9.2}$$

Since dV is an exact differential, the right hand side of this expression satisfies the reciprocity relation (Eq. B.3.3). Thus,

$$-\frac{\partial\beta_T}{\partial T} = \frac{\partial\alpha}{\partial P} \qquad (3.9.3)$$

Consequently, one can obtain the temperature dependence of β_T if the pressure dependence of α is known accurately, and vice versa. When both $\beta_T = f(T)$ and $\alpha = f(P)$ are known experimentally or retrieved from modeling relations that depend on them, the internal consistency of the data should be checked by using Eq. (3.9.3). Unfortunately, all thermodynamic data bases in the literature do not pass this test.

3.10 Summary of Thermodynamic Derivatives

Lumsden (1952) provided a summary of expressions, in a tabular form, of partial derivatives involving P, T, V, U, H, S, G, F in various combinations. These expressions, which are given in Table 3.1, constitute a convenient and useful reference source. Note that the in the left hand set of derivatives, either P or T is held constant, whereas in the right hand, the conjugate extensive quantities, V or S, are held constant.

Table 3.1 Summary of expressions of partial derivatives of thermodynamic quantities

X	Y	Z	$\left(\frac{\partial Y}{\partial X}\right)_Z$	X	Y	Z	$\left(\frac{\partial Y}{\partial X}\right)_Z$
T	V	P	αV	T	P	V	α/β
T	S	P	C_p/T	T	S	V	$C_p/T - \alpha^2 V/\beta$
T	V	P	$C_P - \alpha PV$	T	U	V	$C_P - \alpha^2 VT/\beta$
T	H	P	C_P	T	H	V	$C_P - \alpha^2 VT/\beta + \alpha V/\beta$
T	F	P	$-\alpha PV - S$	T	F	V	$-S$
T	G	P	$-S$	T	G	V	$\alpha V/\beta - S$
P	V	T	$-\beta V$	T	P	S	$C_P/\alpha VT$
P	S	T	$-\alpha V$	T	V	S	$-\beta C_P/\alpha T + \alpha V$
P	U	T	$\beta PV - \alpha VT$	T	U	S	$\beta PC_P/\alpha T - \alpha PV$
P	H	T	$V - \alpha VT$	T	H	S	$C_P/\alpha T$
P	F	T	βPV	T	F	S	$\beta PC_P/\alpha T - \alpha PV - S$
P	G	T	V	P	G	S	$C_P/\alpha T - S$

$\beta \equiv \beta_T$

References

Anderson OL (1995) Equations of state of solids for geophysics and ceramic science. Oxford University Press, New York, Oxford, p 405

Boehler R, Ramakrishnan J (1980) Experimental results on the pressure dependence of the Grüneisen parameter: a review. J Geophys Res 85:6996–7002

Callen HB (1985) Thermodynamics and an introduction to thermostatics. Wiley, New York, Chichester, Brisbane, Toronto, Singapore, p 493

Grüneisen E (1926) The state of a solid body. Handbuch der Physik, vol 1, Springer, Berlin, pp 1–52 Engl. Transl. NASA RE 2-18-59W (1959)

Heine V, Welche PRL, Dove MT (1999) Geometrical origin and theory of negative thermal expansion in framework structures. Am Ceram Soc 82:1759–1767

Hummel FA (1984) A review of thermal expansion data of ceramic materials, especially ultra-low expansion compositions. Interceram 33:27–30

Jeanloz R (1979) Properties of iron at high-pressures and the state of the core. J Geophys Res 84:6059–6069

Lumsden J (1952) Thermodynamics of alloys. Institute of Metals, London

Mary TA, Evans JSO, Vogt T, Sleight AW (1996) Negative thermal expansion from 0.3 to 1050 K in ZrW_2O_8. Science 272:90–92

Poirier J-P (1991) Introduction to the physics of the earth. Cambridge, p 264

Saxena SK (1988) Assessment of thermal expansion, bulk modulus, and heat capacity of enstatite and forsterite. J Phys Chem Solids 49:1233–1235

Saxena SK, Chatterjee N, Fei Y, Shen G (1993) Thermodynamic data on oxides and silicates. Springer, Berlin

Tunell G (1931) The definition and evaluation of the fugacity of an element or compound in the gaseous state. J Phys Chem 35:2885–2913

Welche PRL, Heine V, Dove MT (1998) Negative thermal expansion in beta-quartz. Phys Chem Miner 26:63–77

Third Law and Thermochemistry

4

The bulk of thermodynamics has been developed on the basis of the first two laws. The third law plays a much lesser role in the development of the subject. Its principal application lies in the development of the concept of absolute entropy of a substance, and its calculation through the heat capacity function. The latter, however, constitutes a major step in the development of the field of thermochemistry. The third law was developed primarily by Nernst (1864–1941; Nobel prize: 1920) and proposed in 1905. In this section, we discuss the observational basis and modern statement of the third law and its implication for the calculation of absolute entropy of a substance. This is followed by a general discussion of thermochemistry, and calculation of changes of thermochemical properties of reactions that are required for phase equilibrium calculations discussed in Chaps. 6 and 10.

4.1 The Third Law and Entropy

4.1.1 Observational Basis and Statement

From analysis of available data on changes of Gibbs energy, $\Delta_r G$, and enthalpy, $\Delta_r H$, of reactions involving pure phases, Nernst concluded that at temperatures near absolute zero, $\Delta_r G$ and $\Delta_r H$ of reactions among pure phases do not have any significant temperature dependence. This observation has its root in the early works of Thomsen, Berthelot and Richard around 1900. Thus, since at $T = 0$, $\Delta_r G = \Delta_r H - T\Delta_r S = \Delta_r H$, the $\Delta_r G$ versus T and $\Delta_r H$ versus T curves not only must be essentially flat, but also coincident near $T = 0$, as shown schematically in Fig. 4.1. (It is easy to see that as $T \to 0$, $\partial(\Delta_r H)/\partial T \to 0$ since this derivative equals $\Delta_r C_p$ and C_p of all substances tend to zero as $T \to 0$.) Now since $(\partial G/\partial T) = -S$, Nernst proposed that the vanishingly small slope of $\Delta_r G$ versus T near $T = 0$ implies that for all reactions among pure phases $\Delta_r S \to 0$ as $T \to 0$. This is known as the Nernst postulate. Planck (1858–1947) took Nernst statement a step further by

© Springer Nature Switzerland AG 2020
J. Ganguly, *Thermodynamics in Earth and Planetary Sciences*,
Springer Textbooks in Earth Sciences, Geography and Environment,
https://doi.org/10.1007/978-3-030-20879-0_4

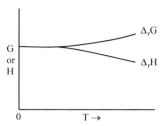

Fig. 4.1 Schematic variation of Gibbs free energy and enthalpy change of a reaction, $\Delta_r G$ and $\Delta_r H$ respectively, as a function of temperature near absolute zero

suggesting that the simplest way this postulate could be valid is to have the entropy of all substance go to zero as $T \rightarrow 0$.

If we now look at the Boltzmann expression of entropy, $S = k_B \ln \Omega$ (Eq. 2.5.1), the statement that the entropy of every substance is 0 at $T = 0$ implies that $\Omega = 1$ at $T = 0$; in other words, there is only **one** microscopic or dynamical state in which the system can exist at $T = 0$. Now consider a solid solution of two components. As we have discussed earlier, there are usually multitude of atomic configurations in which the solid solution of a fixed composition can exist without significantly changing its overall energy. Planck's modification of Nernst statement that $S = 0$ at $T = 0$ implies that there is only one configurational state at this temperature for a solid solution. However, a solid solution may have different configurational states at $T = 0$ that have effectively the same energy. Consequently, a solid solution may have non-zero configurational entropy at $T = 0$. This analysis also applies to a solid of an end-member component if it has defects. This is because defects may be considered as another component in the solid, which leads to a configurational entropy of mixing between the end member component and the defects since a certain number of defects may be distributed within a crystal in a large number of ways even at $T = 0$, without affecting the overall energy of the solid. An example of a solid with lot of defects is the mineral wüstite, which has an ideal defect-free stoichiometry of FeO, but is always found as $Fe_{1-x}O$, with $x > 0$, as a result of vacancies (point defects) in the cation sites. These considerations led Lewis (1875–1946) to restrict the Plank statement of zero entropy of a substance at absolute zero to **pure and perfect** crystalline solids. In summary then, the third law of thermodynamics can be stated as follows:

The entropy of a *pure* and *perfect* crystalline substance vanishes at absolute zero.

There is still a lingering problem with elements that have more than one isotope. It may be possible to have alternative configurational states or distributions of isotopes within the crystal structure of an element that are energetically equivalent even at $T = 0$. In general, it may be possible to have different configurational and vibrational states that are energetically equivalent even at absolute zero. But as noted by Fermi (1956), the number of such equivalent states has to be very large in order to cause a significant deviation from the above statement of the third law, since $S = k_B \ln\Omega$ and $k_B = 1.38 \times 10^{-23}$ J/mol-K. Such a situation may be theoretically possible, but extremely unlikely to exist in nature.

4.1.2   Third Law Entropy and Residual Entropy

The thermodynamic formalisms that developed from the first and second laws always deal with the **relative values** or changes of the thermodynamic potentials with respect to some reference state, instead of their absolute values at any given condition. Thus, for example, we do not speak in terms of absolute value of U or H, but instead speak of ΔU and ΔH. The third law of thermodynamics relieves entropy from this restriction of being a 'relative' quantity, so to speak. Because of the third law one can now calculate the absolute entropy of a substance at a specific condition, as follows.

From the relationship between S and C_P (Eq. 3.7.5), we have

$$S(T') - S(T = 0) = \int_0^{T'} \frac{C_P}{T} \, dT$$

Using now the third law, $S(T = 0) = 0$, we get entropy as an absolute quantity as

$$\boxed{S(T') = \int_0^{T'} \frac{C_P}{T} \, dT} \qquad (4.1.1)$$

Such absolute entropy values, which are calculated from the heat capacity data by invoking the third law, are usually referred to as the **third law entropies**.

One may wish to determine the entropy of a substance at absolute zero without invoking the third law. However, since absolute zero is unattainable (see Sect. 4.4), and measurements at temperatures very close to the absolute zero are very difficult, entropy at absolute zero has to be determined by extrapolation from data at higher temperature. Upon such extrapolation, one may get a positive entropy of a substance at T = 0. This is known as the **residual entropy**, which is due to the persistence of many alternative microscopic or dynamical states of the system up to the lowest temperature of measurement. These states may converge essentially to a unique state or to a limited number of states that are too small to have any significant effect on entropy at absolute zero, but the extrapolation of the higher temperature trend of S versus T relation to T = 0 may not reflect the effect of reduction of the configurational states as a function of temperature.

As an example of the problem of residual entropy, let us consider the entropy of CO that was calculated by Clayton and Giauque (1932) from heat capacity measurements down to 14.36 K. They found that the extrapolated entropy of CO at T = 0 is 1.0 cal/mol-deg. The source of this residual entropy lies in the orientational disorder of carbon monoxide. A carbon monoxide molecule can exist in two different orientations, namely, CO and OC, both of which have almost exactly the same energy. Now since each molecule of carbon monoxide can exist in one of two orientations with effectively equal probability, the molecules in a mole of carbon monoxide can exist in a large number of orientations that is given by 2^L (i.e. $2 \times 2 \times 2 \times 2 \ldots$ up to L terms), where L is the Avogadro's number. Thus, the entropy due to complete orientational disorder is $S(\text{orientation}) = k_B \ln \Omega(\text{orientation}) = k_B \ln 2^L = (k_B L) \ln 2$, where $\Omega(\text{orientation})$ stands for the number of orientational configurations in one mole of carbon monoxide. Now since $L k_B = R$, $S(\text{orientation}) = 1.38$ cal/mol, which is very close to the residual entropy of carbon monoxide.

4.2 P-T Dependence of Heat Capacity Functions

Evaluation of the integral in Eq. (4.1.1) to calculate the third law entropy of a solid requires a knowledge of C_P as a function of temperature from $T = 0$ to T. Usually, C_P is expressed in terms of a polynomial function of T, such as the one given below, that fits the C_P values measured in a calorimeter at different temperatures or retrieved from other sources.

$$C_P = a + bT + c/T^2, \qquad (4.2.1)$$

This polynomial function is known as the **Maier-Kelley** equation (Maier and Kelley 1932) that has been used widely. Use of a C_P function of this form to evaluate the third law entropy according to Eq. (4.1.1) runs into obvious mathematical problem because of the $(a/T)dT$ term in the integral expression. Also, since a polynomial function can have awkward behavior when extrapolated beyond the range of conditions encompassed by the experimental data that are fitted to determine the parameters of the function, it is important to understand the constraints on the behavior of C_P function imposed by physical theory.

   The starting point of any discussion on the heat capacity of solid is the work of Einstein, who derived, just one year after his "miraculous year" of 1905, a relationship between C_V and lattice vibrations, which are the principal mechanisms of heat absorption in solids. He assumed that (a) a crystal is a collection of harmonic oscillators (see Sect. 1.6), and (b) all oscillators vibrate with the same frequency. Debye in 1912 modified Einstein's theory to allow for the fact that atoms in a crystal do not vibrate about their respective mean positions with a single frequency, but instead vibrate with a range or **dispersion** of frequencies (Fig. 1.8), say from ν_1 to a maximum frequency of ν_D that is known as the Debye frequency. The distribution of the frequencies, that is the number of lattice vibrations at each frequency within the range ν_1 to ν_D, depends on the temperature. As the temperature increases, the distribution of frequencies moves to the upper end. According to the Debye model, at some temperature called the **Debye temperature**, θ_D (Eq. 1.6.4), virtually all frequencies will be close to ν_D. The Einstein and Debye models have been discussed in numerous text books on Thermodynamics and Solid Sate Physics (see, for example; Swalin 1962; Denbigh 1981; Kittel and Kroemer 1980). Here we summarize the main results on the limiting behavior of the Debye theory and return to this topic in Sect. 14.4. A comprehensive discussion of these models can be found in Ghose et al (1992).

According to the Debye theory,

(a) As $T \rightarrow 0$, $C_V \propto T^3$, which is known as the **"Debye T to the power third law"**, and

(b) As T becomes very high, $C_V \rightarrow 3R$ per mole of a monatomic solid. If the solid is polyatomic, then $C_V \rightarrow 3nR$, where n is total number of atoms in a molecule of the solid. For example, for the mineral fayalite, Fe_2SiO_4, $C_V = 3(7)$ R = 174.6 J/K per mole of the mineral in the high temperature limit if the lattice vibrators can be considered to be harmonic oscillators. This value agrees almost exactly with that calculated from vibrational data (Ghose et al. 1992). The high temperature limiting behavior of C_V (in the harmonic model) is often referred to as **the Dulong-Petit limit** (the name comes from the observation made by Dulong and Petit in 1819 about the heat capacity values of solid elements).

Since $C_P = C_V + \alpha^2 VTk_T$ (Eq. 3.7.10),

$$\text{as } T \rightarrow 0, C_P = C_V \propto T^3, \tag{4.2.2}$$

while, as T becomes very large,

$$C_P \rightarrow 3(n)R + \alpha^2 VTk_T. \tag{4.2.3}$$

Figure 4.2 shows the measured heat capacity of solid argon versus T^3 below 8 K, which is in excellent agreement with the prediction from Debye model. Figure 4.3 shows a comparison of the measured C_P and C_V data for Cu with the expected limiting behaviors. As discussed in Sect. 1.6, thermal expansion of a solid

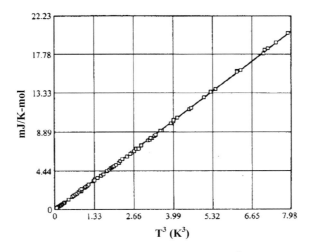

Fig. 4.2 Low temperature heat capacity of solid Argon versus T^3 showing conformation to the prediction from Debye theory. From Kittel and Kroemer (1980)

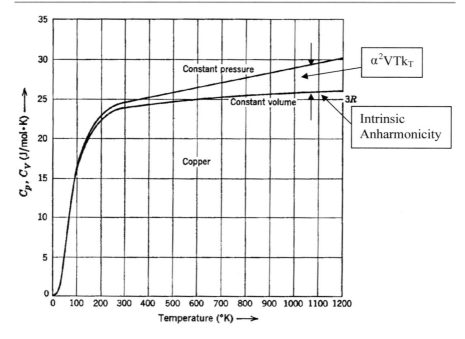

Fig. 4.3 Heat capacity of Cu as function of temperature, illustrating the difference between C_P and C_V, anharmonicity, the Dulong Petit limit (3R), and the form of the heat capacity function. Note the low temperature dependence of heat capacity according to the Debye T^3 law. Modified from Zemansky and Dittman (1981). With permission from Mc-Graw Hill

is a consequence of anharmonicity of lattice vibrations that cause displacements of the mean positions of atoms. This anharmonic effect is incorporated in the relationship between Cp and C_v through the $\alpha^2 V k_T$ term. The departure of C_v from the 3nR or the Dulong-Petit limit is a consequence of the temperature dependence of the vibrational frequencies (v_i) at constant volume that may be determined from spectroscopic data on the pressure and temperature dependencies of vibrational frequencies (Mammone and Sharma 1979; Gillet et al. 1991). The expression of $(\partial v_i/\partial T)_v$ in terms of $(\partial v_i/\partial T)_P$ and $(\partial v_i/\partial P)_T$ is given in Problem 4.3 (the derivation of the expression being left as an exercise to the reader, but with hints). The difference between C_v and the 3nR limit and that between C_P and C_v are sometimes referred to as **intrinsic** and **extrinsic anharmonic effects**, respectively (Gillet et al. 1991).

More often than not, a polynomial fit of experimentally determined C_P versus T data do not satisfy, on extrapolation beyond the temperature range of measurements, the high and low temperature limiting behaviors that follow from the Einstein-Debye theory of lattice vibrations. The extrapolated high temperature behaviors of some of the polynomial functions used in the literature are shown in Fig. 4.4. Several polynomial functions with different degrees of success have been proposed in the geological literature to remedy this problem (see Ganguly and Saxena 1987). One such function, which is due to Fei and Saxena (1987) is as follows.

$$C_P = 3nR(1 + K_1 T^{-1} + K_2 T^{-2} + K_3 T^{-3}) + (A + BT) + \phi, \qquad (4.2.4)$$

where ϕ represents collectively the contributions from anharmonicity, cation disorder and electronic effects (see below). It is retrieved by fitting the C_P versus T data beyond the 3nR limit. The behavior of the above expression for forsterite is also shown in Fig. 4.4.

Born and von Kármán (1912) developed a theory of heat capacity that is much more exact than the Debye theory. (Max Born (1882–1970) received Nobel Prize in 1954 for his pioneering research in quantum mechanics.) However, the application of the Born-von Kármán theory, which is known as **lattice dynamical theory** of heat capacity, was hindered by the fact that it requires knowledge of the phonon density of states (see Sects. 1.6 and 14.4) that had to await major technical advancements (i.e. inelastic neutron scattering) in solid state physics. A compromise between the Debye theory and lattice dynamical theory was developed by Kieffer (1979) that led to the successful prediction of heat capacities of a large number of structurally complex rock-forming minerals from their elastic constants and spectroscopic (infra-red and Raman) data. Direct calorimetric measurements of C_p versus T relations are very demanding and time consuming. Furthermore, these measurements can be carried out only in a very limited number of laboratories

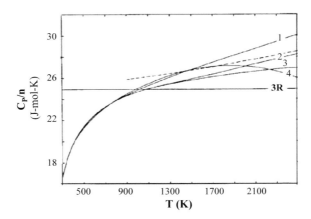

Fig. 4.4 Comparison of the temperature dependence of C_P of forsterite (Mg_2SiO_4) as calculated from different C_P functions that are fitted to calorimetric data within the temperature range ~ 298–1000 K. (1): Maier Kelley equation, (2) Eq. (4.2.5), (3) Berman and Brown equation (1985), (4) Haas and Fisher equation (Robie et al. 1978). The C_P data have been normalized by the number of atoms per formula unit. The dashed line represents $3R + \alpha^2 V T k_T$. From Ganguly and Saxena (1987)

around the world. The **Kieffer model** has, thus, gained wide popularity in the mineralogical literature since it made possible calculation of heat capacities of minerals with much greater success than that from Debye theory, but avoiding the high technical demand of lattice dynamical calculations.

Briefly, Kieffer model is a "hybrid model" in that it combines the formal spirits of Einstein and Debye models to treat the high and low frequency lattice modes, respectively. The high-frequency lattice modes obtained from spectroscopic data are assumed to be dispersionless, as in the Einstein model, whereas the acoustic modes, obtained from the experimentally measured elastic constant data, and the lowest frequency optic modes are assumed to follow specific models of dispersion, which is a Debye-like approach (Fig. 1.8). For a comprehensive review of lattice dynamical and Kieffer theories, and their mineralogical applications, the reader is referred to Ghose et al. (1992).

The pressure dependence of C_p can be derived as follows from the relationship between C_p and H (Eq. 3.7.5) and switching the order of differentiation

$$\left(\frac{\partial C_P}{\partial P}\right)_T = \left[\frac{\partial}{\partial P}\left(\frac{\partial H}{\partial T}\right)_P\right]_T = \left[\frac{\partial}{\partial T}\left(\frac{\partial H}{\partial P}\right)_T\right]_P \qquad (4.2.5)$$

Now differentiating both sides of Eq. (3.1.8) with respect to P and using Maxwell relation (Eq. 3.4.4)

$$\left(\frac{\partial H}{\partial P}\right)_T = T\left(\frac{\partial S}{\partial P}\right)_T + V = -T\left(\frac{\partial V}{\partial T}\right)_P + V \qquad (4.2.6)$$

which, upon substitution in Eq. (4.2.5), yields

$$\left(\frac{\partial C_P}{\partial P}\right)_T = \frac{\partial}{\partial T}\left(-T\left(\frac{\partial V}{\partial T}\right)_P + V\right) = -T\left(\frac{\partial^2 V}{\partial T^2}\right)_P \qquad (4.2.7)$$

For solids with positive thermal expansion, which is the usual case, the thermal expansion increases with increasing temperature (i.e. the second derivative term above is positive) and hence C_p of solids with positive thermal expansion should decrease with pressure (because it involves T, the first term on the right is usually the dominant term). The calculated temperature and pressure effects on the C_p of periclase (MgO), as calculated from the lattice dynamical model of Jacobs et al. (2013) by Tirone (2015), are illustrated in Fig. 4.5 (the behavior of C_v beyond 1000 K at 1 bar seems anomalous).

Fig. 4.5 Pressure and temperature dependence of heat capacity, as calculated by Tirone (2015) from the lattice vibration model of Jacobs et al. (2013). With permission from Oxford University press

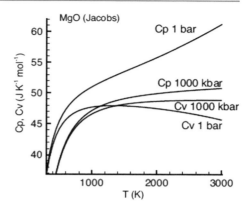

4.3   Non-lattice Contributions to Heat Capacity and Entropy of Pure Solids

4.3.1   Electronic Transitions

Besides lattice vibrations, heat may be absorbed by a solid by means of **electronic** and **magnetic** transitions. Electronic transitions are important for metals, but are usually negligible for non-metals. For metals the electronic transitions become important at high and low temperatures. At high temperatures, sufficient number of electrons may be excited to the conduction band to enable significant heat absorption. At sufficiently low temperatures, the energy absorption by electrons, although small, becomes a significant component of the small amount of total heat absorption by a metal.

Combining the form of temperature dependence of C_V on T due to electronic transition, which follows from quantum theory, with the lattice effect, the C_V versus T for metals at $T \ll \Theta_D$ (where Θ_D is the Debye temperature, Eq. 1.6.4) is given by

$$Cv = \alpha T^3 + \gamma T \qquad (4.3.1)$$

where α and γ are constants. Since at low temperature $C_P \approx C_V$, both α and γ can be simultaneously retrieved by regressing the C_P/T against T^2 at low temperatures. This procedure yields the values α and γ as respectively the slope and intercept of the linear relation between C_P/T versus T^2. An example of C_p/T versus T^2 relation for metal is shown in Fig. 4.6. Anderson (2000) has shown that at the conditions of the Earth's core, the electronic contribution to the heat capacity of Fe, which is the primary constituent of the core, is very significant. His calculated C_v versus T relation is shown in Fig. 4.7.

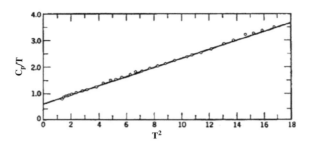

Fig. 4.6 C_p/T versus T^2 (K^2) of metallic silver near T = 0 K. The slope and intercept of the linear relation yield the parameters α and γ in Eq. (4.3.1). The electronic heat capacity is given by the term γT. From Kittel (2005). With permission from Wiley

Fig. 4.7 Vibrational and electronic heat capacities of ε-iron as function of temperature. The temperature at the core-mantle boundary is shown as CMB. From Anderson (2000)

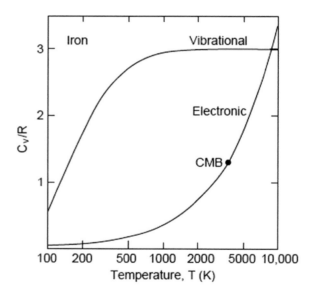

Transition metal ions in silicates may absorb a small, but not insignificant amount of heat by electronic transitions between d-orbitals that split under the influence of a crystal field (see Sect. 1.7). An electron from a lower d-orbital may be excited to a higher orbital provided that it does not violate the Pauli exclusion principle and does not change the number of unpaired electrons, in accordance with the Hund's rule (Sect. 1.7.1). This type of electronic effect on heat capacity has been studied by Dachs et al. (2007) in fayalite, Fe_2SiO_4.

4.3.2   Magnetic Transitions

The magnetic contribution is a quantum mechanical effect arising from the orbital motions and spins of the unpaired electrons. Usually, it is the spin of the unpaired electrons that constitute the dominant magnetic contribution to the entropy of a compound. The magnetic contribution can be significant for crystals with transition metal ions, which have unpaired electrons in the outer d-orbitals. Figure 4.8 shows the λ-shaped magnetic contribution to the heat capacity of Fe-end member of the olivine solid solution series, namely fayalite (Fe_2SiO_4) (Dachs et al. 2007). If C_P of a solid containing a transition metal ion is extrapolated to $T = 0$ K according to Debye relation from measurements above the magnetic transition, then the third law entropy calculated from this C_P versus T relation would be erroneous. In the absence of calorimetric data through the temperature range of magnetic transition, one may, to a very good approximation, estimate the entropy effect due to this transition and add that to the entropy calculated from the extrapolated C_p versus T relation to get a better estimate of the correct third law entropy. The estimation procedure is discussed below.

Electrons in an atom have rotary motion or spin that leads to spin quantum states. Using arguments from quantum theory, it can be shown that the number of spin quantum states (Ω_{spin}) of an atom equals ($2S + 1$) where S is the total spin of all electrons. Since the spin of an electron is $\pm \frac{1}{2}$, and the spin of a paired electron is zero ($+\frac{1}{2} -\frac{1}{2} = 0$), $S = \frac{1}{2}(N_u)$, where N_u is the number of unpaired electrons. Consequently, $\Omega_{spin} = 2(N_u/2) + 1 = N_u + 1$.

When the energy differences among the spin quantum states become small compared to the thermal energy, kT, all spin quantum states become equally probable leading to a (spin quantum) disordering, and thus a contribution to the entropy of the crystal. We may call this spin quantum configurational entropy.

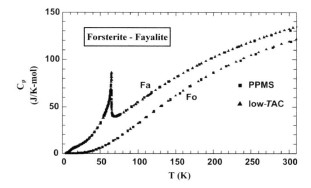

Fig. 4.8 Molar isobaric heat capacities of end-member olivines, forsterite (Fo: Mg_2SiO_4) and fayalite (Fa: Fe_2SiO_4), as determined in a calorimeter. The lamda-shaped feature in the C_p of fayalite is due to magnetic transition of Fe^{2+}. PPMS stands for a commercially designed calorimeter by Physical Properties Measurement System, whereas low-TAC stands for low temperature adiabatic calorimetry. From Dachs et al. (2007). With permission from Elsevier

However, at sufficiently low temperature, the energy difference among the spin quantum states become significant compared to kT, which leads to selective population of the states with lower energies.

The entropy change due to the disordering of a crystal over the available spin quantum states from a completely ordered state can be calculated from the Boltzmann relation. If there are n unpaired ions in a crystal, each with N_u unpaired electrons, then from (Eq. 2.5.1)

$$\Delta S_{mag} = k \ln \Omega_{spin} = k \ln(N_u + 1)^n = nk \ln(N_u + 1) \tag{4.3.2}$$

Thus, if n equals the Avogadro's number L, so that nk = R, then

$$\Delta S_{mag} = R \ln(N_u + 1) \tag{4.3.3}$$

per mol of the ion with unpaired spin.

As an example, let us calculate the ΔS_{mag} for fayalie. Since Fe^{2+} has the electronic configuration $1s^2 2s^2 2p^6 3s^2 3p^6 3d^6$, it has unpaired electron spins only in the 3d orbitals. In the high spin configuration, which is the state of Fe^{2+} except under very high pressure, the distribution of the electrons among the five 3d orbitals is $(\uparrow\downarrow)$ $(\uparrow)$ $(\uparrow)$ $(\uparrow)(\uparrow)$, where an upward pointing arrow indicates a single electron (unpaired spin) and an upward plus downward pointing arrows indicate two electrons with opposite spins in the same orbital. Thus, in the high spin state there are 4 unpaired electrons, and consequently ΔS_{mag} = Rln(5) = 13.38 J/K per mol of Fe^{2+}. However, since a mole of fayalite has 2 mol of Fe^{2+}, ΔS_{mag} = 2(13.38) = 26.76 J/K per mole of fayalite. This constitutes 17.7% of total entropy of fayalite at 298 K, and is only slightly above ΔS_{mag}(calorimetric) which constitutes 17.2% of the total entropy at the same temperature (Dachs et al. 2007). (The electronic entropy that arises from the heat capacity change due to the excitation of electrons from a lower to higher d-orbital, as discussed in the previous section, is 3.3% of the total entropy.) With increasing forsterite content in the olivine solid solution, the transition temperature for magnetic ordering, known as the **Neel temperature**, shifts progressively to lower temperature and is accompanied by a reduction of ΔS_{mag} contribution to the total entropy.

In Lanthanides and actinides, the unpaired d electrons are shielded from interaction with neighboring ions in a crystal structure by completely filled outermost s-orbitals. For these ions, the number of magnetic quantum states is given by 2J + 1, where J is a quantum number representing the total angular momentum vector of the ion. The reader is referred to Ulbrich and Waldbaum (1976) for further discussion of the subject.

4.4   Unattainability of Absolute Zero

Nernst discovered that more a substance cools, the more difficult it becomes to cool it further. From the point of view of the relationship between temperature and entropy or order, it means that more ordered the state of a substance gets, the more difficult it becomes to remove the remaining disorder so that, as proposed by Nernst, the attainment of a state of complete order that characterizes the condition at absolute zero becomes an infinitely difficult task.

To understand the problem associated with the attainment of absolute zero, consider a substance with an entropy $S(T)$ at temperature T in the S-T space, as illustrated in Fig. 4.9. One can think of cooling the substance by a combination of (a) isothermal removal of entropy and (b) isentropic (adiabatic) cooling. These two steps can be achieved, respectively, by isothermal magnetization and adiabatic demagnetization of the substance. (In isothermal magnetization, the electrons are made to spin preferentially in one direction than in the other, thus reducing the entropy of the substance. In adiabatic (isentropic) demagnetization, the magnetic field is removed producing greater disorder of electron spins. The increase of entropy due to spin disorder is compensated by cooling of the material so that the entropy remains constant.) Now, since according to the third law $S = 0$ only at $T = 0$, there is no way to carry out the steps (a) and (b) repeatedly so that the state of the substance intersects either the T-axis or the S-axis at a non-zero value, and then bring the substance to the origin ($S = T = 0$). From the geometric construction shown in Fig. 4.9, the state of $S = T = 0$ cannot be reached by a finite number of

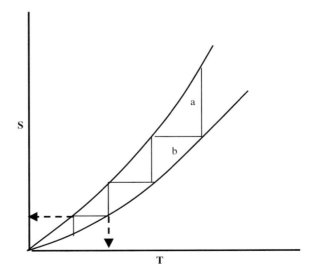

Fig. 4.9 Schematic illustration of cooling of a substance by successive steps involving isothermal removal of entropy and isentropic cooling. The paths illustrated by the dashed lines are impossible, as these violate the third law

steps of the types of (a) and (b). Similar problem arises with any other cyclic process that can be devised to lower the temperature of the substance. Thus, the third law of thermodynamics has led to the notion that *exactly* absolute zero is physically unattainable. However, the third law only precludes attainment of absolute zero by a cyclic process, such as described above. The possibility remains that the thermodynamic barrier to the attainment of absolute 0 may one day be broken by a non-cyclic process.

(The quest for lower and lower temperature has led to the attainment of temperature below 170 nK, thus producing a new state of matter known as the **Bose-Einstein condensate**. In this state the occupancy of the lowest energy state in a system of identical particles that are not subject to the Pauli Exclusion Principle rapidly increases with decreasing temperature below a certain critical temperature. This state was predicted by Albert Einstein on the basis of the seminal work of Satyendra Nath Bose (1894–1974) on quantum statistics of photons. The experimental achievement of Bose-Einstein condensation in 1995, nearly 70 years after Einstein's prediction, led to Nobel prizes to physicists Carl Weinman, Eric Cornell and Wolfgang Ketterle in 2001).

4.5 Thermochemistry: Formalisms and Conventions

4.5.1 Enthalpy of Formation

Since absolute enthalpy cannot be measured, the enthalpy data of an electrically neutral compound are reported as its enthalpy of formation from the constituent oxides or elements, which we will denote as $\Delta H_{f,o}$ and $\Delta H_{f,e}$, respectively. Consider, for example, the mineral calcite, $CaCO_3$. The $\Delta H_{f,o}$ and $\Delta H_{f,e}$ for calcite denote the enthalpy changes of the reactions (a) $CaO + CO_2 = CaCO_3$ and (b) $Ca + C(graphite) + 3/2O_2$-gas $= CaCO_3$, respectively, i.e.

$$\Delta H_{f,o}(CaCO_3) = H(CaCO_3) - [H(CaO) + H(CO_2)]$$

and

$$\Delta H_{f,e}(CaCO_3) = H(CaCO_3) - [H(Ca) + H(C : graphite) + 3/2H(CO_2 - gas)].$$

The enthalpy change of a reaction, $\Delta_r H$, equals the difference between the enthalpy of formation of the product and reactant compounds from either oxides or elements, i.e. $\Delta_r H = \Sigma H_{f,o}(products) - \Sigma H_{f,o}(reactants) = \Sigma H_{f,e}(products) - \Sigma H_{f,e}(reactants)$, as long as a uniform convention is adopted in choosing the forms of elements (for example, graphite instead of diamond for the element C). This is because of the fact that enthalpy is a state function, and therefore the change of enthalpy in going from one state to another is independent of the means by which the change of state is achieved. Consider, for example, the reaction

$$CaO + CO_2 \rightarrow CaCO_3 (calcite : Cc) \tag{4.5.a}$$

for which we can construct the following cyclic process involving decomposition of the oxides to elements and formation of calcite from the latter.

Box (4.5.1)

Because H is a state function, we write $\Delta_r H_a = \Delta H_1 + \Delta H_2 + \Delta H_3 + \Delta H_4$. Since the step 3 does not constitute any reaction, $\Delta_r H_3 = 0$. For the other steps, we can write $\Delta_r H_1 = -\Delta H_{f,e}(CaO)$, $\Delta_r H_2 = -\Delta H_{f,e}(CO_2)$ and $\Delta_r H_4 = \Delta H_{f,e}(CaCO_3)$. Thus,

$$\Delta_r H_a = \Delta H_{f,e}(CaCO_3) - \Delta H_{f,e}(CaO) - \Delta H_{f,e}(CO_2)$$

Values for heats of formation can be found in books of thermochemical properties, such as those referred to in Sect. 4.5.4.

It should be noted in the above example that we have chosen graphite and not diamond, and molecular oxygen and not atomic oxygen to refer the enthalpies of formation of CaO and CO_2 from the elements. The physical states of elements that the enthalpies of formation of compounds are referred to are known as the **reference forms** or **reference states** of the elements. *The ΔH_f of an element in its reference state is taken to be zero.* The choice of the reference states of elements is a matter of convention that is agreed upon by the thermochemists. Except for **phosphorous**, the reference forms of the elements are those forms that are stable at 1 bar and the specified temperature. Earlier choice of reference form of P conformed to this standard practice, that is red triclinic variety from 298.15 to 704 K and ideal diatomic gas from 704 to 1800 K (e.g. Robie et al. 1978), but it is now changed to the metastable white P up to 317.3 K (there is a transition from α to β phase at 195.4 K), liquid P from 317.3 to 1180.01 K and ideal diatomic gas at higher temperature. This change of reference state of crystalline P is due to the fact that **white** P is the most reproducible form. It should be easy to see that as long as a uniform convention is maintained, the choice of reference forms of elements does not have any effect on the calculation of enthalpy change of a reaction, which is ultimately the quantity of common interest in thermochemical calculations.

If an element is not in its reference form at 1 bar, T, then its enthalpy and Gibbs energy of formation are calculated from the reference form. Thus, while the $\Delta H_{f,e}$ of graphite at 1 bar, 298 K or 1 bar, 500 K is zero because graphite is the stable (and hence the reference) form of the element C at these conditions, the $\Delta H_{f,e}$ of diamond at 1 bar, T equals the enthalpy change of the reaction C(graphite) = C(diamond). Thus, at 1 bar, 298 K, $\Delta H_{f,e}$(diamond) = 1.895 kJ/mol. The $\Delta H_{f,e}$ of O_2 and H_2 gases at 1 bar, and, say, 300 K are zero because these are the stable forms of oxygen and hydrogen at these conditions. In thermochemical tables, the reference forms of an element within different temperature ranges are clearly stated, and the enthalpy and Gibbs energy of formation from the elements of these forms are listed as zero.

4.5.2 Hess's Law

Even before the formulation of the first law of thermodynamics, Germain Henri Hess (1802–1850) observed that the heat evolved in a chemical reaction is the same, whether the reaction was carried out directly, or through a series of intermediate steps. We know that this statement must be true for the enthalpy change of the reaction, as discussed above for the decomposition of $CaCO_3$ to CaO and CO_2. The reason that Hess found the heat of a reaction to be independent of the intermediate steps is simply because he conducted his experiments at a fixed pressure of 1 bar, and at constant pressure $\Delta_r Q = \Delta_r H$, which depends only on the initial and final states. It is easy to check the validity of this relation from the application of the first law to chemical reaction, as follows.

For any reaction at a constant pressure, the change in the internal energy is given by $U_p - U_r = (Q_p - Q_r) - P(V_p - V_r)$, where the subscripts p and r stand, respectively, for the products and reactants, regardless of the number of intermediate steps. Thus, at constant pressure $\Delta_r Q = (U_p + PV_p) - (U_r + PV_r)$. But since H = U + PV, we have at constant pressure, $\Delta_r Q = \Delta_r H$.

4.5.3 Gibbs Free Energy of Formation

The Gibbs free energy of formation of a compound from the component elements ($\Delta G_{f,e}$) or oxides ($\Delta G_{f,o}$) is estimated according to thermodynamic relation among G, H and S. Thus, $\Delta G_{f,e} = \Delta H_{f,e} - T\Delta S_{f,e}$, and similarly for $\Delta G_{f,o}$. However, more commonly one utilizes a combination of the enthalpy of formation from the elements or oxides and the third law entropy. For example, $\Delta G_{f,e} = \Delta H_{f,e} - T\Delta S$. This has been called the **apparent** Gibbs free energy of formation (from element or oxide) or sometimes simply just the Gibbs free energy of formation, although the last nomenclature is not quite appropriate. In practice, however, it does not matter if one uses the true or apparent Gibbs free energy of formation of a compound, as long as all compounds in a given system are treated similarly, because the relative Gibbs free energies of different isochemical assemblages are not affected by the choice of either method.

4.5.4 Thermochemical Data

Thermochemical properties of minerals and other substances are measured directly in different types of calorimeters and solid state electrochemical cells, and are also retrieved from experimentally determined phase equilibrium relations that represent consequences of the thermochemical properties of the phases. The modern approach is to combine the directly measured data with the phase equilibrium constraints, and derive by appropriate optimization techniques an internally consistent set of thermochemical data for mineral phases and fluids. By "internally consistent" we mean that the retrieved thermochemical properties are mutually compatible so that the phase relations calculated from these properties are consistent with the available experimental constraints that are considered to be reliable. (In the materials sciences, this is often referred to as the CALPHAD approach, after the name of a consortium and a journal CALPHAD that is an acronym for **Cal**culations of **Ph**ase **D**iagrams.) Using this global approach, several internally consistent data sets have been developed for Earth and Planetary materials such as those by Berman (1988), Johnson et al. (1992), Saxena et al. (1993), Gottscahlk (1997), Holland and Powell (2011), Chatterjee et al. (1998), Fabrichnaya et al. (2004) and Stixrude and Lithgow-Bertelloni (2005a, b, 2011) (the pioneering study being that by Helgeson et al. (1978) which is now superseded by Johnson et al. (1992)). Each data set is internally consistent, but is not necessarily consistent with one another. This means that combination of data from different sets may lead to wrong prediction of new phase relations. The data sets of Saxena et al. (1993), Fabrichnaya et al. (2004), Holland and Powell (2011) and Stixrude and Lithgow-Bertelloni (2011) are especially suited for applications to high pressure phase equilibrium that are appropriate for the Earth's mantle.

The success of a specific data base to correctly produce phase equilibrium relations only implies that the Gibbs free energy values within the data base are internally consistent within a certain P-T range. The Gibbs energy values are typically computed at desired P-T conditions by using their tabulated values at some reference conditions (typically 1 bar, 298 K) and the heat capacity, thermal expansion and compressibility data given within the specific data base. However, the success of a data base with phase equilibrium calculations does not necessarily imply that the individual thermochemical and thermophysical properties (viz. thermal expansivity, compressibility, heat capacity, enthalpy and entropy) are correct. A variety of petrological and geophysical calculations, however, require reliable data for these individual properties and it has been found that some of these properties exhibit behavior at the deep mantle and core conditions of the Earth that are incompatible with physical constraints (see, for example, Stacey 2005; Brosh et al. 2008; Tirone 2015). For example, one may find negative thermal expansion for phases for which such behavior can be excluded, thermal maximum of C_p, increase of C_p with increasing P, in violation of thermodynamic restriction (Eq. 4.2.7), etc. A root cause of these problems can be traced to problems with the adopted equation of states to deal with very high P-T conditions corresponding to the states of the Earth's lower mantle and core (see discussion in Sect. 5.4.2.2).

In addition to the above, there are also empirical and microscopic methods of estimating thermochemical properties, some of which have been discussed in the Appendix C. A detailed discussion of the methods that relate microscopic properties at the atomic level to macroscopic thermodynamic properties is beyond the scope of this book, but the interested readers are referred to Kieffer and Navrotsky (1985), Tossell and Vaughn (1992), Gramaccioli (2002) and the most recent review by Ottonello (2018).

In solid electrochemical cells, one directly measures the free energies of formation of oxides as function of temperature (e.g. O'Neill 1988). Also, from the temperature dependence of $\Delta_f G$ one obtains the entropy and enthalpy values ($\partial G/\partial T = -S$, $H = G + TS$). The basic principles of electrochemical cells have been briefly discussed in Sect. 12.9.

In a calorimeter, one measures the change of heat (i.e. calor in Latin) associated with a change of state of a substance, such as dissolution and phase change or chemical reaction, which is made to take place within a well-insulated chamber. The quantities measured are C_v, C_p, and heats of chemical reactions. The latter may be measured directly, or through a thermochemical cycle when the reaction is too slow to be amenable to direct measurement. The enthalpies of formation of most binary oxides from elements have been determined directly by combustion calorimetry that involves burning metal wires or powders in an oxygen atmosphere. Comprehensive reviews of calorimetric methods, especially in the context of Earth materials, can be found in Navrotsky (1997, 2002) and Geiger (2001).

The idea of a thermochemical cycle that is employed to determine the enthalpy change of a slow reaction can be illustrated by considering the reaction MgO (periclase: Per) + SiO_2 (quartz: Qtz) = $MgSiO_3$ (enstatite: Enst). The enthalpy change of this reaction is the $\Delta H_{f,o}$ of $MgSiO_3$. This reaction is too slow to be amenable to direct calorimetric measurements, but the reaction enthalpy can be determined using the following thermochemical cycle.

$$(1)\ MgO\ (crystal) + Solvent \rightarrow solution \qquad \Delta H_s(1)$$
$$(2)\ SiO_2\ (crystal) + Solvent \rightarrow solution \qquad \Delta H_s(2)$$
$$(3)\ MgSiO_3\ (crystal) + Solvent \rightarrow solution \qquad \Delta H_s(3)$$

where $\Delta H_s(i)$ represents the heat change associated with a specific dissolution step (i) in a calorimeter. (Because the pressure is maintained constant (at 1 bar), the heat change equals ΔH, as shown in Sect. (4.5.2)). The solution is kept very dilute in order to prevent interactions among the dissolved species and the phases are dissolved sequentially in the **same** solvent. The dissolution process may be either exothermic (heat is evolved) or endothermic (heat is absorbed), the latter being the case for most silicates.

Now $\Delta H_{f,o}(MgSiO_3)$ is the enthalpy change of the reaction

$$MgO + SiO_2 = MgSiO_3. \tag{4.5.b}$$

This enthalpy change can be represented by a thermochemical cycle

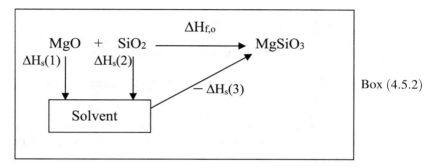

Box (4.5.2)

so that

$$\Delta H_{f,o}(MgSiO_3) = \Delta H_s(1) + \Delta H_s(2) - \Delta H_s(3). \tag{4.5.1}$$

The $\Delta H_{f,e}$ of a ternary compound can be obtained from its $\Delta H_{f,o}$ and the enthalpy of formation from elements of the component binary oxides according to

$$\Delta H_{f,e}(\text{ternary comp.}) = \Delta H_{f,o}(\text{ternary comp.}) + \Sigma \Delta H_{f,e}(\text{binary oxide}) \tag{4.5.2}$$

The enthalpy change of a polymorphic transition can be measured from the difference of heat of solution of the two phases in a solvent. For example, $\Delta_r H$ of the olivine-wadsleyite (α-Mg_2SiO_4 (Ol) = β-Mg_2SiO_4 (Wad)) and of wadsleyite-ringwoodite (β-Mg_2SiO_4 (Wad) = γ-Mg_2SiO_4 (Ring)), which represent major phase transitions within the Earth's mantle at 400 and 400 km depths, was determined by Akaogi et al. (2007) from the difference in the enthalpies of solution of the two polymorphs in a lead borate ($2PbO.B_2O_3$) solvent at 1 bar, 973 K. (Lead borate is a very effective solvent for many rock forming and mantle minerals at a temperature of 973 K (Kleppa 1976), and has been used widely to determine the heats of formation of these phases.) The enthalpy of solution values are: $\Delta H_s(\alpha) = 169.35 \pm 2.38$ kJ/mol, $\Delta H_s(\beta) = 142.19 \pm 2.65$ kJ/mol and $\Delta H_s(\gamma) = 129.31 \pm 1.96$ kJ/mol. These values yield $\Delta_r H$ ($\alpha = \beta$) = $\Delta H_s(\alpha)$ - $\Delta H_s(\beta)$ = 27.2 ± 3.6 and similarly $\Delta_r H$ ($\beta = \gamma$) = 12.9 ± 3.3 kJ/mol. It is left to the reader to figure out the relationship between the $\Delta_r H$ and ΔH_s of the polymorphs by constructing the appropriate thermochemical cycles.

It should be noted that while conventional thermochemical tables (JANAF, Holland and Powell 2011; Fabrichnaya et al. 2004) provide enthalpy data of compounds in terms of formation values from the elements or oxides, usually the former, ab initio quantum chemical calculations yield the enthalpies of both compounds and their constituent elements. Such calculations have now progressed to a level to yield reliable thermochemical values for complex materials such as rock forming minerals (e.g. Ottonello 2018). If the data from the thermochemical tables need to be combined with the results of ab initio calculations to calculate the enthalpy change of a reaction, one must first convert the ab initio enthalpy values to $\Delta H_{f,e}$ (i.e. enthalpy of formation from elements) using the enthalpies of element that have been calculated following the same protocol.

Problem 4.1 The heats of solution of $MgSiO_3$ (orthoenstatite) and component oxides in a lead borate ($2PbO.B_2O_3$) solvent, as determined in a calorimeter at 1 bar, 970 K, are as follows (Charlu et al. 1975).

$$\Delta H_{sol'n}(MgSiO_3) = 36.73 \pm 0.54 \text{ kJ/mol}; \Delta H_{sol'n}(MgO) = 4.94 \pm 0.33 \text{ kJ/mol};$$

$\Delta H_{sol'n}(SiO_2) = -5.15 \pm 0.29$ kJ/mol, where the uncertainties represent 1σ (standard deviation) values.

From these data, calculate the heat of formation from both oxides and elements of orthoenstatite at 1 bar, 970 K, along with their respective standard deviations. Look up the additional data that you may need in a thermochemical Table (e.g. Saxena et al. 1993).

Answer $\Delta H_{f,ox} = -36.94(0.696)$ kJ/mol(1σ); $\Delta H_{f.elem} = -1472.3$ kJ/mol using data from Saxena et al. (1993)

Problem 4.2 Following the derivation for Eq. (4.2.7), show that

$$\left(\frac{\partial C_v}{\partial V}\right)_T = T\left(\frac{\partial^2 P}{\partial T^2}\right)_v$$

*Problem 4.3 The "intrinsic anharmonic parameter" of a vibrational mode i is defined as the dependence of $\ln v_i$ (v_i: frequency) on temperature at constant volume, and may be determined, according to the following relation, from the spectroscopic data on the pressure and temperature dependencies of v_i.

$$\left(\frac{\partial \ln v_i}{\partial T}\right)_V = \alpha k_T \left(\frac{\partial \ln v_i}{\partial P}\right)_T + \left(\frac{\partial \ln v_i}{\partial T}\right)_P$$

Derive this relation.

Hint: Consider $\ln v_i$ to be a function of P and T, and then take its total derivative. Evaluate the term $(\partial P/\partial T)_V$ using the property of implicit function (see Appendix B.4).

References

Akaogi M, Takayama H, Kojitani H, Kawaji H, Atake T (2007) Low-temperature heat capacities, entropies and enthalpies of Mg_2SiO_4 polymorphs, and α-β-γ and post-spinel phase relations at high pressure. Phys Chem Miner 34:169–183
Anderson OL (2000) The Grüneisen ratio for the last 30 years. Geophys J Int 143:279–294

Berman RG (1988) Internally consistent thermodynamic data for minerals in the system N_2O-K_2O-CaO-MgO-FeO-Fe_2O_3-Al_2O_3-SiO_2-TiO_2-H_2O-CO_2. J Petrol 29:445–552

Berman RG, Brown TH (1985) Heat capacity of minerals in the system Na_2O-K_2O-CaO-MgO--FeO-Fe_2O_3-Al_2O_3-SiO_2-TiO_2-H_2O-CO_2: Presentation, estimation and high temperature extrapolation. Contrib Mineral Petrol 89:168–183

Born M, von Kármán T (1912) Über Schwingungen in Raumgittern. Physik Zeit 12:297–309

Brosh E, Shneck RZ, Makov G (2008) Ecplicit Gibbs free energy equation of state for solids. J Phys Chem Solids 69:1912–1922

Charlu TV, Newton RC, Kleppa OJ (1975) Enthalpies of formation at 970 K of compounds in the system MgO-Al_2O_3-SiO_2 from high temperature solution calorimetry. Geochim Cosmochim Acta 39:1487–1497

Chatterjee ND, Krueger R, Haller G, Olbricht W (1998) The Bayesian approach to the internally consistent thermodynamic data base: theory, database and generation of phase diagrams. Contrib Miner Petrol 133:149–168

Clayton JO, Giauque WF (1932) The heat capacity and entropy of carbon monoxide. Heat of vaporization. Vapor pressure of solid and liquid. Free energy to 500 K from spectroscopic data. J Am Chem Soc 54:2610–2626

Dachs E, Geiger C, von Seckendorff V, Grodzicli M (2007) A low temperature calorimetric study of synthetic (forsterite + fayalite) $\{(Mg_2SiO_4 + Fe_2SiO_4)\}$ solid solutions: an analysis of vibrational, magnetic, and electronic contributions to the molar heat capacity and entropy of mixing. J Chem Thermo 39:906–933

Debye P (1912) Zur Theorie der spezifischen wärmen. Ann der Physik 39:789–839

Denbigh K (1981) The principles of chemical equilibrium, 4th Edition, Dover

Fabrichnaya O, Saxena SK, Richet P, Westrum EF (2004) Thermodynamic data, models, and phase diagrams in multicomponent oxide systems. Springer

Fei Y, Saxena SK (1987) An equation of state for the heat capacity of solids. Geochim Cosmochim Acta 51:251–254

Fermi E (1956) Thermodynamics, Dover, p 160

Ganguly J, Saxena SK (1987) Mixtures and mineral reactions. Springer, Berlin, Heidelberg, New York, Paris Tokyo, p 291

Geiger CA (2001) Thermodynamic mixing properties of binary oxide and silicate solid solutions determined by direct measurements: the role of strain. In: Geiger CA (ed) Solid solutions in silicate and oxide systems. EMU notes mineral 3, Eőtvős Univ Press, pp 71–100

Ghose S, Choudhury N, Chaplot SL, Rao KR (1992) Phonon density of states and thermodynamic properties of minerals. In: Saxena SK (ed) Advances in physical geochemistry, vol 9. Springer, New York Berlin Heidelberg, pp 283–314

Gillet P, Richet P, Guyot F (1991) High temperature thermodynamic properties of forsterite. J Geophys Res 96:11805–11816

Gottschalk M (1997) Internally consistent thermodynamic data for rock-forming minerals in the system SiO_2-TiO_2-Al_2O_3-Fe_2O_3-CaO-MgO-FeO-K_2O-Na_2O-H_2O-CO_2. Eur J Miner 9:175–223

Gramaccioli CM (ed) (2002) Energy modelling in minerals. EMU notes in mineralogy, vol 4, Eötvös University Press, Budapest, p 425

Helgeson HC, Delany JM, Nesbitt HW, Bird DK (1978) Summary and critique of the thermodynamic properties of rock-forming minerals. Amer J Sci 278-A:1–229

Holland TJB, Powell R (2011) An improved and extended internally consistent thermodynamic dataset for phases of petrological interest, involving a new equation of state for solids. J Met Geol 29:333–383

Jacobs MHG, Schmid-Fetzer R, van der Berg AP (2013) An alternative use of Kieffer's lattice dynamic model using vibrational density of states for constructing thermodynamic databases. Phys Chem Miner 40:207–227

Johnson JW, Oelkers EH, Helgeson HC (1992) SUPCRIT92: a software package for calculating
    the standard molal thermodynamic properties of minerals, gases, aqueous species, and
    reactions from 1 to 50000 bar and 0 to 1000 C. Comput Geosci 18:889–947
Kieffer SW (1979) Thermodynamics and lattice vibrations of minerals: 1. Mineral heat capacities
    and their relationships to simple lattice vibrational models. Rev Geophys Space Phys 17:1–19
Kieffer SW, Navrotsky A (1985) Editors: Microscopic to macroscopic, Reviews in mineralogy 14,
    Mineralogical Society of America, p 428
Kittel C (2005) Introduction to solid state physics. Wiley, New Jersey, p 704
Kittel C, Kroemer H (1980) Thermal physics. Freeman, San Francisco, p 473
Kleppa OJ (1976) Mineralogical applications of high temperature reaction calorimetry. In:
    Sterns RGJ (ed) The Physics and chemistry of minerals. Wiley, pp 369–388
Maier CG, Kelley KK (1932) An equation for the representation of high temperature heat content
    data. J Am Chem Soc 54:3242–3246
Mammone JF, Sharma SK (1979) Pressure and temperature dependence of Raman spectra of rutile
    structure oxides. Year Book Carnegie Inst Washington 78:369–373
Navrotsky A (1997) Progress and new directions in high temperature calorimetry. Phys Chem
    Miner 24:222–241
Navrotsky A (2002) Thermochemistry, energetic modelling, and systematics. In: Gramaciolli CM
    (ed) Energy modelling in minerals, vol 14. European Mineralogical Union. Eötvös University
    Press, Budapest, Hungary, pp 5–26
O'Neill HStC (1988) Systems Fe-O and Cu-O: Thermodynamic data for the equilibria Fe-"FeO",
    Fe-Fe3O4, "FeO"-Fe3O4, Fe3O4-Fe2O3, Cu-Cu2O, and Cu2O-CuO from emf measurements.
    Am Miner 73:470–486
Ottonello G (2018) Ab initio reactivity of Earth's materials. Rivista del Nuovo Cimento 41:225–
    289
Robie RA, Hemingway BS, Fisher JR (1978) Thermodynamic properties of minerals and related
    substances at 298.15 K and 1 bar (10^5 pascals) pressure and at higher temperatures. US Geol
    Surv Bull 1452, p 456
Saxena SK, Chatterjee N, Fei Y, Shen G (1993) Thermodynamic data on oxides and silicates.
    Springer
Stacey FD (2005) High pressure equations of state and planetary interiors. Rep Prog Phys 68:341–
    383
Stixrude L, Lithgow-Bertelloni C (2005a) Thermodynamics of mantle minerals–I. Physical
    properties. Geophys J Int 162:610–632
Stixrude L, Lithgow-Bertelloni C (2005b) Thermodynamics of mantle minerals–II. Phase
    equilibria. Geophys J Int 184:1180–1213
Stixrude L, Lithgow-Bertelloni C (2011) Thermodynamics of mantle minerals–II. Phase equilibria
    Geophys J Int 184:1180–1213
Swalin RA (1962) Thermodynamics of solids. Wiley, New York, London, Sydney, Toronto, p 387
Tirone M (2015) On the use of thermal equations of state and the extrapolation at high temperature
    and pressure for geophysical and petrological applications. Geophys J Int 202:55–66
Tossell JA, Vaughn DJ (1992) Theoretical geochemistry: application of quantum mechanics in the
    earth and mineral sciences. Oxford, Oxford, New York, p 514
Ulbrich HH, Waldbaum DR (1976) Structural and other contributions to the third-law entropies of
    silicates. Geochim Cosmochim Acta 40:1–24
Zemansky MW, Dittman RH (1981) Heat and thermodynamics. McGraw-Hill, New York, p 543

Critical Phenomenon and Equations of States

5

Consider the familiar phase diagram of H_2O in the P-T space, showing the stabilities of ice, liquid water and water vapor (Fig. 5.1). Formally, a phase is defined to be a substance that is spatially uniform on a macroscopic scale and is physically distinct and separable from the surrounding. Along any of the three lines, the two phases that are on either side coexist in stable equilibrium. However, note that the line separating the fields of liquid water and water vapor ends at a point, C, which has the coordinate of 220.56 bars, 647.096 K (373.946 °C). This point is a **critical end point**. Thus, the critical end point in the phase diagram of H_2O is the terminus of the curve along which liquid and vapor can coexist. As a mixture of liquid and vapor is moved along the coexistence curve to higher temperature and pressure conditions, the properties of the two phases progressively approach one another and the distinction between their properties completely vanishes at the critical end point.

The thermodynamic and transport properties of a fluid changes very rapidly as the P-T conditions approach the critical end point. These sharp changes of properties have important geological and industrial consequences that we would explore in this chapter. In addition, the pressure-temperature condition of the critical point of a fluid has important implications in the formulation of its equation of state.

5.1 Critical End Point

If the P-T condition change between the fields of liquid and vapor along a path that does not intersect the vapor-liquid coexistence curve, such as the curve (p-q-r) in Fig. 5.1, the property of the phase will change **continuously**, and at no point will there be a coexistence of liquid and vapor; there will be either a liquid or a supercritical phase or a vapor. This is unlike the case when the P-T path intersects the liquid-vapor coexistence curve, in which case there is a discontinuous change of properties (e.g. volume). On further compression of a gas after it is compressed to the pressure of the liquid-vapor coexistence curve, there is progressive conversion

© Springer Nature Switzerland AG 2020
J. Ganguly, *Thermodynamics in Earth and Planetary Sciences*,
Springer Textbooks in Earth Sciences, Geography and Environment,
https://doi.org/10.1007/978-3-030-20879-0_5

of gas to liquid without any change of pressure (the effect of compression is compensated by the decrease of volume), thus permitting the coexistence of both liquid and vapor phases. Analogous situation prevails if the temperature is changed at a constant pressure in that when the temperature reaches the coexistence curve, there is progressive conversion of one phase to the other without any change of temperature. The heat that is supplied or withdrawn becomes compensated by the heat change associated with phase transformation.

The P-T space beyond the critical end point is conventionally divided, for descriptive purposes, into domains of supercritical liquid ($P > P_c$, $T < T_c$), supercritical vapor ($T > T_c$, $P < P_c$) and supercritical fluid ($T > T_c$, $P > P_c$), as illustrated in Fig. 5.1, where P_c and T_c denote the pressure and temperature of the critical end point. However, there is no discontinuous transition of properties between any two adjacent domains.

The existence of a critical temperature was discovered by Thomas Andrews in 1869. In the course of determining the effect of temperature and pressure on the properties of carbon dioxide, Andrews found that CO_2 gas would transform to a liquid if the pressure on the gas was increased but only as long as the temperature was kept below 304 K (31 °C). However, beyond this temperature, it was impossible to convert CO_2 gas to liquid by further compression. The critical P-T conditions of CO_2, are now accepted to be 72.8 bar, 304.2 K.

Notice that no critical end point exists on the other coexistence curves in the phase diagram of water. The distinction between the properties of liquid and vapor phases is purely quantitative in nature. In both cases, the water molecules are distributed randomly but the interactions between the molecules in the gas is weaker than those in the liquid. On the other hand, the distinction between liquid water and ice and between ice and vapor is qualitative in nature since ice has a crystal structure with symmetry properties. This observation brings out an

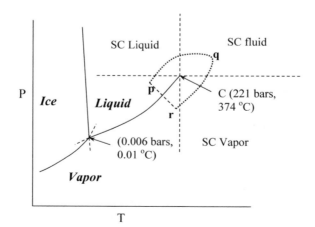

Fig. 5.1 Schematic phase diagram of H_2O showing the critical end point of the vapor-liquid coexistence curve and the domains of supercritical (SC) phases. Note that the horizontal and vertical dashed lines are not phase boundaries, but depict descriptive regimes

important point about the condition for the existence of a critical end point, that is, such a point can exist only on the coexistence curve of two phases that have **quantitative, but not qualitative** difference of properties. Thus, for example, there can be no critical end point on the solid-liquid phase boundary.

Before delving further into the thermodynamics of critical phenomenon, we digress briefly to discuss the nature of fluids in some natural systems. Fluids under conditions of terrestrial metamorphic processes, which are rich in H_2O and CO_2, had been at the supercritical conditions. The critical point of ocean water was experimentally determined to be 403–405 °C, 285–302 bars (Sourirajan and Kennedy 1962; Bischoff and Rosenbauer 1983); it is different from that of pure water because of its salinity. Koschinsky et al. (2008) reported temperatures of fluids emanating from hydrothermal veins on the Mid-Atlantic Ridge system that fall on or above (as high as 407 °C) the critical point of sea water.

The interiors of **Jupiter** and **Saturn** consist of three layers, namely, molecular hydrogen plus helium and metallic hydrogen plus helium envelopes followed by dense cores (ice + rock) (Guillot 1999). With increasing pressure, the molecular gaseous species become a supercritical fluid with progressively increasing density (the critical points of H_2 and He are at 33.2 K, 13.0 bar and 5.2 K, 2.3 bar, respectively), and finally transforms, at pressures above 30 GPa, to a liquid metallic state across a phase transition boundary. Such pressures (and much higher, of the order of Mbar) are attained within the interiors of Jupiter and Saturn.

Let us now consider the P-V diagram showing the isothermal expansions of liquid and gas states, such as those of H_2O (Fig. 5.2). Above the point marked P_c, V_c, which is the critical end point, there is only one phase, the volume of which

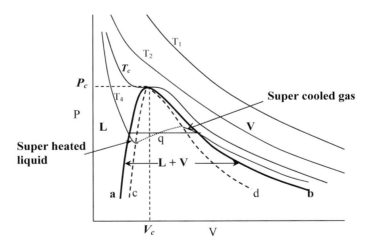

Fig. 5.2 Schematic P-V diagram of a substance with a critical point at P_c, V_c, T_c. T_i indicates an isotherm with T_c as the isotherm at the critical temperature. L and V stand for liquid and vapor, respectively, which are stable at the two sides of the thick solid curve a-b. At any point under the curve a-b, the liquid and vapor phases coexist in stable equilibrium at a fixed temperature, and their volumes are given by the points of intersection of the isotherm with the curve a-b. However, the liquid and solid phases may persist **metastably** up to the points of intersection of an isotherm with the dashed curve c-d. Between c and d, there must be two phases, liquid and vapor, at any temperature

changes continuously with change of pressure. But below this point, there are two phases, liquid and vapor, which are stable on the left and right sides, respectively, of the heavy curve a-b that touches the critical point. The P-V curves for both liquid and gas, which are shown by solid lines, satisfy the inequality $(\partial P/\partial V)_T < 0$, but continue a short distance beyond their respective stability fields, and end at the dashed curve c-d. These distances, which have terminal points at $(\partial P/\partial V)_T = 0$, mark the fields of (metastable) superheated liquid and supercooled gas. The line c-d is the schematic locus of the points at which $\partial P/\partial V = 0$. Note that at any temperature below T_c, the P-V curves of the stable liquid and stable gas intersect the heavy line a-b at the same pressure. This is a pressure on the liquid-vapor coexistence curve (Fig. 5.1) corresponding to the specified temperature. Within the area bounded by the dashed line c-d, which also touches the critical point, a single homogeneous phase is unstable. Instead there are two phases, liquid and gas, the relative proportion of which change with compression. Between a-b and c-d, a homogeneous phase, either gas or liquid is metastable. This means that the "wrong phase" can survive due to kinetic barrier associated with the transformation to the thermodynamically stable phase.

Since the critical point is a point on the curve c-d, it is obvious that at the critical point

$$\left(\frac{\partial P}{\partial V}\right)_T = 0 \tag{5.1.1}$$

Comparing the isotherms T_c and T_4, we find that the critical point is the point of merger of the minimum and maximum on an isotherm. The convergence of a minimum and a maximum on an isotherm denotes the transition from the condition of $\partial^2 P/\partial V^2 > 0$ (condition for a minimum) to $\partial^2 P/\partial V^2 < 0$ (condition for a maximum). Thus, at the critical point, we must have $\partial^2 P/\partial V^2 = 0$. It can also be shown that at the critical point $\partial^3 P/\partial V^3 < 0$ (Landau and Lifshitz 1958).

P-V relations at the critical point:

$$\left(\frac{\partial P}{\partial V}\right)_T = \left(\frac{\partial^2 P}{\partial V^2}\right)_T = 0$$
$$\left(\frac{\partial^3 P}{\partial V^3}\right)_T < 0$$

Box (5.1.1)

One can imagine that a single expression could be found to fit the experimental P-V data of both liquid and gas at a fixed temperature, as shown by connecting the P-V curves of these phases at T_4 by a dotted line inside the curve c-d. The Gibbs energy of the two phases in equilibrium at a fixed P-T condition must be the same. This condition imposes a geometric restriction on the nature of the wavy part of an isotherm.

Consider, for example, the isotherm T_4, which intersects the horizontal line connecting the stable liquid and gas phases at the point q. The requirement that the Gibbs energies of the liquid and gas connected by the horizontal line must be the same implies that the magnitudes of the areas between the wavy part of the isotherm T_4 and the horizontal line on two sides of the point q must be the same. This requirement ensures that the change of Gibbs energy along the wavy line within a-b (which is given by $\int V \mathrm{d}P$ between the end points of the horizontal line through q) is zero.

If we now consider the isobaric T-V relation of liquid and gas, we will have qualitatively the same picture as illustrated in Fig. 5.2, except that the volume would increase with increasing temperature. Thus, at T_C, we would also have

$$\left(\frac{\partial T}{\partial V}\right)_P = 0 \tag{5.1.2}$$

The second and third derivative properties are analogous to those in the P-V relation, as summarized in Box (5.1.1).

5.2   Near- and Super-Critical Properties

5.2.1   Divergence of Thermal and Thermo-Physical Properties

The fact that $\partial P/\partial V = \partial T/\partial V = 0$ at the critical point has important consequences about the behavior of C_P, β_T, and α_T at and near the critical condition. From the definitions of α and β_T (Eqs. 3.7.1 and 3.7.2, respectively), it is easy to see that near the critical end point, both α and β_T tend to $+\infty$. (We discussed in 3.7 that for some solids $\alpha_T < 0$, but there are no critical point bounding the stability fields of such solids). The qualitative behavior of C_P near the critical end point can be derived as follows. From Eq. 3.7.10, the difference between C_P and C_V is given by the term $\alpha^2 VT/\beta_T$, which, on substitution of the expressions of α and β_T yields

$$\frac{\alpha^2 VT}{\beta_T} = -\frac{T\left(\frac{\partial V}{\partial T}\right)_P^2}{\left(\frac{\partial V}{\partial P}\right)_T} \tag{5.2.1}$$

Using now the property of an implicit function (Eq. B.4.4), we have $(\partial V/\partial T)_P = -(\partial V/\partial P)_T/(\partial T/\partial P)_V$, which on substitution in the above equation and rearrangement of terms, yields

$$C_P - C_V = \frac{\alpha^2 VT}{\beta_T} = \frac{T\left(\frac{\partial P}{\partial T}\right)_V^2}{\left(\frac{\partial P}{\partial V}\right)_T} \tag{5.2.2}$$

As $T \to T_C$, the denominator after the second equality tends to zero (Eq. 5.1.1), and consequently, $(C_P - C_V) \to \infty$. Also, since all the terms after the first equality are positive quantities, $(C_P - C_V) \to +\infty$ as the critical temperature is approached. This implies that $C_P \to +\infty$ whereas C_V remains finite, but diverges weakly, as $T \to T_C$.

In summary, **as $T \to T_C$**,

$$\alpha \to +\infty$$
$$\beta_T \to +\infty$$
$$C_p \to +\infty \qquad\qquad \text{Box (5.2.1)}$$
$$C_V\text{: finite}$$
$$\text{(weak divergence)}$$

The divergence of the fluid properties near a critical end point, however, is not limited to those discussed above. Obviously, other fluid properties that are related to one or more of the above divergent properties must also exhibit divergence near a critical point. Thus, for example, sound speed in a fluid must tend to zero as $T \to T_c$ since it is related to β_T according to $C_{sound} = (1/\rho\beta_T)^{1/2}$, where ρ is the fluid density. For comprehensive discussions of the behavior of various fluid properties near the critical end point, the readers are referred to Sengers and Levelt Sengers (1986) and Johnson and Norton (1991). The latter specifically considers the properties H_2O that are of great importance in the study of hydrothermal systems. Sengers and Levelt Sengers (1986) have discussed the criteria for strong and weak divergence of properties near the critical point. According to these criteria, β_T and C_p diverge strongly, in accordance with the above analysis, whereas β_s and C_v diverge weakly.

The divergence of a property near the critical end point is typically expressed in terms of an appropriate exponent of $(T - T_c)$. For example, the divergence of the bulk modulus is expressed as

$$\beta_T = (T - T_c)^{-\gamma} \qquad\qquad (5.2.3)$$

The exponents of $(T - T_c)$ used to describe the temperature dependence of properties near a critical point are known as the **critical exponents**.

5.2.2 Critical Fluctuations

Classical thermodynamics correctly predicts that the properties like C_P, α, β_T should diverge near the critical end point, but it fails to correctly predict the analytical form of the divergence. This problem, however, has no significant consequence except at conditions very near the critical point. There are enormous fluctuations of properties near the critical end point, which cannot be predicted by classical thermodynamics (see Callen 1985, for further discussion). For example, there are huge density fluctuations of water at or very close to the critical end point, as illustrated in Fig. 5.3, which renders water milky or opaque due to the scattering of light. This phenomenon is known as **critical opalescence** and is due to the variation of refractive index of the liquid as a result of its density fluctuations at length scales comparable to the wave length of light. However, a fraction of a degree change of temperature restores the water to its normal transparent state.

The reason behind the fluctuations of properties near the critical end point may be understood by considering the shape of Gibbs energy versus volume curve at points on the coexistence curve at and near this point within the framework of a classical theory of phase transition that is due to the great physicist Lev Landau (1908–1968; Nobel prize 1962), and is commonly referred to as the Landau theory (see Sect. 6.3). According to this theory, the G versus V curve has two equally depressed minima at any point on the vapor-liquid coexistence curve which gradually merge as the critical end point is approached, leading finally to a broad minimum (Fig. 5.4). The two equally depressed minima of the Gibbs energy

Fig. 5.3 Density fluctuations of water near the critical point. From CEA (1998)

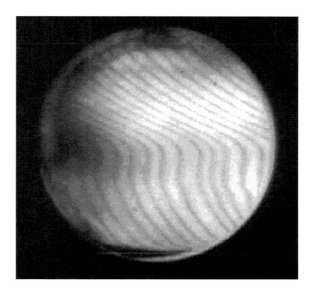

corresponds to the two equally stable physical states of the system. At any point on either side of the coexistence curve, there is one global and one local minimum (that is one minimum of G is lower than the other), the former corresponding to the one stable physical state of the system. The single minimum of G at the critical end point implies that there is only one truly stable state for the system at the critical condition, but since the minimum is very broad, the system can exist in several states of different densities without any significant effect on the overall Gibbs energy. Similar picture is valid at near-critical condition. In other words, the density or the physical state of the system at or very near the critical condition can **fluctuate** over a significantly large length scale because of the broadness of the Gibbs energy minimum. Since the difference in the Gibbs energy between near-equilibrium and exact equilibrium states at or very near-critical condition is very minute, there is not enough driving force to push the near-equilibrium states to the exact minimum of Gibbs energy.

The treatment of the problem of fluctuations near the critical condition is beyond the scope of classical thermodynamics, but the problem was successfully treated, including the long-range correlated behavior of the fluctuations, by Kenneth Wilson, a high energy physicist (Nobel prize, 1982), through what is known as the "renormalization group theory". This theory also correctly predicts the experimental values of critical exponents that classical theory fails to do, and also shows the inter-relationships among these exponents.

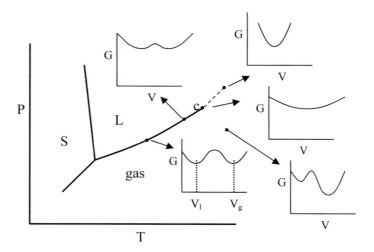

Fig. 5.4 Schematic illustration of the change of form of G versus V curve (**a**) along a coexistence line of liquid and vapor, ending at a critical point, c, (**b**) at metastable extension of the coexistence curve, and (**c**) in the field of stability of gas

5.2.3   Super- and Near-Critical Fluids

Supercritical fluids (SCF) have some properties that make them attractive for use as media to manipulate chemical reactions in industrial processes, and also make their geochemical evolution markedly different from that of the subcritical fluid, as found, for example, in the chemical properties of hydrothermal fluid emanating from some vents on the Mid-Atlantic Ridge system (Koschinsky et al. 2008). A useful review of the subject can be found in Savage et al. (1995). The diffusivity of a species in the supercritical fluid lies between that in a liquid and in a gas. Consequently, reactions that are diffusion controlled in the liquid become faster in the SCF. (The term diffusion controlled means that diffusion is the slowest step in the overall reaction process; hence the reaction rate cannot be faster than the diffusion rate). The solubility of a compound can also substantially increase or decrease in passing from a subcritical to a supercritical condition. The enhanced solubility of reactants could greatly accelerate the reaction rate in the SCF. The partial molar volume of a solute at very high dilution diverges as the critical point is approached. The partial molar properties have been discussed formally in 8.2, but from a physical standpoint we may view partial molar volume of a solute as its *effective* molar volume in a solution. The divergence of partial molar volume is usually towards negative infinity, especially when the size of the solute atom or molecule is smaller than that of the solvent, but the divergence to positive infinity is also possible (Savage et al. 1995). The properties of SCF vary with density, which is very sensitive to temperature and pressure changes near the critical point.

The near-critical (NC) water also has much better solubility than water at lower temperatures for both organic and ionic species. This makes NC-water a good non-pollutant solvent and reaction medium that may be used effectively in industrial processes, and also as a great solvent for scavenging base metals from rocks leading to the formation of ore deposits. The fluctuations of the properties of water near the critical condition also cause sharp fluctuations in the solubility of components, such as $B(OH)_3^=$, which probably explains the oscillatory zoning in minerals like tourmaline, as discussed by Norton and Dutrow (2001), that are found in the carapace of Geysers.

5.3   Near-Critical Properties of Water and Magma-Hydrothermal Systems

In a series of papers, Norton and co-workers (e.g. Norton and Knight 1977; Johnson and Norton 1991; Norton and Dutrow 2001; Norton and Hulen 2001) have discussed the properties of water at and near the critical point, and their implications for the evolution of magma-hydrothermal systems. Norton and Knight (1977) showed that convective heat flux dominates over the conductive heat transfer at permeability values greater than 10^{-18} m^2. The **convective heat flux** by a fluid, J (conv), is given by the product of the mass flux (i.e. mass of fluid crossing a unit area per unit time) and the heat content per unit mass of the fluid. These two quantities are given by $\rho_f \upsilon_f \phi$ and $C'_p T$, respectively, where ρ_f is the fluid density, υ_f is the fluid velocity, C'_p is the specific heat capacity of the fluid and ϕ is the rock porosity. The quantity $\upsilon_f \phi$ is known as the **Darcy velocity** of the fluid. Thus,

$$J(\text{conv}) = (\rho_f \upsilon_f \phi)\left(C'_p T\right) \tag{5.3.1}$$

As discussed by Norton (2002), the magnitude of convective fluid velocity is related directly to the magnitude of lateral gradient of fluid density and inversely to its viscosity, ν. Using chain rule, the lateral gradient of fluid density is given by

$$\frac{\partial \rho_f}{\partial x} = \left(\frac{\partial \rho_f}{\partial T}\right)\left(\frac{\partial T}{\partial x}\right) \tag{5.3.2}$$

The first derivative on the right can be expressed in terms of the coefficient of thermal expansion, α, as $-(\rho_f \alpha_f)$. This relation follows from the definition of α (Eq. 3.7.1) and making the substitutions $V = m/\rho$ and $dV = -(m/\rho^2)d\rho$. Thus,

$$\frac{\partial \rho_f}{\partial x} = -(\rho_f \alpha_f)\left(\frac{\partial T}{\partial x}\right) \tag{5.3.3}$$

Norton and Knight (1977) showed that while α and C_p of water diverge to $+\infty$ near the critical point, the viscosity of water rapidly decreases as the critical point is approached and reaches a minimum near the critical condition, as illustrated in Fig. 5.5. More updated properties of water near the critical condition have been presented by Johnson and Norton (1991), but the general picture remains valid. Thus, because of the large increase in the magnitude of the lateral density gradient as a result of the large increase of α_f, and conspicuous drop of viscosity, the magnitude of convective fluid velocity rapidly increases as the critical point of H_2O is approached. This effect, coupled with the property of divergence of C_p to $+\infty$ at the critical point leads to tremendous enhancement of the **convective heat flux** by fluid as its P-T condition evolves toward the critical point, as should be evident from Eq. (5.3.1). (In a dynamical natural system, however, there are moderation effects to fluid flow and its thermal evolution towards the critical point).

Numerical simulations of heat transfer show that near-critical conditions are indeed realized near the margin of a shallow granitic intrusion in the earth's crust and in portions of the immediately overlying rock (Norton and Hulen 2001). This is illustrated (Fig. 5.6) in a numerical simulation of pressure versus enthalpy evolution of H_2O near the margin and in the lithocap of a granite pluton (Norton and Hulen 2001; Norton, pers.com.). In this simulation, the dots within a trajectory of fluid evolution represent time steps, with the time gap between any two successive dots

Fig. 5.5 Properties of water near the critical point. The near coincidence of the maxima of the coefficient of thermal expansion, α, and isobaric heat capacity, C_p, and the minimum of viscosity, ν, causes rapid increase of convective heat flux by water as the critical point is approached. From Norton and Knight (1977). With permission from American Journal of Science

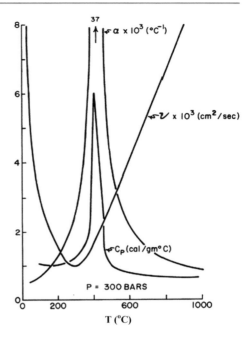

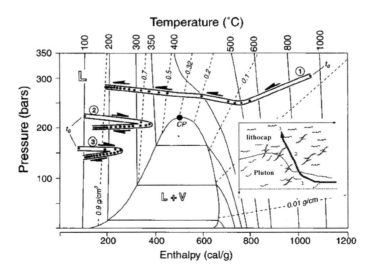

Fig. 5.6 Pressure versus Enthalpy changes of water as a function of time at three locations within and around a granite pluton. The location 1 is within the pluton, and the locations 2 and 3 are in the lithocap, as shown in the inset. The separation between two successive dots within the fluid evolution trajectory represents a 50 ky time step. The dashed lines represent density of H_2O. Note how the temperature of H_2O at locations 1 and 2 evolve towards the critical point, CP. From Norton and Hulen (2001); modified by Norton (pers.com.). With permission from Elsevier

being 50 ky. For the location 1 near the margin of the pluton, the fluid achieves near-critical condition after ~ 250 ky and persists at such condition for ~ 50 ky. During this period, energy is very rapidly advected away by the fluid from the pluton to the overlying rock. Near critical condition also persists for ~ 50 ky at the location 2 in the lithocap. In addition to the very rapid dispersal of energy, the solubilities of various chemical species in fluid are likely to be markedly different during the period over which the fluid remains at super-critical condition than when its P-T condition is significantly removed from it. We also note incidentally, as should be evident from Eq. (5.3.3) and emphasized by Norton (2002), that con-vective flows would initiate along steep margins of a pluton where the magnitude of lateral temperature gradient, and hence of density gradient, is very high (note that buoyancy force depends on the change of fluid density in a horizontal direction).

Figure 5.7 shows the variation of $\alpha(H_2O)$ as function of time at a point slightly to the right of the point 3 in the inset of Fig. 5.6, as calculated by Norton and Dutrow (2001). The strong oscillatory behavior of α between 70,000 and 130,000 years is due to the fluctuations of state conditions as the fluid evolves through conditions near the critical point during the above time frame, and reflects the extreme sensitivity of α to changes of state conditions near the critical point. Monitoring the fluctuations of α in the numerical simulation serves to display the fluctuations of state conditions that result from feedback relations among the thermal transport processes and the force field that drives the fluid velocity (see Norton and Dutrow 2001, for further discussion).

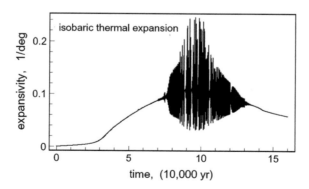

Fig. 5.7 Oscillations in the coefficient of thermal expansion, α, of water near the critical point in a magma hydrothermal system. The oscillations are due to opening and closing of fractures in the rock and extreme sensitivity of α to small variations of pressure. From Norton and Dutrow (2001). With permission from Elsevier

5.4 Equations of State

An equation of state (EoS) usually refers to the relation among V, P and T of a substance. The ideal gas equation, PV = nRT is the simplest example of such an equation. Equations of states have variety of applications in the evaluation of thermodynamic behavior of substances at different state conditions. For example, in order to evaluate the Gibbs energy of a substance at high pressure from a lower pressure datum, one needs to integrate the relation dG = VdP at a constant temperature (recall that dG = −SdT + VdP) for which V needs to be known as a function of pressure at the temperature of interest. This integration is especially important in geological and planetary problems where we need to perform phase equilibrium calculations at high pressure involving solid, melt and gas phases. A polynomial fit of the measured P-V-T relation is good for interpolation of the data, but by its very nature, a polynomial relation can produce physically unacceptable behavior (such as increase of volume with increasing pressure) on extrapolation beyond the range of experimental data. This is illustrated in Fig. 5.8 by fitting the V-P relation of olivine as a function pressure by a polynomial function and extrapolating it beyond the range of experimental data. Therefore, the experimental data need to be fitted by equations that have justifiable theoretical basis. We discuss below the development of some of the equations of state for gas, solid and silicate melt that have been found useful in the treatment of geological and geophysical problems.

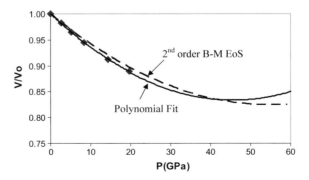

Fig. 5.8 Fits to the experimentally determined P-V data of forsterite (*Source* San Carlos, Arizona) by a polynomial function (solid line), and by 2nd order Birch-Murnaghan equation of state (dashed line), and extrapolations to higher pressures. Note that although the polynomial function fits the experimental data better than the 2nd order B-M equation of state (EoS), it has physically unacceptable extrapolation to higher pressure. The slight mismatch between the experimental data and the fit by 2nd order B-M EoS suggests that the 3rd order B-M equation is a better model for these data. The polynomial function: $V/V_o = 1.0002 - 0.0073P + 8 \times 10^{-5}P^2$ with P in GPa. K_o for B-M EoS: 87.67 GPa. Experimental data are from Robert Downs (pers.com.)

5.4.1 Gas

5.4.1.1 Van der Waals and Reduced Equations of State

The ideal gas EoS, PV = nRT, which followed from the works of Robert Boyle (1627–1691), Jacques Charles (1776–1856), Amedeo Avogadro (1776–1856) and Joseph Louis Gay Lussac (1778–1850), worked well in representing the P-V-T relation of many gases at low pressure, usually up to a few bars. However, it does not hold even in a crude way at significantly higher pressures. This is because ideal gas equation does not consider atomic or molecular interactions within a gas, nor the finite size of these entities. Also, calculation of the P-V relation of an ideal gas does not lead to the qualitative features shown in Fig. 5.2 in that it is devoid of any isotherm like T_c which shows a critical point (where both first and second derivatives of P with respect to volume are zero), and any isotherm like T_4 which shows a domain containing a maximum and a minimum. Lack of these properties implies that an ideal gas would not liquify—it would compress indefinitely. The inter-molecular attraction reduces the pressure that a gas can exert, and the finite size of gaseous molecules or atoms reduce the space available to the gas within an enclosure. However, at sufficiently low pressure, the density of a gas becomes too low for the volume excluded by the gaseous molecules (or atoms) as well the intermolecular attraction to have any significant effect. Thus, all non-ideal gas equations should reduce to the ideal gas form as $P \rightarrow 0$.

The earliest and the simplest form of non-ideal EoS for gas is due to J. D. van der Waals (1837–1923), who derived it as part of his doctoral dissertation, and is known as the van der Waals equation. It accounts, in a simple way, for the effects of molecular interaction and size, and is expressed as

$$P = \frac{nRT}{V - nb} - \frac{n^2a}{V^2}, \tag{5.4.1a}$$

where n is the number of moles, and **a** and **b** are constants. The constant **b** is the volume excluded by the molecules in a mole of gas so that $(V - nb)$ is the effective free volume available to the gas within a container of volume V, whereas the constant **a** is related to the intermolecular attraction. Notice that in the above equation, the intermolecular attraction reduces the pressure that a gas could exert, as expected. The above equation can also be written in terms of molar volume, V_m, as

$$P = \frac{RT}{V_m - b} - \frac{a}{V_m^2} \tag{5.4.1b}$$

The van der Waals equation produces the qualitative features of P-V relation shown in Fig. 5.2. In addition, it is possible to express the constants a and b for a gas in terms of its critical parameters. This is done by setting the first and second derivatives of pressure with respect to volume equal to zero at the critical condition, which yield two relations between the constants **a** and **b** in terms of T_c and V_c that can be solved to yield

$$T_c = \frac{8a}{27bR},$$
$$V_c = 3nb \text{ or } V_{m(c)} = 3b$$

On substitution in Eq. (5.4.1b), these relations yield

$$P_c = \frac{a}{27b^2} \tag{5.4.2a}$$

so that

$$\frac{P_c V_{m(c)}}{RT_c} = \frac{3}{8} \tag{5.4.2b}$$

Substitution of these relations in the van der Waals EoS, and rearrangement of terms yield an equation of state in terms of the ratios P/P_c, V/V_c and T/T_c. These dimensionless variables are called **reduced variables**, P_r, V_r and T_r, respectively.

$$\frac{P}{P_c} = P_r; \quad \frac{T}{T_c} = T_r; \quad \frac{V}{V_c} = V_r \tag{5.4.3a}$$

In terms of these reduced variables, van der Waals equation can be written as

$$\boxed{\left(P_r + \frac{3}{V_r^2}\right)(3V_r - 1) = 8T_r} \tag{5.4.3b}$$

This equation, which is known as the **reduced van der Waals equation of state**, shows that for all gases that follow van der Waals equation, the relationship among the reduced variables is unique since these variables are dimensionless. Thus, using Eq. (5.4.3b), one can specify any two reduced variables for a van der Waals gas to obtain the third one, and from that determine its P-V-T relation if the critical properties are known.

5.4.1.2 Principle of Corresponding States and Compressibility Factor

The representation of the P-V-T relations of gases in terms of the reduced variables is known as the **law of corresponding states**, the implicit idea being that the behavior of all gases expressed in this form should be very similar, as suggested by the reduced van der Waals equation of state, Eq. (5.4.3b). This law was first

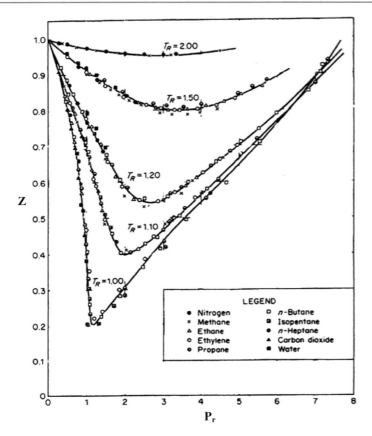

Fig. 5.9 Plot of the dimensionless compressibility factor, Z (=PV_m/RT) versus reduced pressure, P_r, at different values of the reduced temperature, T_r for a number of gases which closely obey van der Waals equation of state. The solid lines show the average behavior of seven hydrocarbons. Modified from Su (1946) by Kondepudi and Prigogine (1998). With permission from American Chemical Society

proposed by van der Waals in 1881. To examine the deviation from ideal gas behavior, and also to see the validity of the corresponding state approach, it is customary to define a dimensionless ratio, Z, as

$$Z = \frac{PV_m}{RT} \tag{5.4.4}$$

This ratio is known as the **compressibility factor**, which obviously has a value of unity for an ideal gas. Figure 5.9, which is modified from Su (1946), shows the Z vs P_r relation at several values of T_r for a number of gases. The solid lines are not least squares fits to the data but the average behavior of seven hydrocarbons. The overall average deviation of the data in Fig. 5.9 from the solid lines is 1%. Thus, it is evident that P-V-T relations of the gases with Z between 1.0 and 0.2 closely follow the law of corresponding states.

5.4.1.3 Redlich-Kwong and Related Equations of State

As more data on the high pressure behavior of gases became available, the reduced van der Waals equation of state failed to adequately represent their behavior. The problem is especially severe at high pressures that are of geological interest. This failure of van der Waals equation is illustrated in Fig. 5.10 by considering the properties of water at 0.1, 1 and 10 kbar pressures. Burnham et al. (1969) determined the P-V-T relation of water up to 10 kbar and 1000 °C. An isothermal combination of P-V data from this study was used to calculate the temperature according to Eq. (5.4.3b) and compared with the experimental temperature. The gross deviation of the actual behavior from the van der Waals type behavior is evident.

Despite its failure to predict P-V-T relation of gases at high pressure, the van der Waals equation forms the basis of a more successful two parameter EoS, which is due to Redlich and Kwong (1949), and is as follows.

$$P = \frac{RT}{V_m - b} - \frac{a}{V_m[V_m + b]\sqrt{T}}, \qquad (5.4.5)$$

A modified form of the Redlich-Kwong (RK) EoS was first introduced by Holloway (1977) in the geological literature, and since then a number of modifications of this EoS have been proposed by different workers to model P-V-T relations of

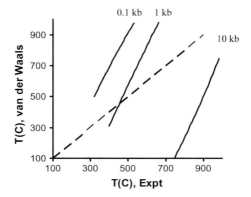

Fig. 5.10 Comparison of temperatures calculated from reduced van der Waals equation of state (Eq. 5.4.3) with the experimental temperatures for H_2O of given volumes at 0.1, 1.0 and 10.0 kb pressures. The dashed line represents the line of perfect agreement between the calculated and experimental temperatures. The P-V-T data for H_2O are from Burnham et al. (1969)

Fig. 5.11 Comparison of the shock wave data of H_2O (small squares) with the prediction from the MRK equation of state developed by Halbach and Chatterjee (1982). From Chatterjee (1991)

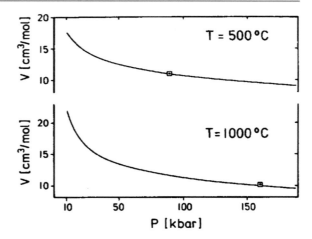

gases to high P-T conditions. Most of these modifications, which are commonly referred to as modified Redlich-Kwong or **MRK** equation of states, treat the a and b terms as specific functions of temperature and pressure, respectively, instead of constants. For example, Halbach and Chatterjee (1982) expressed a and b as

$$a(T) = A_1 + A_2T + \frac{A_3}{T} \qquad (5.4.6a)$$

$$b(P) = \frac{1 + B_1P + B_2P^2 + B_3P^3}{B_4 + B_5P + B_6P^2} \qquad (5.4.6b)$$

where the A and B parameters are constants. They derived values of these constants from the experimentally measured P-V-T data of H_2O up to 10 kb, 1000 °C (Burnham et al. 1969), and thereby predicted the PVT properties of H_2O up to 200 kb, 1000 °C. As illustrated in Fig. 5.11, the predicted volumes of H_2O are in excellent agreement with those derived from shock wave measurements at very high pressure.

Evaluation of the Gibbs free energy of a substance at a certain pressure, P_2, from that at P_1 requires evaluation of the integral of VdP between the limits P_2 and P_1. Since we typically have P as f(V) instead of the reverse, the integration is carried out by the standard method of integration by parts so that $\int VdP$ can be evaluated from $\int PdV$ (evaluation of Gibbs free energy at high pressures using equations of states, P = f(V, T), is discussed in Sect. 6.8). Recognizing the difficulty of analytical integration of PdV when the parameter b in an MRK formulation becomes a complex function of P, Holland and Powell (1991) proposed the following method of treatment of a modified Redlich-Kwong equation of state. They found that if the term b is held constant, then the measured volume diverges from that predicted by an MRK equation above a threshold pressure P_o, and that this divergence can be represented by an equation of the form

$$V_m = V_m^{mrk} + \left[c\sqrt{(P - P_o)} + d(P - P_o)\right] \qquad (5.4.7)$$

where c and d are functions of temperature, and V_m^{mrk} is the volume calculated from an MRK equation of state with constant b. Holland and Powell (1991) called this compensated Redlich-Kwong or **CORK**, and showed that this form works well in the P-T range of 1 bar to 50 kbar and 100–1600 °C.

Holland and Powell (1991) expressed the parameters a, b, c, d in terms of some constants with specific values and the critical temperature, T_c, and critical pressure, P_c, so that these parameters can be used for different types of gases, which conform to the principle of corresponding states.

$$a = a_o \frac{T_c^{5/2}}{P_c} + a_1 \frac{T_c^{3/2}}{P_c}T$$

with $a_o = 5.45963 \times 10^{-5}$, $a_1 = -8.63920 \times 10^{-6}$ with the unit of a in kJ^2 $kbar^{-1}$ mol^{-2}. This relation can be used to explicitly express the parameter a as a function of temperature for any gas by substituting its T_c and P_c values. Evaluation of Gibbs free energy of gas at high pressure using the CORK formulation is discussed in Sect. 6.8.3.

For **gaseous mixtures**, Redlich and Kwong (1949) proposed the following mixing rules:

$$a = \sum_i \sum_j X_i X_j a_{ij} \qquad (5.4.8a)$$

and

$$b = \sum_i X_i b_i \qquad (5.4.8b)$$

where X_i is the mole fraction of the gaseous species i and a_{ij} is a cross coefficient, the nature of which depends on the nature of the gas molecules, whether these are polar (e.g. H_2O) or non-polar (e.g. CO_2). For unlike nonpolar molecules, a_{ij} represents the geometric mean of the **a** parameters, i.e.

$$a_{ij} = \left(a_i a_j\right)^{1/2} \qquad (5.4.9)$$

For a mixture involving polar molecules, the cross coefficient should contain terms that account for the formation of complexes. Thus, for example, for the geologically most important fluids H_2O and CO_2, the cross coefficient can be expressed as (Flowers 1979)

$$a_{ij} = (a_{H_2O}^o a_{CO_2}^o)^{1/2} + 0.5R^2 T^{5/2} K \qquad (5.4.10)$$

where K is the equilibrium constant (to be discussed later in Sect. 10.4) of the reaction

$$H_2O + CO_2 = H_2CO_3$$

5.4.1.4 Virial and Virial-Type EoS

Since all gases must behave ideally at sufficiently low pressures, one may express the compressibility factor Z (Eq. 5.4.4) as a function of P such that Z = 1 as P $\rightarrow$ 0. As an example,

$$Z(P, T) = 1 + BP + CP^2 + DP^3 + \ldots \qquad (5.4.11)$$

This type of equation is known as **Virial** equation of state. The coefficients of P in the above equation have theoretical significance in terms of statistical mechanics in that the terms B, C, D etc. represent successive contributions to N-body interactions (two, three, four etc.) to the deviation from ideal gas behavior. These interactions can, in principle, be calculated from models of molecular interactions, and therein lies one of the appeals of Virial EoS.

The virial equations of state fail for many gases of geological interest, especially at high pressure. Saxena and Fei (1987) found that an EoS of the same form as the virial EoS to be quite successful in fitting the PVT data of gases at high pressures, if the unity in the above equation is replaced by a temperature dependent term A(T), and the coefficients of P (i.e. B, C, D etc.) are treated as functions of pressure.

$$Z(P, T) = A(T) + BP + CP^2 + DP^3 + \ldots \qquad (5.4.12)$$

With EoS of this form, which we would refer to as the **Virial-type** EoS, it is necessary to break up the data for a given fluid species into different pressure regimes, and then treat the data within each pressure regime separately. Saxena and Fei (1987) were successful in treating the data for a number of molecular species by considering three pressure regimes, <1 kbar, 1–10 kbar and >10 kbar, each with its own values of the coefficients. This virial-type EoS will, of course, fail at very low pressures since it does not reduce to the ideal gas law as P $\rightarrow$ 0.

Belonoshko and Saxena (1991, 1992) estimated the volumes of several geologically important fluid species, H_2O, CO_2, CH_4, CO, O_2, H_2O, at P-T conditions up to 1 Mbar and 4000 K by molecular dynamic (MD) simulation. They treated the results from the MD simulation as if these were experimental data ("computer experiments") and combined these with the available (conventional) experimental data at lower P-T conditions to develop an EoS for dense fluids. They found that the entire set of data above 5 kbar can be fitted quite well by a simple virial-type EoS of the form

$$P = \frac{a}{V} + \frac{b}{V^2} + \frac{c}{V^m} \qquad (5.4.13)$$

where

$$a = \left(a_1 + a_2 \frac{T}{1000}\right) \times 10^4$$

$$b = \left(b_1 + b_2 \frac{T}{1000}\right) \times 10^6$$

$$c = \left(c_1 + c_2 \frac{T}{1000}\right) \times 10^9$$

and a_1, a_2, b_1, b_2, c_1, c_2, and m are constants, with the units of a, b and c being bar-cm^3, bar-(cm^3)^2 and bar-(cm^3)^m, respectively. The above equation is an extension of an EoS that was proposed by Tait (1889) and revived, nearly hundred years later, by Spiridonov and Kvasov (1986). The values of the constants in the above equation for fluid species of common geological interest, namely, H_2O, CO_2, CH_4, CO, O_2 and H_2, are given by Belonoshko and Saxena (1992), and are valid within specified range of pressure (5 kbar-1 Mbar) and temperature (700–4000 K for H_2O and 400–4000 K for other species). Use of these values along with V in cm^3/mol yields P in bars. Upon comparing the volumes of dense fluids calculated from their EoS parameters with the available experimental data, these authors concluded that their EoS with the associated parameters reproduce experimental data with a maximum error of 5–6%.

Pitzer and Sterner (1994) developed equations of state for H_2O and CO_2 that are continuously valid over extremely wide range of pressure, 0–10 GPa, and temperature from below the critical temperature to 2000 K. A particular advantage of their equation of state is that, because of the continuity of P-V-T relation over extremely large range, these EoS can be differentiated and integrated to yield other thermodynamic quantities without the need to avoid some specific P-T conditions at which the functions become discontinuous. For example, calculation of fugacity of a species at a pressure P′ requires evaluation of the integral $\int VdP$ between a very low pressure, at which fugacity becomes equal to pressure, to P′ (Eq. 3.6.9). In this case, it is advantageous to have P-V relation that is continuous over the desired range of pressure. If there is a discontinuity, as in the case of Belonoshko-Saxena EoS, then the integral needs to be broken up into different pressure ranges within each of which a P-V relation is valid and continuous. Detail discussion of the Pitzer-Sterner EoS is beyond the scope of this section, but the EoS is linked to an online fugacity calculator for water.

5.4.2 Solid and Melt

The equations of state discussed in this section were initially derived for solids, but have been found to be useful also for melts. We first discuss the Birch-Murnaghan equation of state in detail because of its historical importance in geophysics and the wide ranging applications that have been made for this EoS. This is followed by shorter exposition and discussion of further developments.

5.4.2.1 Birch-Murnaghan Equations

Using the finite strain theory developed by Murnaghan (1937), Birch (1952) showed that the strain, $\in$ and volume V or density ρ are related according to

$$\frac{V_o}{V(P)} = \frac{\rho(P)}{\rho_o} = (1 - 2 \in)^{3/2} = (1 + 2f)^{3/2} \qquad (5.4.14)$$

where the subscript o refers to the zero pressure condition, and f ($= -\in$) was introduced as a more convenient variable, being always positive for compression. For compression, $V_o/V(P) > 0$ and this requires $f > 0$.

In order to relate pressure and volume at constant temperature, we need to deal with Helmholtz free energy, F, which relates these two variables according to $P = - (\partial F/\partial V)_T$ (Eq. 3.1.9). Using chain rule, we then relate P to f as

$$P = -\left(\frac{\partial F}{\partial V}\right)_T = -\left(\frac{\partial F}{\partial f}\right)_T \left(\frac{\partial f}{\partial V}\right)_T \qquad (5.4.15)$$

Rewriting Eq. (5.4.14) as $V/V_o = (1 + 2f)^{-3/2}$ and differentiating both sides at constant temperature, we get

$$\left(\frac{\partial V}{V_o}\right)_T = -\frac{3}{(1 + 2f)^{5/2}} \partial f$$

so that

$$\left(\frac{\partial f}{\partial V}\right)_T = -\frac{1}{3V_o}(1 + 2f)^{5/2} \qquad (5.4.16)$$

We now need to find an expression for the first derivative term on the right hand side of Eq. (5.4.15) to be able to express P in terms of volume or density. Birch (1952) expressed F as a polynomial function of compression, f,

$$F = af^2 + bf^3 + cf^4 + \ldots \qquad (5.4.17)$$

(Note the unusual nature of this power series expression in that it does not contain a constant a constant term and a first order term in f, say F_o and βf respectively. We would show at the end of this section how these two terms vanish). Assuming that the terms higher than second order in f are insignificant, a condition that would prevail for small values of f (i.e. small compression), we have

$$\left(\frac{\partial F}{\partial f}\right)_T = 2af \tag{5.4.18}$$

The next step is to express **a** and **f** in terms of the volume or density of the substance. This is the most tricky step in the derivation of final expression for the equation of state. With this objective in mind, we first rewrite Eq. (5.4.14) as $V(P)/V_o = (1 + 2f)^{-3/2}$ and expand the right hand side in a binomial series[1] to yield, for small values of f,

$$\frac{V}{V_o} = (1 - 3f)$$

so that

$$\frac{\Delta V}{V_o} = -3f \tag{5.4.19}$$

where $\Delta V = V - V_o$.

Now, from the definition of isothermal bulk modulus, k_T, (Eq. 3.7.2, with $k_T = 1/\beta_T$), we have,

$$\lim_{P \to 0} k_T \equiv k_{T,o} = -V_o \left(\frac{\Delta P}{\Delta V}\right)_T$$
$$= -V_o \frac{P}{\Delta V} \tag{5.4.20}$$

where $k_{T,o}$ is the isothermal bulk modulus at zero pressure (in other words, it is the inverse of the slope of the V versus P curve at a fixed temperature at P = 0: see Fig. 3.4). Note that in the above equation, we have replaced ∂P by $\Delta P = P - 0 = P$ and ∂V by ΔV. This is because within a small interval, the derivatives can be represented by finite differences. On substitution of the expression of P, as given by Eq. (5.4.15), into the above equation,

$$k_{T,o} = \frac{V_o}{\Delta V} \left(\frac{\partial F}{\partial f}\right)_T \left(\frac{\partial f}{\partial V}\right)_T \tag{5.4.21}$$

For small compression, the first derivative term on the right is given by Eq. (5.4.18) whereas the second one can be obtained from Eq. (5.4.19) as $(\partial f/\partial V) = -1/3(V_o)$. (Here we are talking about "small compression" since we are dealing with $k_{T,o}$.) Substitution of these relations into Eq. (5.4.21), and replacing f by $-\Delta V/3V_o$ according to Eq. (5.4.19), we obtain the important relation

---

[1]When n is a real number, the term $(1 + x)^n$ can be expressed, according to binomial series, as
$(1 + x)^n = 1 + nx + \frac{n(n-1)}{2!}x^2 + \frac{n(n-1)(n-2)}{3!}x^3 + \cdots$.

$$a = \frac{9}{2}\left(k_{T,o}V_o\right), \tag{5.4.22}$$

(Note that in deriving this equation, we used $(\partial f/\partial V) = -1/3(V_o)$, which is valid only for small compression. The expression of $(\partial f/\partial V)$ in Eq. (5.4.16) is not restricted to small compression, but it obviously reduces to $-1/3(V_o)$ when this restriction is imposed).

Combination of Eqs. (5.4.18) and (5.4.22) yields

$$\left(\frac{\partial F}{\partial f}\right)_T = 9k_{T,o}V_of \tag{5.4.23}$$

Substitution of this equation and Eq. (5.4.16) into Eq. (5.4.15) yields

$$P = 3k_{T,o}f(1 + 2f)^{5/2}, \tag{5.4.24}$$

Expressing now f and $(1 + 2f)$ in terms of ρ/ρ_o according to Eq. (5.4.14), we obtain the desired relation between pressure and density at a fixed temperature:

$$P = \frac{3k_{T,o}}{2}\left[\left(\frac{\rho}{\rho_o}\right)^{7/3} - \left(\frac{\rho}{\rho_o}\right)^{5/3}\right], \tag{5.4.25}$$

This is known as the **second-order** Birch-Murnaghan (B-M) EoS, since it was obtained by truncating the expression of Helmholtz free energy, F, after the second power of compression, f (Eq. 5.4.17). Figure 5.8 shows the fit to the P-V data of olivine using the 2nd order B-M equation of state, and extrapolation to higher pressure.

If the polynomial of F is truncated after the cubic term, then one obtains, after even more tedious manipulations, the so-called **third-order** B-M EoS, as

$$P = \frac{3k_{T,o}}{2}\left[\left(\frac{\rho}{\rho_o}\right)^{7/3} - \left(\frac{\rho}{\rho_o}\right)^{5/3}\right]\left\{1 + \xi\left[\left(\frac{\rho}{\rho_o}\right)^{2/3} - 1\right]\right\}, \tag{5.4.26}$$

where

$$\xi = \frac{3}{4}\left(k'_{T,o} - 4\right),$$

with $k'_{T,o}$ being the pressure derivative of the bulk modulus at $P = 0$ (i.e. $(\partial k_o/\partial P)_T$). Both second and third order EoS have been used extensively in the geophysical literature.

It is easy to see, upon comparing the last two equations, that the third-order B-M equations of state reduces to the second-order form when $k'_{T,o} = 4$. Thus, when a set of compressibility data are fitted by the second-order EoS, one should check the value of k_o' obtained from the fitted relation. If it deviates significantly from 4, then the second order form is an inadequate model to fit the data, even when it fits the experimental data adequately, and is likely to yield erroneous volume data on extrapolation.

The justification of the polynomial expression of the Helmholtz energy (Eq. 5.4.17) used by Birch can be seen as follows. If F is expressed according to a Taylor series expansion (Appendix B.6) around $f = 0$, then

$$F = F_o + \beta f + af^2 + bf^3 + cf^4 + \ldots,$$

or

$$\frac{\partial F}{\partial f} = \beta + 2af + 3bf^2 + 4cf^3 \ldots \qquad (5.4.27)$$

where $F_o = F(f = 0)$ and the coefficients of f are constants (being related to the successively higher order derivatives of F with respect to f). (Note that the above approach of F was also used for G in the development of the celebrated **Landau theory** of phase transition discussed in Sect. 6.3.1). Now for $f = 0$, we have $dF/df = \beta$. However, since F must be at a minimum under equilibrium condition, we must have $\beta = 0$ at equilibrium. Thus, $F - F_o = af^2 + bf^3 + cf^4 + \ldots$. The appearance of an F_o term in the expression of F does not alter the result for (dF/df) that is needed to develop the expression of P in terms of V or ρ (Eq. 5.4.15), since F_o is a constant. However, strictly speaking what Birch wrote as the Helmholtz free energy, F, is in fact $(F - F_o)$.

5.4.2.2 Further Developments for High P-T Conditions

The Birch-Murnaghan EoS does not seem to work well for compressions exceeding 25%. For example, it leads to negative thermal expansion for periclase (MgO), an important constituent of the Earth's lower mantle, at 160 GPa pressure (Belmonte 2017). (Note that while negative thermal expansivity is possible for certain types of structures, as discussed in the Sect. 3.7.1, it can be completely ruled out for periclase). Thus, with the domain of activity of Earth scientists now extending to much higher compression levels to address properties of the deep mantle and the Earth's core, there is an urgent need for an EoS that works better than Birch-Murnaghan at very high P-T conditions and does not lead to physically unacceptable behavior (see Fig. 7.7 to appreciate the P-T regime). Some of the later but by no means exhaustive advancements in this direction are discussed in this section. (Equations of state based on lattice dynamic theories have not been discussed since the development of the theoretical formalisms to appreciate these developments is beyond the scope of this book).

An important later development is the work of Vinet et al. (1986, 1987) who argued that although the nature of interatomic interactions in different types of solids can be widely different, which would seem to preclude a common description of their energetic properties, the isothermal equations of states of various classes of solids can still be expected to follow a universal form. They derived this universal form by considering interactions at the atomic level, and showed that it works better than the popular Birch-Murnaghan EoS, especially when the compression exceeds 25% of the initial volume. Thus, there now seems to be a shift of preference from B-M to Vinet EoS in the geophysical community dealing with very high pressure behavior of rocks and melt (e.g. Anderson and Isaak 2000). Likewise the Birch-Murnaghan EoS, there are also different orders of the EoS due to Vinet et al. However, a third order form seems to be adequate to represent V-P properties of materials of geological and geophysical interest. With $\eta = V/V_o$, the third order Vinet EoS can be expressed as follows.

$$P(\eta) = 3k_o \frac{(1 - \eta^{1/3})}{\eta^{2/3}} \exp\left[\frac{3}{2}(k_o' - 1)(1 - \eta^{1/3})\right] \tag{5.4.28}$$

Here V_o, k_o and k_o' are volume, bulk modulus and pressure derivative of bulk modulus, respectively, at zero pressure and a fixed temperature.

The Vinet equation was derived by using a specific form of internal energy in terms of interatomic energy at zero pressure and interatomic distances at zero and high pressures, and taking its partial derivative with respect to V at T = 0 (recall that dU = TdS − PdV− see Sect. 2.7− so that at T = 0, $\partial U/\partial V = -P$). Vinet et al. (1987) subsequently argued that the above EoS can also be used at higher temperature by making the reference parameters (V_o, k_o and k_o') functions of temperature and showed several successful applications of the EoS to Au, NaCl and Xe to ~ 1300 K.

As discussed above (Eq. 5.4.15), the appropriate thermodynamic potential to use to derive an expression of pressure at T > 0 by differentiating it with respect to volume is the Helmholtz free energy, F. Thus, the Vinet EoS would hold at high temperature as long as F can be adequately expressed according to the form of the expression of internal energy used in its derivation (recall that the Birch-Murnaghan EoS was also derived by assuming a specific form of F, Eq. 5.4.17). The required form of F(V, T) that is analogous to the form of U in Vinet EoS is as follows (Tirone 2015):

$$F(V, T) = F_o(T)[1 + \alpha(T)]e^{-\alpha(T)} \tag{5.4.29}$$

where $F_o(T)$ is the Helmholtz energy at zero pressure at T and $\alpha(T)$ is a constant at T. It remains to be seen to what extent such a form of F remains valid.

Ghiorso (2004) showed that there is a region in the P-T space that is sometimes accessed by silicate melts at which the Vinet EoS suffers from the problem of mathematical singularity. He has, thus, proposed a new EoS that does not suffer from the limitation of singular behavior of the Vinet EoS, and applied that to treat the thermodynamic properties of silicate melt.

Stacey (2005) provided a comprehensive analysis of the problems with the available equations of state of solids and proposed a new EoS for solids that is a modified form of an EoS proposed by Rydberg, and called it "Generalized Rydberg Equation". This EoS seems to have a larger P-T range of validity than the Vinet EoS.

Brosh et al. (2007) proposed a formulation of G(P, T) by combing the pressure effect on G due to cold compression with G(P_o, T), where P_o is a reference pressure such as 1 bar at which Gibbs energy data are available, and some other parameters. The Brosh formulation is not a conventional EoS formulation in the sense that it does not provide an explicit P-V-T relation, but it serves the purpose of such formulation as it yields G(P, T) values for solid phases that are needed for phase equilibrium calculations, and V, S, Cp, α, β can be retrieved from G(P, T) by differentiation. This formulation seems quite promising and Saxena and Eriksson (2015) have used it to calculate phase diagram for Fe-S system at 328 GPa and 3000–6000 K (and discussed its implications for the composition and thermal structure of the inner and outer cores of the Earth) without encountering any unphysical behavior of the derivative properties of G; the experimental constraints on the thermodynamic properties of the phases in the Fe-S system extend up to 200 GPa.

Helffrich and Connolly (2009) carried out a mathematical analysis to understand the reason behind the problem of negative thermal expansion of phases that is encountered at high pressure upon the use of Birch-Murnaghan relation. They found that it is a consequence of the usual assumption of linear decrease of K′ with T that has been used in the high temperature application of the Murnaghan and Birch-Murnaghan EoS. They suggested a remedy to this problem by using a specific form of exponential decrease of K′ with T with an asymptotic limit of K′ = 0. This modification restores the applicability of the widely used Birch-Murnaghan EoS to mid-Earth condition or the core-mantle boundary ($\sim$165 GPa; Fig. 7.7) in that the EoS does not produce any physically unacceptable behavior.

Holland and Powell (2011) used a modified form of the Tait equation of state (see Sect. 5.4.1.4) that is valid for 298 K and added to it what is known as a thermal pressure term to make it suitable for high temperature applications. The concept of thermal pressure has been discussed in Sect. 7.1, but briefly it is the pressure change resulting from change of temperature at constant volume. Thus, the P-V relation in the EoS proposed by Holland and Powell (2011) has the following form: $P(V)_T = P(V)_{T=298\,K}$ (Tait) + $(\Delta P)_{thermal}$, where the last term is given by the integral $\int(\alpha k_T)dT$ (see Eq. 7.1.3) over 298 K to the desired temperature. Holland and Powell (2011) argued that the term αk_T versus T should have the same form as C_v versus T, rising to a constant value above a certain temperature (Fig. 4.3). This expectation seems to be satisfied in several cases (e.g. Fig. 7.1), but Wang and Reeber (1996) argued on theoretical grounds, coupled with analysis of some of the available thermophysical data, that it is $\alpha k_T V$ rather than αk_T that should be expected to remain roughly independent of temperature beyond a threshold value and that αk_T should ultimately decrease with T at high temperature, such as for MgO (Fig. 7.1). Tirone (2015) showed examples of anomalous behavior of the C_p function calculated from the widely used thermodynamic database of Holland and Powell (2011) that was partly built on their modified Tait EoS and has been highly successful for phase equilibrium calculations of common petrological interest (<2 GPa).

Problem 5.1 It was discussed above that the use of Birch-Murnaghan EoS leads to negative thermal expansion for MgO at 160 GPa. What would be the consequence of this anomalous thermal expansion behavior for pressure effect on the entropy of MgO?

Answer: Positive (but why?)

References

Anderson OL, Isaak DG (2000) Calculated melting curves for phases of iron. Amer Mineral 85:376–385

Belmonte D (2017) First principles thermodynamics of minerals at HP-HT conditions: MgO as a prototypical mineral. Minerals 7:1–35

Belonoshko AB, Saxena SK (1991) A molecular dynamics study of the pressure-volume-temperature properties of supercritical fluids: II. CO_2, CH_4, CO and H_2. Geochim Cosmochim Acta 55:3191–3208

Belonoshko AB, Saxena SK (1992) Equations of state of fluids at high temperature and pressure (Water, Carbon Dioxide, Methane, Carbon Monoxide, Oxygen, and Hydrogen) In: Saxena SK (ed) Advances in physical geochemistry, vol 9, Springer, New York, Berlin, Heidelberg, pp 79–97

Birch F (1952) Elasticity and the constitution of the Earth's interior. J Geophys Res 57:227–286

Bischoff JL, Rosenbauer RJ (1983) Critical point and two-phase boundary of the seawater, 200–500 °C. Earth Planet Sci Let 68:172–180

Brosh E, Makov G, Shneck RZ (2007) Application of CALPHAD to high pressures. CALPHAD 31:173–185

Burnham CW, Hollaway JR, Davis NF (1969) Thermodynamic properties of water at 1,000 °C and 10,000 bars. Geol Soc Am Special Paper 132:96

Callen HB (1985) Thermodynamics and an introduction to thermostatics. Wiley, New York Chichester Brisbane Toronto Singapore, p 493

CEA (1998) Supercritical fluids. http:/www-drecam.cea.fr/drecam/spec/publi/rapport98. Commissariat A L'Energie Atomique: Département de Recherche sur l'Etat Condensé les Atomes et les Molécules

Chatterjee ND (1991) Applied mineralogical thermodynamics. Springer, Berlin, Heidelberg, New York, p 321

Flowers GC (1979) Correction to Hollway's (1977) adaptation of the modified Redlich-Kwong equation of state for calculation of the fugacities of molecular species in supercritical fluids of geologic interest. Contrib Mineral Petrol 69:315–318

Ghiorso MS (2004) An equation of state of silicate melts. I. Formulation of a general model. Am J Sci 304:637–678

Guillot T (1999) A comparison of the interiors of Jupiter and Saturn. Planet Space Sci 47:1183–1200

Halbach H, Chatterjee ND (1982) An empirical Redlich-Kwong-type equations of state of water to 1000 °C and 200 kbar. Contrib Mineral Petrol 79:337–345

Helffrich G, Connolly JAD (2009) Physical contradictions and remedies using simple polythermal equations of state. Amer Mineral 94:1616–1619

Holland TJB, Powell R (1991) A compensated-Redlich-Kwong (CORK) equation for volumes and fugacities of CO_2 and H_2O in the range 1 bar to 50 kbar and 100–1600 °C. Contrib Mineral Petrol 109:265–271

Holland TJB, Powell R (2011) An improved and extended internally consistent thermodynamic dataset for phases of petrological interest, involving a new equation of state for solids. J Met Geol 29:333–383

Holloway JR (1977) Fugacity and activity of molecular species in supercritical fluids. In: Fraser DG (ed) Thermodynamics in geology. Reidel, Dordrecht, Holland, pp 161–181

Johnson JW, Norton DL (1991) Critical phenomena in hydrothermal systems: state, thermodynamic, electrostatic, and transport properties of H_2O in the critical region. Am J Sci 291:541–648

Kondepudi D, Prigogine I (1998) Modern Thermodynamics: from heat engines to dissipative structures. John Wiley, New York, p 486

Koschinsky A, Garbe-Schönberg D, Schmidt K, Gennerich H-H, Strauss H (2008) Hydrothermal venting at pressure-temperature conditions above the critical point of seawater, 5°S on the Mid-Atlantic Ridge. Geology 36:615–618

Landau LD, Lifschitz EM (1958) Statistical physics. Pergamon, London, Paris, p 484

Murnaghan FD (1937) Finite deformations of an elastic solid. Amer J Math 59:235–260

Norton DL (2002) Equation of state: H_2O-system. In: Marini L, Ottonello G (eds) Proceedings of the Arezzo Deminar on Fluids Geochemistry, DIPTERIS University of Genova. pp 5–17

Norton DL, Dutrow BL (2001) Complex behavior of magma-hydrothermal processes: role of supercritical fluid. Geochim Cosmochim Acta 65:4009–4017

Norton DL, Hulen JB (2001) Preliminary numerical analysis of the magma-hydrothermal history of the Geysers geothermal system, California, USA. Geothermics 30:211–234

Norton DL, Knight JE (1977) Transport phenomena in hydrothermal systems: cooling plutons. Am J Sci 277:937–981

Pitzer KS, Sterner SM (1994) Equations of state valid continuously from zero to extreme pressures for H_2O and CO_2. J Chem Phys 101:3111–3115

Redlich O, Kwong JNS (1949) On the thermodynamics of solutions. V. An equation of state. Fugacities of gaseous solutions. Chem Rev 44:223–244

Savage PE, Gopalan S, Mizan TL, Martino CN, Brock EE (1995) Reactions at supercritical conditions: applications and fundamentals. Amer Inst Chem Eng J 41:1723–1778

Saxena SK, Fei Y (1987) Fluids at crustal pressures and temperatures I. Pure species. Contrib Mineral Petrol 95:370–375

Saxena SK, Eriksson G (2015) Thermodynamics of Fe-S at ultra-high pressure. CALPHAD 51:202–205

Sengers JV, Levelt Sengers JMH (1986) Thermodynamic behavior of fluids near the critical point. Ann Rev Phys Chem 37:189–222

Sourirajan S, Kennedy GC (1962) The system H_2O-NaCl at elevated temperatures and pressures. Amer J Sci 260:115–141

Spiridonov GA, Kvasov IS (1986) Empirical and semiempirical equations of state for gases and liquids. Tev Thermophys Prop Matter 57(1):45–116 (in Russian)

Stacey FD (2005) High pressure equations of state and planetary interiors. Rep Prog Phys 68:341–383

Su G-J (1946) Modified law for corresponding states for real gases. Ind Eng Chem 38:803–806

Tait PS (1889) On the virial equation for molecular forces, being part IV of a paper on the foundations of the kinetic theory of gases. Proc Roy Soc Edinb 16:65–72

Tirone M (2015) On the use of thermal equations of state and the extrapolation at high temperature and pressure for geophysical and petrological applications. Geophys J Internat 202:55–66

Vinet P, Ferrante J, Smith JR, Ross JH (1986) A universal equation of state for solids. J Phys C Solid State Phys 19:L467–L473

Vinet P, Smith JR, Ferrante J, Ross JH (1987) Temperature effects on the universal equation of state of solids. Phys Rev B 35:1945–1953

Wang K, Reeber RR (1996) A model for evaluating and predicting high-temperature thermal expansion. J Material Res 11:1800–1803

Phase Transitions, Melting, and Reactions of Stoichiometric Phases

6

In this chapter we discuss briefly the phenomenon of solid state phase transitions in which one substance transforms to a different one of the same composition but different symmetry property, with or without redistribution of atoms within the different crystallographic sites or change of ordering state. This is followed by the development of thermodynamic formulations for calculations of equilibrium conditions of heterogeneous reactions involving phases of fixed compositions or stoichiometric phases. The analogous problems involving solutions are discussed later (Chap. 10) after the development of the formalisms that are required to treat the thermodynamic properties of solutions.

6.1 Gibbs Phase Rule: Preliminaries

Any discussion of phase transformation or phase equilibrium should be preceded by a discussion of what is known as the Gibbs Phase Rule, or simply the Phase Rule, which determines the number of intensive variables that can be varied independently in a system, the latter being anything under observation. The derivation of phase rule is presented in Sect. 10.3, but here we state what it is, and emphasize the main points about its properties and applications so that these points are not lost in the formalities of derivation. Besides, an appreciation of phase rule is needed for the concepts developed in this chapter.

A phase is defined to be a substance that is spatially uniform at the macroscopic scale and is physically distinct from its surroundings in a system. A homogeneous liquid, or a gas or a mineral (but not different grains of the same mineral) constitute different phases. In the sense of Phase Rule, the number of components has a special meaning in that it is the *smallest* number of chemical species that is required to express the compositions of all the phases in the system. The choice of the components, however, is not unique, and is usually a matter of convenience, but by definition, the *number* of components in a system is unique. As an example, let us

© Springer Nature Switzerland AG 2020
J. Ganguly, *Thermodynamics in Earth and Planetary Sciences*,
Springer Textbooks in Earth Sciences, Geography and Environment,
https://doi.org/10.1007/978-3-030-20879-0_6

consider an assemblage of three phases, albite ($NaAlSi_3O_8$), jadeite ($NaAlSi_2O_6$) and quartz (SiO_2). Here we need only two components to express the compositions of all three phases. There are, however, three different ways to choose the two components, viz., (a) $NaAlSi_2O_6$ and SiO_2, (b) $NaAlSi_3O_8$ and $NaAlSi_2O_6$, and (c) $NaAlSi_3O_8$ and SiO_2. We can use any of these sets of two components to express the compositions of all three phases by appropriate linear combinations, but two is the minimum number of chemical species that are needed to express the compositions of all three phases. One can, of course, express the compositions of three phases by linear combinations of the oxides Na_2O, Al_2O_3 and SiO_2, but these are not components in the sense of Phase Rule since it involves more species than the minimum needed to express the compositions of all phases in the system.

With the above definition of components, and assuming that (a) the system is at equilibrium (that requires uniformity of temperature and lack of any tendency of chemical mass transfer) and (b) all phases are subjected to a uniform pressure, the Phase Rule is stated as follows:

$$\boxed{F = C - P + 2,}\qquad(6.1.1)$$

where F stands for degrees of freedom in the system, which means the number of *intensive* variables that can be varied independently, P stands for the number of phases and C stands for the number of components. In the assemblage of albite, jadeite and quartz, there are only two intensive variables, namely pressure (P) and temperature (T) that can affect the stability of the phases. But from the phase rule, $F = 1$ (since $P = 3$ and $C = 2$), which means that we can vary either pressure or temperature, but not both, as long as all three phases are present in the system, and are in equilibrium; the variation of one intensive variable depends on that of the other. This is also stated by saying that the reaction albite = jadeite + quartz is **univariant**, and this is why the equilibrium stability fields of albite and of jadeite plus quartz must be separated by a line, and not by a domain, in the P-T space, as long as none of the minerals have additional components in solid solution, such as anorthite ($CaAl_2Si_2O_8$) that dissolves in albite to form plagioclase feldspar or diopside ($CaMgSi_2O_6$) that dissolves in jadeite to form jadeitic clinopyroxene. Figure 6.1 shows the experimentally determined location of the univariant equilibrium. The stabilities of albite and of jadeite plus quartz are confined within two different divariant P-T fields within which both P and T may be varied independently within certain limits. Note that when we are dealing with the stability of albite alone, $F = 2$ because $C = 1$ (we need only one component, $NaAlSi_3O_8$, to express the composition of stoichiometric albite).

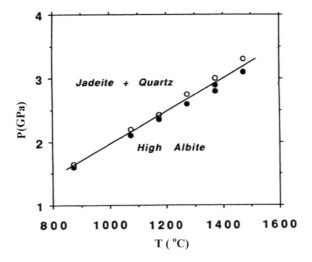

Fig. 6.1 P-T location of the univariant equilibrium High-Albite = Jadeite + Quartz, as determined experimentally by Holland (1980). The filled symbols indicate growth of high-albite at the expense of jadeite plus quartz, whereas the open symbols indicate the reverse reaction. High-albite is a high temperature polymorph of albite in which Al and Si are disordered over four tetrahedral positions

6.2   Phase Transformations and Polymorphism

Transformations of a compound between structures of different crystallographic symmetries or states of matter are known as phase transformations. The phenomenon of the existence of different crystallographic structures for the same compound (e.g. kyanite, sillimanite and andalusite) is known as polymorphism. In this section we would first discuss classification of phase transformations, and then a special class of phase transformation, namely that accompanied by a change of ordering parameters in solid.

6.2.1   Thermodynamic Classification of Phase Transformations

There are different types of classification of phase transformation in solids. A comprehensive review of these classifications can be found in Oganov et al. (2002). We discuss here the classification proposed by Ehrenfest (1933) on the basis of changes of thermodynamic properties at the transition condition. According to this scheme, a phase transformation is said to be first order, second order or higher order depending on whether the discontinuities *first* appear in the first, second or higher derivatives of the Gibbs energy, G, with respect to pressure and temperature at the transition condition. Thus, in a **first order** transformation, discontinuities appear in entropy and volume, which are first derivatives of G with respect to temperature and pressure, respectively ($\partial G/\partial T = -S$; $\partial G/\partial P = V$), whereas in a **second order** transition, the first derivatives are continuous but discontinuities appear in their derivatives (or in the second derivatives of G), that is in C_P, α and β, at the transition condition (Fig. 6.2). Since at equilibrium $\Delta H = T\Delta S$, the discontinuity in entropy in the first-order transition also implies discontinuous change of enthalpy. At the first order transition temperature C_p becomes infinite, since $C_p = \Delta H/\Delta T$, and the absorption or liberation of heat does not cause any change of temperature until the transition is complete. No transformation of higher than second order has yet been established, but possible examples have been discussed by Pippard (1957).

A second order transition may change to a first order transition with the change of state conditions. The point in the P-T space where this transition from the second to the first order behavior takes place is known as the **tricritical** point. An ordinary critical point (see Sect. 5.1) cannot exist in a second order phase transition since it separates phases of two different symmetries, but a tricritical point is somewhat analogous to an ordinary critical point in that the first derivative properties of G (volume, enthalpy and entropy) change discontinuously on one side of the point across the phase boundary, and change continuously beyond the termination of this boundary at the tricritical point.

Most polymorphic transformations that we commonly encounter, such as kyanite/sillimanite/andalusite (Al_2SiO_5), calcite/aragonite ($CaCO_3$), quartz/coesite/stishovite (SiO_2), graphite/diamond (C), are first order in nature. Proven second order transformations in the sense of Ehrenfest's original description of a discontinuity of the second derivative of Gibbs energy at the transition point, while the first derivatives are continuous, are extremely rare. Superconducting transition of tin at zero magnetic field, which shows a finite discontinuity of C_P between normal and superconducting tin (Keesom and van Laar 1938) at the transition temperature, represents an example of this very rare type of phase transition. Carpenter (1980) suggested that the phase transformation of omphacite (sodic clino-pyroxene solid solution) from $P2/n$ to $C2/c$ space group is likely to be second order.

Phase transformations sometimes show a λ-shaped behavior of C_P around the transition temperature, as illustrated in Fig. 6.3 for the transition of SiO_2 from α-quartz (trigonal) to β-quartz (hexagonal). This type of transition is neither first-order nor second-order. Because of the shape of the C_P function, these are called **lambda** transitions. Some other examples of lambda transitions are the onset of ferromagnetism in paramagnetic substances, onset of ferroelectricity and onset of superfluidity in liquid helium.

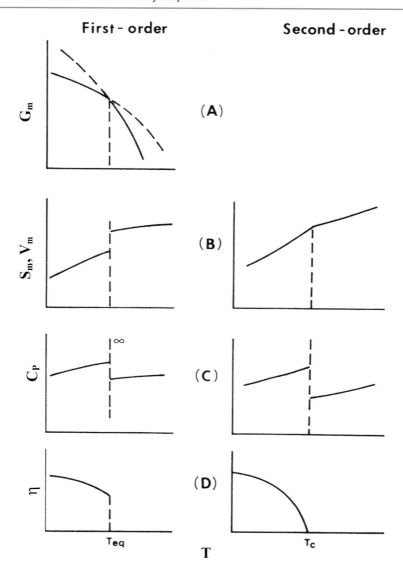

Fig. 6.2 Schematic illustration of the nature of change of some thermodynamic properties as function of temperature in first- and second-order phase transitions. (**A**) Change of Gibbs free energies of two phases related by first-order transition. (**B**), (**C**) Change of first-derivatives of G_m (i.e. S_m and V_m) and of a second derivative of G_m, namely C_P, at the first-order and second-order phase transitions. (**D**) Change of a long range order parameter, η, for the two types of transitions

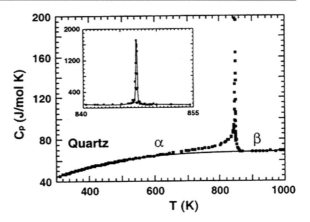

Fig. 6.3 Change of heat capacity, C_P, of quartz through alpha/beta transition showing a λ type behavior. From Richet (2001). With permission from Academic/Plenum publishers

6.3 Landau Theory of Phase Transition

6.3.1 General Outline

In a second order phase transformation, one or more properties of the phases change smoothly between the low temperature and high temperature forms. For example, the state of atomic ordering or the orientation of an atomic group or the unit cell parameters or the atomic positions may change continuously between the low temperature and high temperature phases. Lev Landau (1908–1968, Nobel prize; 1962) in 1937 developed an important phenomenological theory, which is commonly referred to as the Landau theory, to deal with the Gibbs energy change of the phases involved in these types of phase transitions. This theory has been applied to many mineralogical problems, especially by the Cambridge school of mineralogy (e.g. Carpenter 1985; Salje 1990; Putnis 1992). A brief outline of the basic structure of the theory is discussed below. For more expanded treatments, the interested readers should consult the above references and Landau and Lifschitz (1958).

In the Landau theory, one defines an order parameter, η, such that it has a value of zero for the high temperature and a non-zero positive or negative value for the low temperature phase. In a second order phase transition η increases or decreases smoothly from zero as the temperature decreases below the transition temperature, T_{tr}. For example, in an AB alloy, (e.g. CuZn or brass), which becomes completely disordered at T_{tr}, one can define a configurational order parameter to describe the distribution of the atoms among the lattice sites α and β as

$$\eta = X_A^\alpha - X_B^\alpha \tag{6.3.1}$$

where X defines the atomic fraction of the specified species in the lattice site α. Let us say that α is the site preferred by the atom A when the phase is ordered (in which case β is the lattice site preferred by the atom B). If the phase is completely disordered, $X_A^\alpha = X_B^\alpha$, in which case $\eta = 0$. If the phase is completely ordered, then $X_A^\alpha = 1$ and $X_B^\alpha = 0$, so that $\eta = 1$. For an intermediate state of ordering η has values between 0 and 1. Note that η depends on the average compositions of the lattice sites in the crystal, which may be different from the composition within a small domain within the crystal lattice. Thus, η is a **long range** order parameter. Landau theory deals only with long range order parameters, and thus falls in the class of **mean field theory** in that it deals with the average or mean behavior and ignores local fluctuations. It fails very close to the transition condition where local fluctuations are large. This domain of large fluctuations is known as the Ginzberg interval. (Ginzberg received Nobel prize in Physics in 2003 for his theoretical contribution, in collaboration with Landau, to the understanding of the superconducting state. The Ginzberg-Landau theory has its origin in Landau's approach to second order phase transformation).

Landau assumed that near a transition point where $\eta = 0$, the Gibbs energy may be expressed as a power series of η as

$$G(\eta) = G_o + \alpha\eta + A\eta^2 + B\eta^3 + C\eta^4 + D\eta^5 + \ldots \qquad (6.3.2a)$$

where $G_o = G(\eta = 0)$. This expression of Gibbs energy is often referred to as the **Landau potential**. (One would notice that this expression is formally similar to the power series expansion of Helmholtz free energy, F, around the state of zero strain (Eq. 5.4.17) that was used to derive the Birch-Murnaghan equation of state in Sect. 5.4.2). It should be easy to see that the Landau potential represents a Taylor series expansion of G around $\eta = 0$ (Appendix B.6), and thus the coefficients of η represent successively higher order derivatives of G with respect to η (e.g. $\alpha = \partial G/\partial\eta$, $A = \partial^2 G/\partial\eta^2$, and so on). Consequently, since a second order transition point represents a singularity, the above power series expansion of G cannot be carried out to arbitrarily high orders since the n th order coefficient of η requires that G is differentiable up to that order. The power series form of G is also often expressed as

$$G(\eta) = G_o + \alpha\eta + \frac{1}{2}A'\eta^2 + \frac{1}{3}B'\eta^3 + \frac{1}{4}C'\eta^4 + \frac{1}{5}D'\eta^5 + \ldots \qquad (6.3.2b)$$

in order to get rid of the numerical coefficients when G is differentiated with respect to η. For example, $dG/d\eta = \alpha + 2A\eta + 3B\eta^2 + \ldots = \alpha + A'\eta + B'\eta^2 + \ldots$.

Landau derived constraints on the values of the coefficients of η on the basis of the thermodynamic stability criteria of the high temperature ($\eta = 0$) and the low temperature ($\eta \neq 0$) phases, and the stability of the transition point itself. For a detailed discussion of these procedures, the reader is referred to Landau and Lifschitz (1958). We present below some of the important results and later illustrate by simple examples how the property of a coefficient of η could be constrained from the stability conditions, and also how a non-zero value of an odd order coefficient makes second order phase transition impossible.

The **main results** about the properties of the expansion coefficients and their relationship with the order of phase transitions are as follows.

(a) The linear term, i.e. α, must vanish. This conclusion follows from the fact that at $\eta = 0$, $(\partial G/\partial \eta) = \alpha$, but the stability of the completely disordered phase (i.e. minimum of G) requires this derivative to be zero.

(b) The second order coefficient A has the property that it must be positive and negative respectively above and below a specific temperature, T_i, and consequently must vanish at T_i. In other words

$$A = a(T-T_i), \text{ with } a > 0 \qquad (6.3.3)$$

(c) If the expression of G contains an odd order term, i.e. B, D etc., then the phase transition must be **first order** (which is characterized by two values of the order parameter at the transition condition, one being 0).

(d) A **first order** transition is possible even when the expression of G has only even order terms if the second order coefficient $A > 0$, the fourth order coefficient $C < 0$, and there is a sixth order term, $E\eta^6$, with $E > 0$ i.e. $G(\eta) = G_o + A\eta^2 + C\eta^4 + E\eta^6$ with A & $E > 0$, $C < 0$, $E > 0$.

(e) A **second order** transition requires the power series expansion of G to have only even order terms. However, assuming that an expansion of G up to 5 th order is adequate, the fourth order coefficient, C, must be positive for the transition to be of second order, i.e. $G(\eta) = G_o + A\eta^2 + C\eta^4$, $C > 0$.

(f) For the last two cases, a G versus η curves for $\eta < 0$ is a mirror image of that with > 0 with the same values of the expansion coefficients.

(g) If the odd coefficients are zero, and the fourth order coefficient $C = 0$, which represents the intermediate case between the first order $(C < 0)$ and second order $(C > 0)$ transitions, then we have a **tricritical** condition. In this case, a positive sixth order term (i.e. $E > 0$) is required to obtain a minimum of G as a function of η:

$$G(P,T,\eta) = G_o + A\eta^2 + E\eta^6, E > 0 \qquad (6.3.4)$$

(As remarked before (Sect. 6.2.1), a tricritical point marks the **transition** between a first order and a second order transformations).

(h) For **second order** and **tricritical** phase transition, T_i equals the transition temperature, T_{tr}, itself, whereas for first order phase transition, T_i could be greater than T_{tr}.

Although all coefficients in the power series expansion of G are, in principle, function of temperature at a given pressure, it is usually found to be adequate to treat only the second order coefficient, A, as a function of temperature, according to the form of Eq. (6.3.3). In the same spirit, only A is usually considered to be affected by pressure change. There is, however, no theoretical requirement on the nature of pressure dependence of A. The simplest alternative is to assume that A varies linearly with pressure.

A crystal may have more than one type of order parameter. For example, the mineral albite, $NaAlSi_3O_8$, undergoes a displacive transition from a monoclinic ($C2/m$: high temperature) to triclinic ($C\bar{1}$: low temperature) symmetry at low temperature. In alkali feldspar of composition $(Na_{0.69}K_{0.31})AlSi_3O_8$ this transition takes place at ~ 415–450 K (Salje 1988). But the distribution of Al and Si over the four types of tetrahedral sites, namely ($T_1(O)$, $T_1(m)$, $T_2(O)$ and $T_2(m)$), also changes with decreasing temperature, with Al being progressively more ordered to the $T_1(O)$ site. Thus, there are two order parameters involved in two types of processes. However, these two processes are not independent because both affect the unit cell dimension of the crystal. It is possible to incorporate different order parameters describing different types of processes, and their couplings, in the Landau potential. Comprehensive discussions of this topic can be found in Salje (1988) and Putnis (1992). As an example, the Gibbs energy of albite incorporating the displacive and (Al, Si) ordering effects can be expressed as the sum of two power series in terms of two ordering parameters, η_d and η_{conf}, and their couplings, as follows (Putnis 1992)

$$\begin{aligned} G(\eta_d, \eta_{conf}) = G_o &+ \left[A_1 (\eta_{conf})^2 + C_2 (\eta_{conf})^4 + E_1 (\eta_{conf})^6 + \ldots \right] \\ &+ \left[A_2 (\eta_d)^2 + C_2 (\eta_d)^4 + E_2 (\eta_d)^6 + \ldots \right] \\ &+ \lambda (\eta_{cont})(\eta_d) \end{aligned} \qquad (6.3.5)$$

It can be shown (Salje 1988; Putnis 1992) that as a consequence of the order parameter coupling, there is only one phase transition in which both order parameters participate, instead of two phase transitions, and the stability field of the lower temperature triclinic form is expanded relative to what it would have been in the absence of Al-Si disordering.

Using mathematical arguments from group theory, Landau derived some important results on the relationships between the symmetry properties of the low and high temperature phases in a second order transition. The derivation of the results is discussed in Landau and Lifschitz (1958). The main results are summarized below.

(a) For a second order transition, the symmetries of the two phases must be interrelated, with the high temperature phase having all the symmetry elements of the lower temperature phase and some extra elements as well. In group theoretical language, the symmetry group of the low temperature phase must be a sub-group of the symmetry group of the higher temperature phase. There is no requirement of any relation between the symmetry groups of the two phases in the first order transition.

(b) A second order phase transformation is possible when the number of symmetry elements of the low temperature phase is half those of the high temperature phase, and is impossible when the reduction factor is three.

Notice that the symmetry arguments do not tell if a second order phase transformation will take place, but provide the permissive criterion for such phase transformations, and tell us when a second order phase transformation is impossible.

6.3.2 Derivation of Constraints on the Second Order Coefficient

For the purpose of illustration of the point that the thermodynamic stability conditions indeed impose restrictions on the properties of the coefficients in the expansion of G in terms of η (Eq. 6.3.2), let us derive the property of the coefficient A in a solid that undergoes a second order transition. Now for a second order transition there are only even order terms in the expression of G, and the terms higher than the fourth order are negligible. Thus, since G must be at a minimum when equilibrium is achieved at a P-T condition, we have from Eq. (6.3.2a)

$$\left(\frac{\partial G}{\partial \eta}\right)_{P,T} = 2A\eta + 4C\eta^3 = 0 \tag{6.3.6}$$

and

$$\left(\frac{\partial^2 G}{\partial \eta^2}\right)_{P,T} = 2A + 12C\eta^2 > 0 \tag{6.3.7}$$

For the coefficient A, it is easy to see that $A > 0$ for the high temperature phase ($\eta = 0$) in order to satisfy Eq. (6.3.7). For the low temperature phase ($\eta \neq 0$), the first stability condition leads to the relation $2C\eta^2 = -A$, which on substitution in Eq. (6.3.7) leads to—$4A > 0$. Consequently $A < 0$ at $T < T_{tr}$. Now since $A > 0$ at $T > T_{tr}$, and $A < 0$ at $T < T_{tr}$, we must have $A = 0$ at $T = T_{tr}$, thus conforming to Eq. (6.3.3)

6.3.3 Effect of Odd Order Coefficient on Phase Transition

To see how the presence of odd order coefficient in the G versus η relation leads to a first order transition (i.e. two η values at the minimum of G), let us assume, for the sake of simplicity, that expansion up to 4th order is adequate to represent G as a function of η near the transition temperature. At equilibrium, we have at the transition temperature T_{tr},

$$\frac{dG}{d\eta} = 0 = 2A\eta + 3B\eta^2 + 4C\eta^3$$

or

$$4C\eta^2 + 3B\eta + 2A = 0$$

which yields

$$\eta = \frac{-3B \pm (9B^2 - 32AC)^{1/2}}{8C} \tag{6.3.8}$$

Thus, there are two solutions for η at the transition temperature. Since one of the solutions must be $\eta = 0$ (a property required for the high temperature phase), we have $AC = 0$. Consequently, the other solution is $\eta = -3B/(4C)$.

6.3.4 Order Parameter Versus Temperature: Second Order and Tricritical Transformations

It is now instructive to derive the dependence of the order parameter on temperature near the transition temperature in second order and tricritical transitions. Combining Eqs. (6.3.3) and (6.3.6), we have for a **second order** transition (for which $T_i = T_{tr}$)

$$\eta = \left[\frac{a}{2C} (T_{tr} - T) \right]^{1/2} \tag{6.3.9}$$

If the order parameter is so defined that it has a maximum value of unity, we can write $\eta = 1$ at $T = 0$, so that

$$T_{tr} = \frac{2C}{a} \tag{6.3.10}$$

in which case Eq. (6.3.9) reduces to

$$\boxed{\eta = \left(1 - \frac{T}{T_{tr}} \right)^{1/2}} \tag{6.3.11}$$

Now for a **tricritical** transition, we recall that the odd order coefficients and the fourth order coefficient (C) are zero, and $E > 0$ (Eq. 6.3.4). Thus, the minimization of G in Eq. (6.3.2a) with respect to η and substitution of Eq. (6.3.3) yields a $(T - T_{tr}) + 3E\eta^4 = 0$. Consequently,

$$\eta = \left[\frac{a}{3E} (T_{tr} - T) \right]^{1/4} \tag{6.3.12}$$

Again if the value of η is unity in the most ordered state $(T = 0)$, then

$$T_{tr} = \frac{3E}{a} \tag{6.3.13}$$

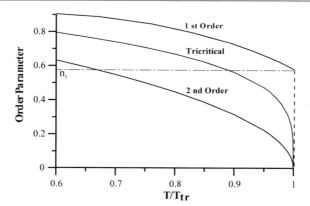

Fig. 6.4 Variation of an order parameter as a function of T/T_{tr} for second order, tricritical and first order phase transitions, where T_{tr} is the transition temperature in the respective cases. The second order and tricritical properties have been calculated according to Eqs. (6.3.11) and (6.3.14), respectively, whereas the variation of the order parameter in the first order transition is schematic, showing two different ordering states, $\eta = \eta_1$ and $\eta = 0$ at the transition temperature. It is assumed that power series expansion of Gibbs energy as a function of η is valid to $T/T_{tr} = 0.6$

in which case

$$\eta = \left(1 - \frac{T}{T_{tr}}\right)^{1/4} \qquad (6.3.14)$$

Within the framework of Landau theory, it is thus possible to determine the nature of the phase transition by following the behavior of the order parameter, η, as function of T. The exponents of $(1 - T/T_{tr})$ in the above expressions of η are known as **critical exponents**. The variation of η as function of (T/T_{tr}) for the second order and tricritical transitions are illustrated in Fig. 6.4. For the first order transition, the illustrated variation is schematic.

6.3.5   Landau Potential Versus Order Parameter: Implications for Kinetics

The variation of $(G - G_o)$ as a function of η for a second order transition is schematically illustrated in Fig. 6.5a. The schematic variations of $(G - G_o)$ for first order transitions are illustrated in Fig. 6.5b, c, the former referring to the case in which there are only even order terms in the expansion of G, whereas the latter referring to that with odd order coefficients (case (e) and (c), respectively, in Sect. 6.3.1). In Figs. 6.5a, b, the parameter η for a phase in a given state of ordering (i.e. compositions of sublattices) can have either positive or negative values of the same magnitude, depending on how it is defined, $\eta = X_A^\alpha - X_B^\alpha$ or the reverse.

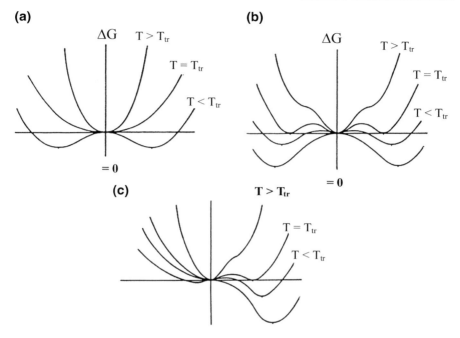

Fig. 6.5 Schematic illustrations of the change of ΔG, defined as $(G - G_o)$, in the second order (**a**) and first order (**b**), (**c**) phase transitions. The symmetry of the curves in (**a**) and (**b**) is due to the presence of only even order terms in the power series expansion of G as function of the order parameter, η, whereas the asymmetry of the curves in (**c**) is due to the presence of odd order terms in the expression of G. In (**a**) and (**b**), the parameter η can have zero and either positive or negative values for a phase in a given state of ordering, depending on how it is defined. In (**c**), $\eta > 0$ for an ordered phase. Note that at the transition temperature (T_{tr}) of a first order transition, (**b**) and (**c**), G has two equally depressed minima, one at $\eta = 0$ and the other at $\eta \neq 0$, implying two equally stable phases, whereas in a second order transition (**a**) G has only one minimum: at $\eta = 0$ for $T \geq T_{tr}$, and at $\eta \neq 0$ for $T \leq T_{tr}$, implying the presence of only one stable phase at all temperatures. From Carpenter (1985). With permission from Mineralogical Society of America

The G versus η curves in Figs. 6.5a, b are, thus, symmetric (i.e. G for $\eta < 0$ is a mirror image of that for $\eta > 0$) because of the presence of only even order terms in the power series expansion of G as function of η (Eq. 6.3.2). Figure 6.5c is, however, asymmetric since it contains only odd order terms in the power series expression of G versus η. In this case η is so defined as to have positive value for an ordered state, and $\eta < 0$ denotes an anti-ordered or physically inaccessible phase.

Note that at the transition temperature of a first order transition (Fig. 6.5b, c), there is a minimum of G at $\eta = 0$ and an equally depressed minimum at $\eta \neq 0$. Thus, there are two stable phases with two different states of ordering at the transition temperature. On the other hand, in the second order transition, there is only one minimum at T_{tr} since there is only one stable phase at the transition temperature. The form of the Gibbs energy curves lead to different kinetic pathways for the two types of transition. To illustrate this point, let us consider that a solid which undergoes first order transition is being cooled from a temperature above T_{tr}, and that $\eta > 0$ at $T < T_{tr}$. If the cooling rate is sufficiently slow for the solid to achieve the equilibrium ordering state as a function of temperature, then the ordering state of the solid at any temperature will be determined by the minimum of G corresponding to that temperature. Now when the solid is cooled to slightly below the transition temperature, between T_{tr} and T_i (Fig. 6.5b, c), it should transform to a new ordered state with η (>0) determined by the minimum of Gibbs energy. However, small fluctuations in the ordering state of the solid will cause the Gibbs energy to rise, which in turn would prevent the fluctuations to grow any further. Only when the fluctuations exceed a critical value would the Gibbs energy decrease by further growth of the fluctuations, and thereby lead to the formation of the equilibrium ordered state. This process is known as homogeneous nucleation. Nucleation can also be heterogeneous by having nuclei of the equilibrium phase derived from an external source or stabilized at discontinuities (see Sect. 13.11 for exposition of nucleation theory). But at any rate, first order phase transformation resulting from sudden or discontinuous change of ordering state requires nucleation and growth. On the other hand, no such Gibbs energy barrier exists in the second order phase transformation by order-disorder process. Therefore, a second order phase transformation does not require nucleation (in a sense these properties of first order and second order phase transformations resulting from changes of ordering state are analogous to nucleation and spinodal decomposition related to phase separation or unmixing, which is discussed in Sect. 8.14).

6.3.6 Some Applications to Mineralogical and Geophysical Problems

The Landau coefficients can be determined from laboratory measurements of a thermodynamic property such as enthalpy, H, as function of order parameter. For example, if G is expressed according to Eq. (6.3.2a) with the coefficient A given by $A = a(T - T_{tr})$ (Eq. (6.3.3)), and other parameters are insensitive to temperature, then

$$S = -\frac{\partial G}{\partial T} = S_o - a\eta^2 \qquad (6.3.15)$$

where $S_o = S(\eta = 0)$. Now if we are dealing with a second order transition, then assuming that expansion of G up to the fourth order term is adequate, as is usually the case, we have, using $H_o = H(\eta = 0)$ and $A = a(T - T_{tr})$

$$\begin{aligned} H &= G + TS = H_o + \left(A\eta^2 + C\eta^4\right) - T(a\eta^2) \\ &= H_o - aT_{tr}\eta^2 + C\eta^4 \end{aligned} \qquad (6.3.16)$$

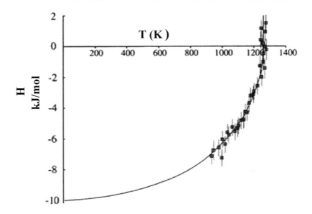

Fig. 6.6 Effect of orientational disorder of CO_3 groups as function of temperature and enthalpy of calcite, as determined by drop calorimetry. ΔH is the enthalpy of calcite with an ordering state ηrelative to that in the completely disordered state ($\eta = 0$). From Redfern et al. (1989)

Now, since $C = aT_{tr}/2$ (Eq. 6.3.10), the above equation reduces to

$$H(\eta) = H_o - aT_{tr}\eta^2 + \frac{a}{2}T_{tr}\eta^4, \qquad (6.3.17)$$

which enables determination of the coefficient 'a' from a knowledge of T_{tr} and experimental data on the variation of H as a function of η, and hence of $(S - S_o)$ from Eq. (6.3.15). The function $(G(\eta) - G_o)$ is then determined from the relation $G(\eta) - G_o = (H(\eta) - H_o) - T(S(\eta) - S_o)$.

As a specific example, let us consider the calorimetric data on the relationship between the enthalpy of calcite and orientational disorder of CO_3 groups (Redfern et al. 1989), which is illustrated in Fig. 6.6. Analysis of the experimental data on η versus T shows that it conforms to Eq. (6.3.14) (Putnis 1992), which means that the order-disorder transition is **tricritical**. Consequently, as discussed above (Sect. 6.3.1), $C = 0$, and the sixth order coefficient E is needed in the power series expansion of G versus η. Thus, since in this case, $E = aT_{tr}/3$ (Eq. 6.3.13), we have following the above procedure

$$H(\eta) - H_o = -aT_{tr}\eta^2 + \frac{a}{3}T_{tr}\eta^6 \qquad (6.3.18)$$

Using the data in Fig. 6.6, we then have

$$H(\eta = 1) - H(\eta = 0) = -10000\,\text{J/mol}$$
$$= -aT_{tr} + \frac{a}{3}T_{tr}$$

where $T_{tr} = 1260$ K. Thus, $a = 11.9$ J mol^{-1} K^{-1}, from which one can retrieve $(S(\eta) - S_o)$ and $(G(\eta) - G_o)$, following the above procedure.

It is possible for the melting curve of a solid to have a tricritical point where the dP/dT slope becomes zero. Thermodynamic analysis of such melting behavior within the framework of Landau theory, yields a quadratic dependence of temperature on pressure, as follows (Aitta 2006).

$$T = T_{tc} - (T_{tc} - T_o)\left(\frac{P}{P_{tc}} - 1\right)^2 \text{ for } P < P_{tc} \qquad (6.3.19)$$

where the subscript tc indicate tricritical condition and T_o is the iron melting temperature at zero ($\sim$ atmospheric) pressure. In the absence of direct determination of the tricritical conditions, the values of P_{tc} and T_{tc} of a solid may be constrained from the available melting data at lower pressure, if the existence of tricritical point can be justified. Aitta (2006) argued that the melting curve for iron, which is the primary component of the core of the Earth (see Sect. 7.3.2) and other rock-iron planetary bodies, may have a tricritical point. And assuming that they do, she modeled a selection of the available data to constrain the tricritical parameters using the above equation to yield T_{tc} = 8500 K and P_{tc} = 800 GPa. Here the good agreement between the model melting curve and twelve experimental data of iron in the range of 0–250 GPa constitutes a permissible criterion of the validity of the assumption for the existence of a tricritical melting condition. The calculated iron melting curve (Aitta 2006) yields a temperature 6290 ($\pm$80) K at the pressure of the inner core boundary of the Earth, which is 329 GPa (Fig. 7.7, in Chap. 7). This should be considered to be the upper limit of temperature at the inner core boundary since there are dissolved impurities in the molten iron in the outer core that have the effect of lowering the solidification temperature of iron at a given pressure (see Sect. 7.3.2). Quite similar iron melting temperature at the inner core boundary pressure was also calculated by Komabayashi (2014) from thermodynamic data of solid and molten iron assessed from the available melting data at 0–200 GPa.

6.4 Reactions in the P-T Space

6.4.1 Conditions of Stability and Equilibrium

Let us consider a univariant reaction among stoichiometric phases. The reaction causes discontinuous changes of S and V. Let $\Delta_r G$, $\Delta_r S$ and $\Delta_r V$ denote the changes of the specified thermodynamic properties caused by the reaction. As a simple example, let us consider the reaction

$$NaAlSi_3O_8(\text{Albite: Ab}) = NaAlSi_2O_6(\text{Jadeite: Jd}) + SiO_2(\text{Quartz: Qtz}),$$

which is illustrated in Fig. 6.1. At any point within the depicted field of stability of albite, G(Ab) < (G(Jd) + G (Qtz)), whereas at any point within the field of stability of Jd + Qtz, G(Ab) > (G(Jd) + G (Qtz)). In general, within the field of stability of the reactants (i.e. the phases that are written in the left hand side of a reaction),

$$\sum_i r_i G(R_i) < \sum_j p_j G(P_j) \tag{6.4.1}$$

where R_i is a reactant phase and r_i is its stoichiometric coefficient, and P_j is a product phase with a stoichiometric coefficient p_j. The reverse relation holds in the field of stability of the product phases (i.e. phases that are written on the right hand side of a reaction). These inequality relations follow from the fact that at a given P-T condition, a system restricted only to P-V work evolves in the direction of lower Gibbs free energy (Eq. 3.2.3).

A schematic G versus P relation at constant temperature of the phases albite and jadeite plus quartz is illustrated in Fig. 6.7. The slope of a G versus P curve must always be positive since $\partial G/\partial P = V$, which is a positive quantity. When albite is in equilibrium with jadeite plus quartz in the P-T space, $G(Ab) = G(Jd) + G(Qtz)$. In general, when **equilibrium** is achieved between the product and reactant assemblages in the P-T space, the following condition must be satisfied.

$$\underbrace{\sum_j p_j (GP_j) - \sum_i r_i G(r_i)}_{\Delta_r G} = 0 \tag{6.4.2}$$

The quantity on the left hand side is called the Gibbs free energy change of a reaction, $\Delta_r G$. Conventionally, it is written as total G of the product minus that of the reactant. If $\Delta_r G < 0$, the reaction proceeds to the right (i.e. in the direction of the product assemblages) and vice versa.

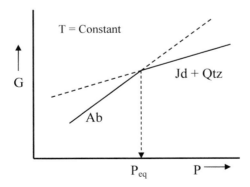

Fig. 6.7 Schematic illustration of the G versus P relation at a constant temperature for albite and jadeite + quartz, conforming to the stability relations shown in Fig. 6.1. P_{eq} is the equilibrium pressure at the chosen pressure. The free energy of a stable assemblage is shown by a solid line

6.4.2 P-T Slope: Clapeyron-Clausius Relation

To deduce the P-T slope of an equilibrium boundary at which there are discontinuous changes of entropy and volume, we write, using Eq. (3.1.10)

$$d\Delta_r G = -\Delta_r S dT + \Delta_r V dP \qquad (6.4.3)$$

Now, since at equilibrium $\Delta_r G = 0$, we have along the equilibrium boundary, $d\Delta_r G = 0$, and consequently,

$$\frac{dP}{dT} = \frac{\Delta_r S}{\Delta_r V} \qquad (6.4.4)$$

This relation is known as the Clapeyron-Clausius relation.

Problem 6.1 Equation (6.4.4) is not applicable to a second-order phase transitions since for that case $\Delta_r S = \Delta_r V = 0$, which causes a 0/0 indeterminacy. A different relation is needed to express the P-T slope of second order transitions. Show that this relation is

$$\frac{dP}{dT} = \frac{\Delta\alpha}{\Delta\beta} \qquad (6.4.5)$$

Hint: Begin by expressing V = f(P, T), and writing the total differential of V.

Similarly, using S instead of V, show that

$$\frac{dP}{dT} = \frac{\Delta_r C_p}{T\Delta_r(V\alpha)} \qquad (6.4.6)$$

These are known as **Ehrenfest relations**, after the physicist Paul Ehrenfest (1880–1933), who first derived these relations.

Problem 6.2 Draw a schematic G versus T diagram at constant pressure, analogous to Fig. 6.7, to depict the relative stabilities of albite and jadeite plus quartz. Pay attention to the thermodynamic restriction on the sign of G-T slope.

Problem 6.3 Using the result from the Problem 2.2, calculate the change in slope of the univariant reaction Albite = Jadeite ($NaAlSi_2O_6$) + Quartz due to the phase transition of albite from low-albite to high-albite that is smeared around 600 °C. Assume that the disordering of Al and Si does not have any significant effect on the molar volume of albite (a justifiable assumption) and the phase transition of albite takes place at a discrete temperature of 600 °C. Sketch a qualitative phase diagram in the P-T space that is consistent with the effect of free energy change of albite (due to low- to high-albite transition) on the phase relations.

Hint: Consider any point on the metastable extension of low-Ab = Jd + Qtz in the high-albite field (i.e. beyond 600 °C), and determine the sign of $\Delta_r G$ for the high-Ab equilibrium at that point.

Problem 6.4 In Eq. (6.4.4), $\Delta_r S$ and $\Delta_r V$ are, in general, functions of P and T. Thus, in order to calculate the slope of an equilibrium boundary at a given P-T condition of equilibrium (say, P′, T′), one needs to calculate $\Delta_r S$ and $\Delta_r V$ at that specific P-T condition. These properties can be calculated from the entropy and volume data at 1 bar, 298 K, and the thermal expansion and compressibility data from following relations:

$$\Delta_r S(P', T') = \Delta_r S^+ + \int_{298}^{T'} \left(\frac{\Delta_r C_p}{T} \right)_{1\text{bar}} dT - \int_1^{P'} \Delta_r (\alpha V)_{T'} dP \qquad (6.4.7a)$$

$$\Delta_r V(P', T') = \Delta_r V^+ + \int_{298}^{T'} \Delta_r (\alpha V)_{1\text{bar}} dT - \int_1^{P'} \Delta_r (\beta V)_{T'} dP \qquad (6.4.7b)$$

where the superscript + denotes the properties at 1 bar, 298 K. Derive the above relations.

6.5   Temperature Maximum on Dehydration and Melting Curves

Devolatilization, in particular dehydration and melting reactions, are of extraordinary importance in the understanding of geological processes at pressures in the lower crust and the Earth's mantle. These played critical role in the release of volatiles which are structurally bound in minerals and the melting behavior of rocks in the Earth's interior. The release of volatiles have many important consequences for the properties of rocks in the Earth's interior such as their rheological and transport properties, and the melting temperatures. In fact, there would be no volcanism in the subduction zone environment (circum-Pacific "ring of fire") if it were not for the dramatic effect of water in lowering the melting temperatures of minerals and rocks.

The entropy change of a dehydration reaction is always positive. However, while the volume change of a dehydration reaction is positive at low pressures, it becomes negative at high pressures. This is because of the fact that the compressibility of H_2O is much greater than the overall compressibility of the reactant solid phases. As a consequence, the volume of the dehydration products becomes less than that of the reactant solids after a threshold pressure is exceeded, as schematically illustrated in Fig. 6.8. This leads to the change in sign of $\Delta_r V$ from positive at low pressure to negative at high pressure, and consequently a "bending backward" phenomenon of the dehydration boundary in the P-T space. The dehydration boundaries of several important minerals, as calculated and experimentally constrained by Bose and

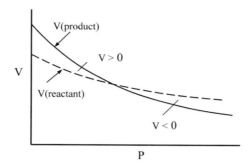

Fig. 6.8 Schematic illustration of the volumetric behavior of the reactant and product phases in a dehydration reaction as a function of pressure. Here V(product) and V(reactant) are the total volumes of the product and reactant phases, respectively, as written in a reaction, i.e. V(product) = $\Sigma v_{pi}P_i$, and V(reactant) = $\Sigma v_{ri}R_i$, with the reaction being of the form $v_{r1}R_1 + v_{r2}R_2 + \ldots \leftrightarrow v_{p1}P_1 + v_{p2}P_2 + \ldots$

Ganguly (1995), are illustrated in Fig. 6.9. The progressive change in the P-T slope of dehydration reactions with increasing pressure causes release of water from hydrous minerals at much shallower depths than it would have been otherwise as these get buried deep into the Earth's interior during geological processes (e.g. subduction of the oceanic lithosphere) with major consequences for such phenomenon as volcanism (Bose and Ganguly 1995) and probably also for seismicity in the subduction zone environment (Hacker et al. 2003).

The melting temperatures of solids are expected to exhibit similar temperature maximum as the dehydration temperatures, and for similar thermodynamic reasons. The entropy change of melting of a solid or a group of solids is always positive, but the volume change of melting is positive at low pressure and negative above a threshold pressure because of the greater compressibility of the melt compared to that of the solid. This property is very well illustrated by the melting behavior of Cs (Fig. 6.10a). The melting behavior of the most predominant rock type in the Earth's upper mantle, namely that of peridotite, is shown in Fig. 6.10b. In the latter case, the change in the slopes of solidus and liquidus[1] is a consequence of the changing composition of the melt and residual matrix as well as their compressibilities. However, it is noteworthy that there is a melting temperature maximum where we must have ΔV_m = V(melt) − V("melting" solid) = 0. (By "melting solid" we imply the portion of solid that underwent melting, since in a multi-phase system the entire assemblage of solid does not melt at a discrete temperature.) As a consequence, melt that might form in the deep interior of the Earth could become heavier than the surrounding mantle, and would thus stay trapped in the interior.

---

[1]In a multiphase system, melting is, in general, spread over a temperature interval. The temperature at which the melting begins is known as the solidus whereas that at which melting goes to completion is known as the liquidus. Between the solidus and liquidus, the composition of melt and residual solid matrix, as well as the assemblage of phases in the latter, change as a function of temperature.

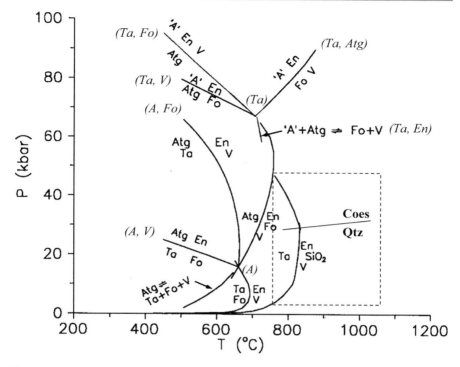

Fig. 6.9 Dehydration boundaries of the minerals talc and antigorite (serpentine) as individual phases and by reaction between them or with other phases. **Ta**: Talc ($Mg_3Si_4O_{10}(OH)_2$); **Atg**: Antigorite ($Mg_3 Si_2O_5(OH)_2$); **En**: Enstatite ($MgSiO_3$); **Fo**: Forsterite (Mg_2SiO_4); **A**: Phase A ($Mg_7Si_2O_8(OH)_6$); **V**: Vapor (H_2O). Note the "back-bending" of the dehydration boundaries at high pressure. The parenthetical phases shown in italics indicate those absent from an equilibrium in the sub-system consisting of the phases Ta, Atg, En, Fo, A and V. Modified from Bose and Ganguly (1995). With permission from Elsevier

It is of incidental interest here to discuss the possibility of presence and entrapment of melt at the base of upper mantle that is defined by the 410 km discontinuity. Presence of water is known to lower the melting temperature of minerals. It has been suggested that enough free water may be present at the base of the upper mantle of the Earth to induce partial melting of the mantle rock (Huang et al. 2005). However, such a melt layer can be gravitationally stable only if the melt has a greater density than the surrounding mantle rock as a consequence of the pressure effect on ΔV_m, as discussed above. To address this problem, Sakamaki et al. (2006) determined the density of anhydrous and hydrous melt derived from partial melting of mantle rock (peridotite) as function of pressure at 1600 °C, which is the inferred approximate temperature at 410 km discontinuity. Their results, which are illustrated in Fig. 6.11, suggest that magma with dissolved water content of less than ~6 wt%, should be gravitationally stable just above 410 km discontinuity. The presence of a melt layer is consistent with seismic anomaly observed above this discontinuity.

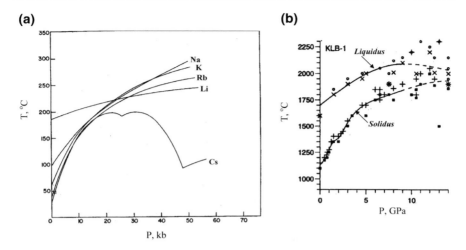

(a)

(b)

Fig. 6.10 Melting temperature as a function of pressure for (**a**) several alkali metals (modified from Newton et al. 1962) and (**b**) a mantle peridotite, known as KLB-1 (from Iwamori et al. 1995, with permission from Elsevier). In (**a**), note the melting temperature maxima of Cs. The rise of temperature after a maximum is due to phase transformation of Cs to a denser polymorph. In (**b**), the melting takes place over a temperature interval between solidus and liquidus. The experimental data are indicated

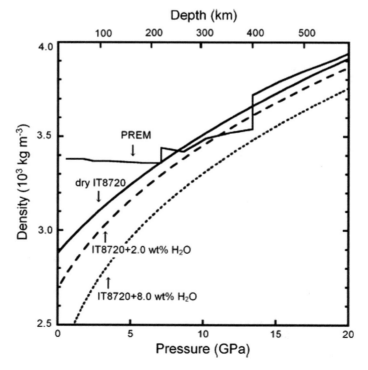

Fig. 6.11 Density versus pressure relations of peridotitic mantle, and of dry and hydrous partial melts with different quantities of dissolved water. PREM stands for the Preliminary Reference Earth Model of Dziewonski and Anderson (1981). From Sakamaki et al. (2006). With permission from Nature

6.6 Extrapolation of Melting Temperature to High Pressures

Because of experimental difficulties associated with the determination of melting temperatures at very high pressures, Earth scientists often have to estimate melting temperature at the desired pressures on the basis of the available data on melting temperatures at lower pressures. Since the melting curve is nonlinear, such extrapolations require theoretical understanding of how melting temperature of a solid is related to its physical properties for which data are available. Several methods have been proposed for extrapolation of melting temperature of stoichiometric solids to high pressures. Two such methods, which are used widely in the Earth science community, are discussed below.

6.6.1 Kraut-Kennedy Relation

Kraut and Kennedy (1966) discovered that for metals, T_m changes linearly if it is plotted against the compression of the solid, $\Delta V_s/V_o$, instead of pressure, where $\Delta V_s = V_o - V_s(P)$, with V_o being the volume of the solid at the ambient condition (Fig. 6.12). This is often referred to as the Kraut-Kennedy melting law, and is stated as

$$T_m(P) = T_m^o + C\frac{\Delta V_s}{V_o} \qquad (6.6.1)$$

where T_m^o is the melting temperature at the atmospheric condition ($P \sim 0$), and C is a constant. Kennedy and co-workers (Kennedy and Vaidya 1970; Leudemann and Kennedy 1968; Akella and Kennedy 1971) subsequently discovered that this linear relation holds for metals except for the soft metals like Pb, which have a relatively large value (~ 3) of the Grüneisen parameter at low pressure (Akella et al. 1972); for the ionic solids T_m versus $\Delta V/V_o$ relation is concave downwards while for the van der Waals solids it is concave upwards, usually when the compression exceeds 6 and 10%, respectively, of the volume at 1 bar (V_o).

Libby (1966) and Mukherjee (1966) independently showed that the Kraut-Kennedy melting law follows from the Clapeyron-Clausius relation as a special case. To see this, let us re-write the Clapeyron-Clausius relation (Eq. 6.4.4) for the pressure dependence of melting temperature as

$$\begin{aligned} dT_m &= \frac{\Delta V_m}{\Delta S_m}dP \\ &= \frac{\Delta V_m}{\Delta S_m}\left(\frac{dP}{dV_s}\right)dV_s \end{aligned} \qquad (6.6.2)$$

where ΔV_m and ΔS_m are the volume and entropy changes of melting, respectively, V_S is the volume of the solid, and dP and dV_S refer to the property changes along the melting curve. If we now assume that these property changes are approximately the same as those under isothermal condition, then from Eq. (3.7.2).

$$\frac{dP}{dV_s} \approx \left(\frac{\partial P}{\partial V_s}\right)_T = -\frac{k_T}{V_o} \tag{6.6.3}$$

Combining the last two equations,

$$dT_m \approx -\left(\frac{\Delta V_m k_T}{\Delta S_m V_o}\right)(dV_S)_T \tag{6.6.4}$$

Assuming the term within the parentheses to be constant, we can write

$$\int_{T_m(P')}^{T_m(P)} dT_m \approx -\left(\frac{\Delta V_m k_T}{\Delta S_m V_o}\right) \int_{V_s(P')}^{V_s(P)} (dV_S)_T \tag{6.6.5}$$

or

$$T_m(P) \approx T_m(P') + \left(\frac{\Delta V_m k_T}{\Delta S_m}\right) \frac{[V(P') - V(P)]_T}{V_o}, \tag{6.6.6}$$

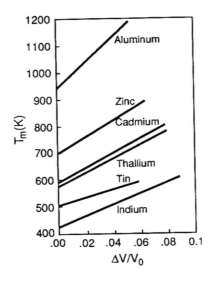

Fig. 6.12 Melting temperature versus compression of some metals showing linear relations. V_o is the solid volume at 1 bar pressure. From Anderson (1995), based on the data by Kennedy and Vaidya (1970)

which is formally the same as the Kraut-Kennedy melting law (Eq. 6.6.1). Kraut and Kennedy specifically set $P' = 1$ bar. However, it is easy to see that the predictive success of this melting law would be better if P' is set to higher pressure, and the term $[V(P') - V(P)]$ is evaluated at as close to the melting temperature as the available compressibility data permit.

The Kraut-Kennedy relation has been used by Boehler (1993) to extrapolate the melting temperature of the Fe-O-S system that is measured up to 2 Mbar in the laboratory to the pressure of the inner-core outer boundary (3 Mbar) of the Earth. The extrapolated melting temperature is used to fix the temperature at the boundary between the solid inner and liquid outer core, as illustrated in Fig. 7.7.

6.6.2   Lindemann-Gilvarry Relation

Lindemann (1910) tried to calculate the Einstein vibrational frequency of solids (see Sect. 1.6) in which the atoms are assumed to behave as harmonic oscillators, and hypothesized that at the melting point the amplitude of vibration of the atoms become so large that they collide with each other. Gilvary (1957) picked up on this hypothesis and assumed that instead of the amplitudes of vibration being large enough to lead to atomic collisions, melting takes place when the root-mean-square amplitude (roughly the average amplitude) of vibration exceeds a critical fraction of the distance separating the atoms. This led to the development of the following melting relation that is referred to as Lindemann-Gilvarry or simply Lindemann melting relation.

$$\frac{dT_m}{dP} = \frac{2T_m}{k_T}\left(\Gamma_{th} - \frac{1}{3}\right) \tag{6.6.7}$$

where Γ_{th} is the thermodynamic Grüneisen parameter (Eq. 3.8.2). The above relation has been applied to treat melting relations of metals with considerable success. Subsequently, Anderson and Isaak (2000) applied it to describe the melting relations of iron in the Earth's core. The interested reader is referred to Poirier (1991) and Anderson (1995) for derivation of the above melting relation and other theoretical melting models.

Problem 6.5 Show that the Lindemann-Gilvarry melting relation can be expressed as

$$\frac{d\ln T_m}{d\ln\rho} = 2\left(\Gamma_{th} - \frac{1}{3}\right) \tag{6.6.8}$$

6.7   Calculation of Equilibrium P-T Conditions of a Reaction

6.7.1   Equilibrium Pressure at a Fixed Temperature

If P and T are the variables that we wish to control, then at equilibrium any reaction must satisfy the relation $\Delta G_r(P_e, T_e) = 0$, where P_e, T_e are the equilibrium pressure-temperature condition of the reaction. As above, we will use the symbol Δ_r to denote the change of a specified property between the reactant and product. Conventionally, one writes $\Delta_r Y = \sum Y(\text{product}) - \sum Y(\text{reactant})$. As an example, let us consider the reaction

$$\begin{array}{cccc} \text{An} & \text{Gr} & \text{Ky} & \text{Qtz} \\ 3CaAl_2Si_2O_8 = Ca_3Al_2Si_3O_{12} + 2Al_2SiO_5 + SiO_2 \end{array} \tag{6.7.a}$$

which is a very important reaction for the determination of pressure of metamorphic rocks (see Sect. 10.11; An: anorthite, Gr: grossularite garnet; Ky: kyanite; Qtz: quartz). As long as the minerals are confined to the stoichiometric end member compositions, as written, the change of Gibbs free energy, $\Delta_r G$, of this reaction at any P-T condition is written as $\Delta_r G(P, T) = G(Gr) + 2G(Ky) + G(Qtz) - 3G(An)$.

Now let us suppose that we know the thermochemical properties of the phases involved in a reaction at a P, T condition at which the reaction is not at equilibrium. In order to calculate the equilibrium pressure (P_e) at T′, we need to develop an expression of $\Delta_r G$ as a function of P at constant temperature, and solve for P_e at T′ by imposing the equilibrium property, i.e. $\Delta G_r(P_e, T') = 0$. To this end, we write, making use of the property $(\partial G/\partial P)_T = V$

$$\Delta_r G(P_e, T') = \Delta_r G(P, T') + \int_P^{P_e} (\Delta_r V)_{T'} dP = 0 \tag{6.7.1}$$

This equation can be solved for P_e if $\Delta_r V$ is known as a function of P at T′. The mathematical strategy for calculating the last integral from an equation of state that expresses P as a function of V (instead of V as a function of P) is discussed in Sect. 6.8. It should be noted, however, that when we deal with a reaction involving only solid phases, such as the reaction (6.7.a) above, $\Delta_r V$ does not change significantly for a change of pressure of several thousand bars. This is because of the fact that minerals have similar compressibilities so that the change of volume with pressure of the products and reactants substantially cancel out. In such cases, we would introduce very little error by taking $\Delta_r V$ out of the integration sign, and solving for P_e at T′.

In thermochemical tables, one commonly finds the enthalpy of formation of compounds from elements or oxides at 1 bar, 298 K, and third law entropies of compounds at 1 bar, 298 K along with their C_P-s as functions of temperature. In order to use these data to solve Eq. (6.7.1), we express $\Delta_r G(P_e, T')$ as

$$\Delta_r G(P_e, T') = \Delta_r G(1, T') + \int\limits_{1}^{P_e} \Delta_r V(P, T')dP$$

$$= \Delta_r H(1, T') - T'\Delta_r S(1, T') + \int\limits_{1}^{P_e} \Delta_r V(P, T')dP \qquad (6.7.2)$$

$\Delta_r H(1, T')$ and $\Delta_r S(1, T')$ are related to their respective 1 bar, 298 K values, denoted by $\Delta_r H^+$ and $\Delta_r S^+$, according to (Eq. 3.7.5).

$$\Delta_r H(1, T') = \Delta_r H^+ + \int\limits_{298}^{T'} \Delta_r C_p dT \qquad (6.7.3)$$

and

$$\Delta_r S(1, T) = \Delta_r S^+ + \int\limits_{298}^{T'} \frac{\Delta_r C_p}{T} dT \qquad (6.7.4)$$

Substituting the last two equations into Eq. (6.6.2), rearranging the terms and imposing the condition of equilibrium, $\Delta_r G(P_e, T') = 0$, yields

$$\boxed{\Delta_r H^+ - T'\Delta_r S^+ + \left[\int\limits_{298}^{T'} \Delta_r C_p dT - T \int\limits_{298}^{T'} \frac{\Delta_r C_p}{T} dT + \int\limits_{1}^{P_e} (\Delta_r V(P, T') \, dP \right] = 0}$$

$$(6.7.5)$$

where the superscript + denotes properties at 1 bar, 298 K. Note that the first two terms do **not** constitute $\Delta_r G^+$ since $\Delta_r S^+$ is multiplied by T' instead of 298.

As discussed in Sect. 4.5.1), the $\Delta_r H^+$ is calculated from the enthalpy of formation (ΔH_f) values of the product and reactant phases from either elements or oxides according to

$$\Delta_r H(P, T) = \Sigma \Delta H_{f,e}(\text{products}) - \Sigma H_{f,e}(\text{reactants})$$
$$= \Sigma \Delta H_{f,o}(\text{products}) - \Sigma H_{f,o}(\text{reactants})$$

Note that Eq. (6.7.5) can also be used to retrieve $\Delta_r H^+$ and $\Delta_r S^+$ from experimental data for a univariant equilibrium. This is done by calculating the term within the square brackets at several P-T conditions along the equilibrium boundary, and regressing it against T. The slope and intercept of the linear regression yield $\Delta_r S^+$ and $-\Delta_r H^+$, respectively.

6.7.1.1 Solved Problem: Depth of Diamond Formation

"Diamonds are for ever", but from what depths within the Earth were the diamonds transported?

The diamonds are recovered from eclogite and peridotites xenoliths that represent fragments of the Earth's mantle, which were ripped off by the CO_2-rich kimberlite magmas during their explosive ascent towards the Earth's surface in early geologic periods (mostly Cretaceous, but some also Pre-Cambrian). The lower limit for the depth of origin of diamond may be estimated by comparing the P-T condition of graphite-diamond transition boundary with the paleo-geotherms in diamond producing areas. Instead of using experimental data for the graphite-diamond transition boundary, here we calculate this boundary, as an illustration of the application of Eq. (6.7.5) to relatively simple phase equilibrium calculation.

We first represent the graphite-diamond transition as a reaction

$$C(graphite) = C(diamond) \qquad (6.7.b)$$

From the thermochemical data in Robie et al. (1978), we have

$$H_{f,e}(D: 1 \text{ bar, } 298 \text{ K}) = 1895 \text{ kJ/mol}$$
$$C_p = a + bT + cT^2 + dT^{-0.5} + eT^{-2} \text{J/K-mol}$$

with

$$a = 98.445, \ b = -3.6554(10^{-2}), \ c = 1.2166(10^6), \ d = -1.6590(10^3),$$
$$e = 1.0977(10^{-5}) \text{ for diamond,}$$

and

$$a = 63.160, b = -1.1468(10^{-2}), c = 6.4807(10^5), d = -1.0323(10^3),$$
$$e = 1.8079(10^{-5}) \text{ for graphite.}$$

Chatterjee (1991) summarized the volumetric data for the two phases as

$$V(D: 1 \text{ bar, T}) = 0.3409 + 0.2015(10^{-5})T + 0.984(10^{-9})T^2 \text{ J/bar-mol}$$
$$V(G: 1 \text{ bar, T}) = 0.5259 + 1.2284(10^{-5})T + 2.165(10^{-9})T^2 \text{ J/bar-mol}$$
$$\beta(D: 298 \text{ K}) = 0.18(10^{-6}) \text{bar}^{-1}$$
$$\beta(G: 298 \text{ K}) = 3.0(10^{-6}) \text{bar}^{-1} (\text{both } \beta \text{ being independent of pressure})$$

Since graphite is the stable form of element at 1 bar, T, it is taken to be the reference form of the element at 1 bar, T, and accordingly, $\Delta H_{f,e}(G: 1 \text{ bar, T}) = 0$. Thus, $\Delta_r H$ of the above the reaction (6.7.b) at 1 bar, 298 K (i.e. $\Delta_r H^+$) is given by $\Delta H_{f,e}(D: 1 \text{ bar,} 298 \text{ K}) = 1895 \text{ kJ/mol}$. The $\Delta_r C_p$ term for this reaction can be written as

$$\Delta_r C_p = \Delta a + \Delta b T + \Delta c T^2 + \Delta d T^{-0.5} + \Delta e T^{-2}$$

where $\Delta a = a(D) - a(G) = 35.285$, and so on.

It is easy to see that the sum of the first four terms in Eq. (6.7.5) yields $\Delta_r G(1,T)$, which we can now evaluate by substituting the values of $\Delta_r H^+$, $\Delta_r S^+$ and carrying out the required integrations of the $\Delta_r C_p$. As an example, this procedure yields

$$\Delta_t G(1 \text{ bar}, \ 1373 \text{ K}) = 7666.83 \text{ kJ}.$$

Using the data for molar volume as function of temperature at 1 bar, we have

$$\Delta_r V(1 \text{ bar}, \ 1373 \text{ K}) = -0.201 \text{ J/bar-mol}$$

Now, since $(\partial V/\partial P)_T = -V(1 \text{ bar}, T)\beta_T$, (Eq. 3.7.2 and follow-up discussion), we have

$$V(P, T') = I - V(1 \text{ bar}, T') \int \beta_{T'}(P) dP \qquad (6.7.6)$$

where I is an integration constant. We now assume that $\beta_{T'}(P) = \beta_{298}(P)$, which is a constant, according to the data summarized above. Thus,

$$V(P, T') = I - V(1 \text{ bar}, T')\beta_{298}P$$

The integration constant is evaluated by substituting $V(1 \text{ bar}, T')$ for $V(P, T')$ and $P = 1$ bar on the right hand side so that

$$V(P, \ T') = V(1, \ T') + V(1, \ T')\beta_{298} - V(1, \ T')\beta_{298}P \qquad (6.7.7)$$

(Note that the second term on the right is multiplied by 1 bar so that it has the unit of cm^3/mol or J/bar-mol). Using the last equation, we have

$$\int_1^P V(P, T') \, dP = [V(1, T') + \beta_{298} V(1, T')]P' - \frac{\beta_{298} V(1, T')(P')^2}{2}$$

so that

$$\int_1^{P_e} \Delta_r V(P, T') \, dP = [\Delta_r V(1, T') + \Delta_r \langle \beta_{298} V(1, T') \rangle]P_e - \frac{\Delta_r \langle \beta_{298} V(1, T') \rangle (P_e)^2}{2}$$

$$(6.7.8)$$

Substitution of this equation into Eq. (6.7.5) and rearrangement of terms yields a quadratic equation

$$-\frac{\Delta_r\langle\beta_{298}V(1,T')\rangle}{2}P_e^2 + [\Delta_r\langle\beta_{298}V(1,T')\rangle + \Delta_rV(1,T')]P_e + \Delta_rG(1,T') = 0$$

$$(6.7.9)$$

where $\Delta_rG(1, T')$ stands for the first four terms in Eq. (6.7.5).

The solution of the last equation by the usual method of solving a quadratic equation yields two values of P_e, one of which is physically unreasonable. For example, at 1300 K, the two values of P_e are 46.6 and 208.5 kb, corresponding, respectively, to the use of minus and plus sign before the square root term of the solution. The larger value corresponds to a depth of ~ 600 km which is too deep for the generation of kimberlite magma and transport of material to the surface. Furthermore, calculations of the P-T condition of xenoliths in kimberlite indicate that these were transported from a depth of no greater than of ~ 200 km (e.g. Ganguly and Bhattacharya 1987). Thus, we accept the lower value of P_e to define the equilibrium graphite-diamond transition. (When the equation of equilibrium involves higher power of P_e due to a more complex dependence of V on P, then the solution of P_e needs to be obtained numerically).

The results for calculation of P_e versus T relation are illustrated in Fig. 6.13 along with the paleogetherm of a diamond bearing locality in the southern part of India. The paleogeotherm was calculated by Ganguly et al. (1995) on the basis of the inferred P-T conditions of xenoliths (crosses) in diamond bearing kimberlites of Proterozoic age (~ 1 Ga) and heat flow data. The geotherm and the graphite-diamond transition boundary intersect at a pressure of 48 kb that corresponds to a depth of ~ 150 km. The geotherm is also quite similar to those in other diamond producing kimberlite localities, especially South Africa (Ganguly and Bhattacharya, 1987). Thus, diamonds were transported from a depth of at least ~ 150 km from within the Earth's interior.

6.7.2   Effect of Polymorphic Transition

The equilibrium (6.7.a) is subject to the effect of polymorphic transition (which is a transition between different crystallographically distinct forms of the same compound) as it is extended to temperature beyond the field of stability of kyanite. Many reactions of geological interest involve phases that undergo polymorphic transitions. Another example of such a reaction is

$$\begin{array}{ccc} \text{Talc} & \text{Enst} & \text{Qtz/Coes} \\ \end{array}$$
$$\text{MgSi}_4\text{O}_{10}(\text{OH})_2 = 3\text{MgSiO}_3 + \text{SiO}_2 + \text{H}_2\text{O} \qquad (6.7.c)$$

The equilibrium boundary of this reaction across the quartz-coesite transition is illustrated in Fig. 6.9. The general procedure for accounting for the effect of polymorphic transition is developed below using the equilibrium (6.7.c) as an example.

Let us suppose that we know the equilibrium boundary of the above reaction in the quartz field, and want to calculate the equilibrium boundary in the coesite field

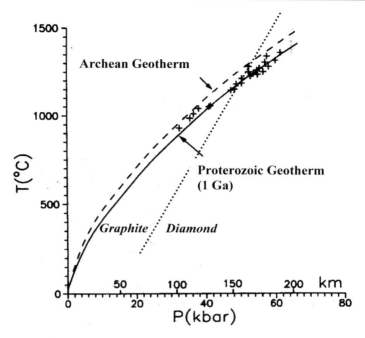

Fig. 6.13 Comparison of graphite/diamond equilibrium boundary with the geotherm calculated by Ganguly et al. (1995) for the southern Indian shield area where diamonds are found in mantle xenoliths brought up by kimberlite eruptions during the proterozoic period. This comparison shows that diamonds form at depths of at least 150 km within the Earth's interior. The crosses represent calculated P-T conditions of mantle xenoliths on the basis of mineral chemistry. The geotherm was calculated to satisfy the xenolith and heat flow data

using the thermochemical data for quartz-coesite transition. The breakdown reaction of talc in the coesite field can be expressed as a linear combination of that in the quartz field and quartz-coesite polymorphism as follows.

$$\text{Talc} = 3 \text{ Enst} + \text{Qtz} + \text{H}_2\text{O} \tag{1}$$

$$\text{Qtz} = \text{Coes} \tag{2}$$

$$\text{Talc} = 3 \text{ Enst} + \text{Coes} + \text{H}_2\text{O} \tag{3}$$

Thus, in general

$$\Delta_r G_3(P, \ T) = \Delta_r G_1(P, \ T) + \Delta_r G_2(P, \ T)$$

Now, let P_1 and P_3 be the equilibrium pressures of reactions (1) and (3), respectively, at a temperature T' (Fig. 6.14). The Gibbs energy change of reaction (3) at P_1, T' is then given by

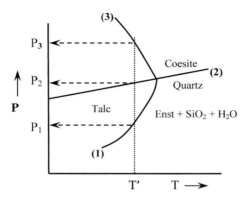

Fig. 6.14 Schematic illustration the P-T parameters defined for calculation of the effect of polymorphic transition on an equilibrium boundary of a reaction. The actual data for the reaction chosen here are illustrated in the inset of Fig. 6.9

$$\Delta_r G_3(P_1, T') = \Delta_r \overset{0}{\underset{}{G_1}}(P_1, T') + \Delta_r G_2(P_1, T') \qquad (6.7.10)$$

$$= \Delta_r G_2(P_1, T')$$

The Gibbs energy change of reaction (3) at P_3, T' can be expressed as

$$\Delta_r G_3(P_3, T') = \Delta_r G_3(P_1, T') + \int_{P_1}^{P_3} (\Delta_r V_3)_{T'} dP = 0 \qquad (6.7.11)$$

The last equality is due to the fact that P_3 represents the equilibrium pressure of the reaction (3) at T'. Combining the last two equations, we finally have

$$\boxed{\Delta_r G_2(P_1, T') + \int_{P_1}^{P_3} (\Delta_r V_3)_{T'} dP = 0} \qquad (6.7.12)$$

This equation can be solved for P_3 at different T' values to obtain the equilibrium dehydration boundary of talc in the coesite field (P_3 vs. T) using the data for Gibbs energy change of quartz to coesite transition, if the volume change for the reaction (3) is known as a function of pressure and temperature. Of course, this procedure is needed when thermochemical data of talc is not available. If it is, then the equilibrium dehydration boundary of talc in the coesite field can be calculated by solving the general equation of equilibrium, Eq. (6.7.5).

In the last expression, the first term on the left may be replaced by the $\Delta_r V$ term for the polymorphic reaction as follows. If P_2 is the equilibrium pressure of the polymorphic transition at T' (Fig. 6.14), then

$$\Delta_r G_2(P_1, T') = \Delta_r G_2(P_2, T') + \int_{P_2}^{P_1} (\Delta_r V_2)_{T'} \, dP = \int_{P_2}^{P_1} (\Delta_r V_2)_{T'} \, dP \qquad (6.7.13)$$

Since volumetric data of solids are more accurate than their Gibbs energy data, it is preferable to take advantage of the last expression to calculate P_3 versus T relation according to Eq. (6.7.12).

Problem 6.6 The volume change of reaction (3) can be expressed as $\Delta_r V_3 = \Delta_r V_3^S + V^{H_2O}$, where $\Delta_r V_3^S$ is the solid volume change of the reaction, i.e. $\Delta_r V_3^S = 3V_{Enst} + V_{Coes} - V_{Tlc}$. Now, assuming that $\Delta_r V_3^S$ is insensitive to pressure change between the limits of integration of Eq. (6.7.13), calculate the equilibrium pressure for the talc dehydration reaction in the coesite field at 800 °C, using the following data: $\Delta_r V_3^S = -1.750 \, J/bar$, $\Delta G°(Qtz \rightarrow Coes) = 4200 \, J$ at 800 °C and the corresponding pressure at the qtz-coesite boundary (29.6 kb) and the virial-type equation of state of water by Belonoshko and Saxena (1992), as given by Eq. (5.4.13). Fitting of the V-P relation of water at 800 °C calculated from this equation yields $V = 39.286(P)^{-0.286}$ where P is kb.

(a) Express $\int_{P^Q}^{P^C} \left(V^{H_2O} \right)_T dP$ (P^C and P^Q: equilibrium pressures at T in the coesite field

and the quartz field, respectively) using the method in Sect. 6.8, and solve for P^C by successive approximations, satisfying Eq. (6.7.12). (b) Also, use the V-P relation above to solve for P^C.

Answer: (a) 40.0 kb and (b) 41.2 kb.

Problem 6.7 Suppose that you have the Gibbs energy change of a reaction, $\Delta_r G$, at 1 bar, 298 K, but not $\Delta_r H$ and $\Delta_r S$, and in addition you have the C_P data of all phases involved in the reaction. The C_P function is expressed as $C_P = a + bT + c/T^2$. Derive an analytical expression to calculate $\Delta_r G$ at any arbitrary temperature at 1 bar, using the available data, and the relation $\partial G/\partial T = -S$. Complete all integrations. (Note: By adding the integral term from Eq. (6.7.1) to your derived expression for $\Delta_r G(1 \, bar, T)$, you obtain an expression for $\Delta_r G(P, T)$, which you can use, instead of Eq. (6.7.2) to calculate equilibrium P at T.

Hint: First express S as a f(T) using the C_P versus T relation.

*****Problem 6.8** In quantum chemical calculations of pressure of first order phase transition at a fixed temperature, it is quite common to see calculation of U versus V relation of the two phases that are convex downwards and take the common tangent to the two curves to determine the transition pressure (Fig. 6.15). What is the (a) rationale behind this practice and (b) restriction on the temperature at which the

Fig. 6.15 Schematic
illustration of U versus V
relation of two polymorphic
phases, I and II

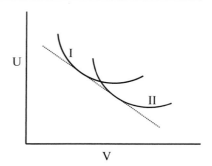

calculated transition pressure holds? Also what thermodynamic potential should be
used in place of U to calculate the transition pressure at any temperature following
the "common tangent" approach?

6.8 Evaluation of Gibbs Energy and Fugacity at High Pressure Using Equations of States

In many Equations of states, such as Redlich-Kwong, Birch-Murnaghan and Vinet,
as discussed above, P is expressed as a function of volume, instead of the reverse,
and these equations cannot be re-arranged to yield V(P) that can be used directly to
calculate the integral $\int V dP$, which relates the Gibbs free energies between two
different pressures. Also, this integral is needed to calculate the change of logf,
where f is the fugacity, for a given change of pressure.

We can use a P(V) EoS to evaluate $\int V dP$ from the relation $d(PV) = P dV + V dP$, so that

$$\int V dP = \int d(PV) - \int P dV \qquad (6.8.1)$$

If we evaluate the first integral between the limits P_1 (lower limit) and P_2 (upper
limit), then the corresponding limits of the second integral are $[P_1, V(P_1)]$ and $[P_2,
V(P_2)]$, and those of the last integral are $V(P_1)$ and $V(P_2)$ (for brevity, we write V_1
for $V(P_1)$ and V_2 for $V(P_2)$). Thus, we have

$$\int_{P_1}^{P_2} V dP = [P_2 V(P_2) - P_1 V(P_1)] - \int_{v_1}^{v_2} P dV \qquad (6.8.2)$$

We derive below the expressions for the integral on the right for some commonly
used EoS for solid and fluid.

6.8.1 Birch-Murnaghan Equation of State

If the P-V relation is described by the third-order Birch-Murnaghan EoS (Eq. 5.4.23), then we can write

$$P = \alpha \left[\left(\frac{V_o}{V} \right)^{7/3} - \left(\frac{V_o}{V} \right)^{5/3} \right] \left\{ 1 + \xi \left[\left(\frac{V_o}{V} \right)^{2/3} - 1 \right] \right\} \qquad (6.8.3)$$

where

$$\alpha = \frac{3k_o}{2}, \text{ and } \xi = \frac{3}{4}(k_o' - 4)$$

(To recapitulate, k_o is the bulk modulus at $P = 0$, T, and k_o' is the pressure derivative of k_o.) From this P-V relation, we obtain

$$\int_{V_1}^{V_2} PdV = \frac{3}{2}\alpha V_o^{5/3}(1 - \xi)\left[\frac{1}{V^{2/3}}\right]_{V_1}^{V_2} - \frac{1}{2}\alpha\xi V_o^3 \left[\frac{1}{V^2}\right]_{V_1}^{V_2}$$
$$+ \frac{3}{4}\alpha V_o^{7/3}(2\xi - 1)\left[\frac{1}{V^{4/3}}\right]_{V_1}^{V_2} \qquad (6.8.4)$$

For the second order B-M EoS, we have the special case of the above expression with $\xi = 0$.

6.8.2 Vinet Equation of State

Writing $c = 3/2(k_o' - 1)$, we can express the Vinet EoS (Eq. 5.4.29) as

$$P = 3k_o \frac{1 - \eta^{1/3}}{\eta^{2/3}} \exp\left[c\left(1 - \eta^{1/3}\right)\right] \qquad (6.8.5)$$

where $\eta = V/V_o$. Now since $dV = V_o d\eta$,

$$\int PdV = 3k_o V_o \int \frac{1 - \eta^{1/3}}{\eta^{2/3}} \exp\left[c\left(1 - \eta^{1/3}\right)\right] d\eta \qquad (6.8.6)$$

Using $y = (1 - \eta^{1/3})$ so that $d\eta = -3\eta^{2/3}dy$.

$$\int PdV = -9k_o V_o \int ye^{cy}dy$$

Definite integral of this expression yields

$$\int_{V_1}^{V_2} PdV = -9k_0 V_0 \left[\frac{e^{cy}}{c^2}(cy - 1)\right]_{y_1}^{y_2} \qquad (6.8.7)$$

in which, summarizing, $y = 1 - \eta^{1/3} = 1 - (V/V_0)^{1/3}$, and $c = 3/2(k_0' - 1)$.

6.8.3 Redlich-Kwong and Related Equations of State for Fluids

Using the P-V relation of Redlich-Kwong equation of state, as given by Eq. (5.4.5), we obtain

$$\int_{V_2}^{V_2} PdV = \left[RT\ln(V - b) + \frac{a}{b\sqrt{T}}\ln\left(\frac{(b + V)}{V}\right)\right]_{V_1}^{V_2} \qquad (6.8.8)$$

where, for brevity, we have used V for the molar volume, which is written as V_m in Eq. (5.4.5). In the modified Redlich-Kwong (MRK) equation of states, a and b are usually treated as functions of temperature and pressure, respectively (e.g. Eqs. 5.4.6a, b). Obviously, the above integral cannot be used when b is a function of pressure. The integration needs to be carried out numerically.

As discussed in Sect. (5.4.1.3), recognizing the problem with the analytical integration of MRK when b is a function of pressure, Holland and Powell (1991) proposed a **compensated Redlich-Kwong (CORK)** equation of state in which b is treated as a constant. They found that the error resulting from treating b as a constant can be accounted for by introducing a correction or compensation term with a general form given by Eq. (5.4.7). Denoting the terms within the square brackets of this equation as a correction volume, V^{cor}, we have

$$\int_{P_1}^{P_2} VdP = \int_{P_1}^{P_2} V^{mrk}dP + \int_{P_1}^{P_2} V^{cor}dP \qquad (6.8.9)$$

The first integral on the right can be evaluated by combining Eqs. (6.8.2) and (6.8.8). The second integral on the right is given by

$$\int_{P_1}^{P_2} V^{cor}dP = \left[\frac{2c}{3}(P - P_0)^{3/2} + \frac{d}{2}(P - P_0)^2\right]_{P_1}^{P_2} \qquad (6.8.10)$$

Thus, for the **compensated Redlich-Kwong (CORK)** equation of state proposed by Holland and Powell (1991), we finally obtain

$$\int_{P_1}^{P_2} VdP = \left[P_2 V_2^{mrk} - P_1 V_1^{mrk}\right] - \left[RT \ln(V^{mrk} - b) + \frac{a}{b\sqrt{T}} \ln\left(\frac{b + V^{mrk}}{V^{mrk}}\right)\right]_{V_1^{mrk}}^{V_2^{mrk}}$$

$$+ \left[\frac{2c}{3}(P - P_o)^{3/2} + \frac{d}{2}(P - P_o)^2\right]_{P_1}^{P_2}$$

$$(6.8.11)$$

where V^{mrk} is the **molar** volume calculated from a modified Redlich-Kwong equation of state with the term a calculated at the temperature of interest and b treated as a constant.

Problem 6.9 Using the equation of state (EoS) of fluids by Belonoshko and Saxena, as given by Eq. (5.4.13), calculate

(a) the Gibbs free energy and fugacity of H_2O at 10 kb, 720 °C, using the experimentally derived values (Burnham et al. 1969) of G (5 kb, 720 °C) = −27,903 J/mol, f(5 kb, 720 °C) = 3964 bars, V(5 kb, 720 °C) = 1.3569 cm³/gm and V(10 kb, 720 °C) = 1.0903 cm³/gm
(b) fugacity coefficient of water at 10 kb, 720 °C
(c) see Problem 8.5.

Partial Answers: G(10 kb, 720 °C) = −17,267 J/mol; f(10 kb, 720 °C) = 14,376 bars).

Hint: To calculate G(10 kb, 720 °C), evaluate $\int$ VdP from the EoS according to the method of integration by parts (Eq. 6.8.2); also see Sect. 3.6 for calculation of fugacity. The units of the constants a, b, and c are bar-cm³, bar-cm⁶ and bar-cm³ᵐ, respectively.

Comment: The experimentally determined values of G and f at 10 kbar, 720 °C are − 17,071 J/mol and 14,723 bars, respectively (Burnham et al. 1969), which are in good agreement with the values derived from the EoS data of Belonoshko and Saxena (1992). These authors have also presented an equation for calculating the fugacity of different geologically important volatile species in the system C-O-H at 5 kb, T, which may be used to calculate the fugacity values at any other pressure using their EoS, which is valid for P > 5 kb).

Problem 6.10 The estimated P-T conditions of formation of the Archean granulites in the southern part of India plot much above the steady state geotherm in the same area that is illustrated in Fig. 6.13. Thus the granulites had formed by some kind of a transient thermal perturbation. One of the proposed mechanisms of such thermal perturbation is basaltic magma under-plating of the terrane, where the heat given off by the cooling magma body from its liquidus condition raised the temperature of the crustal rocks. At 10 kb pressure, the average temperature of the granulites is $\sim 750\ °C$, which is $\sim 400\ °C$ higher than the temperature at the ambient geotherm, and the liquidus temperature of basalt is $1300\ °C$. Assuming that the heat capacity (C_P) of both the crustal rocks and basaltic magma/rock are $\sim 10^3$ J/kg-K, and the heat of fusion (ΔH_f) of basalt is $4(10^5)$ J/kg, calculate the mass of the supposed under-plated basalt relative to that of granulite.

(Presence of under-plated magma body, if it is sufficiently large, can be detected by geophysical methods).

Answer: mass of basalt/mass of granulite ~ 0.42.

6.9 Schreinemakers' Principles

The Schreinemakers' principles constitute a set of rules that enable us to organize the possible equilibria in a system in a self-consistent manner. The principles require systematic application of the Phase Rule, keeping track of some simple facts about self-consistency in the stability analysis of phases, such as an assemblage of the phases A and B must lie within the fields of stability of both A and B.

Let us consider a system consisting of 6 phases, namely, talc (Ta), antigorite (Atg), enstatite (En), forsterite (Fo), phase A (A) and vapor (V). (The phase A is one of several dense hydrous magnesium silicate phases that have been discovered in the laboratory experiments at high P-T conditions. These are given alphabetical names, A, B, C etc., and are generally referred to as DHMS phases.) The univariant reaction relations among these phases in the P-T space, as determined by a combination of experimental data and theoretical considerations, including the Schreinemakers' principles, are illustrated in Fig. 6.9 (Bose and Ganguly 1995). We will discuss Schreinemakers' principles by referring to this Figure. It should, however, be noted at the outset that phase relations may also be depicted using intensive variables other than P and T. Schreinemakers principles are equally applicable to those cases. An example can be found later in the Sect. 12.8.1 (Fig. 12.7). For simplicity, but without losing any generality, we have not included quartz and coesite in the system, although the phase diagram shows equilibria involving these phases (Ta = En + SiO_2 (Qtz/Coes) + V and Qtz = Coes). Only three components, MgO, SiO_2, H_2O are required to describe the compositions of all the phases, assuming that each phase is restricted to their respective pure end-member states.

6.9.1 Enumerating Different Types of Equilibria

If the number of components in a reaction is C, then according to the Phase Rule, the reaction is univariant if it has (C + 1) number of phases. However, sometimes a specific reaction may involve a smaller number of components than there are in the general system. In that case, the reaction becomes univariant with smaller number of phases than those that involve all the components of the general system. An example is the reaction Qtz = Coes. It is univariant since it involves only one component, SiO_2, even though the phases in the general system may require additional components such as MgO and H_2O. Such univariant equilibria that involve lesser number of phases than those that contain all the components of the general system are called **degenerate equilibria**. We will return to the properties of this type equilibrium later.

We now ask the following question. How many univariant equilibria may be found in a system in which the number of phases is P and that of components is C. The question can be easily answered by using combinatorial principle, as follows.

$$\text{Number of (univariant) equilibria} \leq \frac{P!}{(C+1)!(P-C-1)!}$$

with the equality holding when all equilibria are non-degenerate. When P = 6 and C = 3, as in the chosen illustrative system, we have No. (Univariant equilibria) $\leq$ 15.

$$\text{Similarly, the number of invariant points} \leq \frac{P!}{(C+2)!(P-C-2)!}$$

so that there are a maximum of 6 invariant points in the system. All of these invariant points are, of course, not of geological interest. Only two invariant points are shown in Fig. 6.9.

For book keeping purposes in the Schreinemakers analysis, an equilibrium is designated by writing the phases **absent** from the equilibrium within parentheses. Thus, the two invariant points in Fig. 6.9 are labeled (*Ta*) and (*A*). Each univariant equilibrium intersecting at (*Ta*) lacks talc plus an additional phase. Similarly, each univariant equilibrium intersecting at (*A*) lacks the phase A and an additional phase. There are five univariant equilibria radiating from each invariant point. If there are X number of phases at an invariant point, then there are a maximum of X number of univariant equilibria radiating from that point. All univaraint reactions around an invariant point can be written out in chemically balanced forms by using linear algebraic method (e.g. Korzhinskii 1959).

Let us now consider a system of five phases, namely, Ta, En, Qtz, Coes and V. The phase relations are shown within a box in the lower right hand portion of Fig. 6.9. Here again, the overall system consists of three components, MgO, SiO_2, H_2O. Thus, each non-degenerate univariant equilibrium must consist of 4 phases. Two of these are Ta = En + Qtz + V (Coes) and Ta = En + Coes + V (Qtz), which intersect to generate an invariant point consisting of five phases. If all univariant equilibria in the system are non-degenerate, then there must be three additional equilibria radiating from the invariant point, each being characterized by the absence of one phase from the set present at the point. However, the univariant reaction Qtz = Coes is characterized by the absence of three phases, (*Ta*), (*En*) and (*V*), from the set present at the invariant point. Thus, this equilibrium is considered to be **triply degenerate**, which means that for the book keeping purpose, it is taken to be equivalent to three non-degenerate univariant equilibria. Thus, all univariant reactions radiating from the invariant point have been accounted for, and hence there can be no other balanced univariant reaction in this part of the system.

6.9.2   Self-consistent Stability Criteria

If there is a reaction A + B = C + D, then any univariant reaction that involves the phases A and B (in either the same side or the two sides of the reaction) must lie in the half-plane defined by the A + B side of the first reaction. We would refer to this domain as the A-B **half-plane**. The concept of half-plane is illustrated in Fig. 6.16. Consider, for example, the reaction (*Ta, Atg*) in the top right part of Fig. 6.9. The stable parts of all reactions involving the phases A and En, namely the stable parts of the reactions (*Ta, Fo*) and (*Ta, V*), lie in the half-plane defined by the A + En side of the reaction (*Ta, Atg*). Similarly, the stable parts of all reactions involving Fo and V lie in the half-plane defined by the Fo + V side of the reaction (*Ta, Atg*). The arrangement of all reactions in Fig. 6.9 satisfies this "half-plane constraint". The reason behind this topological restriction is a simple stability argument. If the stable part of the reaction (*Ta, Fo*) is on the Fo + V half-plane, then the phase A and En in this reaction would react to form Fo + V. The half-plane concept can be stated in a different and general way as follows.

The phases that are absent from any equilibrium must not appear as products of the reactions bounding the sector which contains the stable part of that equilibrium. Consider, for example, the equilibrium (*Ta, Fo*) in the top left part of Fig. 6.9.

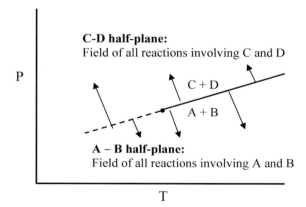

Fig. 6.16 Illustration of half-plane concept in Schreinemakers' analysis. The stable (solid line) and metastable (dashed line) portions of a univariant reaction A + B = C + D divide the P-T space into A-B and C-D half-planes. The dot represents an invariant point. All reactions involving A and B lie in the A-B half-plane. Similarly, all reactions involving C and D lie in the C-D half-plane

The phases Ta and Fo do not appear as products of the reactions (*Ta, V*) and (*Ta, Atg*) that bound the sector containing the stable part of the reaction (*Ta, Fo*).

The fact that an assemblage of phases cannot be stable beyond the field of stability of any of its subsets (self-consistent stability criteria) requires that the stable part of a non-degenerate univariant reaction must be truncated at an invariant point. However, the stable part of a **degenerate** reaction may sometimes extend through an invariant point since such extension does not necessarily violate the self-consistent stability criteria. In the example chosen above, the degenerate reaction Qtz = Coes extends through the invariant point as it does not violate any stability criterion.

If a univariant reaction, say 1, can be expressed as a linear combination of two other equilibria, say 2 and 3, then the metastable extension of the reaction 1 must lie within the field defined by the stable parts of the reactions 2 and 3. Consider, for example, the reaction (*Ta, En*) in Fig. 6.9. It can be expressed as a linear combination of the reactions (*Ta, Fo*) and (*Ta, Atg*). Thus, the metastable extension of the reaction (*Ta, En*) past the invariant point (*Ta*) lies within the field bounded by the stable parts of the reactions (*Ta, Fo*) and (*Ta, Atg*).

6.9.3 Effect of an Excess Phase

In geological problems, we often encounter some phases that are present in excess such that these are not completely consumed by any reaction. An example is quartz in metapelite. Although quartz is involved in numerous reactions, it is never completely consumed. If a phase, say A, is present in excess, then all univaraint equilibria characterized by the absence of that phase, (*A, ...*), become unstable. Consequently, the invariant points to which these equilibria connect also become unstable. However, if an equilibrium characterized by the absence of an excess phase is degenerate, then that equilibrium becomes stable. The proof of this theorem is given by Ganguly (1968).

6.9.4 Concluding Remarks

The configuration of the phase relations derived only by the application of Schreinemakers principle is self- consistent, but not unique, and sometimes meaningless without additional constraints on the slopes and positions of at least some of the equilibria. For example, the reaction topology shown in Fig. 6.9 may be rotated by any arbitrary angle without violating the above principles. One needs to use additional information about the slopes and positions of at least some of the reactions to derive a meaningful configuration. A detailed analysis of possible self-consistent Schreinemakers' topologies in a system of C + 3 phases, with applications to geologically important systems, is given by Cheng and Greenwood (1990).

The main utility of Schreinemakers' principles lies in the fact that once some reasonable constraints about some of the reactions are available, a meaningful topology may be derived leading to new insights about geological and planetary processes. In addition, it provides a "route map" for experimental investigations of a complex system (e.g. Ganguly 1968, 1972). It allows one to choose the critical equilibria for experimental investigation that bring out the important properties of a phase diagram, and to search for conditions where other equilibria, which are yet to be determined experimentally, are to be found as the phase diagram topology gets refined. For example, referring to Fig. 6.9, if the reaction (*Ta, V*) is determined experimentally, then we know where to look for the reaction (*Ta, Fo*) in the experimental studies.

As an example of the practical utility of interactive Schreinemakers analysis, we note that the arrangement of univariant reactions around the invariant point (*Ta*) was derived by this approach along with constraints on the slopes of the reactions that could be deduced from the available thermodynamic data and some experimental information. The reaction (*Ta, Atg*) was determined experimentally by Luth (1995). This reaction was found to intersect the antigorite breakdown reaction Atg = En + Fo + V, as calculated by Bose and Ganguly (1995), at ~70 kbar pressure, generating the invariant point (*Ta*). Since the invariant point (*Ta*) consisted of five phases, it was evident that there must be three additional univariant reactions, if all are non-degenerate, radiating from the invariant point. These univariant reactions were deduced by algebraic method, and their spatial dispositions were inferred by the application of Schreinemakers' principles and thermodynamic constraints on the P-T slopes.

It was found that the P-T profile of the leading age of an old slab (≥ 50 Myr) subducting with a velocity of >3 cm/year intersects the reaction (*Ta, V*). Thus, instead of dehydrating, the mineral antigorite, which is a major water bearing phase on the leading edge of a slab, delivers all the water to the phase A, enabling transportation of water into the deep mantle. This important discovery of the possible transport pathway of water into the deep mantle by Bose and Ganguly (1995) is an example of the important geological insight that may be gained through the application of Schreinemakers' principles, coupled with experimental and thermodynamic data.

References

Aitta A (2006) Iron melting curve with a tricritical point. J Stat Mechanics: Theory and Experiment 2006:P12015

Akella J, Kennedy GC (1971) Melting of gold, silver, and copper—proposal for a new high-pressure calibration scale. J Geophys Res 76:4969–4977

Akella J, Ganguly J, Grover R, Kennedy GC (1972) Melting of lead and zinc to 60 kbar. J Phys Chem Solids 34:631–636

Anderson OL (1995) Equations of state of solids for geophysics and ceramic science. Oxford, New York, p 405

Anderson OL, Isaak DG (2000) Calculated melting curves for phases of iron. Amer Mineral 85:376–385

Belonoshko AB, Saxena SK (1992) Equations of state of fluids at high temperature and pressure (Water, Carbon Dioxide, Methane, Carbon Monoxide, Oxygen, and Hydrogen) In: Saxena SK (ed) Advances in physical geochemistry, vol 9, Springer, New York, Berlin, Heidelberg, pp 79–97

Boehler R (1993) Temperatures in the Earth's core from melting-point measurements of iron at high pressures. Nature 363:534–536

Bose K, Ganguly J (1995) Experimental and theoretical studies of the stabilities of talc, antigorite and phase A at high pressures with applications to subduction processes. Earth Planet Sci Lett 136:109–121

Burnham CW, Hollaway JR, Davis NF (1969) Thermodynamic properties of water at 1,000° C and 10,000 bars. Geol Soc Amer Special Paper 132:96

Carpenter MA (1980) Mechanism of exsolution in sodic pyroxene. Contrib Mineral Petrol 71: 289–300

Carpenter MA (1985) Order-disorder transformation in mineral solid solution. In: Kieffer SW, Navrotsky A (eds) Microscopic to macroscopic: atomic environments to mineral thermodynamics. Rev Mineral 14:187–224

Chatterjee ND (1991) Applied mineralogical thermodynamics. Springer, Berlin, Heidelberg, New York, p 321

Cheng W, Greenwood HJ (1990) Toplogical construction of nets in ternary (n + 3)-phase multisystems, with applications to Al_2O_3-SiO_2-H_2O and MgO-SiO_2-H_2O. Canad Mineral 28:305–320

Dziewonski AM, Anderson DL (1981) Preliminary reference earth model. Phys Earth Planet Inter 25:297–356

Ehrenfest P (1933) Phaseumwandlungen im üblichen und erweiterten Sinn klassifiziert nach de entsperechenden Singularitäten des thermodynamischen Potentials. Proc Amsterdam Acad 36:153

Ganguly J (1968) Analysis of the stabilities of chloritoid and staurolite, and some equilibria in the system FeO-Al_2O_3-SiO_3-H_2O-O_2. Amer J Sci 266:277–298

Ganguly J (1972) Staurolite stability and related parageneses: theory, experiments and appliations. J Petrol 13:335–365

Ganguly J, Bhattacharya PK (1987) Xenoliths in proterozoic kimberlites from southern India: petrology and geophysical implications. In: Nixon PH (ed) Mantle xenoliths, pp 249–266

Ganguly J, Singh RN, Ramana DV (1995) Thermal perturbation during charnockitization and granulite facies metamorphism in southern India. J Metamorphic Geol 13:419–430

Gilvary JJ (1957) Temperatures in the Earth's interior. J Atm Terrestr Phys 10:84–95

Hacker BR, Peacock SM, Abers G, Holloway SD (2003) Subduction factory 2. Are intermediate-depth earthquakes in subducting slabs linked to metamorphic dehydration reactions? J Geophys Res 108:11–15

Holland TJB (1980) Reaction Albite = Jadeite + Quartz determined experimentally in the range 600 to 1200 C. Amer Min 65:129–134

Holland TJB, Powell R (1991) A compensated-Redlich-Kwong (CORK) equation for volumes and fugacities of CO_2 and H_2O in the range 1 bar to 50 kbar and 100–1600 °C. Contrib Mineral Petrol 109:265–271

Huang X, Xu Y, Karato S-I (2005) Water content in the transition zone from electrical conductivity of wadselyite and ringwoodite. Nature 434:746–749

Iwamori H, McKenzie D, Takahashi E (1995) Melt generation by isentropic mantle upwelling. Earth Planet Sci Lett 134:253–266

Kennedy GC, Vaidya SN (1970) The effect of pressure on the melting temperature of solids. J Geophys Res 75:1019–1022

Kesson WH, van Laar PH (1938) Measurements of the atomic heats of tin in the superconductive and non-superconductive states. Physica 5:193–201

Komabayashi T (2014) Thermodynamics of melting relations in the system Fe-FeO at high pressure: Implications for oxygen in the Earth's core. J Geophys Res Solid Earth 119: 4164–4177

Korzhinskii DS (1959) Physicochemical basis of the analysis of the paragenesis of minerals. Consultants Bureau, New York, p 142

Kraut E, Kennedy GC (1966) New melting law at high pressures. Phys Rev 151:668–675

Landau LD, Lifschitz EM (1958) Statistical physics. Pergamon, London, Paris, p 484

Leudemann HD, Kennedy GC (1968) Melting curves of lithium, sodium, potassium and rubidium to 80 kilobars. J Geophys Res 73:2795–2805

Libby WF (1966) Melting points at high compression from zero compression properties through Kennedy relation. Phys Rev Lett 17:423–424

Lindemann FA (1910) Uber der berechnung molecularer Eigenfrequenzen. Zeit Phys 11:609–612

Luth R (1995) Is phase A relevant to the Earth's mantle? Geochim Cosmochim Acta 59:679–682

Mukherjee K (1966) Clapeyron's equation and melting under high pressures. Phys Rev Lett 17:1252–1254

Newton RC, Jayaraman A, Kennedy GC (1962) The fusion curves of the alkali metals up to 50 kilobars. J Geophys Res 67:2559–2566

Oganov AR, Brodholt JP, Price GD (2002) Ab *initio* theory of phase transitions and thermochemistry of minerals. In: Gramaccioli CM (ed) Energy modelling in minerals, European Mineralogical Union Notes in Mineralogy, vol 4, Eötvös University Press, Budapest, pp 83–160

Pippard AB (1957) Elements of classical thermodynamics for advanced students in physics. Cambridge University Press, London

Poirier J-P (1991) Introduction to the physics of the earth. Cambridge, p 264

Putnis A (1992) Introduction to mineral sciences. Cambridge, p 457

Redfern SAT, Salje E, Navrotsky A (1989) High temperature enthalpy at the orientational order-disorder transition in calcite: implications for the calcite/aragonite phase equilibrium. Contrib Mineral Petrol 101:479–484

Richet P (2001) the physical basis of thermodynamics with applications to chemistry. Kulwar Academic/Plenum Publishers, New York, p 442

Robie RA, Hemingway BS, Fisher JR (1978) Thermodynamic properties of minerals and related substances at 298.15 K and 1 bar (10^5 pascals) pressure and at higher temperatures. US Geol Surv Bull 1452:456

Sakamaki T, Suzuki A, Ohtani E (2006) Stability of hydrous melt at the base of the Earths' upper mantle. Nature 439:192–194

Salje EKH (1988) Structural phase transitions and specific heat anomalies. In: Salje EKH (ed) Physical properties and thermodynamic behaviour in minerals, NATO, pp 75–118

Salje EKH (1990) Phase transitions in ferroelastic and co-elastic crystals. Cambridge, UK

Thermal Pressure, Earth's Interior and Adiabatic Processes

7

Unwary readers should take warning that ordinary language undergoes modification to a high-pressure form when applied to the interior of the Earth; a few examples of equivalents follow: certain (high pressure form) – dubious (ordinary meaning), undoubtedly (high pressure form) – perhaps (ordinary meaning) ...

Francis Birch

In this chapter we discuss several applications of the basic thermodynamic relations developed so far to problems relating to the properties and processes in the Earth's interior, from shallow crustal level to the outer core, as well as adiabatic flow processes that have applications to a variety of processes in different natural environments. The most important source of our information about the deep interior of the Earth, which is physically inaccessible, is the velocity of seismic waves as these pass through the materials that constitute the interior. However, interpretation of the seismic wave velocities in terms of the density structure and mineralogical constitution of the Earth's interior requires an understanding of the link between seismic velocities and thermodynamic properties, which is also discussed in this chapter. The processes in the deep interior of the Earth, such as decompression and melting of rocks and rise of mantle plumes, take place under effectively adiabatic condition. These topics are discussed in this chapter from two different viewpoints, namely the adiabatic processes take place under equilibrium condition so that these are isentropic and that the adiabatic decompression is an irreversible process that leads to entropy production.

© Springer Nature Switzerland AG 2020
J. Ganguly, *Thermodynamics in Earth and Planetary Sciences*,
Springer Textbooks in Earth Sciences, Geography and Environment,
https://doi.org/10.1007/978-3-030-20879-0_7

7.1 Thermal Pressure

7.1.1 Thermodynamic Relations

The thermal pressure is defined as the pressure change associated with temperature change at a constant volume. The expression for thermal pressure can be derived as follows. The P-V-T relation of a substance can be expressed in the functional form $f(P,V,T) = 0$. This type of function, which is known as an implicit function, leads to the relation (see Appendix B.4)

$$\left(\frac{\partial P}{\partial T}\right)_V = -\frac{\left(\frac{\partial P}{\partial V}\right)_T}{\left(\frac{\partial T}{\partial V}\right)_P} = -\frac{\left(\frac{\partial V}{\partial T}\right)_P}{\left(\frac{\partial V}{\partial P}\right)_T} \qquad (7.1.1)$$

Using Eqs. (3.7.1) and (3.7.2), the ratio of derivatives on the right equals $-\alpha/\beta_T$. Thus, we have,

$$\left(\frac{\partial P}{\partial T}\right)_V = \frac{\alpha}{\beta_T} = \alpha k_T \qquad (7.1.2)$$

or

$$\Delta P_{th} = \int_{T_1}^{T_2} (\alpha k_T) dT \qquad (7.1.3)$$

where ΔP_{th} stands for thermal pressure

Both α and k_T are affected by temperature changes, but the value of αk_T of a solid seems to be fairly insensitive to temperature change above its Debye temperature, θ_D, in many cases (Anderson 1995; also see Sect. 5.4.2.2 for further discussion about deviation from this relation; the concept of Debye temperature is explained in Sect. 1.6). The behavior of αk_T for a few compounds of geophysical interest is illustrated in Fig. 7.1. Thus, the thermal pressure of a solid due to temperature change above the Debye temperature may be approximated by assuming a constant value of αk_T, as

$$\Delta P_{th} = P(T) - P(\theta_D) = \alpha k_T(T - \theta_D) \qquad (7.1.4)$$

where $P(\theta_D)$ is the pressure at the Debye temperature. We discuss below two examples of the application of the concept of thermal pressure in Geophysical and Geological problems (an additional example has been discussed in Sect. 5.4.2.2).

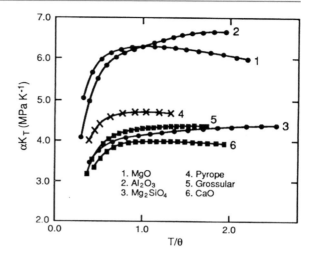

Fig. 7.1 The behavior of αk_T for a few compounds of geophysical interest as a function of temperature normalized to the Debye temperature, θ_D. From Anderson (1995)

Problem 7.1 Making some assumptions, the thermal pressure can be related to the thermodynamic Grüneisen parameter, Γ_{th} (see Sect. 3.8), as follows:

$$\Delta P_{th} = \Gamma_{th} \frac{\Delta U}{V} \qquad (7.1.5)$$

This is the well-known **Mie-Grüneissen equation of state**. Derive this relation starting with Eq. (7.1.3) and using the thermodynamic Grüneisen parameter as defined by Eq. (3.8.2).

7.1.2 Core of the Earth

The Earth's core represents 32% of its mass and is about half of its radius. It is generally accepted from geophysical data and observations of meteorite samples that the core of the Earth is made primarily of iron, with a molten outer and a solid inner components. The radii of the outer limits of the inner and outer cores are at 1221 and 3480 km, respectively, corresponding to depths of 5155 and 2885 km. In this section, we examine if we should expect some light alloying elements to be dissolved in the core of the Earth.

Anderson (1995) calculated the ρ versus P relations of three polymorphs of iron at 300 K, α (bcc: body centered cubic), ε (hcp: hexagonal close pack) and γ (fcc: face centered cubic) for which sufficient data are available, and compared the results with the ρ versus P relation constrained for the core by the **P**reliminary **R**eference **E**arth **M**odel or **PREM** (Dziewonski and Anderson 1981) that is deduced from the geophysical data. The results are shown in Fig. 7.2a.

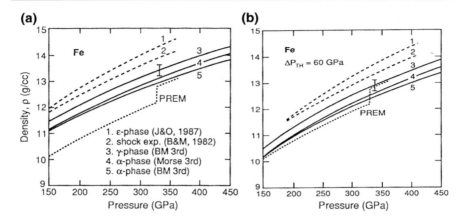

Fig. 7.2 Density versus pressure relation of different polymorphs of iron at (**a**) 300 K and (**b**) after correction for thermal pressure, and comparison with the Preliminary Reference Earth Model (PREM) of Dziewonski and Anderson (1981). The thermal pressure arises from the difference between the temperature at the Earth's core and 300 K. The density jump in the PREM is due to the transition from liquid inner to solid outer core. From Anderson (1995)

Phase diagram for the P-T stabilities of the different iron polymorphs (Anderson 2000; Saxena and Dubrovinsky 2000), developed on the basis of the available experimental data, suggests that of the three polymorphs for which ρ versus P relations are shown in Fig. 7.2a, only ε-iron could be stable at the core pressure. However, its density at 300 K and the pressure of the solid inner core (dashed curves) is much higher than the density of inner core in the PREM. At least a part of this discrepancy is due to the fact that the temperature of the inner core is ~ 5000 K (Fig. 7.7). The adjustment of density of the ε-iron can be carried out if the coefficient of thermal expansion of iron is known at core pressures. Alternatively, we can adjust the pressure at a fixed density by adding the thermal pressure resulting from the difference between the core temperature and 300 K, taking advantage of the fact the αk_T is rather insensitive to temperature above the Debye temperature.

Using Eq. (7.1.3), and assuming a temperature of 5500 ± 500 K for the inner/outer core boundary, Anderson (1995) calculated $\Delta P_{th} \sim 58 \pm 5.2$ GPa. Figure 7.2b was generated by him from Fig. 7.2a, using $\Delta P_{th} = 60$ GPa to illustrate the density versus pressure relation of iron polymorphs at the temperature of Earth's core. It is found that the density of ε-iron is still significantly higher than the PREM density of the inner core. Thus, it seems very likely that there are light alloying elements in the inner core. For several reasons, a primary candidate seems to be S (Anderson 1989; Li and Fei 2003). The phase diagram of iron at the core pressure is still uncertain. If some other polymorph of iron other than the ε (hcp) phase is stable at the core pressure, then its density should be higher than that of the latter, thereby making the case for light alloying elements even stronger.

Box 7.1: A Brief Overview of the Earth's Interior The interior of the Earth is subdivided into four major components, **Crust, Mantle, Outer Core** and **Inner core**, on the basis of the observed discontinuities in the velocities of seismic body waves, which are known as the P (or longitudinal) and S (or shear) waves. The S waves cannot pass through liquid.

The major subdivisions of the Earth's interior are illustrated in Fig. B.7.1. The core consists primarily of iron, and the outer core is known to be in a liquid state as it does not transmit S-waves. The mantle of the Earth, which is essentially made up of minerals belonging to the system $MgO-FeO-CaO-Al_2O_3-SiO_2$, is subdivided into upper and lower mantles, which are separated by a **transition zone**. Within the latter, the velocities of both P and S waves change much more rapidly relative to their changes in the upper and lower mantle.

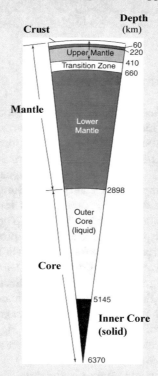

Fig. B.7.1 From Winter (2001)

There is a ~ 200 km thick layer (with irregular upper boundary) above the core-mantle boundary, generally referred to as **D″ (D double prime) layer**, defined by seismic discontinuities and within which the seismic properties are anisotropic.

In the upper mantle, there is a narrow zone between 60 and 220 km, within which both P and S waves slow down relative to the velocities above and below. This is known as the **Low velocity zone**, which may consist of a small amount of melt. The most abundant minerals in the upper and lower mantles are olivine, $(Mg,Fe)_2SiO_4$, and perovskite, $(Mg,Fe)SiO_3$, respectively.

There is a sharp temperature jump from the lower mantle to the outer core (Fig. 7.7) and possibly also between the upper and lower mantle through the transition zone. These narrow zones connecting two different thermal regimes are known as **thermal boundary layers**. The **acceleration due to gravity**, g, is nearly constant at ~ 10 m/s^2 from the surface down to the core/mantle boundary (range 9.8–10.6 m/s^2), after which it decreases almost linearly to 0 at the center. Pressure at the core of the Earth is 3.64 Mbar. At the core/mantle boundary, the density changes from 5.56 to 9.90 g/cm^3.

7.1.3 Magma-Hydrothermal System

We next consider the problem of thermal pressure associated with the cooling and crystallization of a tonalite magma, and its consequences. Tonalite is a granitic rock that is made primarily of quartz and feldspar, but the percentage of plagioclase feldspars, $NaAlSi_3O_8$–$CaAl_2Si_2O_8$ solid solution, is much higher than that of alkali feldspars, $(Na,K)AlSi_3O_8$. This problem was analyzed by Knapp and Norton (1981), who calculated, using experimental data of Burnham and Davis (1971), the thermal pressure that would be generated when a tonalite magma with 4 wt% dissolved H_2O intrudes a country rock and begins to crystallize. All magmas that are rich in quartz and feldspar components have approximately this level of dissolved H_2O when they intrude into rocks in the Earth's crust (The dissolution of H_2O in water involves dissociation according to $H_2O + O^{2-} = 2(OH)^{-1}$, where the O^{2-} is a polymerized oxygen in the magma (see Sect. 8.9). Thus, the specified amount of H_2O represents the amount that the magma had dissolved, and not the amount of molecular water actually present in the magma after dissociation).

The P-T phase diagram of the system is shown in Fig. 7.3. There are four fields: (a) a field of only liquid above the liquidus curve, (b) a field of XLS (crystals) + magma just below the liquidus curve, (c) a narrow field of XLS + magma + fluid, which we would refer to as the three-phase field, and (d) a field of completely crystalline tonalite rock plus H_2O fluid below the solidus curve (A solidus is the temperature at which melting begins in a multiphase assemblage whereas the liquidus is the temperature at which the melting of the assemblage goes to completion). As the magma cools below the liquidus, it begins to crystallize, but without giving off any water until the temperature of the magma falls within the three-phase field. At this point, the residual melt becomes saturated with H_2O because of its diminishing mass while the mass of dissolved H_2O remains constant. Thus, any

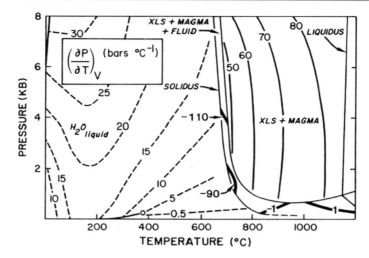

Fig. 7.3 Phase diagram of tonalite with 4 wt% H_2O. The system is completely crystalline at temperatures below the solidus and completely molten at those above the liquidus. Crystallization begins after the magma cools below the liquidus, and is completed when the temperature at fixed pressure decreases to that of the solidus. Below the solidus, the system consists of a tonalite rock and exsolved H_2O. The narrow band just above the solidus defines a domain within which crystals and melt coexists with a H_2O fluid (labeled as XLS + MAGMA + FLUID). Above this narrow band, we have crystals + fluid until the liquidus condition. The solid contours indicate the net thermal pressure, $(\partial P/\partial T)_V$, for the stable assemblage in a given field. The dashed contours are thermal pressures of H_2O. From Knapp and Norton (1981) with modifications by Norton (pers.com)

further cooling leads to the exsolution or expulsion of H_2O from the melt. Complete crystallization of the system takes place when it cools below the solidus temperature.

Figure 7.3 also shows the contours of thermal pressure in each field, as calculated by Knapp and Norton (1981) and Norton (personal communication). Except for the dashed contours below the solidus, which are for H_2O, all thermal pressure contours are for the assemblage of the stable phases in a particular field. It is interesting to note that within the three-phase field, the thermal pressures are negative, and are, on the average around -100 bars per degree. This negative thermal pressure is a consequence of the progressive expulsion of dissolved H_2O from the magma with decreasing temperature, which leads to an increase of volume for the system of XLS + magma + fluid, and thus a negative value of the coefficient of thermal expansion, α, for the composite system.

The P-T paths of ascending magmas of constant volume, with 4 wt% dissolved H_2O (Knapp and Norton 1981), are illustrated in Fig. 7.4. The initial paths within the two phase field show large pressure drop reflecting moderately large positive values of the thermal pressure. Once within the three phase field, there is increase of pressure within the system as it cools because of negative thermal pressure, but the increase is small until a pressure of 1 kb is reached. Beyond this point, there is rapid

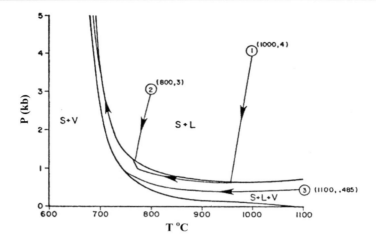

Fig. 7.4 P-T paths (lines with arrows) at constant volume of cooling tonalitic magma systems with 4 wt% water, as determined by the thermal pressure of the system within the fields of stabilities of different phase assemblages (see this figure). Within the two phase field (solid (s) + liquid (l)), cooling leads to sharp pressure drop because of the large positive values of the thermal pressure. As the magma cools into the three phase field, pressure goes up with further cooling because of negative values of the thermal pressure, and especially sharply when the pressure exceeds 1 kb. The parenthetical numbers beside the initial locations of the different magma systems indicate their initial T (°C), P (kb) coordinates. From Knapp and Norton (1981)

increase of pressure due to cooling of the three phase system because of fairly large magnitude of negative thermal pressure (Fig. 7.3). As a consequence, the system is driven away from the solidus curve which delays its complete freezing.

The continued buildup of overpressure within the three phase system, combined with the effect of thermal expansion of trapped water, which is already present in the country rock, as a consequence of heating by the intrusive body leads to the development of extensive fracture networks within the country rock. Magmatic fluids that scavenge ore forming metals from the magma become focused in these fractures (Norton 1978), and thus lead to the formation of ore deposits. Furthermore, intrusion of the magma that is prevented from complete freezing due to the buildup of overpressure leads to the formation of dike and sill like bodies.

7.2   Adiabatic Temperature Gradient

This is a topic of considerable geological and geophysical interest, and provides the framework for our understanding the thermal gradient in the Earth's mantle and core, and in magma chambers. From the second law (Eq. 2.4.6), an adiabatic process ($\delta q = 0$) under **reversible** condition is an isentropic process. Thus, for equilibrium condition, the problem of deriving an expression for the adiabatic temperature gradient reduces to finding an expression for $(\partial T/\partial P)_S$. This is easily achieved by writing the total differential of S in terms of T and P, and imposing the condition that $dS = 0$, as follows.

$$dS = \left(\frac{\partial S}{\partial T}\right)_P dT + \left(\frac{\partial S}{\partial P}\right)_T dP \qquad (7.2.1)$$

The first parenthetical term equals C_P/T (Eq. 3.7.5), whereas, using Maxwell relation (Eq. 3.4.4), the second parenthetical term equals $-(\partial V/\partial T)_P$. Using the definition of the coefficient of thermal expansion, α (Eq. 3.7.1), the last term equals $-\alpha V$. Thus, the above equation reduces to

$$dS = \left(\frac{C_P}{T}\right) dT - (\alpha V) dP \qquad (7.2.2)$$

so that, for isentropic condition ($dS = 0$),

$$\left(\frac{\partial T}{\partial P}\right)_S = \frac{VT\alpha}{C_P} \qquad (7.2.3)$$

Dividing both numerator and denominator by the molecular weight (or the weighted mean molecular weight if the system consists of several phases), M, and noting that $V/M = 1/\rho$, where ρ is the density, and using the symbol C'_p for the **specific** heat capacity C_p/M, we have

$$\left(\frac{\partial T}{\partial P}\right)_S = \frac{T\alpha}{\rho C'_p} \qquad (7.2.4)$$

For a vertical column of material under hydrostatic condition, for which $dP = gdZ$, where g is the acceleration of gravity and Z is the depth (+ve downwards), we then have

$$\left(\frac{\partial T}{\partial Z}\right)_S = \frac{gT\alpha}{C'_p} \qquad (7.2.5a)$$

or

$$\left(\frac{\partial \ln T}{\partial Z}\right)_S = \frac{\alpha g}{C'_p} \qquad (7.2.5b)$$

An expression for $(\partial \ln T/\partial Z)_S$ was derived earlier, Eq. (3.8.9), in terms of the thermodynamic Grüneisen parameter, Γ_{th}. It is left to the reader to verify that the last equation is equivalent to Eq. (3.8.9).

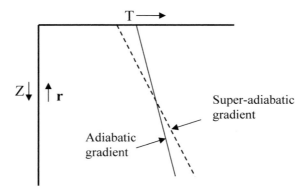

Fig. 7.5 Schematic illustration of adiabatic and super-adiabatic gradients. Z is depth increasing downwards. r is the radius of the Earth increasing upwards. Convection tends to establish an isentropic gradient which is equivalent to the adiabatic gradient since at equilibrium dS = 0 when δq = 0

Equation (7.2.5) is usually used to calculate the temperature change with pressure in an efficiently convecting system. By definition, an isentropic temperature gradient represents the **equilibrium** adiabatic temperature gradient in a vertical column of material. Thus, when the actual temperature gradient in a medium exceeds the isentropic gradient, it tends to convect in order to bring the temperature gradient to the latter condition. If the material properties of the system are such that these offer little resistance to convection, then the actual temperature gradient in the medium should be only slightly superadiabatic (Fig. 7.5). This is because the medium would convect to bring about the equilibrium temperature distribution when there is only a small departure from equilibrium. Thus, the isentropic temperature gradient represents a good approximation of the actual temperature gradient in an *efficiently* convecting medium.

7.3   Temperature Gradients in the Earth's Mantle and Outer Core

7.3.1   Upper Mantle

It is generally accepted that Earth's mantle below ∼200 km depth and the outer core are convecting very efficiently so that the temperature gradients in these regimes are only slightly super-adiabatic. Furthermore, the conductive heat loss from these domains is very slow so that these may be assumed to be under effectively adiabatic conditions. One could, thus, approximate the temperature gradients in the sub-lithospheric mantle (i.e. below ∼200 km depth) and in the

Fig. 7.6 Steady state geotherms in the crust and shallow upper mantle in the oceanic and continental environments, as calculated by Turcotte and Schubert (1982). The temperature profile below ~ 200 km, which is the same in both environments, corresponds to the adiabatic (isentropic) gradient (Eq. 7.2.5). There are, however, significant regional variations of the geotherms

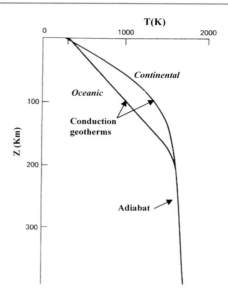

outer core by their respective isentropic gradients. These temperature gradients are often referred to as mantle- and core-**adiabats**, respectively.

Figure 7.6 shows two steady state geotherms in the crust and shallow upper mantle in the oceanic and continental environments, as calculated by Turcotte and Schubert (1982). The temperature profile below ~ 200 km, which is the same in both environments, corresponds to the adiabatic gradient (Eq. 7.2.5). There are, of course, regional variations of the geotherms but the general feature is the same. Efficient mantle convection below 200 km or so maintains the temperature gradient close to the isentropic gradient. The top part, where there is rapid temperature change as a function of depth, represents a thermal boundary layer. Here the dominant mode of heat transfer is conduction. The surface temperature is assumed to have remained fixed at 298 K.

We can calculate an approximate average value of the isentropic or adiabatic gradient in the upper mantle using the following average values of the mantle rocks that consist typically of 60 vol% olivine, 20 vol% orthopyroxene, 10 vol% garnet and 10 vol% clinopyroxene: $\alpha = 5.2 \times 10^{-5}$ K^{-1}, $C'_p = 1.214$ kJ-kg^{-1}-K^{-1}, $g = 10$ m-s^{-2}. To have dimensional compatibility, we first change the unit of C'_p by noting that 1 J = 1 N m = 1 (kg-m-s^{-2}) (m), so that $C'_p = 1214$ m^2-s^{-2}-K^{-1}. Substitution of these values in Eq. (7.2.5) yields

$$(\partial T / \partial Z)s = 0.64 \text{ K/km}$$

at T = 1600 K (The average thermochemical properties are calculated from the data in Saxena et al. 1993).

7.3.2   Lower Mantle and Core

A number of workers have tried to determine the adiabatic temperature profile in the lower mantle. Here we discuss the work of Brown and Shankland (1981) who calculated the isentropic thermal profile using geophysical data. They first calculated entropies of the mantle minerals from seismic velocities, inferred densities at the mantle conditions, and assigned values of temperatures. The relationship among entropy and the other variables are given by a modified version of the Debye theory of lattice vibrations (the general idea of Debye theory is discussed Sects. 4.2 and 14.5). Brown and Shankland (1981) assigned a temperature to the 660 km seismic discontinuity that defines the top of the lower mantle, calculated the entropy, and then found the temperature at a greater depth that yields the same entropy. They calculated several isentropic temperature profiles, but preferred the one based on a 1600 °C temperature at the 660 km discontinuity as this temperature seems most compatible with the P-T condition of the phase transition (spinel to perovskite) that is generally accepted to be the major reason for the observed seismic discontinuity. The upper mantle adiabat calculated by Brown and Shankland (1981) using 1600 °C temperature at 670 km depth is shown in Fig. 7.7.

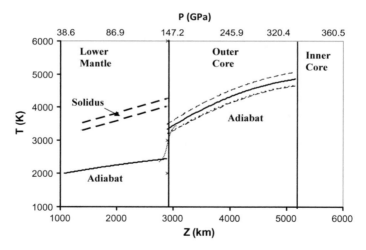

Fig. 7.7 Adiabatic temperature profiles in the Earth's outer core and lower mantle, and solidus temperature of the upper mantle. The mantle adiabat is from Brown and Shankland (1981) using T = 1873 K at the 670 km discontinuity. The upper mantle solidus is from Boehler (1993): upper curve: extrapolation of experimental data in diamond cell up to 2 GPa; lower curve: extrapolation of experimental data in multianvil apparatus up to 1.6 GPa. The outer core profiles are calculated by fixing temperature of 4850 ± 200 K at the inner core/outer core boundary, as suggested by Boehler (1993) on the basis of his experimental data on the melting temperature in the Fe-O system up to 200 GPa. Extrapolations of the adiabats to the core/mantle boundary yield temperature discontinuity of ∼895 ± 140 K. The temperatures of the core-mantle boundary given by the mantle and core adiabats are 2450 and 3345 ± 140 K, respectively. The smoothing of the discontinuity by a dotted line is schematic. The numbers on the upper horizontal axis are pressures in GPa at the depths shown on the lower horizontal axis

We now calculate the adiabatic thermal profile of the outer core using Eq. (7.2.5). For this purpose, we need to account for the variation of g as a function of depth in the outer core, and also for the significant variation of material properties. The g varies from 1068 cm/s^2 at the core-mantle boundary to 440 cm/s^2 at the inner core-outer core boundary. The g versus Z data for the outer core tabulated in the PREM (Dziewonski and Anderson 1981) can be fitted almost exactly by the polynomial relation g = 15.458 − 0.001Z − 2(10^{-07})Z^2, with Z in km and g in m/s^2. Thus, assuming α and C_p' to be constants, integration of Eq. (7.2.5b) yields

$$\ln \text{T}(Z_2) = \ln \text{T}(Z_1) + \frac{\alpha}{C_p'} \left[(15.458)Z - 5(10^{-4})Z^2 - \frac{2(10^{-7})Z^3}{3} \right] \quad (7.3.1)$$

Brown and Shankland (1981) showed that α of outer core varies between 13.2 (10^{-6})/K at the core-mantle (C/M) boundary and 7.9(10^{-6})/K at the outer core-inner core (Oc/Ic) boundary. The C_p' data can be calculated from their inferred values of C_V value of outer core of 27.66 J/mol-K and Gruneisen parameter (Γ_{th}) of the outer core. They found the latter to vary from 1.66 at the top to 0.94 to the bottom of the outer core. The relationship among C_p, C_v and Γ is given by Eq. (3.8.1) as $C_p = C_v(1 + \alpha T \Gamma_{th})$. A finite difference scheme needs to be used to calculate the thermal profile of the outer core according to the above equation in order to allow for the variation of α and C_p'. For this purpose, the outer core is divided into a series of small concentric shells, and the temperature profile within each shell is calculated using the properties at the lower limit of the shell, beginning with the inner core (Ic)-outer core (Oc) boundary.

The temperature at the Ic/Oc boundary is fixed by the melting temperature of iron in the presence of postulated impurities in the core that seem to be primarily O and S. In the absence of direct experimental determination of melting temperature of iron at such high pressure ($\sim$3 Mbar), a number of attempts have been made over many years to estimate the temperature by extrapolation from lower pressure data. Also, there are very significant differences among the lower pressure data because of experimental difficulty of detecting the onset of melting. All these estimates have resulted in a scatter of melting temperature of iron between 4000 and 8000 K at 3 Mbar pressure. Perhaps the most reliable experimental data are by Boehler (1993), who determined the melting temperature in the system Fe-FeO with 30 wt% of the oxide component at static pressure up to 2 Mbar in a laser heated diamond cell assemblage. From extrapolation of his data, Boehler (1993) suggests a temperature of 4850 ± 200 K at the Ic/Oc boundary (The extrapolation between 2 and 3 Mbar pressure is based on the Kraut-Kennedy melting relation that is discussed in Sect. 6.6.1). On the other hand, Saxena and Eriksson (2015) suggested a maximum temperature of $\sim$4200 ($\pm$500) K for the Ic/Oc boundary on the basis of calculated phase relation in the Fe-S system at the pressure of the boundary (3.28 MPa). For the purpose of this discussion of the thermal profile in the outer core, we fix the temperature at the Ic/Oc boundary using Boehler's data for the Fe-O system. Changing the boundary temperature by a few hundred degrees does not significantly change the gradient of the thermal profile. Thus, if a different temperature at the Ic/Oc interface seems to be more appropriate later, then the calculated thermal profile in the outer core may be linked to it if the revised temperature is within a few hundred degrees of what has been used in this calculation.

The adiabatic temperature profile of the outer core calculated by fixing the temperature at the Ic/Oc boundary at Boehler's suggested value is illustrated in Fig. 7.7. The lower mantle and core adiabats intersect the C/M boundary at 2450 and 3345 (±140) K, respectively, causing a temperature jump of 895 (±140) K. A sharp temperature jump at C/M boundary causes major heat flux (which is proportional to the temperature gradient) out of the Earth's core, thereby leading to crystallization and growth of the inner core. The latent heat released by the crystallization, however, helps restore the temperature of the outer core. Heat conduction across the core-mantle boundary would, however, somewhat smoothen the temperature discontinuity, as schematically illustrated in Fig. 7.7.

It has been reported (Murakami et al. 2004; Oganov and Ono 2004) that the perovskite phase, which constitutes the bulk of the lower mantle, undergoes transformation to a denser phase, known as the post-perovskite phase (orthorhombic, space group Cmcm), at ~125 GPa (~2700 km depth) at temperature on the lower mantle adiabat. The transition boundary has a large positive slope which implies that the post-perovskite phase has lower entropy than the perovskite phase (see Sect. 6.4). Consequently, the temperature on the adiabat in the post-perovskite field must increase to restore the isentropic condition. This will reduce the magnitude of temperature jump at the core-mantle boundary. Additionally, dissolution of other light alloying elements such as S and C in the core could reduce the melting temperature at the Ic/Oc boundary (e.g. Saxena and Eriksson 2015) and thus the temperature jump at the core-mantle boundary.

The zone over which temperature changes between two adjacent thermal domains maintained at different temperatures is commonly referred to as a **thermal boundary layer (TBL)**. Sharp thermal boundary layers may develop instabilities, if certain dynamical and physical conditions are satisfied. Seismic evidence and numerical simulations (e.g. Olson 1987; Kellogg 1997) strongly suggest formation of thermal plumes in the TBL at the core-mantle boundary (another TBL that could be a source of mantle plumes is at the boundary between the upper and lower mantle). These thermal plumes ascend nearly adiabatically within the Earth and play very important roles in the mantle dynamics. Extensive volcanisms and "hot spots" (e.g. Iceland, Hawaii) on the surface of the Earth that cannot be directly related to plate-tectonics are most likely the result of intersection of the adiabatic P-T path of the mantle plumes with the mantle solidus (see below) and piercing of the ascending magma through the surface.

Figure 7.7 also shows the solidus of the upper mantle as estimated by Boehler (1993) from extrapolation of experimental data up to 2 Mbar. We find that the solidus temperature at the C/M boundary is at least ~550 K higher than the maximum possible temperature of the mantle. Thus, according to Boehler's data, there should be no melting of lower mantle at the C/M boundary or in the D″ layer.

This conclusion is at variance with that of Zerr et al. (1998) who used Boehler's data, but inferred a temperature of the outer core at the C/M boundary of 4000 ± 200 K.

The thermal profile in the Earth's core has been a highly debated issue, stemming from the uncertainties in our knowledge of (a) the nature and amount of light alloying elements that could bring the density of iron plus the additional elements at the P-T conditions of the core to be in agreement with the PREM density for the core (Dziewonski and Anderson 1981), (b) the experimental data on melting temperature at high P-T conditions and extrapolation of the latter to the pressure of Oc/Ic boundary. The discussion presented above is meant to show the role of thermodynamics in understanding the thermal structure of the outer core and the temperature jump at the core-mantle boundary along with their potential geophysical ramifications.

7.4   Isentropic Melting in the Earth's Interior

One of the most important mechanisms of magma generation in the Earth's interior is the upwelling of the rocks in the upper mantle resulting from local perturbations, such as stretching of the lithosphere, or local variation of density. Because of the low thermal diffusivity of the rocks relative to the upward velocity of mantle material, the upwelling process remains essentially adiabatic.[1] To analyze the melting process in the upwelling mantle material, one could begin by assuming that the entropy production within the decompressing mantle material is negligible. This cannot be strictly true but it provides a useful starting point (some aspect of the effect of entropy production is addressed later). In this case, the temperature of the upwelling material would follow the isentropic gradient which is $\sim 0.5 - 1.0$ C/kb. Figure 7.8, which is reproduced from Iwamori et al. (1995), shows the isentropic P-T trajectories of rocks within the Earth's mantle with different initial temperatures, as calculated from Eq. (7.2.5), and the variation of solidus (beginning of melting) and liquidus (end of melting) temperatures of the mantle peridotite as functions of pressure. Also shown by dotted lines between the solidus and liquidus are the contours with constant mass fractions of the melt. The trajectories within the solidus and liquidus are also constructed to conserve entropy without any loss or segregation of the melt phase. The intersection of the solidus and isentropic P-T trajectory of the mantle rocks roughly defines the depth of magma generation within the Earth's mantle.

---

[1]Whether or not there will be any significant heat loss from an upwelling material depends on the value of a dimensionless parameter, known as the Peclet number (Pe), which is given by Pe = vl/k, where v is the upward velocity, l is the distance traveled and k is the thermal diffusivity of the material. There is no significant heat loss when Pe is significantly greater than unity. For mantle material Pe ~ 30 (McKenzie and Bickle 1988).

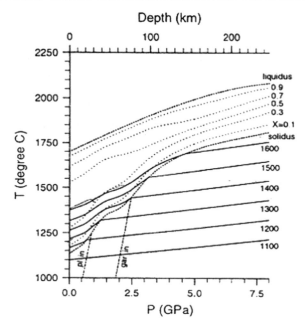

Fig. 7.8 Isentropic P-T trajectories of rocks within the Earth's mantle with different initial temperatures, as calculated from Eq. (7.2.5), and the variation of solidus and liquidus of the mantle peridotite as functions of pressure. The dotted lines between the solidus and liquidus are the contours with constant mass fractions of the melt. The trajectories within the solidus and liquidus are constructed to conserve entropy without any loss or segregation of the melt phase. From Iwamori et al. (1995). With permission from Elsevier

A number on an isentropic trajectory in Fig. 7.8 is the temperature of projection of the trajectory on to the Earth's surface. Following McKenzie and Bickle (1988), these are commonly referred to as the "potential temperature", T_p, of the mantle rocks. As long as there is no significant entropy production due to irreversible processes, rocks at different depths along an adiabat within the mantle have the same potential temperature. These rocks would undergo essentially the same extent of partial melting even though their initial temperatures are different. Only rocks with significantly different T_P values are considered to undergo significantly different extent of partial melting (We would return to this interesting topic in Sect. 7.8 to account for the effect of entropy production associated with irreversible decompression).

An important geological question concerns the melt productivity during the melting process, as the magma plus residual solid moves towards the surface. A geometric analysis of the problem, as presented by Stolper (1996), is simple to follow and is summarized below. For simplicity, consider the melting of a single phase, as illustrated in Fig. 7.9a. The solid that has upwelled following an isentropic gradient meets the melting curve at the point 2. If thermodynamic

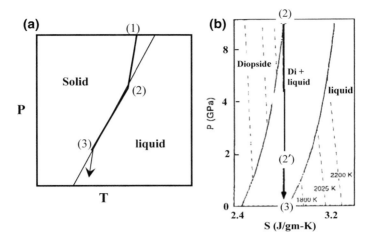

Fig. 7.9 Melting of diopside during decompression to shallower depth. (**a**) Schematic P-T diagram showing the fields of stability of the solid and liquid phases, and the path of decompression, labelled (1)-(2)-(3). (**b**) Progressive melting under isentropic condition. Adapted from Stolper (1996)

equilibrium is maintained closely, then during further upwelling the mixture of solid plus melt would follow the melting curve until all the solid transforms to melt at the point 3 (This is due to the restriction imposed by the Phase Rule that is discussed in Sect. 6.1. According to the Phase Rule, there is one degree of freedom in a system consisting of two phases of the same composition – in this problem a solid and melt of the same composition – so that either pressure or temperature can be varied independently. Here pressure is an independent variable that is changing due to decompression process, and consequently the temperature changes along the melting curve in accordance with the pressure change).

We now define a term '**melt productivity**' as the change in the amount of melt fraction, x_m, as a function of pressure during decompression, and ask the following question: Does the melt productivity decrease or increase or remain the same as the system moves along the univariant melting curve from 2 to 3 under isentropic condition? This question can be addressed by recasting the phase diagram in a pressure-entropy space and following the isentropic decompression path (This type of representation of a phase diagram in the T-S space and their usefulness were first discussed by Thompson, 1970). For this purpose, one needs to first calculate the entropy change of both diopside and liquid along the melting curve. The calculations carried out by Stolper (1996) using the available thermodynamic data are shown in Fig. 7.9b. Now consider an isentropic melting beginning at a pressure of 7 GPa. When the mixture of solid plus melt reaches the point 2′, the masses of solid and melt are equal, as can be determined by application of "lever rule" (see Sect. 10.8.1). At the point 3, the system is completely molten. Notice now that the melt fraction (x) has increased by the same amount from 2 to 2′ as between 2′ and 3, but the pressure change between 2 and 2′ is much greater than that between 2′ and

Fig. 7.10 Illustration of the effect of phase transformation in a one-component system, SiO₂, on isentropic decompression-melting behavior. C: coesite; Q: quartz; L: liquid. From Ghiorso (1997). With permission from Annual Reviews

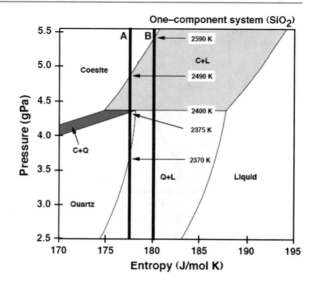

3. In other words, $\Delta x_m/\Delta P$ between 2′ and 3 is much greater than that between 2 and 2′. Thus, we arrive at the interesting conclusion that the melt productivity rapidly increases during isentropic decompression-melting in a single component system. With this insight from a single component system, an elaborate analysis of the problem was carried out for realistic multicomponent and multiphase mantle material, namely peridotite, by Asimov et al. (1997), and the melt productivity was found to increase rapidly with pressure drop as long as there is no subsolidus reaction or phase transformation within the mineral assemblage.

Let us now consider the isentropic decompression-melting behavior in which the solid phases undergo phase transition during the decompression process. For the sake of simplicity, we consider the decompression-melting behavior of the one component system SiO₂ that was analyzed by Asimov et al. (1995), and also discussed by Ghiorso (1997). The phase diagram of the system in a pressure-entropy space is shown in Fig. 7.10. As the high pressure SiO₂ polymorph, coesite, decompresses along the isentropic path A, melting begins at 2490 K, followed by progressive increase of melt fraction that can be determined by lever rule. However, when the system cools to 2400 K, the low pressure polymorph of SiO₂, namely quartz, appears in the system leading to an invariant condition since there are now three phases in the system, viz. coesite, quartz and liquid (see Phase Rule, Sect. 6.1). Further isentropic decompression can only take place after all liquid crystallizes to quartz. Below 2375 K, all of coesite converts to quartz, if equilibrium is maintained. Melting commences again below 2370 K with increase of melt productivity during decompression. Thus, along the isentropic path A, there are two generations of melt separated by a barren zone of ~ 0.6 GPa or ~ 18 km.

Along path B, the system partly freezes at 2400 K since the new phase quartz appearing at this temperature by the transformation coesite $\rightarrow$ quartz has higher entropy than coesite, and this increase of entropy must be compensated by decreasing the amount of liquid. However, after complete conversion of coesite to quartz at 2400 K, the melt productivity increases again as the system decompresses further within the field of quartz plus liquid. From this simple example, we find that phase transformations of solids during decompression impart an episodic character or oscillation to melt productivity if the decompression follows an isentropic path. One can see this qualitatively by noting that during decompression a solid phase or an assemblage of solid phases transforms to a higher entropy assemblage of solids if the transformation boundary has a positive slope in the P-T space (the relationship among entropy change, volume change and P-T slope is discussed in Sect. 6.4.2; since the lower pressure assemblage has a larger volume than the higher pressure one, it must also have a larger entropy than the latter in order that the ratio $\Delta V/\Delta S > 0$). Consequently, there must be a decrease of melt fraction so that the entropy remains constant.

7.5 The Earth's Mantle and Core: Linking Thermodynamics and Seismic Velocities

7.5.1 Relations Among Elastic Properties and Sound Velocities

It is interesting to note that the dynamic properties of a substance like its sound or seismic wave velocities are related to its elastic moduli, which represent static properties. This has important implications in the measurement of elastic properties using sound velocities, and in the interpretation of the seismic wave velocities in the earth's interior in terms of their mineralogical properties.

It can be shown that under isentropic condition, the sound velocity, υ, of a solid is related to its change of density with pressure according to (e.g. Zeldovich and Razier 1966; Kieffer and Delany 1979).

$$\upsilon^2 = \left(\frac{\partial P}{\partial \rho}\right)_s, \tag{7.5.1}$$

Now, from Eq. (3.7.3), we have

$$k_s = \frac{1}{\beta_s} = -V\left(\frac{\partial P}{\partial V}\right)_s$$

But, since from the relation between mass, m, and density, ρ, (i.e. $V = m/\rho$), $dV = -(m/\rho^2)d\rho$, the above equation reduces to

$$k_s = \rho \left(\frac{\partial P}{\partial \rho} \right)_s \qquad (7.5.2)$$

so that from Eq. (7.5.1)

$$\upsilon^2 = \frac{k_S}{\rho}$$

Thus, the adiabatic bulk modulus of a substance can be determined from measurement of its ultrasonic sound velocity. The entropy production during the adiabatic passage of a sound wave through a substance is usually not significant (Kieffer 1977).

In general, a wave velocity in a medium is related to its elastic properties according to the form of the above equation, in which a different elastic modulus or a combination of elastic moduli may take the place of k_S. Specifically, for the velocities of the longitudinal (υ_P) and the shear (υ_S) components of a seismic wave, we have

$$v_p^2 = \frac{k_s + \frac{4}{3}\pi'}{\rho} \qquad (7.5.3)$$

and

$$v_s^2 = \frac{\pi'}{\rho} \qquad (7.5.4)$$

where π' is the shear modulus. Combining these two equations, we have

$$\upsilon_P^2 - \frac{4}{3}\upsilon_S^2 = \frac{k_S}{\rho} \equiv \phi \qquad (7.5.5)$$

Because of its relation with the seismic wave velocities, the ratio k_S/ρ is called the **seismic parameter**, and is commonly represented by the symbol ϕ in the geophysical literature. As we would see later, the development of this connection between the ratio k_S/ρ and the seismic wave velocities constitute an important step in the understanding of the internal constitution of the Earth from the seismic velocities and knowledge of the material properties. Substituting Eq. (7.5.5) into Eq. (3.8.9) that relates temperature gradient to k_S/ρ and the thermodynamic Grüneisen parameter, Γ_{th}, we obtain an expression of adiabatic temperature gradient in terms of the seismic velocities in the Earth's mantle.

$$\left(\frac{\partial \ln T}{\partial Z} \right)_S = \frac{(\Gamma_{th})g}{\phi} = \frac{(\Gamma_{th})g}{\upsilon_p^2 - \frac{4}{3}\upsilon_S^2} \qquad (7.5.6)$$

Upon analyzing the available data for rock forming minerals, Anderson (1989) found a linear relation between $\ln\rho/\overline{M}$ and $\ln\phi$, where $\overline{M}$ stands for the mean atomic weight of a mineral or a rock:

$$\frac{\ln\rho}{\overline{M}} = -1.130 + 0.323\ln\phi \qquad (7.5.7)$$

Combining the last two equations, we obtain an expression of temperature gradient within the Earth's interior in terms of density and mean atomic weight as

$$\boxed{\left(\frac{\partial\ln T}{\partial Z}\right)_S = \frac{(\Gamma_{th})g}{33.06(\rho)^Q}} \qquad (7.5.8)$$

where $Q = 1/(0.323\,\overline{M})$.

7.5.2   Radial Density Variation

7.5.2.1   Williamson-Adams Equation

Williamson and Adams (1923) derived an expression for the density variation in a self-compressing sphere in terms of the seismic parameter that has been found to be useful in the first order discussion of the density structure of the Earth's interior, and served as a basis for further theoretical developments in this area. If the density variation within a spherical body is only due to self-compression under adiabatic condition, then from Eq. (3.7.3).

$$\left(\frac{\partial V}{\partial P}\right)_S = -V\beta_S = -\frac{V}{k_S}$$

Substituting the relations $V = m/\rho$, and $dV = -(m/\rho^2)d\rho$ in the above equation and rearranging the terms

$$\left(\frac{\partial\rho}{\partial P}\right)_S = \frac{\rho}{k_S} = \frac{1}{\phi} \qquad (7.5.9)$$

Assuming that the pressure within the Earth's interior to be in hydrostatic equilibrium, $dP = -\rho g dr$, where r is the radius. Thus, the above equation reduces to

$$\boxed{\left(\frac{\partial\rho}{\partial r}\right)_S = -\frac{\rho g}{\phi}} \qquad (7.5.10)$$

This is known as the **Williamson-Adams equation**. Note that ρ, g and ϕ are functions of r, although g remains appreciably constant to depth near the core-mantle boundary.

Since the Earth is not of uniform composition, the Williamson-Adams formulation is used to determine the density change from seismic velocities within a shell that is of fairly uniform composition, starting with the density at the top of the shell, as shown below.

The last expression can be written as

$$\partial \ln \rho(r) = - \frac{g(r)}{\phi(r)} \partial r$$

so that

$$\ln \rho(r_2) = \ln \rho(r_1) - \int_{r_1}^{r_2} \frac{g(r)}{\phi(r)} dr \qquad (7.5.11)$$

This integral can be evaluated numerically to obtain adiabatic density variation due to self-compression within a spherical shell of uniform composition. As an example, the density profile in the outer core, as calculated from the above equation, is illustrated in Fig. 7.11 The outer core is divided into a number of thin shells and the density at the top of the outer core at 2891 km depth is set equal to 9.90 g, according to the data in the PREM (Dziewonski and Anderson 1981; Anderson 1989). Using g(r) and ϕ(r) data that are also listed in this model, the average density of each shell is calculated successively according to

$$\ln \rho(Z_i) = \ln \rho(Z_{j-1}) + \frac{\bar{g}}{\bar{\phi}}(\Delta Z_j)$$

where (ΔZ_j) is the thickness of the j th shell, and $\bar{g}$ and $\bar{\phi}$ are, respectively, the average density and average seismic parameter within the shell. From Eq. (7.5.5), $\phi = V_p^2$ since the $V_S = 0$ for the liquid outer core.

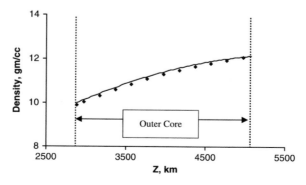

Fig. 7.11 Density variation in the outer core of the Earth. Sold line: calculated numerically from the Williamson-Adams equation. Symbols: data from PREM

Modification by Birch

Since the core and mantle of the Earth are convecting, the temperature distribution within the core and mantle cannot be strictly isentropic. The temperature gradient must be super-adiabatic to some extent, as discussed in Sect. 7.2. If the temperature gradient is not strictly adiabatic (or, more correctly, not isentropic), then, as shown by Birch (1952) and derived at the end of this section, Eq. (7.5.10) modifies to

$$\frac{d\rho}{dr} = -\frac{\rho g}{\phi} + \rho\alpha\tau \tag{7.5.12}$$

where τ denotes the departure of the true temperature gradient from the isentropic gradient (Fig. 7.5) according to

$$\frac{dT}{dr} = \left(\frac{\partial T}{\partial r}\right)_S - \tau \tag{7.5.13}$$

Since $(\partial T/\partial r)_S < 0$ (as T increases downward and r increases upward), $\tau > 0$, so that the temperature gradient is super-adiabatic (i.e. dT/dr is more negative than $(\partial T/\partial r)_S$).

A rough estimate of the effect of the term τ on the calculated density profile may be made by recasting Eq. (7.5.11) in terms of the thermodynamic Grüneisen parameter, Γ_{th}. From Eq. (3.8.2) we have $\alpha = \Gamma_{th}C_p/Vk_S$, which on substitution in Eq. (7.5.12) yields

$$\frac{d\rho}{dr} = -\frac{\rho g}{\phi(r)} + \frac{\Gamma_{th}C_p\tau}{V(k_s/\rho)} = -\frac{\rho g}{\phi} + \frac{\Gamma_{th}C_p\tau}{V\phi}$$

Factoring out $-\rho g/\phi$

$$\frac{d\rho}{dr} = -\frac{\rho g}{\phi}\left(1 - \frac{\Gamma_{th}C_p\tau}{g(V\rho)}\right)$$

Noting that $C_p/(V\rho) = C_p/M$ (M: molar weight) $= C_p'$ (i.e. specific heat capacity), we then have

$$\frac{d\rho}{dr} = -\frac{\rho g}{\phi}\left(1 - \frac{\Gamma_{th}C_p'\tau}{g}\right) \equiv -\frac{\rho g}{\phi}(1 - \sigma) \tag{7.5.14a}$$

or

$$\frac{d\rho}{dZ} = \frac{\rho g}{\phi}(1 - \sigma) \tag{7.5.14b}$$

where $\sigma = \Gamma_{th}C'_p\tau/g$. As discussed in Sect. 3.8, the value of the Grüneisen parameter has a restricted range of variation around a value of 2, whereas C_P of the mantle minerals is typically ~ 1 kJ/kg-K, which equals 1000 Nm/kg-K, or since Newton (force) has the unit of kg-m/s^2, C'_p of the mantle minerals is ~ 1000 m^2/s^2-K. The adiabatic gradient in the Earth's mantle is ~ 0.6 K/km, as discussed in Sect. 7.3.1. Thus, if the true temperature gradient in the earth's mantle is super-adiabatic, and differs from the adiabatic gradient by ~ 0.6 K/km ($\sim 100\%$ deviation), then using $g = 10$ m/s^2, $\Gamma_{th} = 2 \pm 0.5$, we get $\sigma \sim 0.09 - 0.15$, or ~ 9 to 15% of the main term.

Derivation of Eq. (7.5.12) This equation is derived by considering $\rho = f(P,T)$, which leads to

$$\frac{d\rho}{dr} = \left(\frac{\partial\rho}{\partial p}\right)_T \frac{dP}{dr} + \left(\frac{\partial\rho}{\partial T}\right)_p \frac{dT}{dr} \qquad (7.5.15)$$

Expressing the first and second partial derivatives in terms of k_T and α, respectively (Eqs. 3.7.1 and 3.7.2), and assuming hydrostatic relation ($dP = -\rho g dr$), we have

$$\frac{d\rho}{dr} = -\frac{g\rho^2}{k_T} - \rho\alpha\frac{dT}{dr}, \qquad (7.5.16)$$

Substituting (dT/dr) from Eq. (7.5.13) in the above equation along with the expression for adiabatic (isentropic) temperature gradient according to Eq. (7.2.5a), that is $(\partial T/\partial r)_S = -(\partial T/\partial Z)_S = -gT\alpha/C'_p$, and re-arrangement of terms, yields

$$\frac{d\rho}{dr} = -\frac{g\rho^2}{k_T} + \frac{\rho\alpha^2 Tg}{C'_P} + \rho\alpha\tau$$
$$= -\frac{g\rho^2}{k_T}\left(1 - \frac{\alpha^2 Tk_T}{\rho C'_P}\right) + \rho\alpha\tau \qquad (7.5.17)$$

From (3.7.15), the parenthetical term in the above expression equals k_T/k_S. Substituting now the seismic parameter ϕ for k_S/ρ, the expression of $d\rho/dr$ reduces to Eq. (7.5.12).

7.5.3 Transition Zone in the Earth's Mantle

The question as to whether the Earth's mantle is chemically and mineralogically homogeneous or not was first addressed through a thermodynamic analysis by Birch (1952), and his pioneering contribution in this regard has laid the foundation of our current state of knowledge about the Earth's mantle. Birch derived an expression for the change of seismic parameter, ϕ, as function of the radial distance in a homogeneous self-compressing sphere under adiabatic condition. The expression is

$$1 - \frac{d\phi}{gdr} = \left(\frac{\partial k_s}{\partial P}\right)_S + \frac{\alpha k_S \tau}{\rho g}\left[1 + \frac{1}{\alpha k_S}\left(\frac{\partial k_s}{\partial T}\right)_P\right] \qquad (7.5.18)$$

Using the relation $k_s = k_T(1 + \Gamma_{th}\alpha T)$ (Eq. 3.8.3), Birch (1952) cast this equation in terms of k_T, which is better known than k_S, and the thermodynamic Grüneisen parameter, Γ_{th}. Since the latter varies within a narrow range around 2, this manipulation allowed him to approximately predict the values of the left hand quantity, which he called ψ, as a function of depth from the available experimental data of the properties of minerals that are likely to constitute the Earth's mantle.

Using Eq. (7.5.5), one can also calculate the parameter ψ from the available data for the seismic velocities in the Earth's mantle. Birch (1952) found that the values of ψ between 200 and 900 km depth in the Earth's mantle that were calculated from the last equation are far too small compared to those that were calculated from the seismic data (Fig. 7.12). No reasonable adjustment of thermo-physical parameters of mantle phases could reconcile the rapid rise of the observed value of ψ between 200 and 900 km with a model of self-compressing homogeneous sphere. Birch, thus, concluded that this zone of the earth's mantle must be **inhomogeneous** as a result of either mineralogical or chemical change or both. It is now well known from laboratory experimental data on mineral stabilities that a number of mineralogical transformations must take place within ~ 400 to 700 km depth in the

Fig. 7.12 Comparison of the calculated variation (solid line) of the function $\psi = (1 - d\phi/(gdr))$ versus depth (Z) in the Earth's mantle, assuming it to be a homogeneous self-compressing sphere, with the observed data. Note the mismatch between 200 and 900 km depth. From Birch (1952)

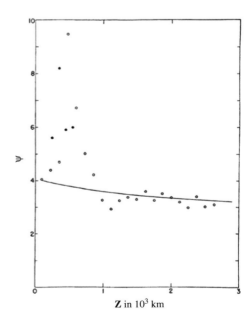

Z in 10^3 km

Earth's mantle, which is commonly referred to as the **transition zone**, and there may be chemical changes as well (The transition zone divides the Earth's mantle into upper and lower mantles, and exerts great influence on the dynamics of the Earth's interior including the subduction of oceanic plates). However, Birch's thermodynamic analysis of the problem laid the groundwork for future investigations, and it is highly instructive to follow his thermodynamic analysis.

Derivation of Eq. (7.5.18) From the relation $k_s = \rho\phi$, we have, $dk_s = \rho d\phi + \phi d\rho$, so that

$$\frac{d\phi}{dr} = \frac{dk_s}{\rho dr} - \frac{\phi}{\rho}\frac{d\rho}{dr} \qquad (7.5.19)$$

Dividing both sides by g, using Eq. (7.5.12) for $d\rho/dr$, and using the hydrostatic relation, $dP = -\rho g dr$, we then have

$$\frac{d\phi}{g dr} = \frac{dk_s}{\rho g dr} - \frac{\phi}{\rho g}\left(-\frac{\rho g}{\phi} + \alpha\rho\tau\right)$$

or

$$1 - \frac{d\phi}{g dr} = \frac{dk_s}{dP} + \frac{\alpha\phi\tau}{g} \qquad (7.5.20)$$

This expression was first derived by Bullen (1949), but Birch (1952) called attention to the temperature effect that is concealed in the dk_s/dP term, which is a total derivative. Since $k_s = f(P,T)$, we have

$$\frac{dk_s}{dP} = \left(\frac{\partial k_s}{\partial P}\right)_T + \left(\frac{\partial k_s}{\partial T}\right)_P \frac{dT}{dP} \qquad (7.5.21)$$

Using $dP = -\rho g dr$, we obtain from Eq. (7.5.13)

$$\frac{dT}{dP} = \left(\frac{\partial T}{\partial P}\right)_S + \frac{\tau}{\rho g}$$

Thus, Eq. (7.5.21) transforms to

$$\frac{dk_s}{dP} = \left[\left(\frac{\partial k_s}{\partial P}\right)_T + \left(\frac{\partial k_s}{\partial T}\right)_P \left(\frac{\partial T}{\partial P}\right)_S\right] + \left(\frac{\partial k_s}{\partial T}\right)_P \frac{\tau}{\rho g} \qquad (7.5.22)$$

It is easy to see from Eq. (7.5.21) that the terms within the square brackets in the last equation constitute the partial derivative $(\partial k_s/\partial P)_s$. Thus, Eq. (7.5.20) reduces to

$$1 - \frac{d\phi}{gdr} = \left(\frac{\partial k_s}{\partial P}\right)_S + \frac{\alpha\phi\tau}{g}\left[1 + \frac{1}{\alpha\rho\phi}\left(\frac{\partial k_s}{\partial T}\right)_P\right] \qquad (7.5.23)$$

Replacing ϕ by k_s/ρ on the right hand side and rearrangement of terms finally yield Eq. (7.5.18).

7.6 Horizontal Adiabatic Flow at Constant Velocity

7.6.1 Joule-Thompson Experiment and Coefficient

The appropriate starting point in the thermodynamic discussion of adiabatic flow processes is the classic experiments by Joule and Thompson (the latter also known as Lord Kelvin) in 1853 on the **irreversible** horizontal flow of gas through a thin but rigid porous plug within an adiabatic enclosure (Fig. 7.13). Here the gas one side (chamber or subsystem 1) of the plug was at a higher pressure than that in the other side (chamber 2), but the pressure on each side was maintained to be uniform. Thus, the entire pressure drop was made to take place within the porous plug. The gas on both sides was at rest at the beginning and end of the experiment so there was no change of kinetic energy of the system. Thermometers placed on two sides of the plug, however, showed measurable temperature difference that depended on the pressure difference between the two sides. Thermodynamic analysis of the above experiment led to the important conclusion that irreversible adiabatic process of the type described by the Joule-Thompson experiment is **isenthalpic**, as opposed to the reversible adiabatic process that is isentropic. This can be shown as follows.

Let m be the mass of gas that has flowed through the porous plug. Then, the work done **on** the gas in chamber 1 is given by

$$W_1^- = -P_1(\Delta V_1),$$

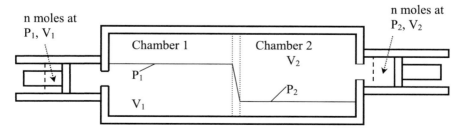

Fig. 7.13 Schematic illustration of the Joule-Thompson experiment. The dashed lines show the initial positions before the transfer of n moles of gas from the left to the right chamber through a porous partition

where ΔV_1 is the volume change of gas in the chamber 1. Using v' as a symbol for specific volume, $\Delta V_1 = -mv'_1$, so that $W_1^- = mP_1v'_1$. On the other hand, the work done **by** the gas in the chamber 2 is given by $W_2^+ = P_2(\Delta V_2) = P_2mv'_2$. Now, since $W_2^+ = -W_2^-$, where W_2^- is the work done **on** the sub-system 2, the net work, W^-, done on the **entire** composite system is given by

$$W^- = W_1^- + W_2^- = m(P_1v'_1 - P_2v'_2) \tag{7.6.1}$$

If u'_i is the specific internal energy of the subsystem i, then the net change of internal energy of the composite system is given by

$$\begin{aligned} \Delta U = \Delta U_1 + \Delta U_2 &= -mu'_1 + mu'_2 \\ &= m(u'_2 - u'_1) \end{aligned} \tag{7.6.2}$$

There is also a small change of internal energy of the thin porous plug, but it is too small compared to that of the rest of the system, and has, therefore, been neglected. Now, since the composite system is under an adiabatic condition, we have (from the first law), $\Delta U = W^-$, so that, using the last two equations, $m(u'_2 - u'_1) = m(P_1v'_1 - P_2v'_2)$ or,

$$U'_2 + P_2V'_2 = U'_1 + P_1V'_1$$

where U'_i and V'_i stand for mu'_i and mv'_i, respectively. However, since the enthalpy H is given by H = U + PV, we arrive at the important conclusion that

$$H'_2 = H'_1 \tag{7.6.3}$$

where H'_i is the total enthalpy of the m moles of gas in the compartment i. In other words, there is no change of enthalpy of the gas (dH = 0) as it is decompressed through a porous plug from a higher to a lower pressure environment, maintaining uniform pressure in both sides. This result is true regardless of the number sub-systems in the composite system, whenever fluid or matter flows through a restriction from a uniformly high to a uniformly low pressure, without any appreciable change of kinetic energy. This type of flow process is often referred to as *throttling process*, which is inherently *irreversible*, and is associated with entropy production.

The expression of change of temperature with that of pressure can now be easily derived by writing the total derivative of H in terms of changes of P and T, and then imposing the condition of constant enthalpy.

$$dH = \left(\frac{\partial H}{\partial T}\right)_P dT + \left(\frac{\partial H}{\partial P}\right)_T dP = 0 \tag{7.6.4}$$

By definition, the first derivative on the right equals C_P, whereas the second derivative on the right can be shown to be equal to $V(1 - T\alpha)$ (which is left as an exercise: Problem 7.2). Thus, we have

$$\left(\frac{\partial T}{\partial P}\right)_H (\equiv \mu_{JT}) = \frac{V(T\alpha - 1)}{C_P} \qquad (7.6.5)$$

(Note that the V and C_P stand for the molar quantities, but the ratio V/C_P does not change if we use values for the respective specific quantities.) Notice that this expression differs from that in the isentropic case (Eq. 7.2.4) by the presence of the term $-V$ in the numerator.

The quantity in the left of Eq. (7.6.5) is known as the **Joule-Thompson coefficient**, and is usually referred by the symbol μ_{JT}. For any substance, $T\alpha$ changes from a value of greater than unity at low pressure to less than unity at higher pressure. The transition point, at which $\mu_{JT} = 0$, is known as the Joule-Thompson inversion point. When $T\alpha > 1$ (i.e. at pressure below that of the inversion point), temperature changes in the same direction as the pressure whereas for $T\alpha < 1$, temperature changes in the opposite direction as pressure. This is shown schematically in Fig. 7.14 In this Figure, X represents the highest pressure and Y represents the highest temperature at which cooling may be obtained by adiabatic expansion of gas through a throttling process (For example, the values for X and Y for N_2, which

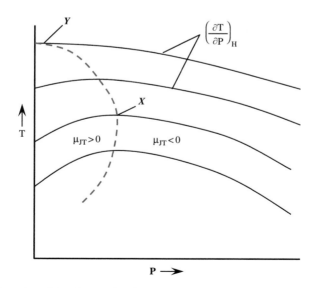

Fig. 7.14 Illustration of the change of Joule-Thompson coefficient as a function of pressure of a hypothetical substance. The points X and Y represents the highest pressure and temperature, respectively, at which the substance can be cooled by irreversible expansion (i.e. decrease of pressure) under adiabatic condition. The dashed line represents the locus of the Joule-Thompson inversion point

Fig. 7.15 Variation of αT
(α: coefficient of thermal
expansion) of H_2O (dashed
lines) and CO_2 (solid lines)
versus pressure. From Spera
(1981)

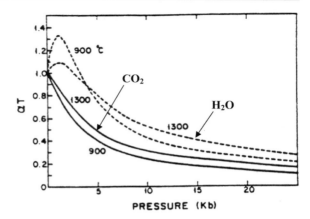

is liquified by adiabatic expansion, are ~ 375 bars (at 40 °C) and 350 °C, respectively). In practical applications of liquefaction of gas (e.g. production of liquid nitrogen and helium), the pressure difference between the low and high pressure side is maintained by a pump instead of a porous plug (The ability to produce liquid He made it possible to study properties of matter at very low temperature and consequent discovery of superconductivity by the Dutch physicist Heike Kamerlingh Onnes (1853–1926); the work was recognized by a Nobel prize in physics in 1911).

Waldbaum (1971) computed values of the Joule-Thompson coefficient, μ_{JT}, for a number of important rock-forming minerals. These μ_{JT} values range from ~ -13 to -30 K/kbar. The variation of the term $T\alpha$ versus P for two most geologically important fluid species, namely H_2O and CO_2, as calculated by Spera (1981), is illustrated in Fig. 7.15 It is clear that μ_{JT} for these fluid species (and also probably other fluid species of geological importance) is negative ($T\alpha < 1$) for a wide range of geologically important conditions.

Spera (1984b) evaluated the effects of salt concentration, temperature and pressure on the μ_{JT} of aqueous solution. The effect of salt concentration is to reduce the value of μ_{JT}, especially at relatively higher temperature. He noted that some fluid inclusions in minerals from porphyry copper deposits have salt concentrations as high as 7.0 molal or even more. Spera (1984b) also predicted that addition of divalent ions would cause a larger decrease of μ_{JT} of an aqueous solution than that of an equivalent amount of univalent ion.

7.6.2 Entropy Production in Joule-Thompson Expansion

Since the Joule-Thompson (J-T) type of adiabatic flow of an inviscid material is an irreversible process, there must be entropy production within the system. To evaluate the latter, we consider the relationship between the changes of enthalpy and entropy in a system, as given by Eq. (3.1.8) or in the Box (3.1.2), viz.

dH = TdS + VdP. This equation was derived for a reversible process in a closed system, but since both enthalpy and entropy are state functions, the change of each of these quantities between the initial and terminal states must be independent of the manner by which the change was achieved. Thus, since $\Delta H = 0$, we have

$$\Delta S = - \int_{P_1}^{P_2} \frac{V}{T} dP \qquad (7.6.6)$$

where P_2 and P_1 are, respectively, the final and initial pressures of the "chambers" 2 and 1. The right hand side could be integrated by expressing V/T as a function of P through the equation of state of the flowing material. For $P_2 < P_1$, the right hand side of the above equation is > 0. Thus, there is an increase of entropy associated with J-T expansion, as it should be for an irreversible process. Note that we can only talk about the entropy change between the final and initial states, but not the entropy production during the expansion of the substance.

Problem 7.2 Show that

$$dH = CpdT - V(T\alpha - 1)dP \qquad (7.6.7)$$

Hint: You know that dH = VdP + TdS (Box 3.1.2). Now express dS in terms of dT and dP using the relation S = f(P,T), and use an appropriate Maxwell relation (Box 3.4.1).

Problem 7.3 Show

$$\mu_{JT} = 0 \quad \text{for a perfect or ideal gas.} \qquad (7.6.8)$$

Hint: Evaluate α using equation of state for an ideal gas.

7.7   Adiabatic Flow with Change of Kinetic and Potential Energies

The conceptually most straightforward way to treat the problems of adiabatic flow processes that involve changes of potential and/or kinetic energies is to consider the overall energy balance of the system, that is the **total energy change of a system**, which is given by the sum of the changes of internal energy, kinetic energy and potential energy, must equal the net **energy absorbed by the system** from outside. We shall treat some relatively simple cases of horizontal flow, in which there is only a change of kinetic energy, and vertical flow in which there is change of both potential and kinetic energies.

7.7.1   Horizontal Flow with Change of Kinetic Energy: Bernoulli Equation

If there is a change of kinetic energy of a system due to a change of velocity, then we can write the following energy balance equation for a horizontal inviscid flow

$$\Delta U + \frac{1}{2} m\Delta(\upsilon^2) = Q + W^-$$
(7.7.1)

(Energy change of the system = Energy absorbed by the system)

where υ is the linear velocity and m is the mass of the system. Note that W^- is the total work absorbed by the system. Now, if there is (a) no heat transfer between the moving body and the surrounding (i.e. $Q = 0$), and (b) the only form of work is the mechanical PV work, then using Eq. (7.6.1)

$$\Delta U + \frac{1}{2} m\Delta(\upsilon^2) = P_1 V_1 - P_2 V_2$$
(7.7.2)

If the fluid is incompressible (i.e. $V_1 = V_2 = V$), then $\Delta U = 0$ (since $dU = \delta q - PdV$), in which case the above equation reduces to

$$\frac{1}{2} m\Delta\upsilon^2 = -V(P_2 - P_1) = -V\Delta P$$
(7.7.3)

or

$$\boxed{\Delta P = -\frac{m\Delta(\upsilon^2)}{2V} = -\frac{\rho\Delta(\upsilon^2)}{2}}$$
(7.7.4)

where ρ is the density. Thus, there is a pressure decrease with increasing horizontal velocity of an inviscid and incompressible fluid under adiabatic condition. This expression is known as the **Bernoulli equation** in fluid dynamics, and was originally derived from purely mechanical arguments. However, derivation of the equation through thermodynamics shows the restrictive conditions under which it is strictly valid.

Although the relation between pressure and velocity will need modification when the effects of viscosity and energy dissipation due to friction are taken into account, the Bernoulli equation provides qualitative understanding of a number of phenomena that are encountered in the real life. For example, Roofs of buildings sometime blow off during a storm because of the high velocity of air passing over the roof that causes a decrease of external pressure. The airplane wings are designed such that the upper surface has an upward curvature while the lower surface is flat. This causes an increase of velocity and consequent decrease of air pressure on the upper surface relative to the pressure on the lower surface, thus giving an upward lift to the airplane (Fig. 7.16). The velocity of a river increases when it flows through a narrow channel. This causes a **drop** of pressure that the river exerts on its

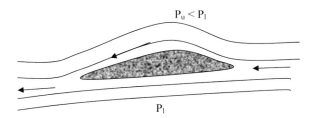

Fig. 7.16 Schematic illustration of streamlines of air around an airplane wing (cross-sectional view). The air pressure at the upper surface, P_u, is lower than that at the lower surface, P_l, resulting in a lifting of the airplane

banks as it flows through a narrow channel – an apparently counterintuitive conclusion. When arteries become constricted by plaques, pressure within the arteries drops because of increased speed of blood flow. This leads to the collapse of the constricted arteries when the pressure falls below a critical limit.

7.7.2   Vertical Flow

7.7.2.1   Change of Temperature with Pressure

If there is a significant change of height over which the flow takes place, as is often the case in geological problems, then there is a change of potential energy of the system, which is given by mgdh, where h is the height (positive upwards). Considering the effects of change of both potential and kinetic energies, the energy balance equation for adiabatic but frictionless vertical displacement of a parcel of material is given by

$$dU + mgdh + \frac{1}{2}m(d\upsilon^2) = \delta\omega^- \qquad (7.7.5a)$$

or

$$\Delta U + mg(\Delta h) + mv(\Delta v) = W^-$$
$$= m\left(P_1 v_1' - P_2 v_2'\right) \qquad (7.7.5b)$$

where the velocity υ is the taken to be positive upwards, v' is the specific volume of the system, as before, and Δ stands for the difference between the states 2 and 1 (e.g. $\Delta \upsilon = \upsilon_2 - \upsilon_1$) (Note that $\frac{1}{2}m(d\upsilon^2) = m\upsilon d\upsilon$). The collection of left hand terms represents the total energy change of the system, internal plus external, whereas the right hand term is the energy absorbed by the system under adiabatic condition.

Rearranging terms in the last equation, we can write

$$(U_2 + mP_2 v_2') - (U_1 + mP_1 v_1') + mg(\Delta h) + m\upsilon(\Delta\upsilon) = 0$$

Now, since $H = U + PV = U + mPv'$, the last equation reduces to

$$\Delta H + mg(\Delta h) + mv(\Delta v) = 0$$

or in a differential form

$$dH + mgdh + m(vdv) = 0 \tag{7.7.6}$$

The first two terms on the left hand side may be collectively viewed as the total derivative of H if we treat H as a function of P, T and h instead of just P and T. In that case, the second term on the left is $(\partial H/\partial h)_{P,T}dh$, and the first term on the left expresses the total change of H due to changes of the changes in P and T, which is given by Eq. (7.6.7).

Combining the last expression with Eq. (7.6.6), we have for a unit mass of material

$$Cp'dT - v'(T\alpha - 1)dP + gdh + vdv = 0 \tag{7.7.7}$$

Differentiating with respect to P and rearranging terms, we finally obtain

$$
\begin{aligned}
\left(\frac{\partial T}{\partial P}\right)_{Q(ir)} &= \frac{(T\alpha - 1)}{\rho C'_p} - \left(\frac{g}{C'_p}\right)\frac{dh}{dP} - \frac{v}{C'_p}\left(\frac{dv}{dP}\right) \\
&= \mu_{JT} - \left(\frac{g}{C'_p}\right)\frac{dh}{dP} - \frac{v}{C'_p}\left(\frac{dv}{dP}\right)
\end{aligned} \tag{7.7.8}
$$

The subscript Q(ir) indicates that it is an adiabatic condition but that the process is irreversible (i.e. $\delta q = 0$, $dS > 0$). We would call this **irreversible adiabatic decompression** or **IAD**. Without the velocity term, this expression was derived earlier by Ramberg (1972). Applications of this equation to the upwelling and melting processes in the Earth's interior are discussed in Sect. 7.8.

7.7.2.2 Geyser Eruption

What is the velocity with which a geyser, such as the **Old Faithful** in the Yellowstone National Park, Wyoming, keeps on erupting through the vent? An approximate answer to this question is provided by the energy balance relation given by Eq. (7.7.6). Rewriting $mvdv$ as $\frac{1}{2}(mdv^2)$, we have

$$dH + mgdh + \frac{1}{2}mdv^2 = 0 \tag{7.7.9}$$

or, replacing d by Δ, and dividing through by m

$$\Delta H' + g\Delta h + \frac{1}{2}(\Delta v^2) = 0 \tag{7.7.10}$$

so that

$$\Delta \upsilon = -\sqrt{2\Delta H' + 2g\Delta h} \qquad (7.7.11)$$

The velocity, υ_2, at the maximum height of the geyser is zero so that $\Delta \upsilon = -\upsilon_1$. Assuming now that the specific enthalpy of the geyser is approximately constant, we have

$$\boxed{\upsilon_1 = \sqrt{2g\Delta h}} \qquad (7.7.12)$$

The last equation was derived earlier by Furbish (1997), who also treated the eruption velocity of Old Faithful. The height of the steam-water eruption column of Old Faithful is ~ 30 m. Thus, we obtain the near vent velocity of the geyser, υ_1, to be ~ 24 m/s. On the other hand, from the conservative estimate of discharge volume of Old Faithful during its initial phase of eruption (6.8 m^3/s) and the cross-sectional area of the vent (0.88 m^2), a conservative estimate of the exit velocity is 7.7 m/s (Furbish 1997).

The Geyser eruption is associated with some heat loss. If ΔQ is the heat change per unit mass between the final and initial state ($\Delta Q = Q_2 - Q_1$), then it must be added to the left hand side of the energy balance equation, Eq. (7.7.10). Consequently, to account for the heat loss, the term $2\Delta Q$ must be added to the term under the square root in the last expression. Since for heat loss $\Delta Q < 0$, the eruption velocity becomes lower than 24 m/s (note that ΔQ has the SI unit of J/kg or N-m/kg; since N = kg-m/s^2, ΔQ can be expressed in the SI unit of m^2/s^2). Thus, the exit velocity, υ_1, of Old Faithful is $7.7 \leq \upsilon < 24$ m/s.

7.8   Ascent of Material Within the Earth's Interior

Equation (7.7.8) yields the temperature change of a substance due to irreversible vertical flow under **adiabatic** condition, but **ignoring** the effect of friction. Since dh/dP < 0, the gravitational effect contributes a positive term. Thus, while negative values of μ_{JT} of minerals and geologically important fluids at moderately high pressure, as discussed above, leads to heating of materials ascending adiabatically from within the Earth's interior, the effect is moderated (that is dT/dP becomes less negative) by the effect of the gravitational field, as was pointed out by Ramberg (1972). In the following subsections, we address two types of problems related to the ascent of materials from the Earth's interior.

7.8.1 Irreversible Decompression and Melting of Mantle Rocks

We now return to the problem of adiabatic decompression and melting within the Earth's mantle that was discussed earlier in Sect. 7.4 from the equilibrium stand-point. Here we consider the adiabatic upwelling of material in a pressure gradient as a series of Joule-Thompson experiments with small sequential decrease of pressure. The formulations presented below, which is due to Ganguly (2005), provide starting points for the treatment of adiabatic upwelling and melting process in the Earth's mantle as irreversible process ($\delta q = 0$, $dS \neq 0$), which represents a shift from the traditional starting point founded on treating these as reversible processes.

Using $dZ = -dh = dP/(\rho_r g)$, where Z is the depth (+ve downwards) and ρ_r is the density of the mantle rock, Eq. (7.7.8) yields

$$\left(\frac{\partial T}{\partial Z}\right)_{Q(ir)} = \frac{\rho_r}{\rho}\left(\frac{gT\alpha}{C_p'}\right) + \frac{g}{C_p'}\left(1 - \frac{\rho_r}{\rho}\right) - \frac{\upsilon}{C_p'}\left(\frac{d\upsilon}{dZ}\right) \qquad (7.8.1)$$

Note that the term within the first parentheses in the right equals the isentropic temperature gradient (Eq. 7.2.5a). Thus,

$$\left(\frac{\partial T}{\partial Z}\right)_{Q(ir)} = \frac{\rho_r}{\rho}\left(\frac{\partial T}{\partial Z}\right)_S + \frac{g}{C_P'}\left(1 - \frac{\rho_r}{\rho}\right) - \frac{\upsilon}{C_p'}\left(\frac{d\upsilon}{dZ}\right) \qquad (7.8.2)$$

The second right hand term can be viewed as the manifestation of entropy production in the system due to the irreversible nature of the process of upward flow. Since $\rho_r > \rho$ (otherwise the material will not move upward), the second term on the right is negative and, thus, somewhat counteracts the effect of the first term. Physically it means that an ascending parcel of material from within the Earth's interior would be hotter than that predicted by the isentropic gradient. This is illustrated in Fig. 7.17a, which is modified from Ganguly (2005). Assuming constant velocity of ascent, the dashed lines are calculated according to the above equation for different values of ρ/ρ_r and typical C_p' value of mantle mineral of 1.2 kJ/kg-K. The density ratio controls the velocity of upward movement ($u \to 0$ as $\rho \to \rho_r$). If it is very close to unity, then the temperature gradient effectively equals isentropic gradient if the adiabatic condition prevails. It is evident from Fig. 7.17a that the ascent of mantle material with 94% density of the surrounding mantle will be essentially isothermal, if it moves with a constant velocity. If the density of the ascending material is lower, then there will be a **net heating** during ascent.

Plumes of ascending mantle material are generally believed to have generated within the Earth's mantle at the thermal boundary layers between the upper- and lower-mantle at ~ 660 km depth and between the lower-mantle and core at ~ 2900 km depth (Box 7.1). Intersection of the adiabatic T-Z trajectories of the plumes with the solidus of the mantle material marks the onset of melting of the plumes. Typically, the adiabatic path of a plume is assumed to be isentropic

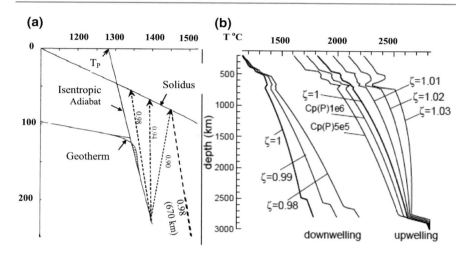

Fig. 7.17 Temperature-depth (T-Z) paths for adiabatic ascents of materials within the Earth's mantle. (**a**) T_P: Potential temperature. The solid line in (**a**) and bold lines in (**b**) depict the isentropic paths. The dashed lines in (**a**) and thin lines in (**b**) illustrate adiabatic paths of mantle materials of different densities relative to the surrounding mantle, taking into account the entropy production due to irreversible expansion of the ascending material. The curve labeled 0.98 (670 km) in (**a**) shows the irreversible adiabatic ascent path of a plume with 98% density of the surrounding mantle, rising from the boundary between upper and lower mantle, which is at the 670 km discontinuity. The numbers on the dashed lines in (**a**) represent ρ/ρ_r (0.90, 0.94, 0.98) whereas in (**b**) $\zeta = \rho_r/\rho$. The profiles labelled by $C_p(P)$ in (**b**) are isentropic trajectories that were calculated by assuming that C_p changes linearly with pressure and attains the Dulong-Petit limit at 10^6 and 10^5 bars (see Sect. 4.2 for discussion of pressure dependence of C_p). (**a**) Modified from Ganguly (2005); (**b**) modified from Tirone (2016)

(e.g. Nicolas 1995). However, there are various types of irreversibility leading to entropy production within the plume. The curve labeled 0.98 (670 km) in Fig. 7.17a shows the adiabatic T-Z trajectory of a plume with 98% density ($\rho/\rho_r = 0.98$) of the surrounding mantle and rising from a depth of 670 km, taking into consideration of the effect of irreversible expansion according to Eq. (7.8.2). The initial temperature of the plume is taken to be defined by the projection of the adiabatic (isentropic) geotherm to 670 km depth. It is found that a plume with 98% density of the mantle would intersect the solidus at a temperature that is $\sim 150\ ^\circ C$ above that defined by its isentropic ascent. Consequently, there would be much larger extent of melt production in the plume as it ascends to shallower depths than that in the case of isentropic ascent. Phase transitions and friction would modify the T-Z trajectory of a plume, but still isentropic T-Z path may lead to substantial underestimation of melting in a plume in some cases.

Tirone (2016) carried out detailed numerical calculations of the T-Z trajectories of ascending and descending mantle materials extending from and to the core-mantle boundary in the CFMAS (CaO-FeO-MgO-Al_2O_3-SiO_2) system that essentially encompasses the entire compositional space of mantle minerals. The results for the calculated T-Z trajectories are illustrated in Fig. 7.17b. The sharp changes in the thermal profiles are due to phase transformation in the mantle. For the upwelling mantle parcel, the pressure, P(Z), at a given Z was calculated from the ambient $P_r(Z)$ using a scaling relation $dP = \zeta dP_r$ and a specified value of ζ. Comparison of the expression derived by Tirone (2016) for $\partial T/\partial Z$ under gravity corrected isenthalpic condition ($dH_g = dH + mgdh = 0$ for constant velocity—see Eq. 7.7.6) with Eq. (7.8.1), which was also derived for the same condition, shows that the scaling factor ζ equals ρ_r/ρ.

The numerical method involves full calculations of (a) compositions and modal abundances of minerals for fixed bulk composition of the system, and (b) total entropy and total enthalpy of the ambient mantle and moving parcel at discrete steps of Z from the bulk mineralogical properties. As discussed later in Sect. (10.3.2) on Duhem's theorem, all properties of a closed system are, in principle, uniquely fixed at specified P-T condition (or, for that matter, by specification of any two variables). The method of Gibbs free energy minimization (G-minimization), as discussed in Sect. (10.13.1), affords a convenient and widely used method for calculation of the modal abundance and composition of minerals in a system of fixed bulk composition at a specified P-T condition. Tirone (2016) determined the initial enthalpy and entropy (H_o and S_o, respectively) of the bulk system on the basis of the calculated mineralogical properties (modal abundance and compositions). For calculation of isentropic and gravity corrected isenthalpic T-Z profiles, the temperature at a fixed depth, and hence fixed P_r, was varied until the respective condition of conservation of S ($S(Z) = S_o$) and H_g ($H_g(Z) = H_{g,o}$) were satisfied, where $H_g(Z) = H(Z) + mgh$.

Tirone (2016) also calculated the effects of entropy production on the T-Z trajectories of moving mantle parcels, including mantle plumes, using full scale dynamic models that involve the effects of viscous dissipation and heat conduction. These T-Z trajectories have been found to be quite similar to the results of much simpler calculations using the gravity corrected JT formulation.

The problem of melt productivity of mantle rocks during irreversible decompression has been analyzed by Ganguly (2005). Beginning with the fundamental energy conservation relation, Eq. (7.7.6), he derived the following expression for the change of extent of melting with pressure.

$$\left[\frac{dx}{dP}\right]_{Q(ir)} \approx \left[\frac{dx}{dP}\right]_S - \left[\frac{x\Delta v'_{(f)} + \upsilon(d\upsilon/dP)}{\Delta H_f + (x\Delta C'_{P(f)} + C'_{P,s})(\partial T/\partial x)_p}\right]$$
$$\approx \left[\frac{dx}{dP}\right]_S - \Lambda \tag{7.8.3}$$

where x is the melt fraction and $\Delta Y'_{(f)}$ is the change of a specific quantity on fusion. Ganguly (2005) evaluated the term Λ, which represents the effect of entropy production, and showed that the melt productivity, or the melt fraction that would develop in an upwelling mantle material after it crosses the solidus (Fig. 7.17a) could be significantly higher, by as much as 50%, than that calculated using the isentropic condition. A potentially important consequence of the increased melt productivity is that melt segregation would take place more quickly than envisaged for the limiting case of isentropic decompression.

7.8.2 Thermal Effect of Volatile Ascent: Coupling Fluid Dynamics and Thermodynamics

To evaluate the temperature change associated with the ascent of material from the Earth's interior, one needs to consider the full set of equations expressing conservation of mass, momentum and energy. The problem was considered by Spera (1981, 1984a, b) in relation to problems of fluid flow in geological problems. The temperature change associated with the steady, one-dimensional flow of single phase fluid of fixed composition in a vertical conduit of diameter d is given by

$$\frac{dT}{dZ} = \mu_{JT}\left(\frac{dP_f}{dZ}\right) - \frac{\upsilon}{C'_p}\left(\frac{d\upsilon}{dZ}\right) + \frac{g}{C'_p} + \frac{4\kappa}{C'_p \dot{m} d}(T - T_w) \qquad (7.8.4)$$

where κ is heat transfer coefficient (J cm^{-2} s^{-1} K^{-1}), $\dot{m}$ is mass flux (gm cm^{-2} s^{-1}), T_w is the temperature along the crack or conduit wall. If the fluid pressure exceeds the lithostatic pressure, P, by a constant amount, $P_f = P + \alpha$ (constant), then $dP_f = \rho_r g dZ$, where ρ_r is the density of the surrounding rock, as in Eq. (7.8.2). It is then easy to see that for adiabatic condition ($\kappa = 0$) the above expression reduces to Eq. (7.8.2) that follows from simple thermodynamic considerations.

Using the conditions of mass and momentum conservation, but neglecting the heat of reaction due to precipitation, hydrolysis and alteration, the above equation transforms to (Wood and Spera 1984)

$$\frac{dT}{dZ} = \frac{\alpha T}{\rho C'_p}\left(\frac{\dot{m}}{\rho}\right)^2 \frac{d\rho}{dZ} + \mu_{JT}\frac{2C_f \dot{m}^2}{\rho d} + \frac{\alpha T g}{C'_p} + \frac{4\kappa}{C'_p \dot{m} d}(T - T_w) \qquad (7.8.5)$$

where C_f is the effective friction coefficient. The first term on the right disappears if the fluid is incompressible.

Spera (1984a) evaluated the temperature changes on decompression of binary H_2O-CO_2 fluid, which is the dominant fluid composition in both metamorphic and magmatic systems, using reasonable range of values of the different parameters in Eq. (7.8.5). He concluded that the combined effects of conductive/convective heat transfer and the temperature change accompanying decompression are often offsetting for metamorphic fluids, making them effective carriers of mantle heat to shallower levels that may produce localized melting within the lithosphere.

It has been concluded by a number of workers that the high temperature granulite facies metamorphic rocks had formed under highly H_2O depleted condition. Newton and co-workers (e.g. Janardhan et al. 1982; Hansen et al. 1987) studied the granulites in southern India and Sri Lanka and concluded that these rocks had formed under highly H_2O depleted condition with $P_{H_2O} < 0.3P_{total}$, and that the flow of CO_2 from either the mantle and/or the lowermost crust was responsible for the depletion of H_2O.

The thermal consequence of the CO_2 flux from the mantle has been examined in detail by Ganguly et al. (1995) in terms of both advective heat transfer and the result of irreversible decompression. It is of interest of here to discuss the latter aspect of their analysis. Using the thermodynamic properties (i.e. α, ρ and C_P) of CO_2 from Bottinga and Richet (1981), and Eq. (7.7.8) with a typical value of dh/dP $\sim$ -3 km/kbar, Ganguly et al. (1995) showed that **adiabatic** irreversible decompression of CO_2 would lead to a net increase in temperature of 15–20 °C/kbar of decompression (which is equivalent to ~ 3 km of ascent) at a constant velocity. The increase in the volumetric heat content of CO_2 due to its irreversible decompression is given by $\rho C_p' \Delta T$, where ΔT is its temperature change due to the decompression process. Ganguly et al. (1995) thus concluded that the adiabatic rise of CO_2 from a depth of 90–20 km (approximate depth of formation of the granulites) in the Earth's interior leads to an increase of the heat content of CO_2 by 594–790 J/cm^3, which is significant compared to the heat derived from the advective heat transfer by the fluid.

References

Anderson DL (1989) Theory of the Earth. Blackwell, Boston, Oxford, London, Edinburgh, Melbourne, p 366

Anderson OL (1995) Equations of state of solids for geophysics and ceramic science. Oxford, New York, Oxford, p 405

Anderson OL (2000) The Grüneisen ratio for the last 30 years. Geophys J Int 143:279–294

Asimow PD, Hirschmann MM, Ghiorso MS, O'Hara MJ, Stolper EM (1995) The effect of pressure induced solid-solid phase transition on the decompression melting of the mantle. Geochim Cosmochim Acta 59:4489–4506

Asimow PD, Hirschmann MM, Stolper EM (1997) An analysis of variations in isentropic melt productivity. Philos Trans Roy Soc Lond A 335:255–281

Birch F (1952) Elasticity and the constitution of the Earth's interior. J Geophys Res 57:227–286

Boehler R (1993) Temperatures in the Earth's core from melting-point measurements of iron at high pressures. Nature 363:534–536

Bottinga Y, Richet P (1981) High pressure and temperature equation of state and calculation of the thermodynamic properties of gaseous carbon dioxide. Am J Sci 281:615–660

Brown JM, Shankland TJ (1981) Thermodynamic parameters in the Earth as determined from seismic profiles. Geohys J Roy Astron Soc 66:579–596

Bullen KE (1949) An Earth model based on compressibility-pressure hypothesis. Monthly notices of the Royal Astronomical Society, 109:72–720

Burnham CW, Davis NF (1971) The role of H_2O in silicate melts: I. P-V-T relations in the system $NaAlSi_3O_8$-H_2O to 10 kilobars, 700 and 1000 °C. Am J Sci 274:902–940

Dziewonski AM, Anderson DL (1981) Preliminary reference earth model. Phys Earth Planet Inter 25:297–356

Furbish DJ (1997) Fluid physics in geology. Oxford, p 476

Ganguly J (2005) Adiabatic decompression and melting of mantle rocks. Geophys Res Lett 32: L06312. https://doi.org/10.1029/2005GL022363

Ganguly J, Singh RN, Ramana DV (1995) Thermal perturbation during charnockitization and granulite facies metamorphism in southern India. J Metamorph Geol 13:419–430

Ghiorso MS (1997) Thermodynamic models of igneous processes. Annu Rev Earth Planet Sci 25:221–241

Hansen EC, Janardhan AS, Newton RC, Prame WKB, Ravindra Kumar GR (1987) Arrested charnockite formation in southern India and Sri Lanka. Contrib Mineral Petrol 96:225–244

Iwamori H, McKenzie D, Takahashi E (1995) Melt generation by isentropic mantle upwelling. Earth Planet Sci Lett 134:253–266

Janardhan AS, Newton RC, Hansen EC (1982) The transformation of amphibolite facies gneiss to charnockite in southern Karnataka and northern Tamil Nadu, India. Contrib Mineral Petrol 79:130–149

Kellogg LH (1997) Growing the Earth's D″ layer: effect of density variations at the core-mantle boundary. Geophys Res Lett 24:2749–2752

Kieffer SW (1977) Sound speed in liquid-gas mixtures—water-air and water-steam. J. Geophys Res 82:2895–2904

Kieffer SW, Delaney JM (1979) Isoentropic decompression of fluids from crustal and mantle pressures. J Geophys Res 84:1611–1620

Knapp R, Norton DL (1981) Preliminary numerical analysis of processes related to magma crystallization and stress evolution in cooling pluton environments. Am J Sci 281:35–68

Li J, Fei Y (2003) Experimental constraints on core composition. In: Carlson RW (ed) The mantle and core, treatise on geochemistry, vol 2. Elsevier, pp 521–546

McKenzie D, Bickle MJ (1988) The volume and composition of melt generated by extension of lithosphere. J Petrol 29:625–679

Murakami M, Hirose K, Sata N, Ohishi Y, Kawamura K (2004) Phase transformation of $MgSiO_3$ perovskite in the deep mantle. Science 304:855–858

Nicolas A (1995) Mid-ocean ridges: mountains below sea level. Springer, p 200

Norton DL, Panichi C (1978) Determination of the sources and circulation paths of thermal fluids: the Abano region, northern Italy. Geochim Cosmochim Acta 42:183–294

Oganov AR, Ono S (2004) Theoretical and experimental evidence for a post-perovskite phase of $MgSiO_3$ in Earth's D″ layer. Nature 430:445–448

Olson P (1987) A comparison of heat transfer laws for mantle convection at very high Rayleigh numbers. Phys Earth Planet Int 48:153–160

Ramberg H (1972) Temperature changes associated with adiabatic decompression in geological processes. Nature 234:539–540

Saxena SK, Dubrovinsky (2000) Iron phases at high pressures and temperatures: phase transition and melting. Am Mineral 85:372–375

Saxena SK, Chatterjee N, Fei Y, Shen G (1993) Thermodynamic data on oxides and silicates. Springer

Saxena SK, Eriksson G (2015) Thermodynamics of Fe-S at ultra-high pressure. CALPHAD 51:202–205

Spera FJ (1981) Carbon dioxide in igneous petrogenesis: II. Fluid dynamics of mantle metasomatism. Contrib Mineral Petrol 77:56–65

Spera FJ (1984a) Adiabatic decompression of aqueous solutions: applications to hydrothermal fluid migration in the crust. Geology 12:707–710

Spera FJ (1984b) Carbon dioxide in igneous petrogenesis: III. Role of volatiles in the ascent of alkali magma with special reference to xenolith-bearing mafic lavas. Contrib Mineral Petrol 88:217–232

Stolper EM (1996) Adiabatic melting of the mantle. Geochem Soc Newslett 92:7–9

Thompson JB Jr (1970) Geochemical reactions in open systems. Geochim Cosmochim Acta 34:529–551

Tirone M (2016) On the thermal gradient in the Earth's deep interior. Solid Earth 7:229–238

Turcotte DL, Schubert G (1982) Geodynamics: applications of continuum physics to geological problems. Wiley, p 450

Waldbaum DR (1971) Temperature changes associated with adiabatic decompression in geological processes. Nature 232:545–547

Williamson ED, Adams LH (1923) Density distribution in the Earth. J Wash Acad Sci 13:413–428

Wood SA, Spera FJ (1984) Adiabatic decompression of aqueous solutions: Applications to hydrothermal fluid migration in the crust. Geology 12:707–710

Zeldovich YB, Razier YP (1966) Physics of shock wave and high temperature hydrodynamic phenomena, vols 1 and 2. Academic press, New York

Zerr A, Diegeler A, Boehler R (1998) Solidus of Earth's deep mantle. Science 281:243–246

Thermodynamics of Solutions

8

> *Diffusion has the reputation of being a difficult subject, much harder than … solution thermodynamics. In fact it is relatively simple. … I can easily explain a diffusion flux. … I suspect that I have never clearly explained chemical potentials to anyone.*
>> E. L. Cussler (1984, Diffusion: mass transfer in fluid systems)

We have, so far, considered thermodynamic potentials for phases that are at a fixed composition. However, when the compositions of the phases become variable, we obviously need to take into account the effect of the compositional changes of the phases on their thermodynamic properties. We have seen earlier (Sect. 3.1) that the extensive thermodynamic potentials H, F and G, which are most useful for practical applications of thermodynamics, are, nonetheless, auxiliary functions, and can be derived by systematic Legendre transformations on the fundamental thermodynamic potential U. For a system with fixed masses of all species and unaffected by a force field, U is completely determined by specifying the extensive properties S and V. If the mole numbers of the different species in the system change, then we should begin by making the appropriate modification to the expression of U, and derive from that the expressions for the auxiliary thermodynamic potentials. The required modification to the expression of U leads to the introduction of a new intrinsic thermodynamic property known as the **chemical potential** of a component in a solution.

8.1 Chemical Potential and Chemical Equilibrium

Gibbs laid the foundation of chemical thermodynamics in his monumental work entitled "On the Equilibrium of Heterogeneous Substances" that was published in 1875 and 1878 in the Transactions of the Connecticut Academy of Sciences

© Springer Nature Switzerland AG 2020
J. Ganguly, *Thermodynamics in Earth and Planetary Sciences*,
Springer Textbooks in Earth Sciences, Geography and Environment,
https://doi.org/10.1007/978-3-030-20879-0_8

(see Gibbs 1961). Here Gibbs argued that if n_1, n_2 etc. are the number of moles of different species in a system that are subject to change by reversible mass exchange with the surrounding, then U must be a function of these mole numbers in addition to S and V, so that the fundamental relation must now be written as

$$U = U(S, V, n_1, n_2, \ldots),$$ (8.1.1)

instead of U = U(S, V), as written in Eq. (2.7.6). The total derivative of U is then

$$dU = \left(\frac{\partial U}{\partial S}\right)_{V,n_i} dS + \left(\frac{\partial U}{\partial V}\right)_{S,n_i} dV + \left(\frac{\partial U}{\partial n_1}\right)_{V,S,n_k \neq 1} dn_1 + \left(\frac{\partial U}{\partial n_2}\right)_{V,S,n_k \neq 2} dn_2 + \ldots,$$
(8.1.2)

where n_i stands for the mole numbers of all components (i.e. n_1, n_2 ...) and n_k stands for the mole numbers of all components except the one appearing in a partial derivative of U with respect to n_i.

We know from the earlier developments that $\partial U/\partial S = T$ and $\partial U/\partial V = -P$. Thus, the partial derivatives of U with respect to the number of moles or masses of specific components are the new partial derivatives in the representation of the total derivative of U. The partial derivative of U with respect to n_i have been called the **chemical potential** of the component i by Gibbs, and is commonly represented by the symbol μ_i. Thus,

$$\mu_i = \left(\frac{\partial U}{\partial n_i}\right)_{S,V,n_j \neq n_i}$$ (8.1.3)

Equation (8.1.2) can now be written as

$$dU = TdS - PdV + \sum_i \mu_i dn_i$$ (8.1.4)

Note that this expression is valid only for a reversible process since the term δq is replaced by TdS. It is easy to see, using the second law (Eq. 2.4.9), that for an irreversible process, dU is less than the right hand quantity.

The relation between U and any of the auxiliary state functions or potentials is not affected by making U dependent on the mole numbers of the components in the system. This is simply because of the fact that the expressions for the auxiliary functions have been derived by *partial* Legendre transforms of U. For example, if we perform the partial Legendre transform of U with respect to V, we obtain $I_v \equiv H = U + PV$ (Eq. 3.1.4), whether or not U depends on the mole numbers of the components. Differentiating now the expressions of H, F and G (i.e. H = U + PV; F = U − TS and G = U − TS + PV: see Box 3.1.1), and writing dU according to Eq. (8.1.4), we obtain

$$\text{Box (8.1.1)}$$

$$dH = VdP + TdS + \sum_i \mu_i dn_i \qquad (8.1.5)$$

$$dF = -PdV - SdT + \sum_i \mu_i dn_i \qquad (8.1.6)$$

$$dG = VdP - SdT + \sum_i \mu_i dn_i \qquad (8.1.7)$$

These expressions are the same as the corresponding expressions for a system of fixed mole numbers of different species (Box 3.1.2) except for the additional term $\sum \mu_i dn_i$.

While Eq. (8.1.3) constitutes the fundamental definition of chemical potential, it is evident from the last three expressions that chemical potential can also be defined in terms of the rate of change of the thermodynamic potentials H, F and G by holding an appropriate combination of variables constant. The different expressions of chemical potential are summarized below.

$$\mu_i = \left(\frac{\partial U}{\partial n_i} \right)_{S,V,n_k \neq n_i}$$

$$\mu_i = \left(\frac{\partial H}{\partial n_i} \right)_{P,S,n_k \neq n_i}$$

$$\text{Box (8.1.2)}$$

$$\mu_i = \left(\frac{\partial F}{\partial n_i} \right)_{V,T,n_k \neq n_i}$$

$$\mu_i = \left(\frac{\partial G}{\partial n_i} \right)_{P,T,n_k \neq n_i}$$

Of these, the last definition of μ_i in terms of G is used most commonly, especially in geological problems, since P and T are the variables of common interest.

Consider now a closed system that is subdivided into two homogeneous parts (I and II) by a membrane, which is permeable to the transfer or diffusion of only one component, i (Fig. 8.1). If the system is held at a constant P-T condition, then the change in the total Gibbs energy of the system as a result of transfer of some amount of i from one part to another equals the sum of the changes of Gibbs energies of the two subsystems. Thus, at constant P and T,

$$
\begin{aligned}
dG &= dG^{I} + dG^{II} \\
&= \mu_i^{I} dn_i^{I} + \mu_i^{II} dn_i^{II} \\
&= dn_i^{II} (\mu_i^{II} - \mu_i^{I})
\end{aligned}
\tag{8.1.8}
$$

The last equality follows from the fact that since the overall system is closed, $dn_i^{I} + dn_i^{II} = 0$.

For spontaneous change at constant P-T condition in a system restricted only to P-V work, $dG \leq 0$ (Eq. 3.2.4), the equality holding only when equilibrium is achieved. Thus, if $\mu_i^{II} > \mu_i^{I}$, then $dn_i^{II} < 0$, that is the component i must flow from the subsystem II to I, that is from its state of higher chemical potential to that of lower chemical potential (Fig. 8.2), so that the overall Gibbs energy of the system is decreased. When $\mu_i^{I} = \mu_i^{II}$, there must be no diffusion of i across the membrane since such a process would not reduce the Gibbs energy of the system. The equality of chemical potential of the component i between the subsystems I and II then defines the condition of chemical equilibrium in the overall system. Following the derivation of Eq. (8.1.8), it may be easily verified that equilibrium under different sets of conditions, namely at constant T and V or constant S and V or constant S and P, for which we minimize the potentials F, U and H, respectively, also lead to the condition of equality of chemical potential of the component i between the two subsystems.

Let us now assume that the overall closed system has an arbitrary number of components and is subdivided into an arbitrary number of compartments or subsystems, which are separated from one another by semi-permeable membranes. In order for the Gibbs energy of the overall system to be at a minimum, it is necessary and sufficient that the chemical potential of each component be the same in all compartments which are open to its diffusive exchange, since, otherwise, diffusion of a component from the state of higher to that of lower chemical potential would reduce the overall Gibbs energy of the system.

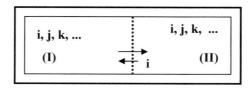

Fig. 8.1 A closed system is subdivided into two parts by a semipermeable membrane that is open to the transfer of only one component, i

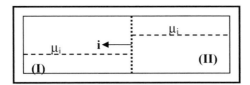

Fig. 8.2   Flow of component i from its state of higher chemical potential to that of lower chemical potential

There are two important points to notice in the above development. First, the condition of chemical equilibrium at a specified condition (constant P-T or T-V or S-V or S-P) requires that the chemical potentials of only those components *that are permitted to flow* be equal in the parts of the system through which the flow can take place. Thus, for example, the equilibrium condition between two minerals $(Fe,Mg)SiO_3$ (orthopyroxene: Opx) and $(Fe,Mg)_2SiO_4$ (olivine: Ol) requires that only $\mu_{Fe}^{Opx} = \mu_{Fe}^{Ol}$, if we ignore charged defects as a component. In the absence of charged defects, the contents of Si and O are fixed in both minerals by the requirement of charge balance, and consequently, these components are not allowed to diffuse between the two minerals.

The second point concerns the flow of matter in a multicomponent diffusion process. For the case of simultaneous diffusion of more than two components, a component may diffuse in the direction of increasing concentration or even chemical potential because of strong cross-coupling with other diffusing components. This cross-coupling is often referred to as the hydrodynamic effect in the multicomponent diffusion theory, and the process of diffusion in the direction of increasing concentration or chemical potential is called **up-hill** diffusion. To be sure, equilibrium is achieved only when the chemical potential of each diffusing species is the same in each part of the system that is open to it, but in general, it is not necessary that each component must always flow in the direction of its lower chemical potential in a multicomponent diffusion process. In such process, the flux of each component depends on the chemical potential gradients of all independent components, not just on its own chemical potential gradient. This is a consequence of irreversible thermodynamics that is discussed in Appendix A. Usually, however, the cross-coupling effects of the other components are not strong enough to change the direction of flux of a component in response to its own chemical potential gradient.

In general, things flow or move from higher to lower values of an appropriate potential. For example, electrical charge flows from higher to lower values of electrical potential or voltage, or matter falls from higher to lower value of the gravitational potential. Hence the name chemical potential since a chemical component should usually flow in the direction of its lower chemical potential.

8.2 Partial Molar Properties

If Y is an extensive thermodynamic property of a system (G, F, H, S or V), then a corresponding partial molar property, y_i, gives the rate of change of Y with respect to a change in the number of moles of the component i when pressure, temperature and the mole numbers of all other components are kept constant. That is

$$\boxed{y_i = \left(\frac{\partial Y}{\partial n_i}\right)_{P,T,n_j \neq n_i}} \qquad (8.2.1)$$

Thus, we define partial molar Gibbs free energy as

$$g_i = \left(\frac{\partial G}{\partial n_i}\right)_{P,T,n_j \neq n_i} \qquad (8.2.2)$$

and similarly for the other partial molar quantities corresponding to F, H, S and V.

Comparing Eq. (8.2.2) and the expressions of μ_i in the Box (8.1.2), we find that the chemical potential of a component i, μ_i, at constant P, T n_j condition is the same as its partial molar Gibbs free energy. It should, however, be obvious that the identity $\mu_i = y_i$ is not valid when the chemical potential is defined in terms of any other state function and y_i represents its partial molar property.

We now seek a relationship between the total or integral value of an extensive quantity and the corresponding partial quantity. Since Y = f(P, T, n_1, n_2, ...), we have at constant P-T condition,

$$dY = \left(\frac{\partial Y}{\partial n_1}\right)_{n_j \neq n_1} dn_1 + \left(\frac{\partial Y}{\partial n_2}\right)_{n_j \neq n_2} dn_2 + \dots$$
$$= y_1 dn_1 + y_2 dn_2 + \dots \qquad (8.2.3)$$

Since a partial molar quantity is an intensive property, its value does not change by changing the size of the system at a constant P-T condition as long as the composition of the system is kept fixed. Since, $n_i = X_i N$, where X_i is the mole fraction of the component i and N is the total number of moles of all components in a system, we have for constant X_i, $dn_i = X_i dN$. Using Eq. (8.2.3), we then have $dY = (y_1 X_1 + y_2 X_2 + \dots)dN$. Integrating this equation, we get

$$Y = (y_1 X_1 + y_2 X_2 + \dots)N + I,$$

where I is an integration constant. Noting that Y = 0 when N = 0 (that is the system does not have an extensive property when it has no content of any component), we get I = 0. Now substituting n_i for NX_i in the last equation, we obtain an important relation between an extensive property and the corresponding partial properties of the components at a constant P-T condition,

$$\boxed{Y = \sum_i n_i y_i} \qquad (8.2.4)$$

Specifically, we note that at constant P-T,

$$G = \sum_i n_i g_i = \sum_i n_i \mu_i \qquad (8.2.5)$$

A molar property of a component is obviously different from its partial molar property, but from Eq. (8.2.4), we may consider a partial molar property of a component as its 'effective' molar property *in solution* such that it yields the corresponding integral property of the solution in the same way that the molar properties of pure substances yield the corresponding integral property of a mechanical mixture. This statement may be illustrated by considering the volumetric properties. For a mechanical mixture, $V = n_1 v_1^o + n_2 v_2^o + \ldots$, where V and v_i^o are, respectively, the total volume of the mixture and molar volume of the pure substance or component i. When the components are in solution, we have an analogous relation between the total volume, V, of the solution, and the partial molar volumes of the components at the specific composition of the solution, viz., $V = n_1 v_1 + n_2 v_2 + \ldots$, which follows from Eq. (8.2.4).

Combination of Eqs. (8.2.5) and (8.1.7) leads to an important relation among the chemical potentials of components in a solution. Differentiating Eq. (8.2.5), we have $dG = \sum n_i d\mu_i + \sum \mu_i dn_i$. However, from Eq. (8.1.7), we also have at constant P, T condition, $dG = \sum \mu_i dn_i$. Comparison of these two expressions for dG shows that at constant P-T condition

$$\sum_i n_i d\mu_i = 0 \qquad (8.2.6)$$

or, upon dividing both sides of this equation by the total mole numbers, N, of the solution,

$$\boxed{\sum_i X_i d\mu_i = 0} \qquad (8.2.7)$$

This is known as the **Gibbs-Duhem relation** that has many applications in the field of solution thermodynamics (see Sect. 8.5 and Problem A.1 for examples). Physically it says that in an n-component system, the chemical potentials of n-1 components are independent.

Problem 8.1 A relation analogous to Eq. (8.2.7) holds for any other type of partial quantity. Show that, in general, at constant P-T condition,

$$\sum_i X_i dy_i = 0 \qquad (8.2.8)$$

Hint: make use of the relation $Y = Y(P, T, n_1, n_2 \ldots)$ and show that at constant P-T, $dY = \Sigma y_i dn_i$. Then follow the logic behind the derivation of Eq. (8.2.7).

We will refer to this equation as the generalized Gibbs-Duhem relation.

*Problem 8.2 Prove the following relations among the partial molar properties:

$$v_i = \frac{\partial \mu_i}{\partial P} ; \quad -s_i = \frac{\partial \mu_i}{\partial T} ; \quad h_i = \mu_i + Ts_i \qquad (8.2.9 : a, b, c)$$

8.3 Determination of Partial Molar Properties

8.3.1 Binary Solutions

If we know the integral property Y of a solution as a function of the mole numbers of the components of the solution, then, of course, determination of the corresponding partial molar quantity of a component of the solution is straightforward as it can be obtained by partial differentiation of Y with respect to n_i, according to the definition of y_i (Eq. 8.2.1). Usually, however, we have the molar properties as functions of the mole fractions or concentrations of the components. In this section we discuss the methods of derivation of a partial molar quantity from the data of the corresponding molar property as a function of mole fractions of the components in a solution. We denote the molar property of a solution as $Y_m = Y/N$ (e.g. molar volume of a solution, $V_m = V/N$, molar Gibbs energy of a solution, $G_m = G/N$, etc., where N is the total number of moles of all components).

Let us first consider a binary solution, for which, according to Eq. (8.2.4), $dY_m = d(X_1 y_1 + X_2 y_2) = (X_1 dy_1 + X_2 dy_2) + (y_1 dX_1 + y_2 dX_2)$. Using the generalized Gibbs-Duhem relation, Eq. (8.2.8), the first parenthetical term of this equation is zero. Thus, we have, at constant P-T condition

$$dY_m = y_1 dX_1 + y_2 dX_2 \qquad (8.3.1)$$

Multiplying and differentiating both sides of this equation by X_1 and with respect to X_2, respectively, and noting that for a binary solution $dX_1 = -dX_2$, we have

$$X_1 \left(\frac{\partial Y_m}{\partial X_2} \right)_{P,T} = -X_1 y_1 + X_1 y_2 \qquad (8.3.2)$$

But from Eq. (8.2.4) $X_1 y_1 = Y_m - X_2 y_2$. Thus,

$$X_1 \left(\frac{\partial Y_m}{\partial X_2} \right)_{P,T} = -Y_m + y_2(X_1 + X_2) = -Y_m + y_2 \qquad (8.3.3)$$

or

$$y_2 = Y_m + X_1 \left(\frac{\partial Y_m}{\partial X_2}\right)_{P,T} \qquad (8.3.4)$$

This equation can be written in two different forms, viz.,

$$y_i = Y_m + (1 - X_i) \left(\frac{\partial Y_m}{\partial X_i}\right)_{P,T} \qquad (8.3.5)$$

and

$$y_i = Y_m + \left(\frac{\partial Y_m}{\partial X_i}\right)_{P,T} - X_i \left(\frac{\partial Y_m}{\partial X_i}\right)_{P,T} \qquad (8.3.6)$$

The usefulness of these alternative forms will be evident when we discuss partial properties in a multicomponent solution.

Equation (8.3.5) has a simple geometric interpretation about the relationship between the partial quantity of a component and the corresponding molar property of the solution. This is discussed with reference to Fig. 8.3 that shows a hypothetical variation of a molar property, Y_m, of a binary solution as a function of the mole fraction of the component 2, X_2. According to Eq. (8.3.4), the partial properties of the two components, y_1 and y_2, at a composition X_2' of the solution is given by the intercepts of the tangent line to the Y versus X curve at X_2' on the vertical lines at $X_2 = 0$ and $X_2 = 1$. In order to prove this geometric interpretation, let us consider the value of y_2 for the solution composition of X_2'. From Fig. 8.3, we have

$$y_2\left(X_2'\right) = Y_m\left(X_2'\right) + Z,$$

Fig. 8.3 Geometric interpretation of a partial molar quantity, y_i, in a binary solution, as defined by Eq. (8.3.5)

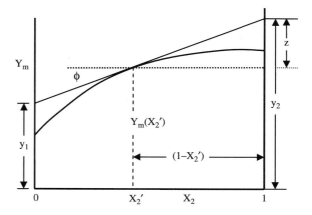

where $Y_m(X_2')$ is the molar value of Y at the solution composition X_2'. However,

$$Z = (1 - X_2') \tan \theta,$$

where $\tan \theta = (\partial Y_m / \partial X_2)$. Thus,

$$y_2(X_2') = Y_m(X_2') + (1 - X_2) \frac{\partial Y_m}{\partial X_2}$$

as in Eq. (8.3.5). It should also be evident from Fig. 8.3 that as $X_i \to 1, y_i \to y_i^o$, where the last quantity is a molar property of a pure component.

(A comment about the pressure dependence of Gibbs energy of a pure species and of its chemical potential in a solution seems appropriate at this point. Since its molar volume is a positive quantity, the Gibbs energy of a pure end member increases with pressure (Eq. 3.1.10; Fig. 6.7). However, the pressure dependence of the chemical potential of the same species, when it is dissolved in a solution, (Eq. 8.2.9a) is not necessarily positive, although it is usually so. This is due to the fact that when the curvature of the line depicting the variation of V_m versus X is convex downwards, it is possible for a component to have a negative partial molar volume in a solution such as for propane and n-butane in propane–methane and n-butane–CO_2 mixtures, respectively (Prausnitz et al. 1986). In some cases the magnitude of the (negative) partial molar volume of a component can be quite large, even exceeding that of its molar volume. This property has interesting applications in chemical engineering and forms the basis of what is known as "supercritical extraction". In this process, the chemical potential of a solute is raised by slightly reducing the pressure near the critical condition of a solvent where the density of the latter changes very rapidly as a function of pressure, thereby leading to the precipitation of solutes. For further discussion of this topic, see Prausnitz et al. (1986). However, no geological example of this type of behavior has yet been reported.

8.3.2 Multicomponent Solutions

For the determination of partial quantities in a **multi-component solution**, one can take two alternative approaches. The first is that due to Darken (1950), who presented a generalization of Eq. (8.3.5) to multi-component solution, and the second is due to Hillert and co-workers (see Hillert 1998), who presented a generalization of Eq. (8.3.6). We present below the final results of their derivations, which we would refer to as Darken and Hillert equations, respectively, and leave the interested readers to consult the original works to understand how these equations were derived.

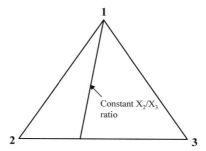

Fig. 8.4 Schematic illustration of the reduction of a ternary solution to a pseudo-binary solution. A line with constant X_2/X_3 ratio defines a pseudo-binary line with 1 as an end-member component

8.3.2.1 Darken Equation

Darken (1950) showed that a partial molar property of a component, y_i, corresponding to the molar property, Y_m, of a multi-component solution can be determined according to

$$y_i = Y_m + (1 - X_i)\left(\frac{\partial Y_m}{\partial X_i}\right)_{P,T,X_j/X_k \ldots X_n/X_k} \qquad (8.3.7)$$

where j or $n \neq k \neq i$. Note that this equation, which is often referred to as the Darken equation, is formally analogous to the expression of a partial molar property of a component in a binary solution, Eq. (8.3.5). By taking the derivative of Y_m at constant relative amounts of all components other than the one for which the partial property is sought, the multi-component solution has been reduced to a pseudo-binary solution. This is easy to understand by considering the case of a ternary solution. For example, if the composition of a ternary solution varies along a straight line connecting the apex 1 to the binary join 2–3 (Fig. 8.4), then the composition of the solution has a fixed value of the ratio X_2/X_3. In this case the ternary solution behaves as a quasi-binary solution. The partial quantity y_i in the ternary solution at any composition along a straight line defining a fixed value of X_j/X_k can be obtained from the Darken equation. The Darken equation has been applied by Sack and Loucks (1985) and Ghiorso (1990) to the problems of multi-component mineral solid solutions.

8.3.2.2 Hillert Equation

The Hillert expression for a partial molar property in a multicomponent system can be written as

$$y_i = Y_m + \left(\frac{\partial Y_m}{\partial X_i}\right)_{P,T,X_{j\neq i}} - \sum_{l=i}^{n} X_l \left(\frac{\partial Y_m}{\partial X_l}\right)_{P,T,X_{i\neq l}} \qquad (8.3.8)$$

This equation has the advantage that the terms for all components within the summation sign are treated in the same way, which makes it simple to handle in computer calculations. There is no need to consider explicitly that in an n component system, there are $n - 1$ independent components because of the stoichiometric constraint $\Sigma X_i = 1$. The reader is referred to Hillert (1998) for derivation of the above equation, but we show below how the binary expression (8.3.6) can be transformed to the above form.

Equation (8.3.6) can be written as

$$y_i = Y_m + \left(\frac{\partial Y_m}{\partial X_i}\right)_{(i+j),P,T} - X_i \left(\frac{\partial Y_m}{\partial X_i}\right)_{(i+j),P,T} \tag{8.3.9}$$

where the subscript $(i + j)$ indicates that the sum of the mole fractions of i and j is held constant (in a binary solution, $X_i + X_j = 1$). For brevity, we will henceforth omit the explicit stipulation in the partial derivatives that P and T are held constant. Now, it can be easily shown that for any function $Z = f(x_i, x_2, x_3 \ldots)$

$$\left(\frac{\partial Z}{\partial x_i}\right)_{(i+j),k} = \left(\frac{\partial Z}{\partial x_i}\right)_{j,k} - \left(\frac{\partial Z}{\partial x_j}\right)_{i,k} \tag{8.3.10}$$

where the subscript k denotes that the mole fractions of all components other than i and j are held constant. (The reader can check the validity of the above relation by considering the function $Y = aX_i + bX_j + cX_k$, and carrying out the operations indicated on the left and right of above equation, both of which will yield $a - b$.) Expanding the derivative terms in Eq. (8.3.9) in the form of the last equation, rearranging terms, and imposing the relation $X_j = 1 - X_i$, we obtain

$$y_i = Y_m + \left(\frac{\partial Y_m}{\partial X_i}\right) - X_i \left(\frac{\partial Y_m}{\partial X_i}\right) - X_j \left(\frac{\partial Y_m}{\partial X_j}\right) \tag{8.3.11}$$

which is the binary form of Eq. (8.3.8).

*Problem 8.3 A molar property of many solutions may be represented as $Y_m = \sum X_i y_i^o + \sum W_{ij} X_i X_j$, where W_{ij} is a constant for the binary join i-j, and y_i^o is the molar property of the pure component i (this type of solutions are known as Regular Solutions; see Sect. 9.2.4). Using Eq. (8.3.8), show that the partial molar property of the component i is given by $y_i = y_i^o + \sum W_{ij} X_j - \sum W_{kj} X_k X_j$, where the first summation is taken over all binary joins involving i, and the last summation is taken over all binary joins. Also show that it reduces to the expression for a binary solution, as derived from Eq. (8.3.4), when the mole fractions of all but two components are set equal to zero.

8.4 Fugacity and Activity of a Component in a Solution

The fugacity of a component in a solution, α, is defined in a formally similar way as that for a pure component, Eq. (3.6.1), by using partial Gibbs energy or chemical potential of the component instead of Gibbs energy of the pure component. Thus,

$$d\mu_i^\alpha = RTdlnf_i^\alpha \qquad (8.4.1)$$

Integrating the above equation at a constant P-T condition between two states of the system of different compositions, we have

$$\mu_i^\alpha(P, T, X) = \mu_i^\alpha(P, T, X^*) + RT\ln\left[\frac{f_i^\alpha(P, T, X)}{f_i^\alpha(P, T, X^*)}\right] \qquad (8.4.2)$$

Using Eq. (8.4.1), the first term on the right can also be expressed as

$$\mu_i^\alpha(P, T, X^*) = \mu_i^\alpha(P', T, X^*) + RT\ln\left[\frac{f_i^\alpha(P, T, X^*)}{f_i^\alpha(P', T, X^*)}\right] \qquad (8.4.3)$$

Combining the last two equations, we have

$$\mu_i^\alpha(P,T,X) = \mu_i^\alpha(P', T, X^*) + RT\ln\left[\frac{f_i^\alpha(P,T,X)}{f_i^\alpha(P', T, X^*)}\right] \qquad (8.4.4)$$

Equations (8.4.2) and (8.4.4) show that the change of chemical potential as a function of pressure and composition can be expressed by the sum of two terms. However, the decomposition of $\mu_i(P, T, X)$ into two terms may be carried out in two different ways. In Eq. (8.4.2), the reference pressure on the right hand side is set to be the same pressure, P, as in the left hand side and is thus a variable, whereas in Eq. (8.4.4), it is fixed to a specific pressure, P'. The conditions assigned to the chemical potential on the right hand side of the two equations, which are the same as those assigned to the corresponding fugacity terms in the denominator, define what is known in thermodynamics as the **standard state**s, whereas the ratio of the fugacity terms is known as the **activity** of the component i at a P, T, X condition, $a_i(P, T, X)$.

The concepts of both fugacity and activity were introduced by Lewis (1907: see Lewis and Randall 1961), leading to the development of thermodynamic formalisms to treat phase equilibria involving non-ideal mixtures of components in solutions and gases. The rationale for introducing the concept of activity lies in the fact that for highly nonvolatile substance, such as a solid, it may be impossible to determine the fugacity values precisely without recourse to highly sophisticated instrumentations. In such cases, it is advantageous to deal with the ratio of fugacity values of a substance between two different states, instead of the individual values at different states.

We can now express $\mu_i^\alpha(P, T, X)$ as either

$$\mu_i^\alpha(P,T,X) = \mu_i^\alpha(P,T,X^*) + RT \ln a_i^\alpha(P,T,X), \qquad (8.4.5)$$

corresponding to Eq. (8.4.2), or

$$\mu_i^\alpha(P,T,X) = \mu_i^\alpha(P', T,X^*) + RT \ln a_i^\alpha(P,T,X) \qquad (8.4.6)$$

corresponding to Eq. (8.4.4). Note that $a_i(P, T, X)$ in these equations are different quantities as these represent relative fugacities with respect to two different standard states. If we set the chemical potentials on the left of the last two equations equal to those of their respective standard states, then we have $RT \ln a_i^\alpha(P, T, X) = 0$ or $a_i^\alpha(P, T, X) = 1$ in both cases. Thus, the activity of a component in its **chosen standard state is unity**, regardless of what that choice happens to be. There is no formal thermodynamic restriction on the choice of this standard state except that it must be at (a) **the temperature of interest**, since the integrations in Eqs. (8.4.2)–(8.4.4) were carried out at a fixed temperature, and (b) **a fixed composition**.

While the final result of a thermodynamic analysis must be independent of the choice of the standard state, a clever choice of standard state could greatly simplify the derivation of the result. This advantage of simplifying the thermodynamic treatment of problem by an appropriate choice of standard state is the reason behind the retention of flexibility in the choice of standard state and outweighs the advantage, as emphasized by Lewis and Randall (1961), of avoiding "the confusion if once for all we should choose for a given substance its standard state." However, as we would see later, certain choices of standard states have proved to be usually convenient in the thermodynamic treatment of solutions. For brevity, we would henceforth indicate the chemical potential and the fugacity at the standard state at T and X^* as $\mu_i^{*,\alpha}(T)$ and $f_i^{*,\alpha}(T)$, respectively, without implying anything about the choice of pressure, if it is the pressure of interest, P, or a fixed pressure, P'. Thus, we restate the last two equations in a general form as

$$\boxed{w\mu_i^\alpha(P, T, X) = \mu_i^{*,\alpha}(T) + RT \ln a_i^\alpha(P, T, X)} \qquad (8.4.7)$$

where

$$\boxed{a_i^\alpha = \frac{f_i^\alpha(P,T,X)}{f_i^*(T)}} \qquad (8.4.8)$$

(Quite often one would hear the statement that the activity of component in its pure state is unity. This, of course, cannot be a generally valid statement unless the standard state has been chosen to be that of pure component at the P-T condition of interest, i.e. $\mu_i^*(T) = \mu_i^o(P, T)$. For reasons discussed in Sect. 6.7, such a choice of standard state, however, happens to be a common practice in problems dealing with non-electrolyte solutions.)

When we deal with a solid or liquid solution, the equilibrium vapor phase above the solution consists of different components. If the vapor phase above the solution behaves as an ideal gas and the components in the condensed solution have very similar energetic properties, then the partial vapor pressure, P_i, of a component would be found to vary linearly with the content of i in the solution. If we use mole fraction of i, X_i, as a measure of the content of i in the solution, then in this simple case we would find a relation $P_i = P_i^o X_i$, where P_i^o is the vapor pressure of the pure component i at the same temperature. An example of such linear relation of the measured vapor pressures of components in a binary solution is shown in Fig. 8.5a. However, when the vapor phase above a solution deviates from the ideal gas behavior (Fig. 8.5b), as is often the case, it is the fugacity (or the corrected vapor pressure) of a component rather than its vapor pressure that is proportional to the mole fraction of the component in the solution. Specifically, we have $f_i = f_i^o X_i$, where f_i^o is the fugacity of the pure component at the P-T condition of interest, if the components in the solution have very similar energetic properties. If the latter restriction is not satisfied, then the fugacity of the component i would follow a relation of the form $f_i = f_i^o X_i \gamma_i$, where γ_i is an adjustable parameter that accounts for effects of the dissimilar energetic properties of the components in solution. To be even more general, we can write

$$f_i^\alpha(T) = f_i^{*,\alpha}(T)\left(X_i^\bullet \gamma_i^\bullet\right)^\alpha \qquad (8.4.9)$$

where $X_i^\bullet$ is some convenient measure or function of the content of i in the solution, and $\gamma_i^\bullet$ is the corresponding adjustment factor. The adjustable parameter γ_i is known as the **activity coefficient** of the component i in the phase α. In addition to the

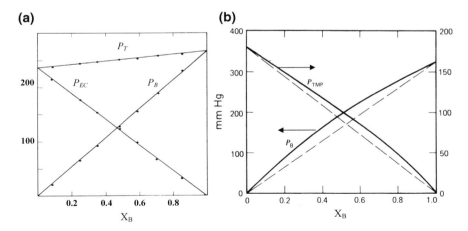

Fig. 8.5 (**a**) Linear and (**b**) nonlinear variation of species vapor pressures and of total pressure (P_T) as function of composition. (**a**) System benzene(B)-ethylene chloride(EC) solution at 49.99 °C, as determined by von Zawidzki (1900: Z. Phys. Chem. **35**, 129); (**b**) system benzene-2,2,4-trimethyl pentane (TMP) at 55 °C. From Sandler (1977)

nature of the compositional function, $X_i^\bullet, \gamma_i^\bullet$ is a function of pressure, temperature and composition. Using now the definition of activity according to Eq. (8.4.8), the above equation can be re-written as

$$a_i^\alpha = \left(X_i^\bullet \gamma_i^\bullet\right)^\alpha \qquad (8.4.10)$$

Consequently,

$$\mu_i^\alpha(P, T, X) = \mu_i^{*,\alpha}(T) + RT \ln X_i^{\bullet,\alpha} + RT \ln \gamma_i^{\bullet,\alpha} \qquad (8.4.11)$$

To conclude this section, we recapitulate that at fixed P, T, X condition,

(a) the fugacity and chemical potential of a component are absolute quantities, while
(b) activity of a component depends on the choice of standard state, and
(c) the value of $\gamma_i^\bullet$ depends on the choice of both standard state and the compositional function $X_i^\bullet$, as should be obvious from the last equation.

8.5 Determination of Activity of a Component Using Gibbs-Duhem Relation

Using Gibbs-Duhem relation, One can determine the activity of a component in a binary system if the activity of the other component is known. Combining Eqs. (8.2.7) and (8.4.5), we have

$$X_1 d \ln a_1 = -X_2 d \ln a_2 \qquad (8.5.1)$$

so that

$$\int_{X_1'}^{X_1''} d \ln a_1 = -\int_{X_1'}^{X_1''} \frac{X_2}{X_1} d \ln a_2$$

or

$$\ln a_1(X_1'') = \ln a_1(X_1') - \int_{X_1'}^{X_1''} \frac{X_2}{X_1} d \ln a_2 \qquad (8.5.2)$$

where X_1' and X_1'' stand for two values of the mole fraction X_1. Thus, if a_2 is known as a function of composition between X_1' and X_1'', and a_1 is known at the composition X_1', then a_1 can be determined at X_1'' by carrying out the integration in the last equation. Graphical integration by plotting X_2/X_1 versus $-\ln a_2$ or numerical integration poses problem of accuracy near the terminal compositions since $X_2/X_1 \rightarrow \infty$ as $X_2 \rightarrow 1$, and $-\ln a_2 \rightarrow \infty$ as $X_2 \rightarrow 0$. As discussed by Darken and Gurry (1953), the second problem can be avoided if we evaluate $\ln\gamma_1$ instead of $\ln a_1$, and then determine a_1 according to $a_1 = X_1\gamma_1$. For this purpose, we first re-write Eq. (8.5.1) as

$$X_1 d\ln X_1 + X_1 d\ln \gamma_1 + X_2 d\ln X_2 + X_2 d\ln \gamma_2 = 0$$

Noting now that $X_i d\ln X_i = dX_i$ and $dX_1 + dX_2 = 0$, we get from the above relation

$$\ln \gamma_1(X_1'') = \ln \gamma_1(X_1') - \int_{X_1'}^{X_1''} \frac{X_2}{X_1} d\ln \gamma_2 \qquad (8.5.3)$$

Since γ_2 is always finite, tending to a constant value as $X_2 \rightarrow 0$ (Henry's law: Sect. 8.8), the graphical evaluation of the above integral near $X_2 = 0$ does not pose any problem. Figure 8.6 illustrates the evaluation of γ_{Pb} in the binary Cd–Pb system by graphical integration according to the last expression. With the aid of a desktop computer, it is, however, more appropriate now to carry out the integration numerically.

Fig. 8.6 Determination of log γ_{Pb} from experimental data of log γ_{Cd} in the Cd-Pb binary system at 500 °C by graphical integration, according to Eq. (8.5.3). Change of log γ_{Pb} between two compositions is given by the area under curve between the compositions. From Darken and Gurry (1953). With permission from Mc-Graw Hill

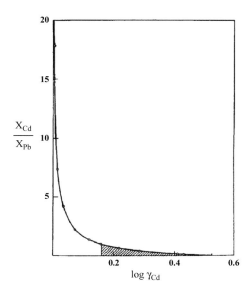

8.6 Molar Properties of a Solution

8.6.1 Formulations

Using Eq. (8.2.5), the molar Gibbs energy of a solution, G_m, is given by

$$G_m = \frac{G}{N} = \frac{1}{N}\left(\sum_i n_i\mu_i\right) = \sum_i X_i\mu_i, \qquad (8.6.1)$$

which, upon substitution of Eq. (8.4.7), yields

$$G_m = \sum_i X_i\mu_i^* + RT\sum_i X_i\ln a_i \qquad (8.6.2)$$

or, decomposing the activity term according to Eq. (8.4.10)

$$G_m = \underbrace{\sum_i X_i\mu_i^*}_{\text{mechanical mixture}} + \left(\underbrace{RT\sum_i X_i\ln X_i^\bullet}_{\Delta G_m^{\text{ideal}}} + \underbrace{RT\sum_i X_i\ln\gamma_i^\bullet}_{\Delta G_m^{\text{xs}}}\right) \qquad (8.6.3)$$

The first right hand term denotes the Gibbs energy per mole of a **mechanical** mixture of the standard state components, whereas the parenthetical term on the right denotes the **chemical** effect of mixing, and will be denoted by ΔG_m^{mix} (the decomposition of ΔG_m^{mix} into ideal and an excess components is explained later). These definitions are illustrated in Fig. 8.7, by choosing, as standard states, the pure end member components at the P-T condition of interest.

Equation (8.6.3) is the **fundamental equation** from which expressions of other molar properties of a solution are derived. Thus, the expressions for the molar entropy (S_m) and volume (V_m) of a solution are obtained from using the identities $S = -(\partial G/\partial T)_P$ and $V = (\partial G/\partial P)_T$, which yield

$$S_m = \sum_i X_i S_i^* + \underbrace{\left(-R\sum_i X_i\ln X_i^\bullet - R\sum_i X_i\ln\gamma_i^\bullet - RT\sum_i X_i\frac{\partial\ln\gamma_i^\bullet}{\partial T}\right)}_{\Delta S_m^{\text{mix}}}$$

$$(8.6.4)$$

and

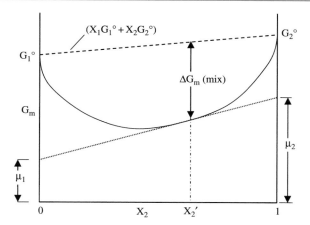

Fig. 8.7 Illustration of Gibbs free energy of mixing in a (stable) binary solution (convex downward solid line). The dashed line represents the Gibbs energy of a mechanical mixture. $\Delta G_m(mix)$ represents the mixing energy at a composition X_2'. μ_1 and μ_2 are the partial molar Gibbs energies or chemical potentials of the components 1 and 2, respectively, at the composition X_2' of the solution

$$V_m = \sum_i X_i V_i^* + \underbrace{\left(RT \sum_i X_i \frac{\partial \ln \gamma_i^\bullet}{\partial P} \right)}_{\Delta V_m^{mix}} \qquad (8.6.5)$$

Also, since $H = G + TS$, we have

$$H_m = \sum_i X_i H_i^* + \underbrace{\left(-RT^2 \sum_i \frac{\partial \ln \gamma_i^\bullet}{\partial T} \right)}_{\Delta H_m^{mix}} \qquad (8.6.6)$$

Let us now illustrate the above concepts by considering the olivine solid solution, $(Fe,Mg)_2SiO_4$. As we would see later (Sect. 9.1), the activity of an end member component, I_2SiO_4 (I: Mg or Fe) should be expressed according to

$$\boxed{a_{I_2SiO_4} = (X_I \gamma_I)^2} \qquad (8.6.7)$$

where the exponent 2 accounts for the fact that there are two moles of cations per formula unit of the solid solution. Comparing this expression with Eq. (8.4.10), in which we have expressed activity of a component in a general form as $a_i^\alpha = (X_i^\bullet \gamma_i^\bullet)$, we note that $X_{Fo}^\bullet = (X_{Mg})^2$ and $\gamma_{Fo}^\bullet = (\gamma_{Mg})^2$ where Fo stands for forsterite (Mg_2SiO_4). Similar expression holds for the activity of the fayalite component (Fe_2SiO_4). Combining Eqs. (8.6.7) and (8.6.2), the Gibbs energy of mixing per mole of the solid solution, $(Fe,Mg)_2SiO_4$, is given by

$$G_m = \sum_i X_i \, \mu_i^* + RT \sum_i X_i \ln(X_i \gamma_i)^2$$

$$= \left(X_{Mg} \, \mu_{Mg_2SiO_4}^* + X_{Fe} \, \mu_{Fe_2SiO_4}^* \right)$$

$$+ \left[2RT(X_{Mg} \ln X_{Mg} + X_{Fe} \ln X_{Fe}) + 2RT(X_{Mg} \ln \gamma_{Mg} + X_{Fe} \ln \gamma_{Fe}) \right]$$

$$(8.6.8)$$

If we choose pure forsterite, Mg_2SiO_4, and pure fayalite, Fe_2SiO_4, at the P-T condition of interest as the standard states, the first parenthetical term equals $\left(X_{Mg}G^oFo + X_{Fe}G^oF_a \right)$.

8.6.2 Entropy of Mixing and Choice of Activity Expression

At this point it is useful to recall the expression of entropy of mixing that was derived in Chap. 2 from the Boltzmann equation. It was shown earlier (Eq. 2.6.6) that for **random** distribution of the mixing units in a solution, $S_{conf} = -\nu R \sum X_i \ln X_i$, where ν stands for the number of moles of the mixing units per mole of the solution. This equation yields the first term within the square bracket of Eq. (8.6.8), which is the contribution from the ideal part of entropy of mixing of the end member components, $T\Delta S^{mix}$(ideal), noting that random distribution is a necessary condition for ideality. This simple analysis provides the rationale for the above choice of activity expression for the end member components in $(Fe,Mg)_2SiO_4$ solid solution. In general, the activity expression should be such that it yields the entropy of mixing obtained from the Boltzmann equation when the activity coefficient terms are neglected.

8.7 Ideal Solution and Excess Thermodynamic Properties

8.7.1 Thermodynamic Relations

A solution is defined to be **thermodynamically** ideal, if the activity coefficient of each component, which has been chosen to describe the properties of the solution, is unity. Thus, from Eq. (8.4.11) we have for an **ideal** solution

$$\mu_i^\alpha(P, T, X) = \mu_i^{*\alpha}(T) + RT \ln X_i^{\bullet;\alpha} \qquad (8.7.1)$$

for each component, i.

The molar properties of an ideal solution follow easily from Eqs. (8.6.3) to (8.6.6) by equating $\gamma_i = 1$. It is obvious that for a thermodynamically ideal solution, molar enthalpy (H_m) and molar volume (V_m) are given simply by the linear combination of the corresponding standard state properties. Such linear combinations are often referred to as **mechanical mixtures**. Thus, for example, if the standard states are chosen to be the states of pure components at the P-T condition of interest, then H_m and V_m of a solution would be given by linear combination of the respective end member properties, if the solution behaves ideally with respect to mixing of the end-member components. However, note from Eqs. (8.6.3) and (8.6.4) that there is a non-zero Gibbs free energy of mixing and entropy of mixing **even for** a thermodynamically ideal solution. The ideal and non-ideal parts of the Gibbs energy of mixing are shown in Eq. (8.6.3).

The molar properties of **ideal solution** are summarized below.

$$G_m = \sum_i X_i \mu_i^* + RT \sum_i X_i \ln X_i^{\bullet}$$

$$S_m = \sum_i X_i S_i^* - R \sum_i X_i \ln X_i^{\bullet}$$

$$V_m = \sum_i X_i V_i^*$$

$$H_m = \sum_i X_i H_i^*$$

Box (8.7.1)

Note that since $X_i < 1$, the ideal Gibbs free energy of mixing is always less than zero whereas ideal entropy of mixing is always greater than zero. The ideal volume and enthalpy of mixing are obviously zero.

The difference between a thermodynamic property of a solution and the corresponding ideal solution property is defined to be an **excess** (**xs**) thermodynamic property. For example, from Eq. (8.6.8), we have for one mole of olivine solid solution, $(Fe,Mg)_2SiO_4$

$$\Delta G_m^{xs} = 2RT\left(X_{Mg} \ln \gamma_{Mg} + X_{Fe} \ln \gamma_{Fe}\right), \tag{8.7.2}$$

Again, the term 2 in these equations is due to the fact that there are two moles of (Fe + Mg) per mole of olivine of the chosen formula representation. The ideal and excess Gibbs free energies of mixing for a solution with one mole of mixing units, such as $(Fe,Mg)Si_{0.5}O_2$, is illustrated in Fig. 8.8. The excess part of the molar Gibbs energy is indicated in Eq. (8.6.3). In all expressions of molar properties of a solution, Eqs. (8.6.3)–(8.6.6), the terms containing the activity coefficient, γ_i, constitute the excess thermodynamic quantities.

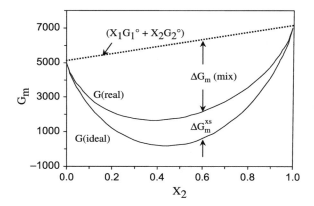

Fig. 8.8 Illustration of molar Gibbs free energy of mixing, $\Delta G_m(mix)$, and excess molar Gibbs free energy of mixing, ΔG_m^{xs}, of a binary solution. The diagram is calculated with the following values of the different parameters: $G_1^o = 5000$, $G_2^o = 7000$, $W_{21} = 7400$ and $W_{12} = 5200$ J/mol. The last two are subregular parameters for nonideal mixing, as explained in Sect. 9.2.2. In this illustration, $\Delta G^{xs} > 0$

By definition, the term $RT\ln\gamma_i$ represents the excess chemical potential of the component i. But since chemical potential represents the partial molar Gibbs free energy, $RT\ln\gamma_i$ represents the excess partial molar Gibbs free energy of the component i in a solution. Consequently, from the definition of a partial molar property (Eq. 8.2.1)

$$RT\ln\gamma_i \equiv \mu_i^{xs} = \left(\frac{\partial \Delta G^{xs}}{\partial n_i}\right)_{P,T,n_j \neq n_i} \tag{8.7.3}$$

Using Eq. (8.3.5), we can express $RT\ln\gamma_i$ in terms of the excess Gibbs free energy per mole, ΔG_m^{xs}, of a binary solution as

$$RT\ln\gamma_i = \Delta G_m^{xs} + (1 - X_i)\left(\frac{\partial \Delta G_m^{xs}}{\partial X_i}\right) \tag{8.7.4}$$

Extensions of this relationship to multicomponent solution are easily obtained from Eqs. (8.3.7) and (8.3.8). The latter expression is more convenient for computer calculation of activity coefficients in a multicomponent system.

8.7.2 Ideality of Mixing: Remark on the Choice of Components and Properties

From the standpoint of calorimetric measurements, a solution is said to have zero enthalpy of mixing if the heat of formation varies linearly between those of the

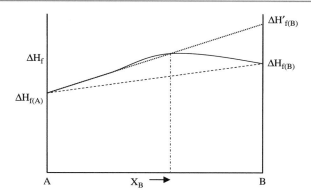

Fig. 8.9 Reduction of a nonideal enthalpy of mixing between two components A and B of a binary solution to an ideal mixing behavior within a limited compositional range, $0 \leq X_B \leq 0.6$, by assigning a hypothetical enthalpy of formation $\Delta H'_f$ to the component B. The real variation of ΔH_f as function of composition is illustrated by the solid line

end-member components. However, even if the heat of formation and other properties of a solution have non-ideal behavior with reference to the properties of the end-members, one can make a solution behave ideally or less non-ideally over a certain range of composition by ascribing hypothetical properties to the end member components, as illustrated in Fig. 8.9. In this illustration, the mixing of A and B is associated with a positive enthalpy of mixing. However, the mixing behavior within the compositional range of $X_B \sim 0$–0.6 may be treated as ideal by assigning a **hypothetical** heat of formation, $\Delta H'_{f(B)}$, to the component B. It is completely legitimate, and sometimes advantageous to choose hypothetical properties of components or hypothetical components as standard states, and generate their thermodynamic properties by extrapolation of the behavior of the properties from actual measurements or in some other way such that the nonideal behavior within the compositional range of practical interest is minimized.

8.8 Solute and Solvent Behaviors in Dilute Solution

The thermodynamic behavior of solute and solvent in dilute solutions plays important role in our understanding of the appropriate choice of standard state that simplifies the activity-composition relations, speciation of a solute such as that of H_2O in a melt that affects the melt properties, and trace element geochemistry of melt. The limiting behaviors of solute and solvent in a dilute solution are embodied in the statements of Henry's law and Raoult's law, respectively. Although these laws were formulated independently from experimental observations, one is a consequence of the other. We would state Henry's law first, which was proposed before Raoult's law, and develop Raoult's law as a consequence of the Henry's law.

As discussed later (Sect. 11.2), Henry's law plays a central role in the understanding of the equilibrium behavior of trace elements in melts as a function of the degree of melting and the nature of the source region. An useful compilation of the concentration limits over which different elements of geochemical interest obey Henry's law in hydrothermal and magmatic systems can be found in Ottonello (1997).

8.8.1 Henry's Law

It was discovered by William Henry (1774–1836) that when the mole fraction of a solute becomes very dilute, its partial vapor pressure becomes proportional to mole fraction. This is known as the **Henry's law**. In modern statements, the partial vapor pressure is replaced by fugacity to account for the nonideal behavior of the vapor phase. Thus, we state **Henry's law** as

$$\text{Limit } X_i \to 0, \quad f_i = K_H X_i \tag{8.8.1}$$

where X_i is the mole fraction of i, and K_H is known as the Henry's law constant. If the molality, m_i, instead of mole fraction is chosen as the measure of the content of i in the solution, as is the practice in the field of electrolyte solution (Chapter 12), then

$$\text{Limit } m_i \to 0, f_i = K_H^* m_i, \tag{8.8.1'}$$

where K_H^* is the Henry's law constant in the molalilty representation (molality is defined as the number of moles per kg of solvent). Now since fugacity is proportional to activity ($a_i = f_i/f_i^*$, where f_i^* is the standard state fugacity, Eq. 8.4.8), we also have

$$\text{Limit } \quad X_i \to 0, \quad a_i \propto X_i \tag{8.8.2}$$

and similarly in the molality representation, where the proportionality constant is the Henry's law constant divided by f_i^*.

Comparing Eqs. (8.8.2) and (8.4.10), it should be obvious that the proportionality constant in the former equals the activity coefficient, γ_i. Thus, within the domain of validity of the Henry's law, γ_i **is independent** of X_i (or m_i), but it **depends** on P, T and solvent composition.

The solute **i** in the last three expressions is an **actual** solute in the solution, and not a solute which dissociates or associates in the solution. The Henry's law behavior of an **actual** solute in a solution is illustrated in Fig. 8.10. However, the law, as stated above, does not hold between the fugacity (or activity) and undissociated mole fraction of a solute such as HCl, which actually dissociates in a dilute aqueous solution, but holds individually for the products of dissociation, namely, H^+ and Cl^-. Instead of obeying the relations (8.8.1) or (8.8.2), the fugacity of HCl

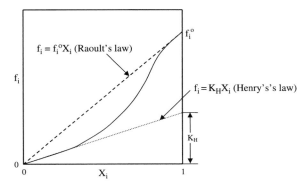

Fig. 8.10 Illustrations of the fugacity versus composition relation of a component in a solution, showing the Henry's law behavior at the dilute end and Raoultian behavior at the concentrated end

will be proportional to $(X_{HCl^\bullet})^2$, as $X_{HCl^\bullet} \to 0$, where $X_{HCl^\bullet}$ is the nominal mole fraction of HCl in the aqueous solution, i.e.

$$X_{HCl^\bullet} = \frac{n_{HCl^\bullet}}{n_{HCl^\bullet} + n_{H_2O}}, \tag{8.8.3}$$

with $n_{HCl^\bullet}$ indicating the number of moles of HCl in the solution had there been no dissociation. We would refer $X_{i^\bullet}$ as the nominal mole fraction of the solute $i^\bullet$. In general, the fugacity of an almost completely dissociated solute will obey the relation

$$\text{Limit} \quad X_i \to 0, \quad f_{i^\bullet} = K_H(X_{i^\bullet})^n, \tag{8.8.4}$$

where n is the number of moles of species into which a mole of the solute $i^\bullet$ dissociates in the solution, and $X_i^\bullet$ is the apparent or nominal mole fraction of the solute. Dividing both sides by the standard state fugacity, f_i^*, of the solute, we have

$$\boxed{\text{Limit } X_i^\bullet \to 0, \quad a_{i^\bullet} = K_H'(X_{i^\bullet})^n,} \tag{8.8.5}$$

where $K_H' = K_H/f_i^*$.

To verify the above statement about the nature of the exponent n, let us consider in detail the behavior of HCl in an aqueous solution. In a dilute solution, it will dissociate almost completely according to

$$HCl(aq) = H^+(aq) + Cl^-(aq) \tag{8.8.a}$$

Using a result derived later (Sect. 10.4), we have, at equilibrium

$$K = \frac{(a_{H^+})(a_{Cl^-})}{a_{HCl}} \qquad (8.8.6)$$

where K is a constant at a fixed P-T condition (and is known as the equilibrium constant), and **a** stands for the activity of the specified species in the aqueous solution. Since activity of a component is proportional to its fugacity (Eq. 8.4.8), we can write from the above relation

$$f_{HCl} \propto (f_{H^+})(f_{Cl^-}) \qquad (8.8.7)$$

Now if both H^+ and Cl^- obey Henry's law, then using Eq. (8.8.1) for each species, we have, as $X_{HCl} \to 0$, $f_{H^+} \propto X_{H^+}$ and $f_{Cl^-} \propto X_{Cl^-}$, so that from the last equation

$$f_{HCl} \propto (X_{H^+})(X_{Cl^-}) \qquad (8.8.8)$$

where

$$X_{H^+} = \frac{n_{H^+}}{n_{H^+} + n_{Cl^-} + n_{H_2O}} \qquad (8.8.9)$$

and

$$X_{Cl^-} = \frac{n_{Cl^-}}{n_{H^+} + n_{Cl^-} + n_{H_2O}} \qquad (8.8.10)$$

If HCl is almost completely dissociated in the solution, then from the stoichiometry of the reaction (8.8.a) we have $n_H^+ = n_{Cl}^- \approx n_{HCl^\bullet}$. Thus,

$$X_{H^+} = X_{Cl^-} \approx \frac{n_{HCl^\bullet}}{2n_{HCl^\bullet} + n_{H_2O}} \qquad (8.8.11)$$

For a dilute aqueous solution of HCl, the denominator in the last expression essentially equals $(n_{HCl}^\bullet + n_{H2O})$, and consequently $X_{H^+} = X_{Cl^-} \approx X_{HCl^\bullet}$. Thus, as $X_{HCl^\bullet} \to 0$, we have from Eq. (8.8.8),

$$f_{HCl} \propto (X_{HCl^\bullet})^2 \qquad (8.8.12)$$

8.8.2 Raoult's Law

It was discovered by Francois-Marie Raoult (1830–1901) from experimental studies that as the mole fraction of a solvent (s) approaches unity, its vapor pressure is given by $P^\circ_s X_s$, where P°_s is the vapor pressure of the pure solvent (Fig. 8.5b). The value of X_s at which such behavior becomes valid depends on the system. Like the modern version of Henry's law, that of Raoult's law is also expressed in terms of fugacity instead of vapor pressure to account for the nonideal behavior of the vapor phase. We show below that this property of the solvent is a consequence of the Henry's law behavior of the solutes in a dilute solution.

Let us consider a solution with an arbitrary number of solutes, each of which obey Henry's law within a certain dilute compositional range. According to the Gibbs-Duhem relation (Eq. 8.2.7), we have at constant P-T condition,

$$\sum_i X_i d\,\mu_i + X_j d\,\mu_j = 0$$

where i stands for a solute and j stands for the solvent. Now, using $d\mu_i = RTdlnf_i$ (Eq. 8.4.1)

$$\sum_i X_i d\,\ln f_i + X_j d\,\ln f_j = 0, \tag{8.8.13}$$

If each solute obeys the Henry's law, then

$$X_1 d\,\ln X_1 + X_2 d\,\ln X_2 + X_3 d\ln X_3 + \ldots X_j d\,\ln f_j = 0 \tag{8.8.14}$$

Differentiating both sides with respect to X_1,

$$X_1 \frac{d\,\ln X_1}{dX_1} + \sum_{i \neq 1} X_i \frac{d\,\ln X_i}{dX_1} + X_j \frac{d\,\ln f_j}{dX_1} = 0 \tag{8.8.15}$$

and using the relation dX/X = dlnX

$$1 + \sum_{i \neq 1} \frac{dX_i}{dX_1} + X_j \frac{d\,\ln f_j}{dX_1} = 0 \tag{8.8.16}$$

Now, differentiating the stoichiometric relation $X_1 + X_2 + X_3 + \ldots X_j$ (solvent) = 1 with respect X_1,

$$1 + \sum_{i \neq 1} \frac{dX_i}{dX_1} + \frac{dX_j}{dX_1} = 0 \tag{8.8.17}$$

Combining the last two equations

$$X_j \frac{d\,\ln f_j}{dX_1} = \frac{dX_j}{dX_1} \tag{8.8.18}$$

so that

$$d\ln f_j = d\ln X_j \tag{8.8.19}$$

Note that since we have imposed Henry's law behavior on the solutes, this expression is valid for the compositional range of the solvent (j) within which the solutes obey the Henry's law. Now, integrating the last expression between $X_j = 1$ and X_j', where the latter is a solvent composition within the domain of Henry's law behavior of the solutes, we have

$$\ln \frac{f_j(X_j')}{f_j^o} = \ln X_j'$$

where $f_j(X_j')$ and f_j^o are the fugacities of the solvent at the composition X_j' and at the pure state at the P-T condition of interest. Thus, since $X_j \rightarrow 1$ as $X_i \rightarrow 0$, we can write

$$\boxed{\lim X_j \rightarrow 1, \quad f_j(X_j) = f_j^o X_j} \tag{8.8.20}$$

If we now choose the pure state of a component at the P, T condition of interest as its **standard state**, then $f_f(X_j)/f_i^o = a_j(X_i)$, in which case the Raoult's law can be stated as

$$\boxed{\lim_{X_j \rightarrow 1} a_j = X_j} \tag{8.8.21}$$

A schematic activity versus composition relation of a component in a solution, referred to a standard state of pure component at the P-T condition of interest, is illustrated in Fig. 8.11. Here the component is chosen to show a negative deviation from ideality.

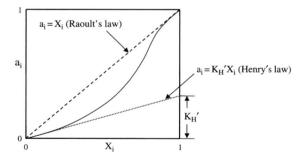

Fig. 8.11 Hypothetical activity versus composition relation of a component (i) in a binary solution showing the limiting behaviors at the dilute and concentrated ends when the standard state of the component is chosen to be its pure state (real or hypothetical) at the P-T condition of interest. $K_H' = K_H/f_i^o$, where K_H is the Henry's law constant in the fugacity representation (Eq. 8.8.1), and f_i^o is the fugacity of pure i at the P-T condition of interest

Problem 8.4 Construct an activity versus composition relation of a component in a solution for which the standard state is chosen to be the hypothetical state obtained by extrapolating the Henry's law behavior of fugacity to $X_i = 1$ (Fig. 8.10).

8.9 Speciation of Water in Silicate Melt

Burnham and Davis (1974) determined the fugacity of H_2O (w) dissolved in a melt of albite ($NaAlSi_3O_8$) composition, and made the interesting discovery that f_w is proportional to $(X_{w.})^2$, where $X_{w.}$ stands for the apparent or nominal mole fraction of H_2O in the melt (that is the mole fraction if H_2O did not dissociate), up to quite high nominal water content of the melt. The observed f_w^m versus $(X_{w.})^2$ relation at 800 °C is illustrated in Fig. 8.12. From this observation, they concluded that H_2O completely dissociates in an albite melt to two hydroxyl ions. This led to further development of ideas of the solubility mechanism water in silicate melt that have important petrological implications.

The fundamental topological variation of the structure of silicate minerals is based on different schemes of sharing (or polymerization) of oxygen among neighboring SiO_4 tetrahedra. Each tetrahedron consists of a central Si atom and four oxygen atoms at the apices. A silicate melt also consists of SiO_4 tetrahedral groups but in a random array and with a degree of sharing of oxygens between neighboring tetrahedra that depends on the extent of sharing in the structure of the mineral from which the melt had formed. For example, there is no shared or polymerized oxygen among the SiO_4 tetrahedra in a melt formed from frosterite, Mg_2SiO_4, which is an orthosilicate and is devoid of any polymerized oxygen in the crystalline state, whereas there are large number of shared oxygens in the melt formed from an albite, which forms a three dimensional network of SiO_4 tetrahedra as a result of sharing of four oxygens in each tetrahedron with four neighboring tetrahedra.

It was hypothesized that water dissolves in a silicate melt by reacting with the polymerized or bridging oxygens (O_b) according to

$$H_2O(m) + O_b(m) \rightarrow 2(OH)^-(m) \tag{8.9.a}$$

This process reaps apart two polymerized tetrahedra, and each $(OH)^-$ sticks at the two tetrahedral apices that were initially connected by a bridging oxygen (Fig. 8.13).[1] This solubility mechanism has a number of interesting implications

---

[1]This solubility mechanism may be rationalized in terms of Pauling's electrostatic valence rule (Pauling 1960). According to this rule, a Si^{4+} ion coordinated to four oxygens contributes a single positive charge to each oxygen atom. Thus, a polymerized O^{2-} ion is charge satisfied as it receives two positive charges from the two Si^{4+} ions in the two shared tetrahedra. Replacing this bridging oxygen by two $(OH)^-$ ions and depolymerizing the tetrahedra keep the charge balance in tact since each hydroxyl group receives one positive charge from a central Si^{4+} ion.

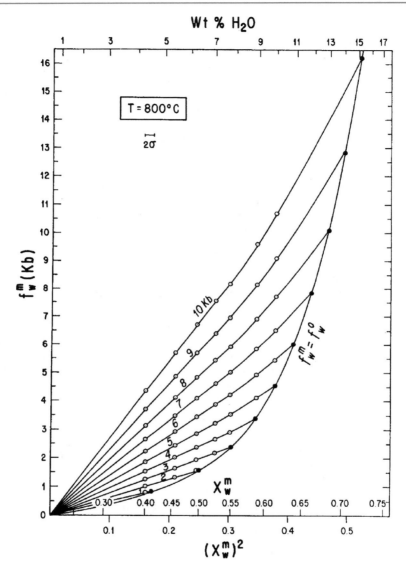

Fig. 8.12 Fugacity of H_2O in $NaAlSi_3O_8$–H_2O melt versus mole fraction, X_w^m, and square of the mole fraction of nominal H_2O in the melt at 800 °C and total pressure of 2–10 kb. The curve $f_w^m = f_w^o$ is the saturation boundary at 800 °C, where f_w^o stands for the fugacity of pure water in the vapor phase. From Burnham and Davis (1974). With permission from American Journal of Science

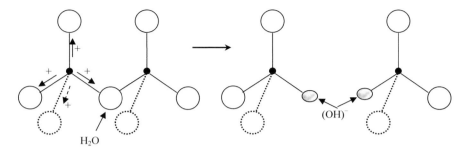

Fig. 8.13 Dissolution mechanism of H_2O in silicate melt by reacting with a bridging oxygen and forming two $(OH)^-$ groups, thereby breaking the linkage between the tetrahedra. The central Si^{4+} ion (small filled circle) contributes one +ve charge to the ligands (oxygen and $(OH)^-$) that are located at the apices of a tetrahedron. Thus, the bridging oxygen (left panel) and $(OH)^-$ groups (shaded) are completely charge satisfied

with respect to the effect of water on the physico-chemical properties of silicate melts, and on differential depression of melting temperatures of minerals in the presence of water that influences the compositions of silicate melts formed by the partial melting of rocks (e.g. Philpotts 1990).

It is shown above (Sect. 8.8.1) that if a mole of solute dissociates completely, then within the domain of validity of Henry's law for the actual solutes, the fugacity of the nominal solute becomes proportional to some power, n, of its mole fraction (Eqs. 8.8.5 and 8.8.12), where n equals the number of moles of the dissociated solutes that formed from one mole of the nominal solute, **only when the mole fraction of the latter tends to zero**. As shown below, the fact that $f_W.$ is proportional to $(X_W.)^2$ in the dilute range clearly implies that H_2O dissociates almost completely to two $(OH)^-$ ions as $X_W. \rightarrow 0$. But, it should be evident from the derivation of Eq. (8.8.12) that the relation $f_W. \propto (X_W.)^2$ cannot hold at high values of $X_W.$, if H_2O continues to dissociate almost completely to two $(OH)^-$. On the contrary, as pointed out earlier by Stolper (1982a), the linear relationship between $f_W.$ and $(X_W.)^2$ up to high $X_W.$ is a strong indication that water does **not** completely dissociate to two $(OH)^-$ ions in a silicate melt except when $X_W.$ is sufficiently dilute so that the activity of the bridging oxygens, $a(O_b)$, in the melt remains effectively constant.

The relationship between $f_W.$ and $(X_W.)$ can be derived as follows. For the reaction (8.9.a), we have at equilibrium

$$K(P,T) = \frac{\left(a_{(OH)^-}^m\right)^2}{\left(a_{H_2O}^m\right)\left(a_{O_b}^m\right)} \tag{8.9.1}$$

where K(P, T), known as the equilibrium constant, is a constant at a fixed P-T condition (the concept of equilibrium constant is developed in Sect. 10.4). Since fugacity is proportional to activity at a fixed P-T condition, we can write from the above relation

$$f_{H_2O}^m \propto \frac{1}{a_{O_b}^m} \left(f_{(OH)^-}^m \right)^2 \qquad (8.9.2)$$

As $X_{W\bullet} \to 0$, $X_{(QH)-} \to 0$ so that, according to Henry's law

$$f_{(OH)^-}^m \propto X_{(OH)^-}^m \qquad (8.9.3)$$

If $X_{(OH)^-}$ is so small $(X_{W\bullet} \to 0)$ that the formation of the hydroxyl ions according to the reaction (8.9.a) does not significantly affect the content of $O^b(m)$, then $a_{O_b}^m$ is effectively a constant at a fixed P-T condition. Thus, as $X_{W\bullet} \to 0$, combination of the last two equations yields

$$f_{H_2O}^m \propto \left(X_{(OH)^-}^m \right)^2 \qquad (8.9.4)$$

But, since according to the reaction (8.9.a), $X_{(OH)^-}^m = 2\, X_{W\bullet}$, and at equilibrium, $f_{H_2O}^m = f_{H_2O}^v$, we finally obtain that as $X_{W\bullet} \to 0$,

$$f_{H_2O}^v = f_{H_2O}^m \propto \left(X_{w\bullet}^m \right)^2 \qquad (8.9.5)$$

Stolper (1982a, b) demonstrated by infrared spectroscopy of quenched silicate glasses and from thermodynamic calculations that speciation of H_2O in the melt changes as a function of the nominal H_2O content of the melt. As illustrated in Fig. 8.14, the spectroscopic data show water dissociates to hydroxyl ions almost completely in the dilute range, as expected from Eq. (8.9.5), but the content of molecular H_2O in the melt increases with increasing nominal water ($H_2O^\bullet$) content, and exceeds the hydroxyl content of the melt when $H_2O^\bullet > 4.5$ wt%.

8.10 Standard States: Recapitulations and Comments

Even at the expense of redundancy, I have chosen to summarize here some of the above discussions on standard states.

(a) The temperature of the standard state must always be the temperature of interest, but there is freedom to choose the composition and pressure of the standard state. This freedom should be exercised carefully.

(b) Let us consider a species **i** in a solution. It is convenient to have the species satisfy the relation $a_i = X_i$, as $X_i \to 1$ at all P-T conditions. As we have seen above (Eq. 8.8.21), this behavior can be realized only if the state of the pure species **i** at the P-T condition of interest is chosen as the standard state. With

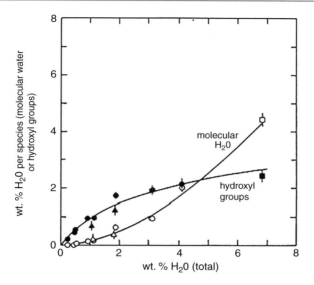

Fig. 8.14 Measured concentrations (wt%) of molecular water groups (open symbols) and water present as hydroxyl groups versus nominal (total) water content in silicate glasses, as determined by infra-red spectroscopy. Circles: rhyolitic glasses; triangles: basaltic glasses; square: albite glass. From Stolper (1982b)

this choice of standard state, $a_i \rightarrow k_H X_i$ as $X_i \rightarrow 0$, where k_H is a constant (Fig. 8.11).

(c) When one deals only with dilute components in solution, it is sometimes convenient to choose a standard state such that $a_i = X_i$, as $X_i \rightarrow 0$ or $m_i \rightarrow 0$, where m_i is the molality of i in the solution, depending on the adopted measure of the content of i in the solution. This property is satisfied by choosing the standard state to be the hypothetical state obtained by extrapolation along the "Henry's law line" to $X_i \rightarrow 1$ or $m \rightarrow 1$ (Problem 8.4). We would return to this choice of standard state in Sect. 12.4 that deals with electrolyte solutions.

(d) Quite often one chooses the standard state of a condensed component to be the state of pure component at 1 bar, T. For this choice of standard state, $a_i \rightarrow X_i$ only at 1 bar, T. In this case, the activity of the pure component at higher pressure is calculated according to

$$RT \ln a_i^o(P', T) = G_i^o(P', T) - G_i^o(1\,\text{bar}, T) = \int_1^{P'} V_i^o dP \qquad (8.10.1)$$

The first equality follows from Eq. (8.4.7).

(e) For gaseous species, it is sometimes convenient to have its activity numerically equal its fugacity. In that case, the standard state of a gaseous species must be chosen to be at that pressure at which the fugacity of the pure gaseous

Fig. 8.15 Choice of standard
state of a gas at unit fugacity.
The pressure of the gas at the
chosen standard state need not
be unity

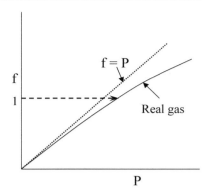

species is unity (Fig. 8.15). Often it is assumed that at $P = 1$ bar, $f_i = P = 1$,
because at sufficiently low pressure, all gaseous species must behave ideally.
However, 1 bar need not be sufficiently low to ensure ideal gas behavior of all
gases, but the error introduced by this assumption is usually not significant,
especially in the treatment of natural processes in which there are always much
larger sources of uncertainties.

Problem 8.5 (continuation of Problem 6.9) Using the results from problem 6.9,
calculate activity of pure water at 10 kb, 720 °C using two different choices of
standard state: (i) state of pure water at 10 kb, 720 °C, and (ii) state of pure water at
1 bar, 720 °C, assuming that $f = P$ at $P \leq 1$ bar.

Answers: (i) a = 1; (ii) a = 14,376.

8.11 Stability of a Solution

A solution may become unstable with respect to phase separation because of
intrinsic instability of the solution to compositional fluctuations and growth of
isostructural stable nuclei of new phases within the solution, or with respect to
decomposition to phases of different structures even though the solution is stable
with respect to fluctuations. Following Mueller (1964), we would call the two types
of instabilities of a solid solution as **intrinsic** and **extrinsic** instabilities, respectively. An example of intrinsic instability is the familiar separation of an alkali
feldspar, $(Na,K)AlSi_3O_8$, upon cooling into two isostructural alkali feldspar phases,
one Na-rich and the other K-rich, whereas an example of extrinsic instability is the
breakdown of clinopyroxene, $Ca(Fe,Mg)Si_2O_6$, which has a monoclinic structure,
into a relatively Mg-rich clinopyroxene and a Mg-poor orthopyroxene, which has
an orthorhombic structure. These breakdown processes may be represented as

$$v_1(\mathrm{Na_xK_{1-x}})\mathrm{AlSi_3O_8} \rightarrow v_2(\mathrm{Na_yK_{1-y}})\mathrm{AlSi_3O_8} + v_3(\mathrm{Na_zK_{1-z}})\mathrm{AlSi_3O_8}$$

monoclinic   monoclinic   monoclinic   (8.11.a)

with y > x, z < x, and

$$v_1'\mathrm{Ca(Mg_{x'}Fe_{1-x'})Si_2O_6} \rightarrow v_2'\mathrm{Ca(Mg_{y'}Fe_{1-y'})Si_2O_6} + v_3'(\mathrm{Mg_{z'}Fe_{1-z'})SiO_3}$$

monoclinic   monoclinic   orthorhombic

(8.11.b)

With y' > x'and z' < x'. We discuss below the thermodynamic aspects of the two types of instabilities.

8.11.1   Intrinsic Stability and Instability of a Solution

At the terminal regions, the free energy of a solution must decrease with the increasing dissolution of a component. To prove this statement, let us consider a binary solution for which

$$\Delta G_m^{mix} = RT(X_1 \ln a_1 + X_2 \ln a_2)$$

For the sake of simplicity, let us assume that the activity is expressed in terms of mole fraction according to the form $a_i = X_i\gamma_i$. Now, according to the laws of dilute solution, as discussed above, as $X_2 \rightarrow 0$, $a_2 = K_H'X_2$ and $a_1 = X_1$, where K_H' is the Henry's law constant (Fig. 8.11). Thus, as $X_2 \rightarrow 0$,

$$\Delta G_m^{mix} = RT(X_1 \ln X_1 + X_2 \ln X_2 + X_2 \ln K_H')$$

Upon differentiating both sides with respect to X_2 at a fixed temperature, and noting that $dX_1 = -dX_2$ (since $X_1 + X_2 = 1$) and $d\ln X_i = dX_i/X_i$, we have

$$\frac{\partial \Delta G_m^{mix}}{\partial X_2} = RT\left(\ln \frac{X_2}{X_1} + \ln K_H'\right) \tag{8.11.1}$$

Thus, as $X_2 \rightarrow 0$, $\partial \Delta G_m^{mix}/\partial X_2 \rightarrow -\infty$. One can easily see by substituting $-dX_1$ for dX_2 in the last equation that also as $X_1 \rightarrow 0$, $\partial \Delta G_m(mix)/\partial X_1 \rightarrow -\infty$. Consequently, since $G_m = X_iG_i^o + (1 - X_i)G_j^o + \Delta G_m^{mix}$, we have, as $X_i \rightarrow 0$

$$\frac{\partial G_m}{\partial X_i} = (G_i^o - G_j^o) + \frac{\partial \Delta G_m(mix)}{\partial X_i} = -\infty \tag{8.11.2}$$

It can be easily verified that the above result is not restricted to the chosen form of activity-composition relation, but is a generally valid result (the reader could try the relation $a_i = (X_i\gamma_i)^\nu$).

Because of the above relation, the Gibbs energy of a solution must decrease with the addition of a very small amount of an additional component. This is why pure minerals are virtually absent in natural environments since dissolution of a small amount of an additional component makes it more stable. When the G_m versus X relation of the solution is convex downwards (i.e. toward the X axis), the molar Gibbs energy, G_m, of the solution is lower than that of the system in an unmixed state. For example, referring to Fig. 8.16a, consider a bulk composition X* and the G versus X relation given by the curve (1). The G_m of the homogeneous solution is given by the point **f**. However, the overall G_m of a combination of unmixed phases of the same bulk composition must always lie above the point **f**, since it is given by the intersection of the line connecting the molar Gibbs free energies of the unmixed (metastable) phases with the vertical line defining the bulk composition, such as the point e in Fig. 8.16a. Now, since $G_m(\mathbf{e}) > G_m(\mathbf{f})$, the unmixed phases are unstable with respect to the homogeneous solution of the same bulk composition. Thus, we conclude that if the G_m versus X relation is convex downwards, or in other words

$$\left(\frac{\partial^2 G_m}{\partial X^2}\right)_{P,T} > 0 \qquad\qquad (8.11.3)$$

(which means that the slope of the G_m vs. X curve increases with X), the solution is stable with respect to unmixing to isostructural phases.

Now consider that an intermediate segment of the G versus X curve is convex upwards, such as shown by the curve curve 2 in Fig. 8.16a. Within this "hump", G_m of a homogeneous solution is greater than the overall G_m of the unmixed phases As an example, the G_m of the homogeneous solution for the bulk composition X* is given by the point **a** on curve 2, which is higher than the overall G_m of any arbitrary combination of unmixed phases, such as given by the points **b**, **c** and **d**. However, the lowest Gibbs energy state for the system is given by the combination of unmixed phases for which the overall G_m is at the point **d**.

This point **d** has the special property that it lies on the **common tangent** to the convex downward segments of the G_m versus X curve. By virtue of this property, the chemical potential of a component in one unmixed phase, α, equals that in the other unmixed phase, β, i.e. $\mu_i^\alpha = \mu_i^\beta$, which is the condition that must be satisfied at equilibrium. This statement may be understood by considering the common tangent as a line of coincidence of two tangents to the G_m versus X curve, one at the α and the other at the β side of the hump, as illustrated in Fig. 8.17. Following the procedure of derivation of chemical potential of a component in a phase, say μ_1^α, from the G_m versus X curve by the method of intercept (Fig. 8.7), it should be easy to see from Fig. 8.17 that at the compositions of the unmixed phases α and β defined by the common tangent to the G_m versus X curve, μ_1^α (i.e. the intercept at $X_1 = 1$ by the tangent on the α side) $= \mu_1^\beta$ (i.e. the intercept at $X_1 = 1$ by the tangent

Fig. 8.16 (a) Schematic Gibbs free energy versus composition relation in a binary system. At the condition P_1-T_1, at which the free energy curve is always convex downward, any arbitrary homogeneous composition, such as f, has lower free energy than that of any combination of unmixed phases of the same bulk composition. At P_2-T_2, any arbitrary composition within the free energy hump attains a lower free energy by unmixing. For composition X*, the lowest free energy state is that of a two-phase combination with compositions X(1) and X(2). (b) Qualitative activity-composition relation corresponding to the condition P_2-T_2; the stable compositions follow the solid lines. From Ganguly and Saxena (1987)

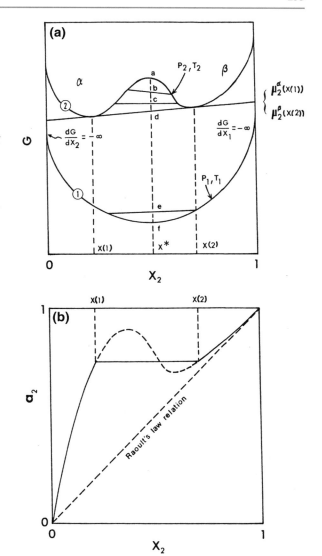

on the β side) and, similarly, $\mu_2^\alpha = \mu_2^\beta$. The qualitative nature of the activity versus composition relation of the solution corresponding to the curve (2) in Fig. 8.16a is illustrated in Fig. 8.16b. Since the unmixed phases are isostructural, each component is referred to the same standard state in both phases, and therefore has the same activity in the unmixed phases at equilibrium.

The two points that are defined by a common tangent to a G_m versus X curve with a 'hump' in a binary system are sometimes referred to as the **binodes**. There is also another special pair of points within the hump on the G_m-X curve, as shown in

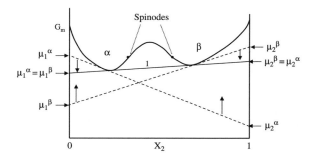

Fig. 8.17 Schematic illustration of the equilibrium property defined by common tangency to the stable portions of G_m versus X curve showing a miscibility gap, and of spinodes. The common tangent line (1) can be viewed as the line of coincidence of two tangents (dashed lines) on the two sides of the G_m versus X curve

Fig. 8.17. These two points mark the transition of the G_m versus X curve from a concave upward to concave downward form. These are known as **spinodes**, and by definition, are inflection points. The spinodes play a special role in the kinetics of unmixing of a solution. We would return to this point in the Sect. 8.13.

8.11.2 Extrinsic Instability: Decomposition of a Solid Solution

As a result of intersection of its G_m versus X curve with that of the other phases, a solid solution may become unstable with respect to a combination of two phases, one of which belongs to a different G_m versus X curve. This is illustrated in Fig. 8.18, which shows the G_m versus X curves for two hypothetical solid solutions A and B of different structures. Each solid solution is intrinsically stable with respect to phase separation. However, the molar Gibbs free energy of either solid solution within the domain defined by the vertical lines is higher than that of a combination of phases of compositions corresponding to those lines. These compositions are defined by a common tangent to the $G_m(A)$ and $G_m(B)$ curves so that these satisfy the equilibrium condition, $\mu_1(A) = \mu_1(B)$ and $\mu_2(A) = \mu_2(B)$.

The geologically important system $MgMgSi_2O_6$–$CaMgSi_2O_6$ (enstatite (En)-diopside (Di)) and its iron counterpart (ferrosilite–hedenbergite) show interesting combination of intrinsic and extrinsic instabilities of solid solutions. The phase diagram for the En-Di system at 1 bar pressure is illustrated in Fig. 8.19. The schematic G_m versus X_{Ca} relations above and below 1320 °C, which is the temperature of coexistence of En, Pig and Di solid solutions, are illustrated in Fig. 8.20. Pyroxene solid solution of monoclinic structure (clinopyroxene: CPx) within this binary join develops a hump in the G_m versus X relation below ~ 1460 °C, and thus decomposes into two monoclinic phases, pigeonite (Pig) and diopside (Di_{ss}; ss: solid solution). At Mg-rich composition, the G_m versus X relation of CPx_{ss} is

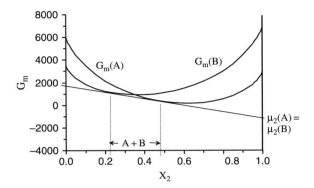

Fig. 8.18 Molar Gibbs free energy versus composition of two binary solid solution phases A and B which have the same range of bulk compositions. Both solid solutions are intrinsically stable with respect the unmixing or phase separation as their G_m versus X curves are always convex downwards ($\partial^2 G_m/\partial X^2 > 0$). However, any solution with composition falling within the range indicated by the vertical dotted lines is unstable. Instead a combination of solid solutions A and B with the compositions indicated by the dotted lines would form with relative amounts of the two phases being determined by the bulk composition

intersected by that of orthopyroxene solid solution within the same binary join leading to the coexistence of enstatite (En_{ss}) and pigeonite solid solutions (Fig. 8.20a). However, as illustrated in Fig. 8.20b, below 1320 °C the common tangent between the G_m versus X_{Ca} curves for OPx and CPx solid solutions (curve b) falls below that (curve a) to the two sides of the hump of CPx solid solution. This leads to the transformation of Pig to En solid solution and an expansion of the compositional gap between the two stable pyroxene compositions. Figure 8.19 shows the metastable solvus (dashed line) below the temperature of Pig to Opx transformation.

8.12   Spinodal, Critical and Binodal (Solvus) Conditions

8.12.1   Thermodynamic Formulations

As illustrated in Fig. 8.21, the size of a hump on a G_m versus X curve of a binary solution changes smoothly as a function of temperature (usually increases with decreasing temperature). The locus of the coexisting equilibrium compositions in the T-X space defines a domain within which the solution is unstable with respect to phase separation. This locus is known as the **solvus** or **binodal**, and the included compositional domain is known as a **miscibility gap**. (Thus, the compositional gap between En_{ss} and Di_{ss} in Fig. 8.19 is not a solvus.) The locus of the two spinodes is known as the **spinodal**. The temperature at which the compositions of the two unmixed phases become identical is known as the **critical or consolute**

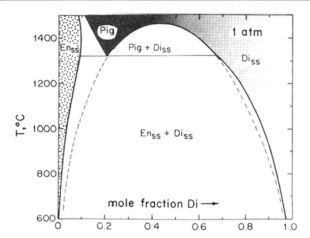

Fig. 8.19 Phase relations in the system $Mg_2Si_2O_6$ (En)–$CaMgSi_2O_6$ (Di) at one atmosphere pressure (ss: solid solution). At high temperature some of the relations may be metastable with respect to other phases. From Lindsley (1983). With permission from Mineralogical Society of America

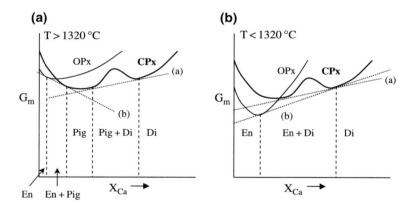

Fig. 8.20 Schematic Gibbs free energy versus composition relations of clinopyroxene (CPx: bold line) and orthopyroxene (OPx: light line) (**a**) above and (**b**) below 1320 °C. Pig: Pigionite (CPx), Di: Diopside (CPx), En: Enstatite (OPx)

temperature of solution. We would denote it by the symbol $T_c(sol)$ to distinguish it from the critical end point in a phase diagram that we have designated by the symbol T_c (Fig. 5.2). However, both critical points have similar significance in that the difference in the properties of the two phases vanishes at these points.

Since a spinode represents a transition between a concave upwards, $\partial^2 G_m/\partial X^2 > 0$, to concave downwards, $\partial^2 G_m/\partial X^2 < 0$, configurations of the G_m versus X curve, we must have $\partial^2 G_m/\partial X^2$ (or G_{XX} for brevity) = 0 at a spinode.

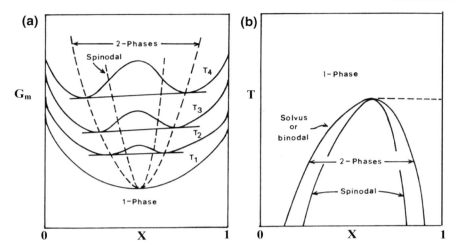

Fig. 8.21 (a) G_m versus X and (b) corresponding T versus X relations in a non-ideal binary system with positive deviation from ideality leading to unmixing or phase separation. From Ganguly and Saxena (1987)

Now, note from Fig. 8.21a that the critical point of a solution represents the point of convergence of the two spinodes on the G_m versus X curve. Thus, a $T_c(sol)$ must also satisfy the condition $G_{XX} = 0$. In addition, the critical temperature must satisfy another condition that is given by the behavior of the third derivative of G_m with respect to X, as discussed below.

The qualitative behaviors of the second and third derivatives of G_m with respect to X at and around the spinodes are illustrated in Fig. 8.22. At the left spinode of a binary system, $\partial^3 G_m / \partial X^3 < 0$ whereas at the right spinode it is $\partial^3 G_m / \partial X^3 > 0$. Therefore, at the critical condition, where the two spinodes meet, we must have $\partial^3 G / \partial X^3 = 0$. In summary, the following thermodynamic conditions must be satisfied at the **spinode** and **critical condition**.

$$\frac{\partial^2 G_m}{\partial X^2} = 0: \text{both spinode and } T_c(\text{sol}) \tag{8.12.1}$$

$$\frac{\partial^3 G_m}{\partial X^3} = 0: \text{Only } T_c(\text{sol}) \tag{8.12.2}$$

The above conditions can be used to develop expressions of ΔG_m^{xs} in terms of compositions that are useful for the calculation of critical and spinodal conditions in a binary system. This procedure is illustrated below using a binary solution that involves one mole of an exchangeable species per mole of an end member component, e.g. **Fe**SiO_3–**Mg**SiO_3 system (exchangeable ions are in bold).

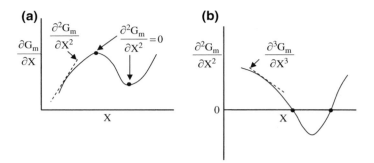

Fig. 8.22 Schematic illustration of the change of the **a** first, second and **b** third derivatives of G_m with respect to X through the spinodal compositions in a binary system. The spinodes are shown by black dots. The first derivative property follows from the G_m versus X curve illustrated in Fig. 8.17

The molar Gibbs energy of the solution is given by

$$
\begin{aligned}
G_m &= [XG_1^o + (1 - X)G_2^o] + \Delta G_m^{\text{ideal}} + \Delta G_m^{\text{xs}} \\
&= [XG_1^o + (1 - X)G_2^o] + RT[X \ln X + (1 - X) \ln(1 - X)] + \Delta G_m^{\text{xs}}
\end{aligned}
\tag{8.12.3}
$$

where, for brevity, we have used X for X_1. Imposing the second derivative condition that must be satisfied at **both** spinode and T_c(sol) (Eq. 8.12.1), we then obtain

$$
\frac{\partial^2 \Delta G_m^{\text{xs}}}{\partial X^2} = -\frac{RT}{X(1 - X)},
\tag{8.12.4a}
$$

Equation (8.12.2) yields an additional relation for T_c(sol), viz.

$$
\frac{\partial^3 \Delta G_m^{\text{xs}}}{\partial X^3} = -\frac{RT(2X - 1)}{X^2(1 - X)^2}
\tag{8.12.4b}
$$

The last two relations can be reduced to various special forms by substituting expressions of ΔG_m^{xs} according to different solution models, which are discussed later in Chap. 9. As an illustration, let us consider a simple class of non-ideal solution that is known as **"Simple Mixture"** or **"Regular Solution"** (see Sect. 9.2 for exposition of the Regular Solution and Sub-Regular solution models). For this type of solutions, ΔG_m^{xs} follows a parabolic form, $\Delta G_m^{\text{xs}} = W^G X(1 - X)$, where W^G is an energetic parameter that, in principle, depends on both temperature and pressure according to the form $W^G = W^H - TW^S + PW^V$ (W^H, W^S and W^V are known as enthalpic, entropic and volumetric interaction parameters, respectively, whereas W^G is known as the free energy interaction parameter). The W^H, W^S and W^V terms are constants. Substitutions of this special expression of ΔG_m^{xs} in the last two equations and carrying out the required differentiations, we obtain the following relations for the spinodal and critical conditions of a **"simple mixture"**.

$$\left. \begin{array}{l} \textbf{Spinodal}: \quad 2\,W^G = \dfrac{RT}{X(1-X)} \\[4mm] \text{and} \qquad\qquad 0 = \dfrac{RT(2X-1)}{X^2(1-X)^2} \end{array} \right\} \textbf{Critical condition}$$

$$\text{(8.12.5a)}$$
$$\text{(8.12.5b)}$$

Equation (8.12.5a) is satisfied by a spinodal, whereas both equations above are satisfied at the critical temperature. The last equation yields $\mathbf{X = 0.5}$ for the critical composition, which on substitution in Eq. (8.12.5a) yields the following expression for the critical temperature of a "**simple mixture**"

$$2RT_{c(sol)} = W^G = W^H - T_{c(sol)}W^S + PW^V$$

or

$$\boxed{T_{c(sol)} = \frac{W^H + PW^V}{2R + W^S}} \qquad\qquad (8.12.6)$$

Unlike the spinodal, the binodal or solvus describes the equilibrium compositions of two coexisting phases. Thus, the binodal curve is calculated by satisfying the thermodynamic condition of equilibrium between the phases. Hence, for a binary system, we have at the **binodal** condition

$$\begin{array}{l} \mu_1^\alpha = \mu_1^\beta \\ \mu_2^\alpha = \mu_2^\beta \end{array} \qquad\qquad (8.12.7)$$

Again various relations between the equilibrium compositions of the coexisting phases on the binodal can be developed from the above equation by substituting expressions for chemical potentials according to different solution models. For the "simple mixture" that we have treated above, the chemical potential of a component is given by

$$\mu_i^\alpha = \mu_i^o + RT \ln X_i^\alpha + W^G(1 - X_i^\alpha)^2$$

and similarly for the phase β. Substitution of the expressions for chemical potentials into Eq. (8.12.7) and rearrangement of terms yield the following expression for the T-X relation along a binodal in a binary system that behaves as a "**simple mixture**"

$$\boxed{\dfrac{RT}{W^G} = \dfrac{1 - 2X_i^\alpha}{\ln\left(\dfrac{1 - X_i^\alpha}{X_i^\alpha}\right)}} \qquad\qquad (8.12.8)$$

The equations governing T-X relations of spinodal and solvus in a binary simple mixture (Eqs. 8.12.5a and 8.12.8, respectively) show that the **pressure effect** on these two relations is manifested through its effect on W^G, which depends on the excess volume of mixing, ΔV_m^{xs} (from the relationship between ΔG_m^{xs} and W^G for simple mixture, as given above, $\partial W^G/\partial P = \partial \Delta G_m^{xs}/\partial P = \Delta V_m^{xs}$). This conclusion, however, is not restricted to simple mixture, but is generally valid. For mineral solid solutions, ΔV_m^{xs} is usually positive, but small. Thus, usually pressure leads to a small expansion of solvus and spinodal. An example of the calculated pressure effect on the solvus in the pyrope-grossular ($(Mg,Ca)_3Al_2Si_3O_{12}$) join of garnet solid solution (Ganguly et al. 1996) is shown in Fig. 8.23. The mixing properties are asymmetric to composition (sub-regular), and there is no excess volume of mixing at the pyrope rich composition.

For **multicomponent** system, one would need to write equations analogous to Eq. (8.12.7) for the other components and then solve for the equilibrium compositions of the two coexisting phases. This exercise, however, has to be carried out in a computer using numerical procedure. For a ternary system, the calculation of spinodal is much simpler than the solvus or the binodal, and is amenable to analytical solution for simple mixture type of solutions. Thus, calculation of spinodal affords an estimate of the miscibility gap in a relatively straightforward way (the miscibility gap must enclose the spinodal but the two must touch one another at the critical temperature).

Fig. 8.23 Calculated Solvus and spinodal in the pyrope-grossular and pyrope-spessartine joins of garnet and calculated pressure effect on the solvus in the pyrope-grossular join. From Ganguly et al. (1996)

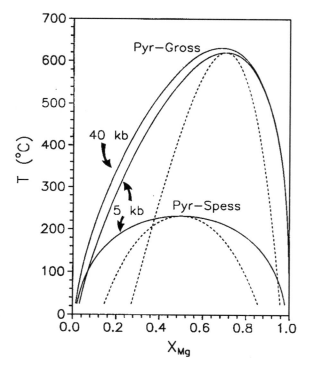

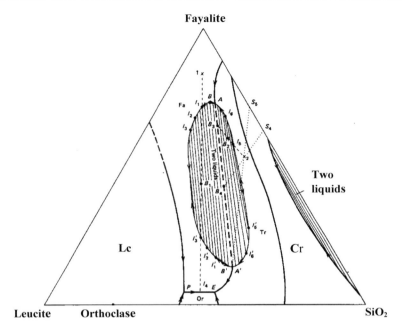

Fig. 8.24 System leucite (KAlSiO$_4$)—fayalite (Fe$_2$SiO$_4$)—silica showing an isolated miscibility gap. The lines within the binary and ternary miscibility gaps join the compositions of the two liquids coexisting in equilibrium. Lc: leucite, Cr: cristobalite. From Philpotts (1990) constructed with data from Roedder (1951)

The procedure for the calculation of ternary spinodal is discussed in Ganguly and Saxena (1987). Analysis of the expression of a spinodal in a **ternary system** in which the bounding binaries behave as simple mixtures shows that for certain combination of the values of the binary W^G parameters, a spinodal may exist as an isolated loop within the ternary compositional space. In other words, a ternary system may have an isolated miscibility gap even when all the binary compositions are stable. Such isolated ternary miscibility gap was first discovered in the system Au-Ni-Cu by Raub and Engel (1947, quoted in Meijering 1950; also see Ganguly and Saxena 1987). Isolated ternary miscibility gap in the field of melt is also displayed by the geologically important system KAlSiO$_4$(leucite)–Fe$_2$SiO$_4$(fayalite)–SiO$_2$ (Fig. 8.24).

Problem 8.6 Show that the following conditions are satisfied at the critical condition of a Sub-Regular solution for which ΔG_m^{xs} is given by $\Delta G^{xs}m = \left(W_{12}^G X_2 + W_{21}^G X_1\right) X_1 X_2$:

$$W_{12}^G(6X_1 - 4) - W_{21}^G(6X_1 - 2) = -\frac{RT_{c(sol)}}{X_1(1 - X_1)}$$

$$6(W_{12}^G - W_{21}^G) = -\frac{RT_{c(sol)}(2X_1 - 1)}{X_1^2(1 - X_1)^2}$$

(8.12.9)

8.12.2 Upper and Lower Critical Temperatures

The miscibility gap most commonly increases with decreasing temperature, such as illustrated in Figs. 8.19 and 8.23. However, there are solutions in various organic systems (e.g. diethylamine-water, benzene-nicotine) for which the miscibility gap decreases with decreasing temperature leading to a lower critical temperature (Fig. 8.25a). It is also possible to have a solution with both upper and lower critical temperatures resulting in a closed miscibility loop in the T-X space, such as shown by the binary solutions of dimythilepyridine-water and m-toluidine-glycerol (Fig. 8.25b). We discuss below the thermodynamic properties that lead to the development of upper and lower critical points, and the possibility of existence of lower T_c in **magmatic** systems.

A solution becomes stable for all compositions either above the upper critical temperature (UCT) or below the lower critical temperature (LCT). Since $\partial^2 G_m/\partial X^2$ (G_{XX} for brevity) = 0 at a critical temperature, and is positive for a stable solution (Eq. 8.11.3), it must increase when the temperature is raised above an UCT and lowered below a LCT. Thus, $\partial G_{XX}/\partial T > 0$ at UCT and <0 at LCT. Now since the order of differentiation is immaterial,

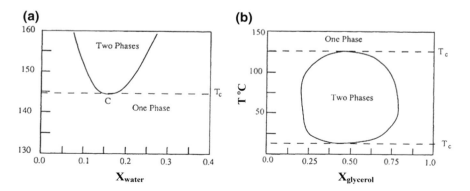

Fig. 8.25 (a) Lower critical temperature in the system diethylamine–water and (b) upper and lower critical temperatures in m-toluidine and glycerol (from Kondepudi and Prigogine 1998)

$$\underbrace{\frac{\partial}{\partial T}\left(\frac{\partial^2 G}{\partial X^2}\right)}_{\partial G_{XX}/\partial T} = \frac{\partial^2}{\partial X^2}\left(\frac{\partial G}{\partial T}\right) = \underbrace{-\frac{\partial^2 S}{\partial X^2}}_{-S_{XX}} \tag{8.12.10}$$

Consequently, $-S_{XX} > 0$ or $S_{XX} < 0$ at an UCT, and $-S_{XX} < 0$ or $S_{XX} > 0$ at the LCT. Also since $G_{XX} = 0$ at the critical temperatures, and consequently $H_{XX} = TS_{XX}$ (because $G = H - TS$), we have the following thermodynamic conditions at upper and lower critical temperatures (Hess 1996).

$$H_{XX} = TS_{XX} < 0 \quad \text{at UCT} \tag{8.12.11a}$$

$$H_{XX} = TS_{XX} > 0 \quad \text{at LCT} \tag{8.12.11b}$$

Now since $H = XH_1^o + (1 - X)H_2^o + \Delta H^{mix}$, where X is the mole fraction of the component 1, we have $H_{XX} = \left(\Delta H^{mix}\right)_{XX} = (\Delta H^{xs})_{XX}$, the last equality following from the fact $(\Delta H^{mix})_{ideal} = 0$ (recall that a mixing property of a solution is the sum of the corresponding ideal and excess properties). Thus, we have the following properties of enthalpy of mixing at the two critical temperatures.

$$\left(\Delta H^{mix}\right)_{XX} = (\Delta H^{xs})_{XX} < 0 \quad \text{at UCT} \tag{8.12.12a}$$

$$\left(\Delta H^{mix}\right)_{XX} = (\Delta H^{xs})_{XX} > 0 \quad \text{at LCT} \tag{8.12.12b}$$

Following Hess (1996), the geometric interpretations of these properties are shown in Fig. 8.26. This figure shows that if there is no inflection point (which implies that ΔH^{mix} versus X curves do not have the wavy features shown in the lower and upper panels of Fig. 8.26a, b, respectively), then (a) ΔH^{mix} must be positive in systems showing upper critical temperature so that $(\Delta H^{mix})_{XX} < 0$, and (b) negative in systems showing lower critical temperature so that $(\Delta H^{mix})_{XX} > 0$. Furthermore,

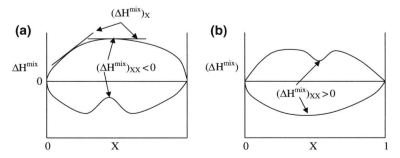

Fig. 8.26 Schematic illustration of ΔH^{mix} versus X relations in a binary solution that lead to (a) $(\Delta H^{mix})_{XX} < 0$ and (b) $(\Delta H^{mix})_{XX} > 0$. The subscripts X and XX denote, respectively, the first and second derivatives of ΔH with respect to X

since for unmixing we must have $\Delta G^{xs} = \Delta H^{mix} - T\Delta S^{xs} > 0$, it follows that $\Delta S^{xs} < 0$ for the existence of LCT in a system characterized by monotonic change of H versus X slope.

Solvus with LCT has not yet been reported from any geologically relevant system. However, Navrotsky (1992) argued that such solvus may be present in oxide melts in that negative ΔH^{mix} was reported in (Na,K) aluminosilicate glasses. Hess (1996) presented a detailed analysis of the problem for silicate melts and suggested that ΔS^{xs} in olivine-rich melt is likely to be negative, and thus a two-liquid field with a LCT may exist in peridotite melts.

8.13   Effect of Coherency Strain on Exsolution

In solid state exsolution process, the lattice planes of the two phases sometimes remain partly or fully continuous or coherent across the interface between the phases when the exsolved lamellae are very fine (Fig. 8.27). Since the lattice spacing in the two phases is different, the continuity of lattice planes across the interface introduces elastic strain energy. Consequently, the exsolution process in the presence of coherency strain cannot be treated simply in terms of ΔG^{mix}, which only accounts for the strain-free chemical interactions, but must also take into account the effect of coherency strain. This problem was treated by Cahn (1962) and Robin (1974). Their analyses show that a **coherent solvus** (i.e. the solvus in the presence of coherency strain) must lie within the strain free or **chemical solvus** (Fig. 8.28).

The elastic strain energy is given by the mechanical work needed to bring, at constant temperature, a chosen mass of a phase to its non-hydrostatically stressed state from its state at the hydrostatic condition. Cahn (1962) defined a new energy function of a solid solution subject to coherency strain as

Fig. 8.27 Semi-coherent lattice planes between two crystals A and B

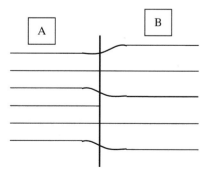

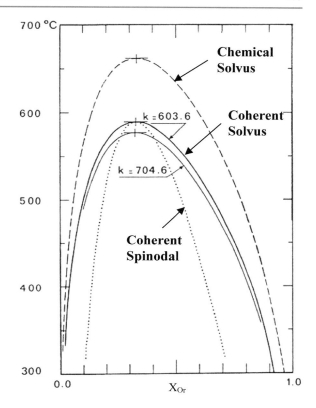

Fig. 8.28 Solvus and spinodal in the system NaAlSi$_3$O$_8$ (Ab)–KALSi$_3$O$_8$ (Or). Coherent solvus and spinodal are due to the combined effects of chemical mismatch and coherency strain that develops when the lattice planes of the two phases remain continuous through the interfaces From Robin (1974). With permission from Mineralogical Society of America

$$\phi_m = G_m + k(X_i - \bar{X}_i)^2 \qquad (8.13.1)$$

where the last term represents the strain energy, and $\bar{X}_i$ is the average value of the mole fraction of the component i in the bulk crystal. This new energy function is referred to as the **Cahn function**. The compositions of the coexisting phases in coherent exsolution process is given by the common tangent to the ϕ_m versus X plot just as the compositions of the phases in strain free exsolution are given by the common tangent to the G_m versus X plot (Fig. 8.17).

From the known values of elastic constants of alkali feldspar, Robin (1974) calculated k = 603.6–704.6 cal/mol, and used these to calculate the coherent solvus in the alkali feldspar solid solution. Figure 8.28, which is from Robin (1974), shows a comparison between the chemical and coherent solvi in the alkali feldspar system. The coherent spinodal is calculated in the same manner as chemical spinodal, but using the Cahn function instead of the Gibbs function.

Lamellar exsolutions are known in many natural minerals. If the exsolved lamella retain coherency of the lattice planes, which can be revealed by transmission electron microscopic studies, then the temperature of exsolution should be calculated from the coherent solvus instead of chemical solvus. (There is, however, no coherency of the lattice planes when the exsolution proceeds to the stage of grain separation.) An example of very fine scale lamellar exsolution is **cryptoperthites**, which are mixtures of very fine scale lamella of K-rich and Na-rich alkali feldspars. Laboratory experiments on the homogenization temperatures of cryptoperthites yield much lower temperature than that predicted from the chemical solvus, as one would expect from the effect of coherency strain in shrinking the size of the solvus.

If one or both exsolved phases with coherent lattice planes undergo phase transformations involving a change of geometry upon cooling, then there is additional strain that affects the free energy and morphological or textural development of the intergrowth. A detailed discussion of the effect of coherency strain on the development of exsolution microstructures of alkali feldspars during cooling of natural rocks in different types of environments (from volcanic ejecta to granulites) can be found in Parsons and Brown (1991).

Problem 8.7 The molar Gibbs energy of mixing of alkali feldspar solid solution, $(Na,K)AlSi_3O_8$ can be expressed according to

$$\Delta G_m^{mix} = RT(X_1 \ln X_1 + X_2 \ln X_2) + (W_{12}X_2 + W_{21}X_1)X_1X_2$$

where the last term represents ΔG_m^{xs} according to a Sub-Regular solution model. Let $1 \equiv Ab$ ($NaAlSi_3O_8$) and $2 \equiv Or$ ($KAlSi_3O_8$). Using the following values of the subregular W parameters, and k = 603 cal/mol, calculate ΔG_m^{mix} and ϕ_m^{mix} versus X_2 at 590 °C, and determine the compositions of coexisting phases in the presence and absence of coherency strain.

$$W_{12} = 6420 - 4.632T \text{ cal/mol}$$
$$W_{21} = 7784 - 3.857T \text{ cal/mol}$$

8.14   Spinodal Decomposition

The spinodal plays a special role in the kinetics of exsolution process. As illustrated in Fig. 8.29, a homogeneous phase with composition within the spinodes (e.g. composition d) is unstable with respect to spontaneous fluctuations as these lead to the development of two phases with an overall free energy that is lower than that of the homogeneous phase. In contrast, a homogeneous phase with composition between a spinode and binode (e.g. composition a) is stable with respect to small fluctuations of compositions since it has a lower free energy than the bulk free energy of the phases developed by these fluctuations. Thus, for composition a, the exsolution process requires formation of stable nuclei of the new phases.

The exsolution that takes place by small fluctuations around an average composition within the spinodal is known as spinodal decomposition. It leads to very fine scale modulated structures without sharply defined boundaries (Fig. 8.30). This

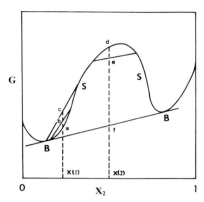

Fig. 8.29 Schematic Gibbs free energy versus composition plot in a binary system illustrating change of free energy of systems with bulk compositions of X(1) and X(2) due to fluctuations. S: spinodes, B: binodes. From Ganguly and Saxena (1987)

type of decomposition is characteristic of **rapidly cooled** environment. The kinetic theory of spinodal decomposition was developed by Cahn (1968). A comprehensive discussion of the subject and its applications to the interpretation of thermal history of rocks can be found in Ganguly and Saxena (1987). As discussed earlier, a spinodal does not define the equilibrium compositions of two unmixed phases. It is a kinetic boundary, and not a phase boundary. This is why spinodal textures such as the one illustrated in Fig. 8.30 cannot survive in a slowly cooled environment that allows enough time for the attainment of equilibrium.

The spinodal decomposition in solids is controlled by the coherent spinodal (Fig. 8.28) and not by chemical spinodal, since the modulated fine scale lamella without sharply defined boundaries are structurally coherent with each other. The

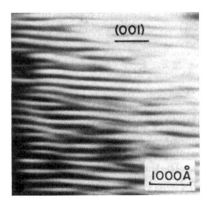

Fig. 8.30 Modulated exsolution microstructure in a subcalcic diopside formed by spinoidal decomposition (McCallister and Nord, quoted in Buseck et al. 1980). With permission from Mineralogical Society of America

stable equilibrium state of the system is achieved when the exsolved compositions shift from the coherent to the chemical solvus by removing the coherency strain that requires "snapping" of the lattice planes at the interface or formation of new grains.

8.15   Solvus Thermometry

Compositions of coexisting minerals that are related to a solvus or a compositional gap provide useful constraints on the temperature of formation of the host rocks if data on the solvus (compositional gap) compositions as function of temperature and pressure are available. The mineral pairs orthopyroxene-clinopyroxene, calcic-olivine–ferromagnesian olivine, calcite-dolomite, alkali feldspar-plagioclase feldspar constitute geological examples of "solvus thermometers". For a comprehensive review of the subject, the reader is referred to Essene (1989). The compositional gap between orthopyroxene and clinopyroxene in the binary system $Mg_2Si_2O_6$–$CaMgSi_2O_6$ at 1 bar pressure is illustrated in Fig. 8.19. Note that in this diagram, the compositional gap between pigeonite and diopside, which is metastable below 1320 °C constitutes a true solvus (see Sect. 8.12.1). However, that between enstatite and diopside is not a true solvus, but is often referred to, albeit loosely, also as a solvus. For brevity, we would refer to both compositional gaps as "solvus" in this section.

In order to determine the temperature of formation of rocks on the basis of the compositions of coexisting minerals on the two limbs of a "solvus", one needs to account for the effects of additional components in natural assemblages on the position of "solvus" at a fixed pressure. For example, both ortho- and clino-pyroxene incorporate significant amount of ferrous components, $Fe_2Si_2O_6$ (ferrosilite: Fs) and $CaFeSi_2O_6$ (hedenbergite: Hd) in solid solutions. Thus, the compositional gap between ortho- and clino-pyroxenes must be treated in terms of at least the quadrilateral components, En-Fs-Di-Hd. The compositional gap is described by a surface in the quadrilateral system with the phase relations shown in Fig. 8.19 representing the terminal section at the magnesian end. The projections of the temperature contours on the compositional gap at 5 kb, which dips towards the Fe-end, are illustrated in Fig. 8.31. In this figure, which is taken from Lindsley (1983), the triangles connect the equilibrium compositions of orthopyroxene, augite and pigeonite at different temperatures. If there are only two coexisting pyroxenes, augite and orthopyroxene, the compositions should ideally fall on the same isotherm on the two sides. However, this is not always the case because of the close spacing of the orthopyroxene isotherms and analytical problems. Many examples of the application of the two pyroxene thermometer (e.g. Lindsley 1983; Sengupta et al. 1999; Schwartz and McCallum 2005) may be found in the literature.

Davidson and Mukhopadhyay (1984) presented the solvus relation between calcic and ferromagnesian olivines. The calcic components are $CaMgSi_2O_4$ (monticellite) and $CaFeSi_2O_4$ (kirschsteinite), whereas the ferromagnesian components are Mg_2SiO_4 (forsterite) and Fe_2SiO_4 (fayalite). Thus, the solvus relation in an olivine

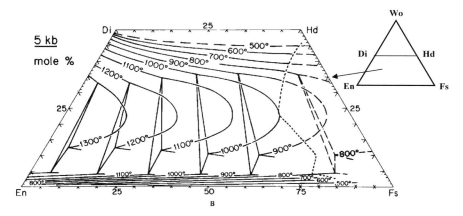

Fig. 8.31 Projection of temperature contours on the 5 kb solvus in the pyroxene quadrilateral, Di $(CaMgSi_2O_6)$–Hd $(CaFeSi_2O_6)$–En $(Mg_2Si_2O_6)$–Fs $(Fe_2Si_2O_6)$. Wo (wollastonite): $Ca_2Si_2O_6$. The triangles connect the equilibrium compositions of augite, orthopyroxene and pigeonite. From Lindsley (1983). With permission from Mineralogical Society of America

quadrilateral in the ternary system Mg_2SiO_4–Fe_2SiO_4–Ca_2SiO_4 is analogous to the pyroxene quadrilateral in the ternary system $Mg_2Si_2O_6$–$Fe_2Si_2O_6$–$Ca_2Si_2O_6$ that is illustrated in Fig. 8.31. The olivine solvus relation is formally similar to that in the pyroxene quadrilateral except for the absence of a third phase (pigeonite in the pyroxene) at specific T-X conditions.

Computer programs are available to define the solvus isotherm that provides the best match to the compositions of coexisting minerals in some of the above systems (e.g. Sack and Ghiorso 1994). If compositions are measured for host mineral and exsolved lamella within the host, it may be necessary to account for the coherency strain on the solvus, as discussed in Sect. 8.13. Robin and Ball (1988) evaluated the effect of coherency strain on the solvus in the pyroxene quadrilateral. They concluded that the effect of coherency strain is significant in the Mg–Ca binary, leading to a depression of the critical mixing temperature by 47 °C. However, this effect progressively decreases as the composition moves into the quadrilateral, becoming virtually negligible for $Fe/(Fe + Mg) \approx 0.6$.

8.16   Chemical Potential in a Field

8.16.1   Formulations

In the presence of a field, such as electrical, magnetic and gravitational fields, the chemical potential of a species is affected by its position in the field. As long as the field potential is uniform, we need not bother about this effect since it does not lead to any change of chemical potential with the change of position of a system in the field. However, if there is a significant change of the field potential, the effect of the field must be taken into account.

Let us start with the effect of gravitational field. If a species i is present in two phases, α and β, which are located at two heights separated by Δh that causes a sufficient change in the gravitational potential energy, then the energy conservation equation is

$$\Delta U + m_i g \Delta h = Q + W^- \qquad (8.16.1)$$

where m_i is the mass of the species and g is the acceleration due to gravity. The left hand side represents the total (internal plus external) energy change of the system whereas the right hand side represents the total energy absorbed by the system (we have already used this equation in connection with the problem of adiabatic flow in Sect. 7.7.2, setting Q = 0.). As a consequence, the condition of chemical equilibrium at constant temperature and volume is no longer given by the constancy of the chemical potentials of the components that are free to move, but by that of the sum of chemical and gravitational potentials of those components (Gibbs 1875, see Gibbs 1961), i.e.

$$\mu_i + m_i g h = \text{constant} \qquad (8.16.2a)$$

so that

$$\mu_{i(1)} + m_i g h_1 = \mu_{i(2)} + m_i g h_2$$

or

$$\mu_{i(1)} - \mu_{i(2)} = m_i g (h_2 - h_1) \qquad (8.16.2b)$$

where the subscripts 1 and 2 indicate locations in a vertical column.

If there is an electrical field, then as shown by Guggenheim (1933: see Guggenheim 1967b), the condition of equilibrium is given by

$$\mu_i^I(o) + (F'Z_i)\varphi^I = \mu_i^{II}(o) + (F'Z_i)\varphi^{II} \qquad (8.16.3)$$

where φ^I and φ^{II} are the electrical potentials acting on the species i at the locations I and II, respectively, F' is the Faraday constant, z_i is the valence of the species i and $\mu_i(o)$ is the chemical potential in the absence of the electrical field at the specified locations. (F' = eL = 9.6485×10^4 Coulombs/mol = 5511.5 J/V-mol, where e is the electronic charge and L is the Avogadro's number.) At this stage, the formal resemblance between the last two equations should be obvious. The quantity φ is the equivalent of gh, both being potentials associated with the field, and Fz_i is the equivalent of m_i, both representing molar properties.

Sometimes the chemical potential in the absence of a field is called the **internal** chemical potential, $\mu_{(int)}$, and the added term due to the effect of the field, that is terms like $Fz\varphi$ and mgh, are called the **external** chemical potentials, $\mu_{(ext)}$. The sum of $\mu_{(int)}$ and $\mu_{(ext)}$ then represents the total chemical potential of the system, $\mu_{(tot)}$. Guggenheim (1933) called $\mu_{(int)} + \mu_{(ext)}$ in an electrical field as the electrochemical potential. In general, it is the **gradient of $\mu_{(tot)}$** that must vanish at equilibrium.

8.16.2   Applications

8.16.2.1   Variation of Pressure and Composition in the Earth's Atmosphere

One of the simple applications of Eq. (8.16.2) is in the derivation of expression relating the variation of pressure with height in the atmosphere. If we set the height at the Earth's surface to zero, and h as the height above it, then from Eq. (8.16.2b),

$$\mu_i(h) + m_i gh = \mu_i(0) \tag{8.16.4}$$

Assuming now that the Earth's atmosphere behaves as an ideal gas, and the atmospheric temperature is uniform, we have from Eq. (8.4.1)

$$\mu_i^*(T) + RT \ln P_i(h) + m_i gh = \mu_i^*(T) + RT \ln P_i(0) \tag{8.16.5}$$

or

$$\boxed{P_i(h) = P_i(0)e^{-m_i gh/RT}} \tag{8.16.6}$$

The above expression is known as the **barometric formula**.

The temperature in the Earth's atmosphere varies between 220 and 300 K. The composition of dry air at sea level is 78 mol% N_2 and 21 mol% O_2. These two species accounts for 99% of the composition of air. For an ideal gas, $P_i = P_T X_i$, where P_T is the total pressure and X_i is the mole fraction of i. Thus, at the surface of the Earth ($P_T = 1$ bar), $P(N_2) = 0.78$ and $P(O_2) = 0.21$ bars.

Figure 8.32 shows the variation of partial pressure of N_2 and O_2 as function of height in the Earth's atmosphere, as calculated according to Eq. (8.16.6) at an approximate average temperature of 260 K. The vertical dotted line represents a typical flight altitude of a plane. The height of Mount Everest is 8,948 km (29,028 ft). The inset of the figure shows the mol% of these two species versus height, as calculated from the relation $P(N_2) + P(O_2) \approx P(total) = P_i/X_i$, where P_i and X_i are respectively the partial pressure and mole fraction of the species i. The data collected by rocket flights between 10 and 40 km (Kittel and Kroemer 1980) are in good agreement with the results in Fig. 8.32.

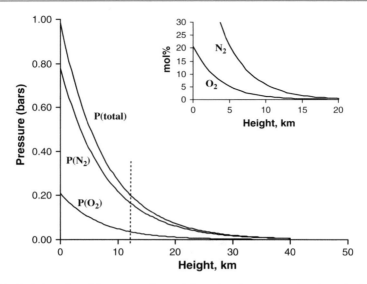

Fig. 8.32 Variation of partial pressure of N_2 and O_2 and total pressure as function of height in the Earth's atmosphere at an approximate average temperature of 260 K. The inset shows the variation of mol% of N_2 and O_2 as a function of height. The vertical dotted line indicates a typical flight altitude of airplane. The height of Mount Everest above the sea level is 8848 m (29,028 ft)

8.16.2.2 Solution in a Gravitational Field

The problem of equilibrium distribution of species in the Earth's gravitational field was treated by Brewer (1951) but has received very little attention from the Earth scientists. The treatment presented below closely follows the formal aspects of Brewar's analysis. The total derivative of the chemical potential of a component at a constant temperature is given by

$$d\mu_1 = \left(\frac{\partial\mu_1}{\partial P}\right)dP + \sum_1^{n-1}\left(\frac{\partial\mu_1}{\partial X_i}\right)dX_i$$
$$= v_1 dP + RT \sum_1^{n-1}\left(\frac{\partial \ln a_1}{\partial X_i}\right)dX_i$$

(8.16.7)

where v_i is the partial molar volume of the component i and n is the number of components. The summation is carried over $n - 1$ components since the mole fractions of only $n - 1$ components are independent. Also, the mole fractions of $n - 2$ components are held constant in the partial derivatives $\partial\mu_1/\partial X_i$ since one of the $n - 1$ independent components is used in the partial derivative.

At hydrostatic equilibrium, $dP = -\rho g dh$, where ρ is the density of the solution. Thus,

$$d\mu_1 = -(v_1 \rho\, g dh) + RT \sum_1^{n-1} \left(\frac{\partial \ln a_1}{\partial X_i} \right) dX_i \qquad (8.16.8)$$

Using $a_1 = X_1\gamma_1$, we obtain

$$\frac{d \ln a_1}{dX_1} = \frac{d \ln X_1}{dX_1} + \frac{d \ln \gamma_1}{dX_1} = \frac{1}{X_1} + \frac{d \ln \gamma_1}{dX_1}$$

so that

$$RT \sum_1^{n-1} \frac{\partial \ln a_1}{\partial X_i} = RT \frac{\partial \ln a_1}{\partial X_1} + RT \sum_{i \neq 1}^{n-2} \frac{\partial \ln a_1}{\partial X_i}$$

$$= \frac{RT}{X_1} + \frac{RT \partial \ln \gamma_1}{\partial X_1} + RT \sum_{i \neq 1}^{n-2} \frac{\partial \ln a_1}{\partial X_i} \qquad (8.16.9)$$

Substitution of this expression into Eq. (8.16.8) yields

$$d\mu_1 = -v_1 \rho\, g dh + RT d \ln X_1 + \left(\frac{RT \partial \ln \gamma_1}{\partial X_1} \right) dX_1 + RT \sum_{i \neq 1}^{n-2} \left(\frac{\partial \ln a_1}{\partial X_i} \right) dX_i$$

$$(8.16.10)$$

From the condition of equilibrium in a vertical column of material that is subjected to a significant change of height in the gravitational field (Eq. 8.16.2a), we have at constant temperature and g

$$\frac{d\mu_1}{dh} + m_1 g = 0 \qquad (8.16.11)$$

$$\boxed{\frac{d \ln X_1}{dh} = (v_1 \rho - m_1) \frac{g}{RT} - \left(\frac{\partial \ln \gamma_1}{\partial X_1} \right) \frac{dX_1}{dh} - \sum_{i \neq 1}^{n-2} \left(\frac{\partial \ln a_1}{\partial X_i} \right) \frac{dX_i}{dh}} \qquad (8.16.12)$$

This is the general condition of equilibrium distribution of a species in a vertical column in a gravitational field at constant temperature and constant acceleration of gravity.

An instructive special case is that of a binary solution, for which the last summation term in the above equation is zero. If the component 1 is sufficiently dilute to follow the Henry's law, or the solution behaves ideally, then the second term on the right also vanishes. Thus, the sign of dX_1/dh is given by that of the first term on the right, which is the same as that of the term $(\rho - m_1/v_1)$. For brevity, we denote this parenthetical term as $\Delta \rho'$.

If the binary solution behaves ideally, then $v_1 = V_1^o$, where V_1^o is the molar volume of the pure component 1. In that case, $\Delta\rho' = \rho - \rho_1^o$, so that if $\rho_1^o > \rho$, $dX_1/dh < 0$, which implies that the component 1 must sink, and vice versa. This result conforms to our common experience. However, Eq. (8.16.12) shows that this is **not** a general behavior. The non-ideality of a solution may compensate for the effect of density difference between a solute and solvent in a multicomponent system. For example, as discussed by Brewar (1951), even though uranium is a heavy element and would thus be expected to increase in concentration with depth, its strong non-ideal interaction with oxygen in a silicate melt makes $\partial\ln a_U/\partial O \ll 0$. Thus, since $dO/dh > 0$, the term

$$\left(\frac{\partial \ln a_U}{\partial O}\right)\frac{dO}{dh} \ll 0$$

Consequently, the effect of non-ideal interaction with oxygen counteracts the tendency of U to sink. Brewar (1951) showed that other analogous terms within the summation sign of Eq. (8.16.12) are relatively unimportant and that the magnitude of the "oxygen effect" term indicated above should be expected to be large enough to make U concentrate towards the top despite its higher density.

8.16.2.3 Variation of Isotopic Ratios with Height

The second and third terms in Eq. (8.16.12) for one isotope must be the same for all practical purposes to the respective terms for another isotope of the same element. Thus, we have in a field of constant value of g

$$\int_{h_1}^{h_2} d\ln\left(\frac{I}{I'}\right) = \frac{(m_I' - m_I)g}{RT}\int_{h_1}^{h_2} dh \tag{8.16.13}$$

where (I/I') stands for the ratio of the two isotopes. Thus,

$$\left(\frac{I}{I'}\right)_{h_2} = \left(\frac{I}{I'}\right)_{h_1}\exp\left[\frac{(m_I' - m_I)g\Delta h}{RT}\right] \tag{8.16.14}$$

where $\Delta h = h_2 - h_1$. Using this relation, we can now calculate the variation of the isotopic ratio of an element with height in a single phase (e.g. melt), and from that the correlation between the isotopic ratios of two elements under equilibrium condition in a gravitational field. However, the efficiency of gravitational effect depends critically on the time scale of equilibration relative to that of the transport process mediated by diffusion and convective circulation. The problem remains to be explored.

8.16.2.4 A Case Study of Equilibrium Distribution in a Gravitational Field

Giunta et al. (2017) presented the first set of data for the variation of isotopic ratios of Cl and Br ($^{37}Cl/^{35}Cl$ and $^{81}Br/^{79}Br$, respectively) as a function of height in a sedimentary aquifer in Illinois basis, USA, that conforms very well to the variation predicted by Eq. (8.16.14). (Unaware of its derivation in the first edition of the book (2008), Giunta et al. (2017) derived it from Boltzmann distribution of non-interacting particles that shows that the fractional occupancy of a species at a given energy level is proportional to its Boltzmann factor that is given by $\exp(-\in_i /k_B T)$ where $\in_i$ is the energy of the level i (see Sect. 14.1), which is mgh_i for the gravitational potential energy.) Their results are illustrated in Fig. 8.33. The solid lines are calculated fractionation behavior versus height at 50 and 130 °C that were inferred to be the plausible limits of temperature over which the fractionation had taken place.

The isotopic variations as a function of height have been expressed as variation of $\delta^{37}Cl$ and $\delta^{81}Br$, where $\delta^{\#}X$ represents $(\Delta R)10^3$, with ΔR standing for the difference between the isotopic ratios at h_2 and h_1, normalized with respect to the ratio at h_1. The relationship between ΔR and h follows easily from Eq. (8.16.14) upon division of both sides by the isotopic ratio at h_1 and subtracting 1 from both sides. Thus, writing r(h) as the isotopic ratio (I/I') at h, we obtain

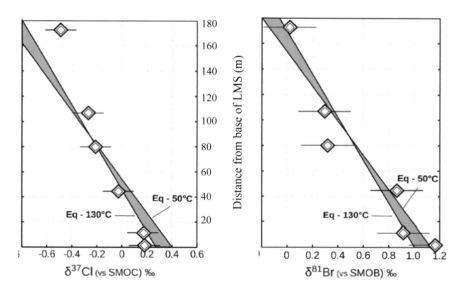

Fig. 8.33 Variations of isotopic ratios of Cl and Br with height in a sedimentary aquifer, LMS (Lower Mount Simon), from the Illinois basis, USA. The isotopic ratios are normalized with respect to standard mean ocean (SMO) values and reported in the conventional del notation (see text). The solid lines represent calculated equilibrium fractionation in an a gravitational field at 130 and 50 °C. From Giunta et al. (2017). With permission from European Association of Geochemistry

$$\Delta R = \frac{r(h_2) - r(h_1)}{r(h_1)} = \exp\left[\frac{(m_1' - m_1)g\Delta h}{RT}\right] - 1 \qquad (8.16.15)$$

Using Boltzmann distribution, Giunta et al. (2017) also derived an expression to describe the equilibrium variation of a species with height, and successfully applied it to model the observed variation of Cl and Br ions in the sedimentary aquifer in the Illinois basin. The relation derived by Giunta et al. (2017) represents the special case of the Eq. (8.16.12) for which all right hand terms except the first one is zero. This requires that the activity coefficient of the solute species of interest is constant (ideal solution or Henry's law behavior in the dilute limit) and the independence of the concentration of the solute on the change in concentration of other species.

Giunta et al. (2017) also calculated the time scale for the relaxation of an initially uniform concentration profile of Cl^- to that required for equilibrium in a gravitational field. They found that gravity-diffusion equilibrium should have been fully established in the LMS unit (Fig. 8.33) after ~ 20 Myr. Thus, the deep ground waters in the aquifer had remained stagnant and isolated at least over this time span.

8.17 Osmotic Equilibrium

8.17.1 Osmotic Pressure, Reverse Osmosis

Consider a U-shaped tube that is fitted with a semi-permeable membrane at the bottom (Fig. 8.34). The membrane is impermeable to the solutes that are dissolved in a solvent, and thus separates the pure solvent from the solution with dissolved solutes. An example would be pure water separated from an aqueous solution of NaCl by a membrane that is permeable only to H_2O. Let us say that the pure solvent is at the left side (side I) and the solution is at the right side (side II) of the U-tube. It will be found that the solvent would flow through the membrane to the right (side II). This is because of the fact that dissolution of solutes lowers the chemical potential of the solvent component in the side II below that in side I that contains only the pure solvent. This flow process would continue until the pressure difference resulting from the unequal column heights reaches a critical value. This equilibrium pressure difference between the two sides separated by a semi-permeable membrane is known as the **osmotic pressure**, which can be calculated from thermodynamics as follows.

The chemical potential of the solvent (j) in the right side of the U-tube is given by

$$\mu_j^{II}(P^{II}, T, X_j) = \mu_j^o(P^{II}, T) + RT \ln a_j^{II}(P^{II}, T, X_j) \qquad (8.17.1)$$

whereas that in the left side is given by $\mu_j^o(P^I, T)$, where μ_j^o stands for the chemical potential of the pure solvent, and P^I and P^{II} are the pressures in the sides I and II, respectively. At equilibrium, the chemical potential of the solvent in two sides of the tube must be the same, in which case, $\mu_j^o(P^I, T) = \mu_j^{II}(P^{II}, T, X_j)$. Thus,

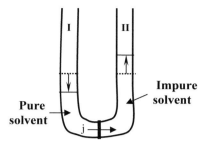

Fig. 8.34 Flow of a solvent (j) through a semi-permeable membrane from a pure side (I) to an impure side (II). The dashed and solid horizontal lines indicate, respectively, the initial and final heights on the two sides

$$\mu_j^o(P^I, T) = \mu_j^o(P^{II}, T) + RT \ln a_j^{II}(P^{II}, T, X_j)$$

so that

$$-RT \ln a_j^{II} = \mu_j^{o,II}(P^{II}, T) - \mu_j^{o,I}(P^I, T) = \int_{P^I}^{P^{II}} V_j^o dP \qquad (8.17.2)$$

where V_j^o is the volume of the pure solvent (the last equality follows from the relation $\partial G/\partial P = V$). If the activity of the solvent in the side II and the molar volume of the pure solvent are known, the above equation can be solved to obtain the osmotic pressure, $P^{osm} = P^{II} - P^I$. If the solvent volume does not change significantly between P^{II} and P^I, which is the typical case, then we may assume V_j^o to be constant, and thus obtain

$$\boxed{P^{osm} = -\frac{RT \ln a_j^{II}}{V_j^o}} \qquad (8.17.3)$$

It is now interesting to note that by exerting a pressure in excess of P_{osm} on the impure side (side II), and thereby making $\mu_j^{II} > \mu_j^I$, the solvent could be made to flow from the impure to the pure side. This is called **reverse osmosis** – a process that is often employed to purify water.

8.17.2 Natural Salinity Gradients and Power Generation

Natural salinity contrasts, such as when fresh water meets sea water, provide a source of osmotic pressure that could, in principle, be harnessed to generate hydroelectric power without causing significant environmental problem. The idea was initially developed by Norman (1974) but got very little attention until the oil

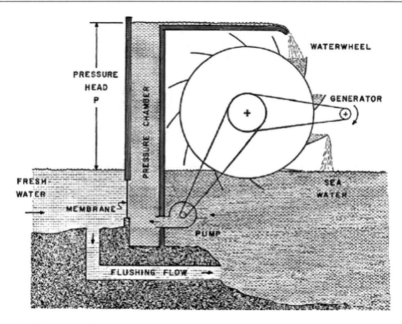

Fig. 8.35 Schematic illustration of the concept of hydraulic power generation using osmotic pressure due to salinity difference at the contact of fresh water and ocean water. From Norman (1974). With permission from the American Association for the Advancement of Science

price increased dramatically around 2005. Figure 8.35, which is reproduced from Norman (1974), illustrates the general concept of how power may be extracted from natural flow of fresh or low-salinity water (feed water) to ocean or high-salinity water (draw water). Because of chemical potential drop of water from the low-salinity to the high-salinity side, the water flowing through a semipermeable membrane rises in the high-salinity side, depending on the difference between their salinities or that between their osmotic pressures with respect to pure water. The rising water column could be made to spill over before the pressure on the draw solution at the height of membrane becomes enough to equalize the chemical potential of water in both sides, and thus drive a wheel to generate hydraulic power. A later version of this concept is known as **pressure retarded osmosis** or PRO, in which a pressure is maintained on the draw water less than that needed to reach equilibrium with the feed water. The pressurized water is allowed to pass through a channel and drive a hydro-turbine to generate power.

An important parameter in the consideration of PRO design is the power density, $\theta(P)$, or the power per unit membrane area. It can be shown to be related to the osmotic and hydraulic pressure differences, $\Delta\pi$ and ΔP, respectively, on the two sides and the water permeability coefficient, A, of the membrane, according to (Achilli and Childress 2010).

$$\theta(P) = A(\Delta\pi - \Delta P)\Delta P \qquad (8.17.4)$$

The permeation of water through the membrane ceases when $\Delta\pi = \Delta P$. It is easy to see that $\theta(P)$ is maximum when $\Delta P = \Delta\pi/2$. The relationship between extractable work and $\theta(P)$ has been discussed by Yip and Elimelech (2012). They showed that maximum work is not necessarily extracted at $\theta(P)_{max}$.

Several technical problems need to be mitigated before natural salinity gradients may be exploited to generate power in cost effective manner. A thorough discussion of the subject can be found in Straub et al. (2016). Two important factors that reduce $\theta(P)$ are (a) contamination of the membrane assembly, which includes a porous layer facing the feed solution and a support structure behind it, by salt and other solute particles that are invariably present in the natural feed solutions, and (b) progressive reduction of the salinity of the draw solution by mixing with the feed solution, thereby reducing the $\Delta\pi$. Figure 8.35 shows partial solutions to these problems by flushing part of the fresh water into the sea water and pumping the sea water back into the pressure chamber utilizing the power generated by the waterwheel.

A major emphasis of current research is to develop more effective membrane, such as that fabricated from porous single layer graphine (Gai et al. 2014), to greatly minimize membrane contamination. Also, to be competitive, there is a need to find combination of feed and draw solutions that have greater salinity contrasts than between river water ocean waters. It was, thus, suggested to combine PRO with reverse osmosis, RO, and use the high salinity residual water from the latter operation to serve as the draw solution for PRO.

A pilot plant based on PRO was installed in 2009 by the Norwegian energy company Starkraft. However, further development was discontinued presumably because the pilot project did not prove to be cost effective.

8.17.3 Osmotic Coefficient

For an ideal solution, Eq. (8.17.3) becomes

$$P^{osm}(\text{ideal}) = -\frac{RT \ln X_j^{II}}{V_j^o} \qquad (8.17.5)$$

The ratio P^{osm} to its ideal value is known as the **osmotic coefficient**, and is commonly denoted by the symbol ϕ. Combining the expressions for P^{osm} (Eq. 8.17.3) and $P^{osm}(\text{ideal})$, ϕ is given by

$$\boxed{\phi = 1 + \frac{\ln\gamma_i}{\ln X_i}} \qquad (8.17.6)$$

The expression for chemical potential of a component in terms of osmotic coefficient can be derived as follows.

$$\mu_i = \mu_i^* + RT \ln X_i + RT \ln \gamma_i$$

Substituting the expression of $\ln \gamma_i$ from Eq. (8.17.6),

$$\mu_i = \mu_i^* + RT \ln X_i + RT[(\phi - 1) \ln X_i],$$

which yields

$$\boxed{\mu_i = \mu_i^* + \phi RT \ln X_i} \tag{8.17.7}$$

It is easy to see that if one uses molality instead of mole fraction as a measure of concentration, μ_i is given by the same expression as above with X_i replaced by m_i.

But why bother about osmotic coefficient since we already have activity coefficient to represent the departure from ideality of mixing of a solution? The answer lies in the fact that in a dilute aqueous solution, a solute may show substantial departure from ideality of mixing, but the activity coefficient of water may still be so close to 1 that it may convey a wrong impression of essentially ideal mixing between the solvent and solute. A classic example from Robinson and Stokes (1970) will clarify the point. The activity coefficient of KCl in a 2 molal aqueous solution ($X(H_2O) = 0.9328$) at 298 K is found to be 0.614, indicating substantial departure from ideal mixing behavior. However, $\gamma(H_2O) = 1.004$ that conveys a false impression of near-ideal behavior. This problem is circumvented by the use of osmotic coefficient to represent the non-ideal behavior, as can be appreciated by calculating $\phi(H_2O)$ from $\gamma(H_2O)$ using Eq. (8.17.6) that yields $\phi(H_2O) = 0.943$.

8.17.4 Determination of Molecular Weight of a Solute

For a binary solution, $X_j = 1 - X_i$, where X_i is the mole fraction of the solute. Using series expansion of $\ln(1 + x)$, we now write

$$\ln X_j = \ln(1 - X_i) = -X_i - \frac{X_i^2}{2} - \frac{X_i^3}{3} - \ldots$$

For a dilute solution of i ($X_i \ll 1$), the quadratic and higher order terms of X_i in the above equation can be neglected, in which case we have $\ln X_j = -X_i = -n_i/n_j$. Also, by the properties of dilute solution, $a_j = X_j$. Thus, for a dilute solution, Eq. (8.17.3) yields

$$P^{osm} = \frac{RT X_i^{II}}{V_j^o} \tag{8.17.8}$$

Now since $X_i^{II} = n_i^{II}/n_j^{II}$ (since the solution is very dilute the total number of moles of the solution is effectively the same as those of the solvent j), we finally obtain

$$P^{osm} = \frac{RTn_i^{II}}{V_j^{II}} \qquad (8.17.9)$$

where V_j^{II} is the **total** volume of the solvent in the impure side, which is effectively equal to the total volume of the impure dilute solution. Thus, by measuring the osmotic pressure of a dilute solution with a known weight of the solute, one can determine the number of moles of the solute, and hence its molecular weight (MW) since $(MW)_i = (\text{total weight})_i/n_i$.

References

Achilli A, Childress AE (2010) Pressure retarded osmosis: From the vision of Sidney Loeb to the first experimental installation – Review. Desalination 261:205–211

Brewar L (1951) The equilibrium distribution of elements in the earth's gravitational field. J Geol 59:490–497

Burnham CW, Davis NF (1974) The role of H_2O in silicate melts: II. Thermodynamic and phase relations in the system $NaAlSi_3O_8$-H_2O to 10 kilobars, 700° to 1100 °C. Am J Sci 274:902–940

Buseck PR, Nord GL Jr., Veblen DR (1980) Subsolidus phenomena in pyroxenes. In Prewitt CT (ed) Pyroxenes. Reviews in mineralogy, vol 7, pp 117–204

Cahn JW (1962) Coherent fluctuations and nucleation in isotropic solids. Acta Metall 10:907–913

Cahn JW (1968) Spinodal decomposition. Trans Metall Soc AIME 242:166–180

Darken LS (1950) Application of Gibbs-Duhem relation to ternary and multicomponent systems. J Amer Chem Soc 72:2909–2914

Darken LS, Gurry RW (1953) Physical chemistry of metals. McGraw-Hill, New York

Davidson PM, Mukhopadhyay D (1984) Ca-Fe-Mg olivines: phase relations and a solution model. Contrib Mineral Petrol 86:256–263

Essene EJ (1989) The current status of thermobarometry of igneous rocks. In: Daly JS, Cliff RA, Yardley RWD (eds) Evolution of metamorphic belts. Geol Soc London Pub No. 43, pp 1–44

Gai J, Gong X, Wang W, Zhang X, Kang W (2014) Ultrafast water transport forward osmosis membrane: porous graphene. J Materials Chem A 2:4023–4028

Ganguly J, Saxena SK (1987) Mixtures and mineral reactions. Springer, Berlin, p 291

Ganguly J, Cheng W, Tirone M (1996) Thermodynamics of aluminosilicate garnet solid solution: new experimental data, an optimized model, and thermometric applications. Contrib Mineral Petrol 126:137–151

Ghiorso MS (1990) Application of Darken equation to mineral solid solutions with variable degrees of order-disorder. Am Min 75:539–543

Gibbs JW (1961) The Scientific papers of J Williard Gibbs, vol 1. Dover, New York, p 434

Giunta T, Devauchelle O, Ader M, Locke R, Louvat P, Bonifacie M, Métivier F, Agrinier P (2017) The gravitas of gravitational isotope fractionation revealed in isolated aquifer. Geochem Perspect Lett 4:53–58

Guggenheim EA (1967a) Theoretical basis of Raoult's law. Trans Faraday Soc 33:151–159

Guggenheim EA (1967b) Thermodynamics. North Holland Publishing Co., Amsterdam

Guggenheim EA (1933) Modern thermodynamics, Metheun, London

Hess P (1996) Upper and lower critical points: thermodynamic constraints on the solution properties of silicate melts. Geochim Cosmochim Acta 60:2365–2377

Hillert M (1998) Phase equilibria, phase diagrams and phase transformations: their thermodynamic basis. Cambridge University Press, Cambridge

Kittel C, Kroemer H (1980) Thermal physics. Freeman, San Francisco, p 473

Kondepudi D, Prigogine I (1998) Modern thermodynamics: from heat engines to dissipative structures. Wiley, New York, p 486

Lewis J (1970) Venus: atmospheric and lithospheric composition. Earth Planet Sci Lett 10:73–80

Lewis GN, Randall M (1961) Thermodynamics. McGraw-Hill, New York, p 723

Lindsley DH (1983) Pyroxene thermometry. Am Mineral 68:477–493

Meijering JL (1950) Segregation in regular ternary solutions, part I. Philips Res Rep 5:333–356

Mueller RF (1964) Theory of immiscibility in mineral systems. Mineral Mag 33:1015–1023

Navrotsky A (1992) Unmixing of hot inorganic materials. Nature 360:306

Norman R (1974) Water salination: a source of energy. Science 186:350–352

Ottonello G (1997) Principles of geochemistry. Columbia University Press, p 894

Parsons I, Brown WL (1991) Mechanism and kinetics of exsolution – structural control of diffusion and phase behavior in alkali feldspars. In: Ganguly J (ed) Diffusion, atomic ordering and mass transport. Advances in physical geochemistry, vol 8. Springer, Berlin, pp 304–344

Pauling L (1960) The nature of the chemical bond. Cornell University Press, p 644

Philpotts AR (1990) Principles of igneous and metamorphic petrology. Prentice Hall, New Jersey, p 498

Prausnitz JM, Lichtenthaler RN, de Azevedo EG (1986) Molecular thermodynamics of fluid phase equilibria. Prentice-Hall, New Jersey, p 600

Robin P-YF (1974) Stress and strain in crypthperthite lamallae and the coherent solvus of alkali feldspars. Am Mineral 59:1299–1318

Robin P-YF, Ball DG (1988) Coherent lamellar exsolution in ternary pyroxenes: a pseudobinary approximation. Am Mineral 73:253–260

Robinson RA, Stokes RH (1970) Electrolyte solutions. Butterworths, London

Roedder E (1951) Low temperature liquid immiscibility in the system K_2O-FeO-Al_2O_3-SiO_2. Amer Min 36:282–286

Sack RA, Ghiorso MS (1994) Thermodynamics of multicomponent pyroxenes II. Applications to phase relations in the quadrilateral. Contrib Mineral Petrol 116:287–300

Sack RA, Loucks RR (1985) Thermodynamic properties of tetrahedrite-tennantites: constraints on the independence of the $Ag \leftrightarrow Ca$, $Fe \leftrightarrow Zn$, $Cu \leftrightarrow Fe$, $Ag \leftrightarrow Sb$ exchange reactions. Am Mineral 71:257–269

Sandler SJ (1977) Chemical engineering thermodynamics. Wiley, p 587

Schwartz JM, McCallum IS (2005) Comparative study of equilibrated and unequilibrated eucrites: Subsolidus thermal histories of Haraiya and Pasamonte. Am Mineral 90:1871–1886

Sengupta P, Sen J, Dasgupta S, Raith MM, Bhui UK, Ehl J (1999) Ultrahigh temperature metamorphism of metapelitic granulites from Kondapalle, Eastern Ghats Belt: implications for Indo-Antarctic correlation. J Petrol 40:1065–1087

Stolper E (1982a) The speciation of water in silicate melts. Geochim Cosmochim Acta 46:2609–2620

Stolper E (1982b) Water in silicate glasses—an infrared spectroscopic study. Contrib Mineral Petrol 81:1–17

Straub AP, Deshmukh A, Eilmelech M (2016) Presssure-retarded osmosis for power generation from salinity gradients: is it viable? Energy Environ Sci 9:31–48

Yip NY, Elimelech M (2012) Thermodynamic and energy efficiency analysis of power generation from natural salinity gradeints by pressure retarded osmosis. Environ Sci Tech, 46:5230–5239

Thermodynamic Solution and Mixing Models: Non-electrolytes

9

Solution properties of minerals and melts form the link between laboratory experimental data in simplified model systems and complex natural and other systems which motivate the experimental investigations. The solution properties need to be expressed as a function of composition, temperature and pressure using forms that are able to represent the data over a sufficiently large range of conditions, and can also be extrapolated well beyond the range of experimental data. The purpose of thermodynamic solution models is the analytical development of these forms. The solution models provide useful rational expressions of the activity of a component in different types of solutions, whereas the mixing models deal with the excess thermodynamic functions. In this chapter we would deal with a variety of thermodynamic solution and mixing models that have been developed over many years, using both theoretical and empirical approaches, and have been applied to model geologically important solutions with different degrees of success.[1]

9.1 Ionic Solutions

An ionic solution is the one in which individual ions or specific ionic complexes constitute the mixing units. As an example, consider a binary olivine solid solution $(Fe,Mg)_2SiO_4$. In this the mixing units are Fe^{2+} and Mg, while the complex $(SiO_4)^{4-}$ constitutes an inert framework. This is an example of a single-site ionic solid solution. A solid solution such as garnet, $^{VIII}(Fe,Mg,Ca,Mn)_3^{VI}(Al,Cr,Fe^{3+})_2Si_3O_{12}$, is an example of two-site ionic solution (the left-hand superscripts indicate the oxygen coordination numbers of the cations). Similarly, there can be multi-site ionic solutions involving substitutions in several sites which are internally charged balanced. When there are substitutions in more than one site, the solution also has a

---

[1]Much of the material in this section was previously published in EMU notes in Mineralogy, v. 3 (Ganguly 2001).

© Springer Nature Switzerland AG 2020
J. Ganguly, *Thermodynamics in Earth and Planetary Sciences*,
Springer Textbooks in Earth Sciences, Geography and Environment,
https://doi.org/10.1007/978-3-030-20879-0_9

reciprocal property arising from interactions between individual sites. Thus, solid solutions involving internally charge-balanced substitutions within more than one site are commonly referred to as **reciprocal solutions**.

It should be noted at the outset that all expressions for the activity of a component are equivalent as long as these are based on the same standard state of the component. But explicit expression of the activity coefficient of a component depends on the form of the activity expression. The ionic solution model provides a rational approach towards the development of such expressions.

9.1.1 Single Site, Sublattice and Reciprocal Solution Models

From a microscopic point of view, ideality of mixing implies random distribution of the mixing units. Thus, in general the activity of an end-member component in a solution should be expressed as a function of composition such that the ideal part of the molar entropy of mixing, as derived from the thermodynamic expression of $\Delta G_m(mix)$, corresponds to the expression of $\Delta S_m(mix)$ derived from the Boltzmann relation, assuming random distribution (Eq. 2.6.6). As shown in Sect. 8.6.2, for a single site solution of the type $(A, B, \ldots)_m F$, the desired property is satisfied if the activity of an end-member component, $A_m F$, is expressed as

$$a_{A_m F} = (x_A \gamma_A)^m \tag{9.1.1}$$

where x_A is the atomic fraction of A within its specific site, and γ_A is the activity coefficient of the ion A reflecting non-ideal interactions with other ions within the **same** site. The term γ_A may be viewed as the activity coefficient of the component $AF_{1/m}$ (e.g. $MgSi_{0.5}O_2$ in the olivine solid solution). Note that X_A equals the mole fraction of the molecular component $A_m F$ (e.g. $X_{Mg} = X_{Fo}$ in olivine where Fo stands for the forsterite component Mg_2SiO_4). It can be shown that with a choice of pure component standard state, $a_{A_m F} = (X_A)^m$, as $X_A \to 1$ at the chosen pressure of that state (to recapitulate, the temperature of the standard state must be the temperature of interest).

Let us now consider a two site (I and II) reciprocal solid solution such as $^I(A, B)_m {}^{II}(C, D)_n P$ in which there is no stoichiometric relation between the substitutions in the two sites, that is the ratio A/B is independent of the ratio C/D. The molar Gibbs energy of such a two site binary reciprocal solution may be expressed using the reference surface illustrated in Fig. 9.1 according to

$$G_m = \left[x_A x_C G^o_{A_m C_n P} + x_B x_C G^o_{B_m C_n P} + x_A x_D G^o_{A_m D_n P} + x_B x_D G^o_{B_m D_n P} \right]$$
$$+ \left[mRT \left(\sum_i x_i \ln x_i \right)^I + nRT \left(\sum_i x_i \ln x_i \right)^{II} \right] + \Delta G^{xs}_m \tag{9.1.2}$$

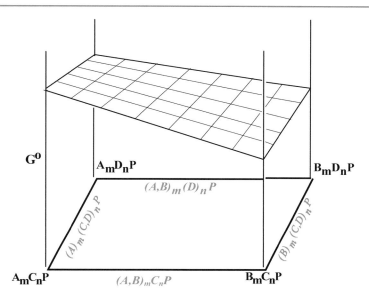

Fig. 9.1 Illustration of the Gibbs free energy surface defined by mechanical mixtures of end-member components in a two-site binary reciprocal solid solution. The bounding binaries define the free energies of mechanical mixtures of end-members in one-site solution. The ruled surface is non-planar

in which x_i is the atomic fraction of i in its specific site indicated by the left hand superscript. The collection of terms within the second square brackets represents $-T\Delta S^{mix}$(**ideal**). The last term, ΔG_m^{xs}, represents the excess molar Gibbs energy of mixing due to non-ideal interactions within the **individual sites**.

The above approach of representing G_m of a reciprocal solution with reference to the Gibbs energies of the end-member compounds, whether these compounds are real or hypothetical, has been called the **compound energy model** by Hillert and co-workers (e.g. Hillert 1998). In general, the method of representing the mixing property of a solution and the activity-composition relations of the macroscopic end-member components in terms of the compositions and properties of the individual lattice sites is often referred to as the **sublattice model**.

Assuming that the interactions within each site are ideal, in which case ΔG_m^{xs} in the last equation is zero, the chemical potential of an end-member component in a binary reciprocal solution is given by

$$\mu_{A_mC_nP} = \mu^o_{A_mC_nP} + \Delta G^o_{rec}\left[{}^I(1-x_A){}^{II}(1-x_C)\right] + RT\left[m\ln{}^I(x_A) + n\ln{}^{II}(x_C)\right]$$

$$(9.1.3a)$$

or

$$\mu_{A_mC_nP} = \mu^o_{A_mC_nP} + \Delta G^o_{rec}\left[{}^I(1-x_A)\,{}^{II}(1-x_C)\right] + RT\ln\left[{}^I(x_A)^{m\,II}(x_C)^n\right] \quad (9.1.3b)$$

where $\mu^o_{A_mC_nP}$ is the Gibbs energy of the pure component A_mC_nP at the P–T condition of interest and ΔG^o_{rec} is the Gibbs energy change of the homogeneous reciprocal reaction

$$A_mC_n + B_mD_n = A_mD_n + B_mC_n \qquad\qquad (9.1.a)$$

This equation was first derived by Flood et al. (1954). The term within the last square brackets of Eq. (9.1.3b) represents the activity of the macroscopic component A_mC_nP when the mixing within individual sites is ideal, and the effect of reciprocal interaction is negligible. In other words,

$$a_{A_mC_nP}(\text{ideal}) = \left[{}^I(x_A)^{m\,II}(x_C)^n\right], \qquad\qquad (9.1.4a)$$

The second term on the right of Eq. (9.1.3) yields

$$RT\ln\left(\gamma_{A_mC_nP}\right)_{rec} = {}^I(1-x_A)\,{}^{II}(1-x_C)\Delta G^o_{rec} \qquad (9.1.4b)$$

For a solution with mixing in more than two sites, the ideal site part of the a-X relation can be written in the same form as Eq. (9.1.4a). The last equation may be written in a general form as

$$RT\ln(\gamma_{i_mj_nP})_{rec} = \pm^I(1-x_i)\,{}^{II}(1-x_j)\Delta G^o_{rec}, \qquad (9.1.5)$$

in which the positive sign holds when i_mj_nP is a reactant component, and the negative sign holds when it is a product component of the reciprocal reaction of the type represented by Eq. (9.1.a).

Equation (9.1.3), and its extension to multisite-multicomponent solution, can be derived in a systematic way by using Eq. (8.3.8), carrying out the indicated differentiations for each sublattice (Wood and Nicholls 1978; Sundman and Ågren 1981; Hillert 1998). For the specific case of two-site solution considered above, Eq. (8.3.8) yields

$$\mu_{A_mC_nP} = G_m + \frac{\partial G_m}{\partial x_A} + \frac{\partial G_m}{\partial x_C} - {}^I\left(\sum_i x_i\frac{\partial G_m}{\partial x_i}\right) - {}^{II}\left(\sum_j x_j\frac{\partial G_m}{\partial x_j}\right), \quad (9.1.6)$$

which, for a binary solution, leads to Eq. (9.1.3).

Equation (9.1.3) highlights an important property of a reciprocal solution. Because of the presence of the term ΔG^o_{rec}, a reciprocal solution behaves non-ideally (in the sense that the chemical potential of a component cannot be determined completely from a knowledge of the composition of the solution), even when the interactions within the individual sites are ideal.

Comparing the statistical-mechanical and thermodynamic derivations of the activity of a component in a reciprocal solution, Førland (1964) suggested that the entropy change of a reciprocal reaction should be very small, which implies that the $\Delta G^{\circ}_{\text{rec}}$ should be quite insensitive to temperature change. This fact was utilized by Liermann and Ganguly (2003) to model the (reciprocal) effect of the variation of Al/Cr ratio in spinel on the (Fe,Mg) fractionation between orthopyroxene, $(Fe,Mg)SiO_3$, and spinel $(Fe,Mg)(Al,Cr)_2O_4$. We will return to the topic of reciprocal solution effect on element fractionation in Sect. 11.1.

When the sites behave non-ideally, the overall activity coefficient of a component has to be expressed by a combination of Eq. (9.1.5) and additional terms reflecting the non-ideal interactions within the individual sites and their mutual interdependence. For the case of a solid solution, the reason for the cross interactions between the sites may be appreciated by noting that the bond distance within one site may be affected by a change of composition of a different site. When the interactions within one site are independent of the composition of the other site, we could write

$$\gamma_{i_m j_n P} = \left[{}^{I}(\gamma_i)^m \, {}^{II}(\gamma_j)^n\right]\left(\gamma_{i_m j_n P}\right)_{\text{rec}}$$

or, substituting Eq. (9.1.5)

$$\gamma_{i_m j_n P} = \left[{}^{I}(\gamma_i)^m \, {}^{II}(\gamma_j)^n\right] \exp\left[\pm {}^{I}(1-x_i)\, {}^{II}(1-x_j)\left(\Delta G^{\circ}_{\text{rec}}/RT\right)\right] \qquad (9.1.7a)$$

and

$$a_{i_m j_n P} = \left[{}^{I}(x_i)^m \, {}^{II}(x_j)^n\right]\gamma_{i_m j_n P} \qquad (9.1.7b)$$

where the γ_i and γ_j reflect the non-ideal interactions of i and j within the specified sites, and x stands for the site-atomic fractions. The sign convention for the exponential term, which represents $\left(\gamma_{i_m j_n P}\right)_{\text{rec}}$, is the same as in Eq. (9.1.5). The interdependence of the mixing properties in the two sites would require additional terms, or may be absorbed in some cases within the site-activity coefficient terms.

As in the case of the ideal single-site expression, Eq. (9.1.1), the above ideal part of the a-X relation in a multi-site solution also follows from a comparison of the expression of entropy of mixing according to the general thermodynamic formulation and that for the statistical thermodynamic formulation for the special case of random distribution within individual sites that is required for ideal mixing. To show this, let us consider a two component solution with two sublattice sites, ${}^{I}(A, B)_m \, {}^{II}(C, D)_n P$, in which the end-member components are $A_m C_n P$ and $B_m D_n P$.

According to the general thermodynamic formulation, we can write

$$
\begin{aligned}
\Delta S_m^{mix} &= -R \sum_i X_i \ln a_i \\
&= -R(X_{A_m C_n P} \ln a_{A_m C_n P} + X_{B_m D_n P} \ln a_{B_m D_n P})
\end{aligned}
\tag{9.1.8}
$$

where X_i stands for the mole fraction of the specified macroscopic end-member component.

If we assume random distribution of atoms within the individual sublattices, then from Eq. (2.6.9)

$$
\Delta S_m^{mix} = -R\left[{}^I\left(x_A \ln x_A^m + x_B \ln x_B^m\right) + {}^{II}\left(x_C \ln x_C^n + x_D \ln x_D^n\right)\right]
\tag{9.1.9}
$$

Now from the stoichiometry, ${}^I x_A = {}^{II} x_C = X_{A_m C_n P}$, and ${}^I x_B = {}^{II} x_D = X_{B_m D_n P}$. Thus, the above equation reduces to

$$
\Delta S_m^{mix} = -R\left[\left(X_{A_m C_n P} \ln\left({}^I x_A^m \,{}^{II} x_C^n\right) + X_{B_m D_n P} \ln\left({}^I x_B^m \,{}^{II} x_D^n\right)\right)\right]
\tag{9.1.10}
$$

Comparing Eqs. (9.1.8) and (9.1.10), we have for ideal mixing or random distribution of atoms within the two sublattices,

$$
a_{A_m C_n P} = \left[{}^I(x_A)^m \,{}^{II}(x_C)^n\right],
$$

and an analogous expression for the other end member component.

9.1.2 Disordered Solutions

Consider now a solid solution of the type ${}^I(A, B)_m \,{}^{II}(A, B)_n P$ in which the species A and B disorder or fractionate between the structural sites I and II. An example of this type of solid solution is orthopyroxene, ${}^{M1}(Fe,Mg)^{M2}(Fe,Mg)\,Si_2O_6$, in which Fe and Mg disorder between the two non-equivalent octahedral sites, M1 and M2 (Ghose 1982). (We note incidentally that the state of the Fe–Mg disordering in orthopyroxene is an important indicator of the cooling rates of the host rocks, e.g. Ganguly et al. 1994.) The disordered solid solutions can also be viewed as reciprocal solutions. Thus, the activity of a component in a two-site disordered solid solution should be expressed according to Eq. (9.1.7). For example, the activity of the component $Mg_2Si_2O_6$ in the orthopyroxene solid solution should be expressed as

$$
\begin{aligned}
a_{Mg_2Si_2O_6} &= {}^{M1}\left(x_{Mg}\gamma_{Mg}\right) {}^{M2}\left(x_{Mg}\gamma_{Mg}\right) \\
&\quad \times \exp\left[{}^{M1}\left(1 - x_{Mg}\right) {}^{M2}\left(1 - x_{Mg}\right) \Delta G_{rec}^o / RT\right]
\end{aligned}
\tag{9.1.11}
$$

where ΔG_{rec}° is the standard state Gibbs energy change of the reciprocal reaction

$$\left(^{M1}Mg^{M2}Mg\right)Si_2O_6 + \left(^{M1}Fe^{M2}Fe\right)Si_2O_6 = \left(^{M1}Fe^{M2}Mg\right)Si_2O_6$$
$$+ \left(^{M1}Mg^{M2}Fe\right)Si_2O_6$$

In the absence of adequate data on the thermodynamic mixing properties, it has been a common practice, however, to express the activity of an end-member component in such a disordered two site solid solution in terms of what has been known as **two-site ideal model**, i.e.

$$a_{A_mA_nP} = \left[^I(x_A)^m {}^{II}(x_A)^n\right] \qquad (9.1.12)$$

where ^Ix_A and $^{II}x_A$ represent the atomic fraction of A within the sites I and II, respectively. By comparing this expression with Eq. (9.1.7), it should be obvious that the two-site ideal model not only implies that $^I\gamma_A = {}^{II}\gamma_B = 1$, but also that $\Delta G_{rec}^{\circ}/RT = 0$. Furthermore, the model also implies, as shown below, negative deviation from ideality for the macroscopic behavior of the solid solution, i.e. $a_{A_mA_nP} < X_A$, where X_A is the macroscopic atomic fraction of A, except in the limiting case of complete disorder.

If p and q are the fractions of the sites I and II, respectively, among the total number of sites participating in the order-disorder process, i.e. $p = m/(m + n)$ and $q = (1 - p)$, then $X_A = p(^Ix_A) + q(^{II}x_A)$. Thus, $X_A \geq (^Ix_A)^m(^{II}x_A)^n$, the equality holding only in the limiting case of complete disorder when $^Ix_A = {}^{II}x_A$ (the reader may easily verify this numerically by taking specific values to the site atomic fractions and site fractions). Consequently, for the two-site ideal model, we have according Eq. (9.1.9), $a_{A_mA_nP} \leq X_A$. Evidently, this model must not be used if there are indications that $a_{A_mA_nP} \geq X_A$. As an example, for the orthopyroxene solid solution, $^{M1}(Fe,Mg)\ ^{M2}(Fe,Mg)Si_2O_6$, one should not write, $a(Mg_2Si_2O_6) = (^{M1}x_{Mg})$ $(^{M2}x_{Mg})$ since the available thermodynamic data seem to indicate a near-ideality or slightly positive deviation from ideality for the mixing of the macroscopic components, $FeSiO_3$ and $MgSiO_3$ (Stimpfl et al. 1999).

9.1.3 Coupled Substitutions

There are many solid solutions that require coupled substitutions of ions in order that the macroscopic electrical neutrality can be preserved. An example is plagioclase feldspar, which has the end-member components $NaAlSi_3O_8$ (albite: Ab) and $CaAl_2Si_2O_8$ (anorthite: An), involving the coupled substitution $(Na^+Si^{4+}) \leftrightarrow (Ca^{2+}Al^{3+})$. When local electroneutrality or charge balance is maintained in the solution, a replacement of Na^+ by Ca^{2+} is accompanied by a replacement of the nearest neighbor Si^{4+} by Al^{3+} in the tetrahedral site. In this case, the expression for the activity of an end-member component (e.g. $NaAlSi_3O_8$) in terms of the ionic solution model should consider X in Eq. (9.1.1) as the mole fraction of the coupled species (e.g. X_{NaSi}), and equate the exponent **m** to the number of moles such species per mole of the solid solution. Thus, recasting the formula for plagioclase solid solution as $(NaSi,CaAl)(AlSi_2O_8)$, we have

$$a(Ab) = X_{NaSi}\gamma_{NaSi} \equiv X_{Ab}\gamma_{Ab}. \qquad (9.1.10)$$

The local charge balance will be destroyed, at least partly, when the thermal agitation overcomes the coulombic forces within the solution. In that case, a different formulation of the activity-composition relation could yield a closer approach to the "ideal" solution behavior. It is, however, interesting to note that in the plagioclase and aluminous clinopyroxene $(CaMgSi_2O_6–CaAl_2SiO_6)$ solid solutions, both of which involve coupled heterovalent substitutions, the activities of the end-member components closely approach their respective molecular fractions even when the calorimetric and structural data indicate significant charge imbalance within the solid solution (Wood et al. 1980; Newton et al. 1980). This implies that the destruction of local charge balance causes similar increases in the ΔH^{mix} and $T\Delta S^{mix}$ terms so that ΔG^{mix} remains nearly the same as in the ideal behavior in a local charge balance (LCB) model. The LCB model, therefore, provides a simple approach for deriving convenient activity expressions in solid solutions involving heterovalent substitutions.

9.1.4 Ionic Melt: Temkin and Other Models

Temkin (1945) proposed that fused salts are completely dissociated into cations and anions and that these two types of ions form two distinct, even though inseparable, sublattices so that there is no intermixing between cations and anions. As a result of strong Coulombic forces, the cations are surrounded by anions and vice versa. Thus, the structure of a molten salt may be viewed as composed of two interpenetrating cation and anion sublattices, as illustrated in Fig. 9.2. Consequently, if there is charge balanced substitution of ions within individual sublattices, then the ideal part of the activity-composition relation of a solution of fused salts can be expressed according to the form Eq. (9.1.4a). In this case, m and n are the dissociation products of the two types of ions. For example, in a solution of $MgCl_2$ and CaF_2, the sublattice representation of the solution is $[^I(Mg^{2+},Ca^{2+})]\,[^{II}(Cl^-,F^-)_2]$ if the two salts completely dissociate in solution according to $X^{2+}Y_2^- \rightarrow X^{2+} + 2Y^-$. This approach of treating ionic melt in terms of cation and anion sublattices is known as the Temkin model. Indeed, from a historical perspective, the sublattice or compound energy models that we have discussed above represent extension of the Temkin model.

Hillert and co-workers (e.g. Hillert 2001 and references therein) extended the Temkin model to treat ionic melts involving heterovalent substitutions in the cation and anion sublattices, including substitutions of neutral species. For example, they modelled the molten solution in the system $Ca–CaO–SiO_2$ in terms of $^I(Ca^{2+})_p\,^{II}(O^{2-}, SiO_4^{4-}, Va^{2-}, SiO_2°)_2$ that permits coverage of the whole range of

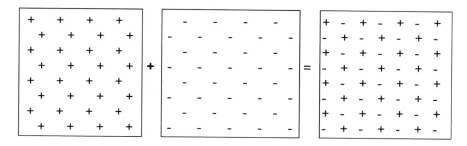

Fig. 9.2 Illustration of Temkin model in which a fused salt or ionic melt is viewed in terms of two interpenetrating sublattices, one for cation and the other for anion

the compositional triangle. In this sublattice formalism, Va^{2-} and SiO_2^o stand for divalent anion vacancy and neutral SiO_2 molecule, respectively. The $CaO–Al_2O_3$ liquid was modelled as $(Ca^{2+}, Al^{3+})_p (O^{2-}, AlO_{1.5}^o)_q$. In these schemes, the coefficients p and q have variable magnitudes that are dictated by the requirement of preserving electrical neutrality of the solution with changing composition. The approach of Hillert and co-workers may be expanded and adapted to treat multicomponent silicate melts that are of great importance to understand magma generation by partial melting of rocks.

9.2 Mixing Models in Binary Systems

The mixing models deal with the different types of representation of the excess thermodynamic quantities as functions of composition. The mixing models described in this section apply to solid solutions in which the substitutions are restricted to one site or in which the substitutions in different sites are coupled, such as in the case of the plagioclase solid solution discussed above. When the solid solution involves internally charge balanced multiple site substitutions, such as in garnet $(Fe,Mg,Mn,Ca)_3(Al,Fe^{3+},Cr)_2Si_3O_{12}$, the mixing properties within each site may also be treated in terms of the models described in this section. The fundamental expression is that of ΔG^{xs} as a function of composition, from which all other excess thermodynamic properties can be derived through standard thermodynamic operations (see Sect. 8.6). We would first deal with binary solutions and then ternary and higher order solutions.

9.2.1 Guggenheim or Redlich-Kister, Simple Mixture and Regular Solution Models

Guggenheim (1937) suggested that the molar excess Gibbs energy of mixing of a binary solution may be represented by the polynomial expression,

$$\Delta G_m^{xs} = X_1 X_2 \left[A_0 + A_1(X_1 - X_2) + A_2(X_1 - X_2)^2 + \ldots \right], \qquad (9.2.1)$$

where the A's are constants at a fixed P–T condition. This polynomial satisfies the requirement that ΔG^{xs} must vanish at the terminal compositions (i.e. $X_1 = 0$ or $X_2 = 0$).

Now recall that $RT \ln \gamma_i = \mu_i^{xs}$, and the latter is related to ΔG_m^{xs} according to Eq. (8.7.4) as

$$\mu_i^{xs} = \Delta G_m^{xs} + (1 - X_i) \left(\frac{\partial \Delta G_m^{xs}}{\partial X_i} \right) \qquad (9.2.2)$$

Using this operation, Eq. (9.2.1) yields

$$RT \ln \gamma_1 = X_2^2 [A_0 + A_1(3X_1 - X_2) + A_2(X_1 - X_2)(5X_1 - X_2) + \ldots] \quad (9.2.3a)$$

and

$$RT \ln \gamma_2 = X_1^2 [A_0 - A_1(3X_2 - X_1) + A_2(X_2 - X_1)(5X_2 - X_1) + \ldots]. \quad (9.2.3b)$$

These expressions for the activity coefficients were first derived by Redlich and Kister (1948) and are usually referred to as **Redlich–Kister relations**. Somehow, even the Guggenheim polynomial is often referred to as the Redlich–Kister expression of excess free energy, which does not seem justified (this is possibly due to the fact that these authors recommended an extension of the Guggenheim polynomial to the ternary system, which is discussed below).

When the A constants with odd subscripts, A_1, A_3 etc. are zero, the ΔG_m^{xs} becomes symmetric with respect to composition. There are, however, various types of symmetries, depending on the number of A terms with even subscripts that are retained in the expression of ΔG_m^{xs}. Following Guggenheim (1967a, b), these types of solutions have been collectively called as **symmetric solutions**. The simplest functional form of a nonideal solution is the one in which all but the first constant in Eq. (9.2.1) is zero. In this case, ΔG_m^{xs} has a **parabolic** symmetry with respect to composition. Guggenheim (1967b) called this type of solution a **Simple Mixture**, as it represents the simplest form of deviation from ideality. Conventionally, A_0 is replaced by the symbol W or W^G when the solution behaves as a Simple Mixture so that

$$\Delta G^{xs} = W^G X_1 X_2, \qquad (9.2.4)$$

and, according to Eq. (9.2.3)

$$RT \ln \gamma_i = W^G (1 - X_i)^2, \qquad (9.2.5)$$

The dependence of W^G on P and T is given by

$$\left(\frac{\partial W^G}{\partial P}\right)_T = \frac{1}{X_1 X_2}\left(\frac{\partial \Delta G^{xs}}{\partial P}\right)_T = \frac{\Delta V^{xs}}{X_1 X_2}, \tag{9.2.6}$$

$$\left(\frac{\partial W^G}{\partial T}\right)_P = \frac{1}{X_1 X_2}\left(\frac{\partial \Delta G^{xs}}{\partial T}\right)_P = -\frac{\Delta S^{xs}}{X_1 X_2}. \tag{9.2.7}$$

Hildebrand (1929) introduced the term **Regular Solution** for the type of solutions which obey Eq. (9.2.4), but in which the interaction parameter W^G is independent of P and T. Thus, a Regular Solution is a special case of a Simple Mixture with ideal volume and entropy of mixing. However, this distinction is not strictly followed in modern usage in that any solution for which ΔG^{xs} conforms to the functional form of Eq. (9.2.4) is often referred to as Regular Solution. We would also use the term Regular Solution in the sense of Simple Mixture. (Historically, the idea of Regular Solution preceded that of Simple Mixture so that it was not introduced to describe a special case of the latter.) Regular solution model holds a special place in the historical development of ideas of solution thermodynamics since the formal nature of the model follows from statistical mechanical consideration of the mixing of non-polar molecules of simple shapes. (From a microscopic point of view, a situation of $\Delta S^{xs} = 0$ not only implies random distribution of the mixing units in a solution, but also negligible contribution to the excess entropy from the vibrational property such as that would prevail if the vibrational spectrum is independent of the atomic arrangements at a given composition of the solution; see Ganguly and Saxena (1987), Sects. 2.II and 2.V, for further discussions.)

Following Thompson (1967), W^G is commonly decomposed into enthalpic (W^H), entropic (W^S) and volumetric (W^V) terms according to

$$W^G(P,T) = W^H(1\,bar, T) - TW^S(1\,bar, T) + \int_1^P W^V dP \tag{9.2.8}$$

It is easy to see that the form of this decomposition of W^G follows that of G in terms of H, S and V. Assuming W^V to be independent of P, the last term in the above equation is often written as a PW^V term in geological literature, since usually $P \gg 1$ bar under geological conditions. The temperature dependence of W^H and W^S is related to the excess heat capacity of mixing, which is due to non-linear change of vibrational properties as a function of composition. Due to an extreme paucity of heat capacity data for solid solutions, the W^H and W^S terms are almost invariably assumed to be constants. However, as shown by Vinograd (2001) from analysis of spectroscopic data in Pyr–Grs and Diop–CaTs solid solutions, there could be significant temperature dependence of these parameters (Pyrope (Pyr): $Mg_3Al_2Si_3O_{12}$; Grossularite (Gros): $Ca_3Al_2Si_3O_{12}$. Diopside (Diop): $CaMgSi_2O_6$; Calcium Tschermak (CaTs): $CaAl_2SiO_6$).

When the thermodynamic mixing properties of solid solutions show symmetric behavior, the data are usually fitted by Regular Solution models. However, the data are rarely good and sufficient enough to permit determination if the symmetry is truly parabolic in nature. Stimpfl et al. (1999) carried out a detailed study, by single crystal X-ray diffraction, of the distribution of Fe^{2+} and Mg between the non-equivalent octahedral sublattices, M1 and M2, in essentially binary orthopyroxene solid solution, $(Fe,Mg)SiO_3$, as a function of temperature. From these data, they calculated the ΔS^{mix} of Fe and Mg, assuming that the distribution is random within each sublattice. Their results show that the ΔS^{mix} is essentially symmetric with respect to composition, but the relation is not parabolic. Instead, the best fit to the data requires two even parameters, A_0^s and A_2^s, in Guggenheim's polynomial expression, where the superscript s denotes terms that are related to the expression of ΔS_m^{xs}, when cast in the form of Eq. (9.2.1) (i.e. $\Delta S_m^{xs} = X_1 X_2 [A_0^s + A_2^s (X_1 - X_2)^2].$).

9.2.2 Subregular Model

This is the simplest model for asymmetric solutions, and has been used most extensively in the petrological and mineralogical literature. It represents a simple extension of the Regular Solution model by making the parameter W^G in Eq. (9.2.4) a simple function of composition as

$$W^G(SR) = W_{21}^G X_1 + W_{12}^G X_2$$

so that

$$\Delta G_m^{xs}(SR) = \left(W_{21}^G X_1 + W_{12}^G X_2\right) X_1 X_2 \qquad (9.2.9)$$

and

$$RT \ln \gamma_i(SR) = \left[W_{ij} + 2X_i\left(W_{ji} - W_{ij}\right)\right] X_j^2 \qquad (9.2.10)$$

where SR implies subregular, and W_{ij}^G is a function only of P and T. The expression for $RT\ln\gamma_i$ in the subregular model is obtained by substituting the expression for $\Delta G_m^{xs}(SR)$ in Eq. (8.7.4) and carrying out the required operations. It reduces to that in the Regular Solution model, Eq. (9.2.5), when $W_{ij} = W_{ji}$.

It is obvious that as $X_1 \to 1$, $W^G(SR) \to W_{21}^G$, and as $X_2 \to 1$, $W^G(SR) \to W_{12}^G$. Thus, the subregular model is simply a weighted average of two Regular Solution models fitted to the data near the two terminal segments of a binary solution (Fig. 9.3). Each Sub-Rgular W_{ij}^G may also be decomposed into enthalpic, entropic and volumetric terms according to Eq. (9.2.8).

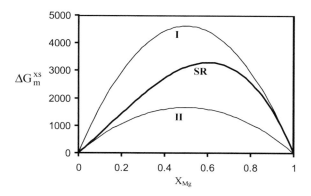

Fig. 9.3 Illustration of the subregular mixing behavior (heavy curve) as a weighted average of the regular solution mixing behaviors (light curves) fitted to the terminal regions. The parameters used for this calculation are those for the pyrope-grossular binary at 600 °C, as given by Ganguly et al. (1996): $W_{CaMg}^G = 18,423$ J/cation-mol and $W_{MgCa}^G = 6630$ J/cation-mol. The sub-regular curve represents $\Delta G_m^{xs} = W^G(SR)X_{Mg}X_{Ca}$ where $W^G(SR) = X_{Mg}W_{CaMg}^G + X_{Ca}W_{MgCa}^G$. The curves I and II are calculated according to $\Delta G_m^{xs} = W_{CaMg}^G X_{Mg}X_{Ca}$ and $\Delta G_m^{xs} = W_{MgCa}^G X_{Mg}X_{Ca}$, respectively

The subregular formulation follows from Guggenheim's polynomial expression for ΔG_m^{xs}, Eq. (9.2.1), by truncating it after the second term and using the identity $A_0 = A_0(X_1 + X_2)$ (since $X_1 + X_2 = 1$), which yields

$$\Delta G^{xs}(SR) = [(A_0 + A_1)X_1 + (A_0 - A_1)X_2]X_1X_2 \qquad (9.2.11)$$

On substitution of W_{21}^G and W_{12}^G for the collection of constants within the first and second set of parentheses, respectively, the above equation reduces to the standard subregular form, Eq. (9.2.9).

Figure 9.4 shows the calorimetric data of the excess heat of mixing in the binary pyrope-grossular ($Mg_3Al_2Si_3O_{12}$–$Ca_3Al_2Si_3O_{12}$) solid solution, as determined by Newton et al. (1977a). These data can be adequately modeled by a sub-regular form, Eq. (9.2.10) (with ΔH^{xs} replacing ΔG^{xs}), as illustrated in the figure. However, the fitted lines in the figure are associated with subregular parameters that represent optimization of both calorimetric and phase equilibrium data. (The phase equilibrium data pertain to the displacement of equilibrium pressure of the reaction Grossular + 2 Kyanite + Quartz = 3 Anorthite as a function of garnet composition at a fixed temperature. The relationship between thermodynamic mixing properties and displacement of equilibria involving solid solution phases are discussed in Sect. 10.12.)

9.2.3 Darken's Quadratic Formulation

Darken (1967) pointed out that when the activity coefficient of a solvent component (say component 1) obeys the Regular Solution relation, as given by Eq. (9.2.5),

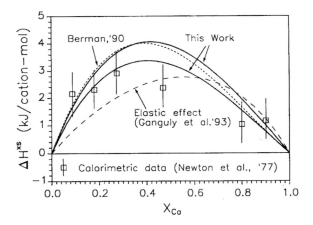

Fig. 9.4 Excess enthalpy of mixing, ΔH^{xs}, in the binary pyrope-grossular join. The squares indicate calorimetric data with $(\pm 1\sigma)$ of Newton et al. (1977a). "This work" refers to the fit to the data according to Sub-Regular model by Ganguly et al. (1996). The model parameters represent an optimization that fit both calorimetric and experimental phase equilibrium data. The upper limit represents the "preferred model" that has the following sub-regular parameters: $W^H_{CaMg} = 21,627, W^H_{MgCa} = 9834$ kJ/cation-mol. "Berman model" illustrated by short-dashed line indicates a sub-regular model fit using similar optimization, but less extensive phase equilibrium data, by Berman (1990). The dashed line indicates theoretical calculation of ΔH^{xs} by Ganguly et al. (1993) using molar volume and elastic property data (see Appendix: C.3.2). With permission from Mineralogical Society of America

then the Gibbs–Duhem relation (Eq. 8.2.7) only requires that the activity coefficient of the solute component must obey the relation

$$RT \ln \gamma_2 = W^G (1 - X_2)^2 + I, \tag{9.2.12}$$

where I is an integration constant. In order that the component 2 conforms to Raoultian behavior, i.e. $a_2 = X_2$ as $X_2 \to 1$ (Eq. 8.8.21), the integration constant must be zero when γ_1 obeys the regular behavior over the entire range of composition. If γ_1 conforms to the Regular Solution property over a restricted compositional range only near the terminal region 1, then γ_2 will conform to the above expression over the same compositional range with $I \neq 0$. In this case, ΔG^{xs}_m near the terminal region 1 is given by,

$$\Delta G^{xs}_m = W^G X_1 X_2 + I X_2 \tag{9.2.13}$$

This is known as **Darken's Quadratic Formulation** (DQF).

Using the above expression, the molar Gibbs free energy of a solution in the terminal region 1 is given by

$$G_m = X_1 G_1^o + X_2 G_2^o + \Delta G_m^{mix}(\text{ideal}) + (W^G X_1 X_2 + I X_2)$$
$$= X_1 G_1^o + X_2 (G_2^o + I) + \Delta G_m^{mix}(R) \qquad (9.2.14)$$

where $\Delta G_m^{mix}(R)$ is given by

$$\Delta G_m^{mix}(R) = \Delta G_m^{mix}(\text{ideal}) + W^G X_1 X_2$$

which is the molar Gibbs energy of mixing in a Regular Solution (see Eq. 9.2.4). Using G_2' for $(G_2^o + I)$, Eq. (9.2.14) can be written as

$$\Delta G_m = X_1 G_1^o + X_2 G_2' + \Delta G_m^{mix}(R) \qquad (9.2.15)$$

Thus, as noted by Powell (1987), a solution obeying DQF in the terminal region 1 may be viewed as a Regular Solution between the real end member 1 and a hypothetical end member, 2′, whose Gibbs free energy, G_2', is given by that of the end member 2 plus the value of the integration constant, I.

By analyzing the experimentally determined activity coefficient data for a number of liquid binary alloys, especially those in which Fe was the solvent, Darken (1967) showed that while the solvent (1) followed Regular Solution behavior up to a certain level of addition of the solute component (2), the latter followed the relation described by Eq. (9.2.12), with $I \neq 0$, over the same range of concentration. From these observations, Darken (1967) suggested that it may be possible to treat many solutions in terms of the Quadratic Formulation in the two terminal regions, each with characteristic values of W and I. The behavior of the intermediate compositional region would be more complex since it has to make the transition from the quadratic properties of one terminal region to those of the other. Powell (1987) analyzed the available molar volume data of several binary mineral solid solutions, and showed that the data in the two terminal segments are better described by DQF than by Regular Solution model.

For those solutions which conform to DQF in the two terminal regions, the intermediate region could follow a relation that is a weighted average of those of the terminal regions in much the same way as the expression of ΔG_m^{xs} of a Sub-Regular solution represents a weighted average of the Regular Solution expressions in the terminal regions (Eq. 9.2.9 and Fig. 9.3). In that case, ΔG_m^{xs} of the intermediate segment of a solution conforming to DQF in the terminal regions is given by

$$\Delta G_m^{xs}(1-2) = X_1 \left(W_{21}^G X_1 X_2 + I_{21} X_2 \right) + X_2 \left(W_{12}^G X_1 X_2 + I_{12} X_1 \right)$$
$$= X_1 X_2 \left(W_{21}^G X_1 + W_{12}^G X_2 + I_{21} + I_{12} \right) \qquad (9.2.16)$$

where the subscript ij represents the property of the terminal region j. Note that when $X_i \rightarrow 1$, the ΔG^{xs} is given by only the $X_i(\ldots)$ term after the first equality.

In applying DQF to treat the mixing property data, one needs to be careful about the quality of the data. Since the data are divided into three segments, two terminal regions and one central region, there is greater flexibility in fitting the data, which permits better conformity of relatively poor quality data to DQF than to the sub-regular model, as illustrated by Ganguly (2001).

9.2.4 Quasi-chemical and Related Models

In the classic Regular Solution model, the distribution of species in a solution has been considered to be random, even though the pair-potential energies are different. However, this cannot be strictly correct since a species would tend to be preferentially surrounded by the ones with which it has a relatively stronger potential energy of interaction. The atomic distribution would be effectively random at high temperature when the thermal energy per mole, RT, is sufficiently high to prevent clustering of such species. Guggenheim (1952) sought to remedy this logical problem with the Simple Mixture model by considering that in a binary solution, the distribution of the 1–1, 2–2 and 1–2 pairs is related to the energy change of the homogeneous chemical reaction

$$1 - 1 + 2 - 2 = 2(1 - 2). \tag{9.2.a}$$

The resultant thermodynamic mixing model is known as the **Quasi-chemical (QC) Model** because of its appeal to a chemical reaction among the different pairs in the solution, and representation of the equilibrium concentration of these pairs in much the same way as that of components by an equilibrium constant of a chemical reaction (Sect. 10.4).

From consideration of the total potential energy of a lattice consisting of 1–1, 2–2 and 1–2 pairs, and neglecting the effect of long range forces, Guggenheim (1952) introduced an interchange energy, W_{QC}, according to

$$W_{QC} = LZ[\Gamma_{12} - 1/2(\Gamma_{11} + \Gamma_{22})], \tag{9.2.17}$$

where L is Avogadro's number, Z is the nearest neighbor coordination number of the atom or ion 1 or 2, and Γ_{ij} is the potential energy of interaction between these species. Here the term Z represents the coordination number of an atom within its specific sublattice instead of the usual polyhedral coordination number around a central atom. For example, in the solid solution between NaCl and KCl, each alkali atom has six nearest neighbor Cl atoms (and vice versa). However, in the above equation, Z must be taken as 12, which represents the number of nearest neighbor alkali atoms surrounding a central alkali atom in the crystal structure. W^H in the Simple Mixture model is exactly the same as the interchange energy defined above. It should be noted that although Guggenheim neglected the effects of long range forces in the derivation Eq. (9.2.17), inclusion of these forces leads to a similar expression except that the $Z[...]$ term is replaced by $\sum Z^{(k)} [\Gamma_{12}^{(k)} - 1/2 (\Gamma_{11}^{(k)} + \Gamma_{22}^{(k)})]$, where the summation is carried out over k-nearest pairs (i.e. 1st nearest, 2nd nearest, 3rd nearest, and so on) (Vinograd 2001). The contribution of the distant pairs may be important even though the energy of interaction decreases rapidly with distance since $Z^{(k)}$ could increase rapidly with distance.

In order to account for the mixing of molecules or atoms of different sizes, Guggenheim (1952) also introduced parameters known as contact factors, q, which represent the geometrical relation of an atom to another atom of different type in a nearest neighbor site. The contact factors have the property that $q_1/q_2 \rightarrow 1$ as either contact factor tends to unity. (One can think of a number of relations between q_1 and q_2 that would satisfy this limiting property. For example, Green (1970) assumed $q_1q_2 = 1$, which satisfies the required relation as either contact factor tends to unity.) Guggenheim (1952) showed that the deviation from a random distribution of the species is given by a parameter β, which is defined as

$$\beta = \left\langle 1 - 4\theta_1\theta_2 \left[1 - \exp\left(\frac{2W_{QC}}{ZRT}\right) \right] \right\rangle^{\frac{1}{2}}, \qquad (9.2.18)$$

where θ is related to the contact factors according to

$$\theta_1 = 1 - \theta_2 = \frac{X_1q_1}{X_1q_1 + X_2q_2} \qquad (9.2.19)$$

As the solution approaches a random distribution, i.e. $W/RT \rightarrow 0$ so that $\beta \rightarrow 1$, whereas for positive (W > 0) and negative (W < 0) deviations from ideality, $\beta > 1$ and <1, respectively.

With the above framework, Guggenheim (1952) derived the following QC expression for the molar excess Gibbs energy of mixing in a binary solution.

$$\frac{\Delta G_m^{xs}}{RT} = \frac{Z}{2} \left\{ \left[\frac{X_1q_1 \ln(\beta + \theta_1 - \theta_2)}{\theta_1(\beta + 1)} \right] + \left[\frac{X_2q_2 \ln(\beta + \theta_2 - \theta_1)}{\theta_2(\beta + 1)} \right] \right\}. \qquad (9.2.20)$$

Using $H = \partial(G/T)/\partial(1/T)$, one then has

$$\Delta H_m^{xs} = \left[\frac{4X_1X_2W_{QC}}{\beta(\beta + 1)} \right] \exp\left(\frac{2W_{QC}}{ZRT} \right) \cdot \left[\frac{X_1q_1\theta_2}{\beta + \theta_1 - \theta_2} + \frac{X_2q_2\theta_1}{\beta + \theta_2 - \theta_1} \right]. \qquad (9.2.21)$$

The ΔS_m^{xs} can be derived from the relation $G = H - TS$.

The unknown parameters, q and W_{QC}, for a binary system can be retrieved from phase equilibrium or enthalpy of mixing data, if the quasi-chemical model provides an adequate analytical representation of the data. From an analysis of NaCl–KCl solvus data, Green (1970) showed that q_{Na^+}/q_{K^+} nearly equals the ratio of the cationic radii or molar volumes of the two end members. However, Fei et al. (1986) failed to find any such relation between the ratio of the retrieved contact factors and molar volumes of the end members in the pyrope–grossular $(Mg_3Al_2Si_3O_{12}-Ca_3Al_2Si_3O_{12})$, diopside–CaTs $(CaMgSi_2O_6-CaAl_2SiO_6)$ and diopside–enstatite $(CaMgSi_2O_6-Mg_2Si_2O_6)$ solid solutions.

In the classic QC theory, the extrema in both enthalpy and entropy of mixing appearing for the negative deviation from ideality (which favors the formation of 1–2 pairs) are at $X_1 = X_2 = 0.5$. In a binary system, ΔH^{mix} exhibits negative deviation from ideality with a "V" shaped form, whereas ΔS^{mix} shows an inverted "W" form in which the sagging of the central portion depends on the degree of ordering (Fig. 9.5). However, in real systems such extrema often occur at compositions other than $X = 0.5$. Several modifications and extensions of the quasi-chemical theory have been suggested by a number of workers in order to better describe the behavior of real solutions and remedy the problem with the location of extrema. The interested reader is referred to Ganguly (2001) and Ottonello (2001) for a discussion of these models.

The Simple Mixture or the Regular Solution model follows in a straightforward way as a special case of the QC formulation when W_{QC}/RT is small and the species 1 and 2 are sufficiently alike in shape and size that their contact factors become similar. However, the magnitude of W_{QC}/RT and the dissimilarity of the contact factors must be intrinsically related in that W_{QC}/RT cannot be a small quantity unless the mixing units are sufficiently alike.

The simple mixture (or Regular Solution) and QC models follow as zeroth and first order approximations, respectively, of a powerful approach developed by Kikuchi (1951), which is known as the **Cluster Variation** method. This method was applied by Burton and Kikuchi (1984a, b) to treat order-disorder in $CaCO_3$–$MgCO_3$ and Fe_2O_3–$FeTiO_3$ solid solutions. Vinograd (2001) discussed in detail the cluster variation method and applied it to garnet and pyroxene solid solutions.

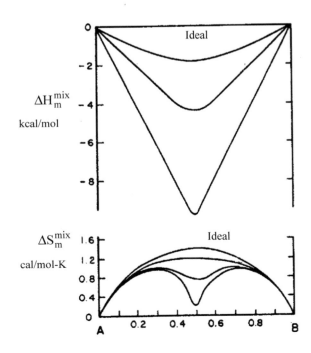

Fig. 9.5 Enthalpy and entropy of mixing of a binary system showing quasi-chemical mixing behavior with degrees of ordering. From Pelton and Blander (1986)

9.2.5 Athermal, Flory-Huggins and NRTL (Non-random Two Liquid) Models

The statistical thermodynamic study of Fowler and Rushbrooke (1937) and calorimetric measurements of Meyer and co-workers (e.g. Meyer and van der Wyk 1944) showed that molecules of different size and shape mix with significant non-random distribution or non-ideal entropy effect even when $\Delta H^{mix} = 0$. This type of solution is known as **Athermal Solution**. Athermal behavior is closely approximated by several polymer solutions in which the components differ in size but have very similar energetic properties. **Analcime**, which is one of the most common rock forming zeolites and forms under wide range of P-T conditions in the Earth's crust, shows athermal mixing behavior (Neuhoff et al. 2004; see Problem 9.1). However, athermal behavior is uncommon among mineral solid solutions since substitutions of atoms of different sizes usually leads to nonideal enthalpic effects owing to the distortion of the lattice and nonlinear change in the bonding energies. Nonetheless, athermal solution model offers a starting point for the development of some other related models that have been successfully used to treat mineral solid solutions.

It was shown independently by Flory (1941, 1944) and Huggins (1941) that the entropy of mixing resulting from the non-energetic solution of a polymer component (2) in a monomer solvent (1) is given by

$$\Delta S_m^{mix} = -R(X_1 \ln \Phi_1 + X_2 \ln \Phi_2), \qquad (9.2.22)$$

where Φ_1 and Φ_2 are the fraction of sites occupied by the solvent and the polymer, respectively. If there are N_1 molecules of the solvent and N_2 molecules of the polymer, and there are p segments in a polymer molecule, then

$$\Phi_1 = \frac{N_1}{N_1 + pN_2}, \quad \Phi_2 = \frac{pN_2}{N_1 + pN_2}, \qquad (9.2.23)$$

where $N_1 + pN_2$ are the total number of sites in the solution. It is assumed that each lattice (or quasi-lattice) site is occupied by either a solvent molecule or a polymer segment. When a lattice site is occupied by a polymer segment, the adjacent sites are occupied by the rest of the segments so that each polymer molecule occupies p lattice sites.

Wilson (1964) extended the Flory–Huggins formulation to include the mixing of molecules which differ not only in size but also in their energetic properties. This extension involved calculation of the relative probabilities of finding molecules of the two components around a central molecule or atom, say of the type i, taking into account the energies of interaction of the i–j and i–i pairs, and from that deriving expressions for the local volume fractions of the components around the central component. Wilson assumed that the ratio of the "local mole fractions" of the components i and j around a central component of i (i.e. χ_{ii} and χ_{ji}, respectively) is given by

$$\frac{\chi_{ji}}{\chi_{ii}} = \frac{X_j \exp(-E_{ji}/RT)}{X_i \exp(-E_{ii}/RT)}, \tag{9.2.24}$$

where E_{ij} is the molar interaction energy between i and j. The local volume fraction of a component around a central component of the same type is then given by

$$\xi_i = \frac{V_i\chi_{ii}}{V_i\chi_{ii} + V_j\chi_{ji}}, \tag{9.2.25}$$

where V_i and V_j are the molar volumes of the components i and j, respectively. Wilson used these local volume fractions in place of the overall site fractions in the Flory–Huggins expression (Eq. 9.2.22). This procedure leads to

$$\Delta G_m^{xs} = -RT[X_1 \ln(X_1 + \Lambda_{12}X_2) + X_2 \ln(X_2 + \Lambda_{12}X_2)] \tag{9.2.26}$$

with

$$\Lambda_{12} = \frac{V_2}{V_1}\exp\left[-\frac{E_{12} - E_{11}}{RT}\right] \tag{9.2.27a}$$

and

$$\Lambda_{21} = \frac{V_1}{V_2}\exp\left[-\frac{E_{12} - E_{22}}{RT}\right] \tag{9.2.27b}$$

It should be noted that the local volume fractions in Wilson's formulation do not always add up to unity (Prausnitz et al. 1986). Also the Wilson expression has no rigorous theoretical justification, but is rather an intuitive extension of the Flory–Huggins formulation to account for the energetic effects on mixing. However, it has been successfully applied to many binary systems (Orye and Prausnitz 1965), and seems to have some appeal in the treatment of multicomponent solutions as discussed later. On the other hand, there are two important formal limitations of the Wilson expression (Wilson 1964; Prausnitz et al. 1986). First, it cannot produce a maximum in the $\ln\gamma$ versus X relation. Second, no values for the parameters Λ_{12} and Λ_{21} can be found that produce phase separation or unmixing, that is produce a 'hump' (or a convex upwards segment) in the G_m versus X curve (which is always convex downwards near the terminal regions). In other words, there are no values of these parameters that could lead to the condition $\partial^2 G_m/\partial X_i^2 < 0$. Thus, if the solution has a miscibility gap, the application of the Wilson equation must be restricted to the P–T–X domain where the solution is continuous.

Renon and Prausnitz (1968) modified Wilson's formulation so that it can produce phase separation by introducing a correction factor, α_{12}, as a multiplier of the energy terms. This model, which is known as the **Non Random Two Liquid Model** (NRTL), leads to the following expression for ΔG_m^{xs}

$$\Delta G_m^{xs} = X_1 X_2 \left[\frac{(E_{12} - E_{22})G_{21}}{X_1 + X_2 G_{21}} + \frac{(E_{21} - E_{11})G_{12}}{X_2 + X_1 G_{12}} \right], \tag{9.2.28}$$

where

$$G_{ij} = \exp \left[-\frac{\alpha_{12}(E_{ij} - E_{jj})}{RT} \right]. \tag{9.2.29}$$

Expressions for the other thermodynamic excess functions can be derived from Eq. (9.2.29), and are given in Prausnitz et al. (1986).

When $\alpha_{12} = 0$ (that implies completely random mixing), $G_{ij} = 1$, and also the ratio of the local mole fraction reduces to that of the bulk mole fractions (Eq. 9.2.24). Under this condition, the ΔG_m^{xs} in the NRTL model reduces to

$$\Delta G_m^{xs} = \Delta E(X_1 X_2), \tag{9.2.30}$$

where, using $E_{12} = E_{21}$,

$$\Delta E = 2E_{12} - (E_{11} + E_{22}) \tag{9.2.31}$$

The above equation is formally similar to the Regular Solution expression, Eq. (9.2.4). However, ΔE reduces to the expression for W in terms of pair potential energies, as given by Eq. (9.2.17), only if $E_{ij} = Z/2(L\Gamma_{ij})$, where Z is the number of nearest neighbors of the components i and j around a central component of i or j. Thus, E_{ij} should be treated as a quantity proportional to the pair potential energy between i and j. By comparing the NRTL and QC models, Renon and Prausnitz (1968) suggested that α_{12} should be similar to $1/Z$. Consequently, α_{12} should be <1. However, the value of α retrieved from experimental mixing property data on mineral solid solutions sometimes depart very significantly from the expected value of $1/Z$.

9.2.6 Van Laar Model

The oldest and one of the most successful of the two constant expressions for binary systems is the one derived by van Laar (1910) on the basis of van der Waal's equation of state and ideal entropy of mixing. The van Laar model can be expressed in the following form.

$$\Delta G^{xs} = \frac{\Omega(a_1 a_2 X_1 X_2)}{(a_1 X_1 + a_2 X_2)(a_1 + a_2)} \tag{9.2.32}$$

where Ω is a term that is related to the interaction energy between the species 1 and 2, and a_1 and a_2 are constants. Using Eq. (8.7.4), the above expression for ΔG^{xs} yields

$$RT \ln \gamma_i = \frac{\Omega X_j^2 a_i (a_j)^2}{(a_i + a_j)(X_i a_i + X_j a_j)^2} \qquad (9.2.33)$$

Ganguly and Saxena (1987) showed that the van Laar model follows as a special case of the Quasi-chemical Model when the interaction energy is small compared to RT. The derivation of van Laar model and its relationship with the QC model suggest that it should be applied to solutions with small or moderate non-ideality, which implies that mixing units in the solution should be quite similar and non-polar. However, despite the severe limitations of its microscopic view, the *form* of the van Laar model has been surprisingly successful, and perhaps more successful than Margules or other two constant formulations (see, for example, Prausnitz et al. 1986). As an example, the van Laar model fits the activity coefficient vs composition data of a mixture of benzene and isooctane, which differ appreciably in size, as illustrated in Fig. 9.6. It has been also been successfully used to treat binary fluid mixtures in the system C-H-O-S that involve both polar (H_2O) and non-polar molecules, and is the most important fluid system in geological processes.

The physical significance of the van Laar constants is unclear, especially when it is applied to treat relatively complex mixtures. However, Saxena and co-workers (Saxena and Fei 1988; Shi and Saxena 1992) have been fairly successful in their treatment of C-H-O-S system by equating the a_1 and a_2 constants with the molar volumes of the end members 1 and 2, respectively. Their lead was pursued by Aranovich and Newton (1999) to treat the activity-composition relation of H_2O and CO_2

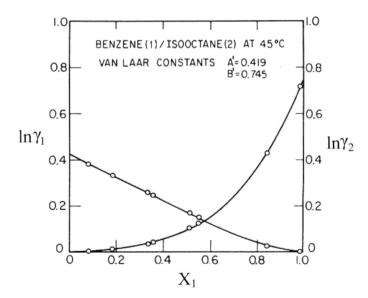

Fig. 9.6 Modeling of the species activity coefficients in a mixture of benzene and isooctane, which differ appreciably in size, by the van Laar model. From Prausnitz et al. (1986)

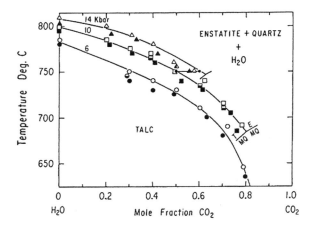

Fig. 9.7 Experimental data on the dehydration equilibrium of talc as a function of fluid composition in the CO_2-H_2O system at 6, 10 and 14 kb pressures, and the model fit of the data using the van Laar mixing model for the fluid phase. Open symbols indicate breakdown of talc to enstatite + quartz, whereas the filled symbol indicate the reverse reaction. From Aranovich and Newton (1999). With permission from Mineralogical Society of America

in the binary system at high pressure and temperature conditions, 6–14 kb, 600–1000 °C. They experimentally determined the effect of change of fluid composition on the equilibrium conditions of decarbonation and dehydration reactions, viz. $CaCO_3$ (calcite) + SiO_2 (quartz) = $CaSiO_3$ (wollastonite) + CO_2, $MgCO_3$ (magnesite) + $MgSiO_3$ (enstatite) = Mg_2SiO_4 (forsterite) + CO_2, and $Mg_3Si_4O_{10}(OH)_2$ (talc) = 3 $MgSiO_3$ + SiO_2 + H_2O. From the known compositions of the fluid phase and their effect on the equilibrium conditions of these reactions (Fig. 9.7), they retrieved the activity coefficients of H_2O and CO_2 in the binary system, assuming that the parameter Ω is a function of P and T according to the form $\Omega = (A + BT)[1 - \exp(-20P)] + CPT$, where P is pressure in kb, and A and B are constants. The term within the square bracket assures ideal gas behavior as $P \rightarrow 0$. (The method for relating the thermodynamic mixing properties of phases to equilibrium conditions of reactions is developed in Sect. 10.12). The values of the constants, as retrieved from experimental data on equilibrium P-T condition versus fluid composition, are A = 12,893 J, B = −6.501 J/K, and C = 1.0112 J/(K-kb). The equilibrium temperature versus mole fraction of CO_2 in the vapor phase at 14 kb, as calculated using these parameters, is shown by solid line in Fig. 9.7.

Note that Eq. (9.2.33) does not permit an extremum of activity coefficient as a function of composition. Thus, the van Laar formulation would be obviously inapplicable where such extremum is known to exist. Recently, Holland and Powell (2003) have developed an extension of van Laar formulation and applied it successfully to treat several asymmetric mineral solid solutions and H_2O-CO_2 mixtures.

9.2.7   Associated Solutions

Associated solutions are those in which there is intermediate compound formation due to negative deviation from the ideality of mixing among the end-member components. The model postulates formation of molecule type associates among the species in the solution. The interaction among these associates and the free ions, if any left after the formation of the associates, are described using different types of solution models, such as the Regular Solution model, as required to model the experimental data. A geologically important system that shows this behavior is the sulfide liquid system O-S-Fe. Figure 9.8, which is reproduced from Kress (2000), shows the variation of the logarithm of sulfur fugacity, $\log(fS_2)$, as a function of the apparent mole fraction of S in the S-Fe binary. By apparent mole fraction, we mean the mole fraction that one would calculate in ignorance of the formation of any intermediate compound. Thus, the apparent mole fraction of S equals $S/(S + Fe)$ where S and F stand for the number of moles of sulfur and iron that were mixed to prepare a solution. Figure 9.8 shows a rapid change of $f(S_2)$ at an intermediate compositional range in the S-Fe binary liquid. This rapid change of $\log(fS_2)$, which implies a rapid change of the chemical potential of S_2 (since $d\mu_i = RTd\ln f_i$), cannot be treated by a model of nonideal interaction between just S and Fe atoms. The different broken curves in Fig. 9.8 illustrate the failure of model predictions based on the non-ideal interactions between these two species. The rapid change of $\mu(S_2)$ implies sharp valley in the G-X diagram at the intermediate composition, and therefore possible presence of intermediate compound(s).

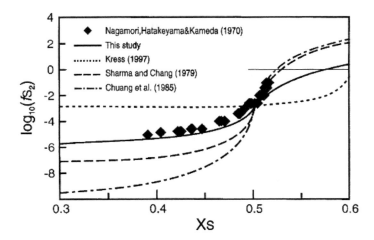

Fig. 9.8 Variation of sulfur fugacity with the mole fraction of sulfur in S-Fe melt, as determined by Nagamori et al. (1970) at 1 bar, 1200 °C. The lines are model predictions by different workers. The most successful prediction is by the associated solution model treatment of Kress (2000), which is labeled as "This study", and shown as a solid line. It considers the melt to be made of Fe, S and FeS (associated compound) with regular solution type interactions. From Kress (2000)

Historically, associated solution model began with the work of Dolezalek (1908) who suggested that the **actual** species in a solution obey Raoult's law or ideal mixing behavior, the only problem being the correct identification of the actual species. An ideal associated solution necessarily leads to negative deviation from ideality of the activity-composition relation of an end-member component if one uses the apparent mole fraction of the component as its true mole fraction. To illustrate this point, I use an example from Hildebrand and Scott (1964), and consider a solution made from a mixture of n_A moles of A and n_B moles of B, with $n_B < n_A$, and suppose that all the moles of B have combined with A to form an intermediate compound AB. In that case, the solution consists of n_B moles of AB, $n_A - n_B$ moles of A and zero moles of B, and consequently the total number of moles, N, in the solution is given by $(n_A - n_B) + n_B = n_A$. Thus, the mole fraction of the species A, y_A, that is actually present in the solution is given by

$$y_A = \frac{n_A - n_B}{n_A} = 1 - \frac{n_B}{n_A} = 1 - \frac{X_B}{X_A} \qquad (9.2.34)$$

where X represents the apparent mole fraction of A or B. For an ideal associated solution $a_A = y_A$, but from the above relation $y_A < X_A$. Thus, the activity of A in an ideal associated solution will be less than its apparent mole fraction. The variation of the activity of A as a function of its apparent mole fraction in an ideal associated solution, as defined by the above equation, is illustrated in Fig. 9.9. The diagonal line represents the other limit in which there is no intermediate compound AB, and A and B mix ideally.

When there is partial combination of A and B, the a_A versus X_A relation will lie between the two limits shown in Fig. 9.9. The calculation of the intermediate case

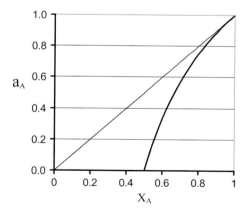

Fig. 9.9 Illustration of the variation of the activity of a dominant component, A, as a function of its apparent mole fraction in an ideal associated solution. The curved line represents the case in which all of the subordinate component, B, combines with A to form an intermediate compound AB. The diagonal straight line represents the case of an ideal solution of A and B without the formation of any intermediate compound. The relationship for a partial combination of B with A will lie between the two limits

requires a knowledge of the equilibrium constant of the homogeneous compound forming reaction (the concept of equilibrium constant is discussed in Sect. 10.4, but a reader with an elementary background of chemistry is expected to be familiar with the concept). As an illustration, let us assume that only a part of the species B has combined to form n_{AB} moles of AB. Then, we have in the solution, $(n_A - n_{AB})$ moles of A, and $(n_B - n_{AB})$ moles of B. Thus, the total number of moles in the solution is $n_{AB} + (n_A - n_{AB}) + (n_B - n_{AB}) = n_A + n_B - n_{AB}$. Assuming that the actual species in the solution mix ideally, we have

$$a_A = y_A = \frac{n_A - n_{AB}}{n_A + n_B - n_{AB}} \qquad (9.2.35)$$

Analogous relations hold for the activities of the two other species, which should be obvious. Assuming ideal mixing of A, B and AB, the equilibrium constant of the reaction A + B = AB is given by

$$K = \frac{y_{AB}}{y_A y_B} = \frac{n_{AB}(n_A + n_B - n_{AB})}{(n_A - n_{AB})(n_B - n_{AB})} \qquad (9.2.36)$$

Since n_A and n_B are known, we can determine n_{AB} from a knowledge of K and substitute that in Eq. (9.2.35) to calculate a_A. Similar calculations can, obviously, be carried out for the other species.

A number of workers (e.g. Blander and Pelton 1987; Zeng and Nekvasil 1996) have successfully modeled the mixing properties of liquids by using the ideal associated models. However, one may also need to assign non-zero interaction energies among the species in the solution to successfully model the property of the solution. For example, Kress (2000) modeled the behavior of the O-S-Fe liquid by considering Regular Solution type interactions among the species in the solution. He considered the existence of the species FeO in the Fe-O binary, FeS in the Fe-S binary and in addition FeO-S in the O-S-Fe ternary system. His model prediction for the S-Fe binary is shown by solid line in Fig. 9.8. It fits the experimental data very well, and much better than predictions from other models which did not consider formation of intermediate compounds in the solution.

The associated solution concept has been incorporated in the widely used MELTS software package to deal with the problem of mixed $H_2O–CO_2$ fluid saturation of rhyolite melt (Ghiorso and Gualda 2015). The dissolved species in melt have been assumed to be hydroxyl melt species, molecular carbon dioxide and carbonate ion, the latter complexed with calcium. The results were applied to several interesting geological problems such as estimation of pressures from composition of H_2O-CO_2 bearing glass inclusions found in quartz phenocrysts of the Bishop Tuff, the partitioning of H_2O and CO_2 between melt and fluid during fluid saturated crystallization, geobarometry etc.

As discussed by Hillert and Sundman (2001), the associated solution model fails to predict miscibility gaps in reciprocal solutions. The sublattice model, on the other hand, over predicts the tendency for unmixing. They have discussed possible modification of the two-sublattice model so it correctly predicts the unmixing properties of a reciprocal solution.

Problem 9.1 Analcime solid solution, which shows athermal mixing behavior, may be treated as a binary solution of the aluminous $(Na_{0.755}Al_{0.755}Si_{2.25}O_6 \cdot 1.125H_2O)$ and siliceous $(Na_{1.05}Al_{1.05}Si_{1.95}O_6 \cdot 0.975H_2O)$ end members. Calorimetric heat of solution data do not show any significant enthalpy of mixing in this binary join, whereas there is a an excess entropy of mixing that is given by

$$\Delta S_m^{xs} = (\Psi + 3R)\left(\sum_k X_k \ln X_k\right)$$

where k stands for an end member and $\psi = -2.20 \; (\pm 75)$ J/mol-K (Neuhoff et al. 2004). Using these data, derive an expression for the activity coefficient, γ_k, of an end member component.

Hint: See Sect. 8.7.

9.3 Multicomponent Solutions

An important practical problem in solution thermodynamics lies in the formulation of method of successful prediction of the properties of multicomponent solutions from those of the bounding binaries. There have been two common approaches in the development of multicomponent excess Gibbs energy models. Following Cheng and Ganguly (1994, 1996), we would call these "**power series multicomponent models**" and "**projected multicomponent models**". In the first approach, one begins with an appropriate power series expression of ΔG_m^{xs} of the multicomponent solution in terms of mole fractions of the components, and then truncate it after a certain number of terms. After some algebraic manipulations, the ΔG_m^{xs} of the multicomponent solution could be expressed to show the nature of the bounding binaries (regular, sub-regular etc.) and the way these are combined to yield the truncated power series of ΔG_m^{xs}. This scheme may then be used as a rational scheme of combining specific types of binaries. The second approach is to combine the binary excess free energies according to certain empirical schemes.

We would refer to a multicomponent solution by the nature of its most asymmetric binary. For example, a subregular multicomponent solution is the one for which the most asymmetric binary has subregular behavior. We would first discuss solution models that deal with mixing within a single structural site, and then a method of combination of mixing within the individual sites to express the multisite mixing properties.

9.3.1 Power Series Multicomponent Models

One of the earliest and most successful multicomponent model that was derived from an expansion of ΔG_m^{xs} in terms of a power series of mole fractions of components is that due to Wohl (1946, 1953). Upon truncating the power series after the third degree terms (that is terms that contain three-body interactions), and algebraic manipulations, Wohl showed that the ΔG_m^{xs} of a ternary solution can be expressed as

$$\Delta G_m^{xs} = \sum_{i \neq j} X_i X_j (W_{ij}^G X_j + W_{ji}^G X_i) + X_i X_j X_k \left[\frac{1}{2} \sum_{i \neq j} (W_{ij} + W_{ji}) + C_{ijk} \right] \quad (9.3.1)$$

where W-s refer to the binary subregular parameters and C_{ijk} represents a ternary interaction term. Subsequently, power series ternary and quaternary expressions were developed by several workers in the geochemical literature. However, Cheng and Ganguly (1994) showed that all these expressions are either equivalent to Wohl's ternary expression or represent its extension to quaternary solution. These authors also developed a power series quaternary expression following Wohl's ternary formulation, and showed that upon truncating it after third degree terms, the excess Gibbs energy of mixing can be expressed as

$$\Delta G_m^{xs} = \sum_{i \neq j} X_i X_j (W_{ij}^G X_{ji} + W_{ji}^G X_{ij}) + \sum_{i \neq j, \neq k} X_i X_j X_k C_{ijk}, \quad (9.3.2)$$

where X_{ji} and X_{ij} are the projected mole fractions of the components j and i, respectively, in the binary join i–j. These **binary** mole fractions are obtained by the normal projection of the multicomponent composition onto that join. Analytically, X_{ij} is given by $\frac{1}{2}(1 + X_i - X_j)$. It is interesting to note that the first right hand term in the above expression represents a summation of ΔG_m^{xs} of the subregular bounding binaries at compositions that are at the shortest distance from the multicomponent composition. Furthermore, a quaternary or higher order solution does not involve a quaternary or higher order term when the binaries have Sub-Regular behavior (this conclusion was also independently reached by Jordan et al. 1950; Helffrich and Wood 1989; and Mukhopadhyay et al. 1993). Using the above equation, and the relation $RT \ln \gamma_i = (\partial \Delta G^{xs}/\partial n_i)$ (Eq. 8.7.3), one can obtain the expression for the activity coefficient of a component in a multicomponent Sub-Regular solution. Note that in this equation G^{xs} refers to the total excess Gibbs energy, not the molar quantity that is expressed by Eq. (9.3.2). The two quantities are related according to $\Delta G^{xs} = N \Delta G_m^{xs}$, where N is the total number of moles in the solution. The expression of $RT \ln \gamma_i$ obtained from Eq. (9.3.2) by the above thermodynamic operation is given by Cheng and Ganguly (1994).

A special case of Eq. (9.3.2) is a multicomponent solution with regular binaries, $W_{ij} = W_{ji} = W_{i-j}$. The activity coefficient of a component in a ternary regular a solution is given by

$$RT \ln \gamma_i = \sum_{j \neq i} W_{i-j} X_j^2 + X_j X_k \left[W_{i-j} + W_{i-k} - W_{j-k} - C_{ijk}(1 - 2X_i) \right] \quad (9.3.3)$$

Effects of additional components that have Regular Solution type interactions with other components can be accounted for by adding terms like $X_j X_k [\ldots]$ on the right hand side of the equation.

Redlich and Kister (1948) utilized Guggenheim's polynomial for binary solution (Eq. 9.2.1) to express ΔG_m^{xs} of multicomponent solution. Their expression, which is commonly referred to as the Redlich–Kister model, involves summation of the binary ΔG^{xs} plus multicomponent correction terms, i.e.

$$\Delta G_m^{xs} = \sum X_i X_j \Delta(G_m^{xs})_{ij} + \text{multicomponent correction terms}, \quad (9.3.4)$$

where $\Delta(G_m^{xs})_{ij}$ is calculated at X_i and X_j from Eq. (9.2.1) (note that these are atomic or mole fractions of the components in the multicomponent system, not the projected binary mole fractions discussed above). The Redlich–Kister model has enjoyed popularity in the metallurgical literature. However, it should be noted that it is equivalent to the Wohl model as long as the bounding binaries follow regular or subregular behavior (Cheng and Ganguly 1994), i.e. when the constant terms higher than A_1 are zero in the polynomial expression of the binary excess free energies.

9.3.2   Projected Multicomponent Models

Several schemes have been proposed for combining the binaries to predict the multicomponent behavior with or without involving multicomponent interaction terms. The popular methods, which are due to Kohler (1960), Colinet (1967), Muggianu et al. (1975) and Bonnier and Caboz (1965, quoted in Hillert 1998), are illustrated in Fig. 9.10. Their expressions for the ΔG_m^{xs} in a ternary solution have been summarized by Hillert (1998). The method of shortest distance was suggested independently by Muggianu et al. (1975) and Jacob and Fitzner (1977). We will, thus, refer to this model as the Muggianu–Jacob-Fitzner model. The expression for ΔG_m^{xs} given by Bonnier and Caboz (1965) was modified later by Toop (1965) in order that the ternary ΔG_m^{xs} appears as a summation of the ΔG_m^{xs} in the binaries when the latter have Regular Solution behavior, i.e. $\Delta G_m^{xs}(\text{ternary}) = \sum X_{ij} X_{ji} \Delta(G_m^{xs})_{ij}$ where X_{ij} and X_{ji} are the shortest distance binary compositions from the ternary compositional point. The modified Bonnier–Caboz formulation is often referred to as Toop's method in the literature. The primary motivations behind the different projected multicomponent formulations is the prediction of the multicomponent behavior from *only* the binary properties, that is to somehow 'absorb' the effects of multicomponent interactions within the scheme of combination of the binaries.

Toop's method is an asymmetric formulation in that it treats one component (component 1 in Fig. 9.10d) differently from the other two. Thus, this method ought to be applied only to ternary systems where one component has a distinctly different

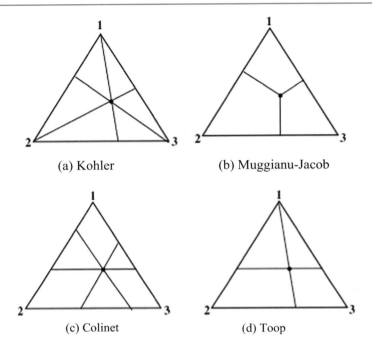

Fig. 9.10 Schematic illustration of the "Projected Multicomponent Models" for a ternary system. The ΔG_m^{xs} of a ternary solution is calculated by combining the ΔG_m^{xs} of the terminal binaries at the projected compositions. From Ganguly (2001)

property from the other two. For example, Pelton and Blander (1986) used a modified QC formulation for the silicate slag system SiO_2–CaO–FeO in which the method of combination of binaries is analogous to that of Toop's method. They chose SiO_2 as the special component 1, since it is an acidic component while the other two are basic components. The predicted ternary properties from combination of the binary data were found to be in good agreement with the experimental data. For solid solutions, one may also be able to identify a component which behaves quite differently from the others. For example, in aluminosilicate garnet, $(Fe,Mg,Ca)_3Al_2Si_3O_{12}$, Ca is the most nonideally mixing component, while Fe and Mg mix nearly ideally (e.g. Ganguly et al. 1996). Thus, Ca may be treated as the unique component in the asymmetric formulation.

9.3.3 Comparison Between Power Series and Projected Methods

The expression of ΔG_m^{xs} obtained for a multicomponent Sub-Regular solution using the power series approach of Wohl (1946) involves a combination of binary excess Gibbs energies at compositions that are at the shortest distance from the multicomponent composition (Eq. 9.3.2). This is exactly the method of combination of the binaries suggested in the Muggianu–Jacob projected multicomponent model (Fig. 9.10b). Thus, there seems to be an independent theoretical justification in the scheme of the combination of binaries in the Muggianu–Jacob model. It was found (Jacob and Fitzner 1977; Jacob, personal communication) that for metallic systems the shortest distance method predicts the ternary ΔG_m^{xs} somewhat better than the other methods, when the ternary interactions are neglected. However, the quality of agreement between the predicted and measured ternary values becomes worse with increasing nonideality of the binaries, especially when a binary ΔG_m^{xs} exceeds 15 kJ/mole, implying increasing importance of the higher order terms.

9.3.4 Estimation of Higher Order Interaction Terms

In principle, it is impossible to determine the multicomponent interaction terms from only the binary data. The higher order interactions specific to a given model can only be determined from comparison of the multicomponent behavior predicted from the binary data with the multicomponent properties determined experimentally. It is, however, doubtful if experimental data are going to be sufficient for such purpose in the foreseeable future, at least for systems of geological interests. A viable alternative would be to determine the binary and multicomponent enthalpic properties from the crystallographic data and pair potential energies of the ions participating in the solid solutions (e.g. Ottonello 1992). Since there are a large body of crystallographic data for a variety of rock-forming mineral solid solutions in both binary and multicomponent systems, it may be possible to use this approach to at least approximately evaluate the relative magnitudes of the higher order terms in the different multicomponent formulations. One may, thus, determine the effectiveness of the different approaches in predicting the multicomponent behavior from only binary data.

From theoretical analysis, Cheng and Ganguly (1994) developed a method of approximation of the C_{ijk} term (Eq. 9.3.2) in the special case that one of the binaries, j–k, behaves nearly ideally. Ilmenite, $(Fe,Mn,Mg)TiO_3$, and garnet, $(Ca, Fe,Mg,Ca)_3Al_2Si_3O_{12}$, are examples of this type of solid solution. In both cases, the first two components mix nearly ideally (Shibue 1999; Ganguly et al. 1996). Indeed, Fe^{2+} and Mg mix with small deviation from ideality in all ferromagnesian silicates for which the thermodynamic mixing properties are known. Thus, this approximation scheme should be applicable to ternary joins of rock-forming minerals involving Fe^{2+}–Mg as one of the subsidiary binaries. The method is as follows.

$$C_{ijk} \approx \sum \sigma_{ij}\sigma_{ik}\left[(W_{ij} - W_{ji})\frac{X_j}{X_j + X_k} + (W_{ik} - W_{ki})\frac{X_k}{X_j + X_k}\right], \qquad (9.3.5)$$

where $\sigma_{ij} = 0$ when $i \equiv j$, and $\sigma_{ij} = 1$ when $i \neq j$.

9.3.5 Solid Solutions with Multi-site Mixing

Hillert (1998) suggested the following expression to represent the ΔG^{xs} of a two site binary reciprocal solution, $^I(A, B)_b{}^{II}(C, D)_c$, for which ΔG_m is given by Eq. (9.1.2)

$$
\begin{aligned}
\Delta G_m^{xs} = x_A x_B x_C (I_{AB:C}) + x_A x_B x_D (I_{AB:D}) + x_C x_D x_A (I_{CD:A}) \\
+ x_C x_D x_B (I_{CD:B}),
\end{aligned}
\tag{9.3.6}
$$

where x_A is the atomic fraction of the species A in the site I, $I_{AB:C}$ is the interaction parameter between A and B in the site I when site II is completely filled by C, $I_{CD:A}$ is the interaction parameter between C and D in the site II when site I is completely is filled by A, and so on. The above expression allows different behavior of ΔG_m^{xs} within a specific sublattice depending on the nature of the species occupying the other sublattice, or in other words, different ΔG_m^{xs} on opposite sides of the compositional square illustrated in Fig. 9.1. For example, when $x_C = 1$, $\Delta G^{xs} = x_A x_B (I_{AB:C})$, but when $x_D = 1$, $\Delta G^{xs} = x_A x_B (I_{AB:D})$. Each site parameter, $I_{AB:C}$ and so on, may be expressed according to the Guggenheim or the so-called Redlich–Kister form, that is, by the expression within the square bracket in Eq. (9.2.1), truncating it after the appropriate number of terms, as demanded by the data. Extensions of this approach to multiple sublattices and multiple components have been discussed by Hillert (1998).

9.3.6 Concluding Remarks

I have summarized above the physical ideas and the basic theoretical structure for a variety of solution models that have been used to treat the thermodynamic properties of mineral solid solutions. These models are also applicable to melts. It may be possible to fit a limited body of data by more than one-solution model equally well, but ΔH^{mix} and ΔS^{mix} predicted by the different models could differ quite significantly. It is, therefore, important to examine the theoretical basis of the solution models in such cases and to ensure, as much as possible, compatibility of the adopted solution model with the known microscopic properties of the solid solution.

From the results of the comparative studies, as discussed above, Guggenheim's polynomial (Eq. 9.2.1) or the so-called Redlich–Kister formulation seems to offer a simple and flexible model for binary solid solutions, although in some specific cases another model, especially QC model when there is short range order, may work better. Use of the Guggenheim polynomial for the binaries affords an additional advantage in the treatment of reciprocal solid solutions in terms of the form suggested by Hillert (1998), Eq. (9.3.6), because the ΔG_m^{xs} in this expression reduces to the form $x_i x_j (I_{ij})$ for the terminal binaries. Binary solutions obeying DQF (Sect. 9.2.3) can also be incorporated in this scheme by defining a solution between a real component and a hypothetical component (Eq. 9.2.15). Finally, if one is to use only the binary terms to predict the multicomponent properties, then the "shortest distance method" of combining the binaries (Fig. 9.10b) is probably the overall best method because of the theoretical justification, as discussed above. The asymmetric Toop method could be advantageous where there is a component with a distinctly different property.

References

Aranovich LY, Newton RC (1999) Experimental determination of CO_2-H_2O activity-composition relations at 600–1000°C and 6–14 kbar by reversed decarbonation and dehydration reactions. Am Mineral 84:1319–1332

Berman RG (1990) Mixing properties of Ca-Mg-Fe-Mn garnets. Am Mineral 75:328–344

Blander M, Pelton AD (1987) Thermodynamic analysis of binary liquid silicates and prediction of ternary solution properties by modified quasichemical equations. Geochim Cosmochim Acta 51:85–95

Burton B, Kikuchi R (1984a) Thermodynamic analysis of the system $CaCO_3$-$MgCO_3$ in the single prism approximation of the cluster variation method. Am Mineral 69:165–175

Burton B, Kikuchi R (1984b) The antiferromagnetic-paramagnetic transition in α-Fe_2O_3 the tetrahedron approximation of the cluster variation method. Phys Chem Mineral 11:125–131

Cheng W, Ganguly J (1994) Some aspects of multicomponent excess free energy models with subregular binaries. Geochim Cosmochim Acta 58:3763–3767

Cheng W. Ganguly J (1996) Erratum to Some aspects of multicomponent excess free energy models with subregular binaries. Geochim. Cosmochim Acta 60:3979

Colinet C (1967) Relation between the geometry of ternary phase diagrams and the thermodynamic properties of liquid solutions. Duplôme d'études supérieures, Univ Grenoble, France (Ph.D. Thesis in French)

Darken LS (1967) Thermodynamics of binary metallic solution. Met Soc AIMS Trans 239:80–89

Dolezalek F (1908) On the theory of binary mixture and concentrated solutions. Zeitschrift Physikalishes Chemie –Stoichiometrie und Verwandtschsftslehre 64:727–747

Fei Y, Saxena SK, Eriksson G (1986) Some binary and ternary silicate solid solutions. Contrib Mineral Petrol 94:221–229

Flood H, Forland T, Grjotheim K (1954) Uber den Zusammenhang zwischen Konzentration und Aktivitäten in geschmolzenen Salzmischungen. Z Anorg Allg Chem 276:290–315

Flory PJ (1941) Thermodynamics of high polymer solutions. J Chem Phys 9:660–661

Flory PJ (1944) Thermodynamics of heterogeneous polymers and their solutions. J Chem Phys 12:425–538

Førland T (1964) Thermodynamic properties of fused salt systems. In: Sundheim BR (ed) Fused salts. McGra-Hill, New York, pp 63–164

Fowler RH, Rushbrooke GS (1937) An attempt to extend the statistical theory of perfect solutions. Trans Faraday Soc 33:1272–1294

Ganguly J (2001) Thermodynamic modelling of solid solutions. In Geiger CA (ed) Solid solutions in silicate and oxide systems. EMU Notes Mineral 3, Eötvös University Press, pp 37–70

Ganguly J, Saxena SK (1987) Mixtures and mineral reactions. Springer, Berlin, p 291

Ganguly J, Cheng W, O'Neill HStC (1993) Syntheses, volume, and structural changes of garnets in pyrope-grossular join: implications for stability and mixing properties. Am Mineral 78:583–593

Ganguly J, Yang H, Ghose S (1994) Thermal history of mesosiderites: quantitative constraints from compositional zoning and Fe-Mg ordering in orthopyroxenes. Geochim Cosmochim Acta 58:2711–2723

Ganguly J, Cheng W, Tirone M (1996) Thermodynamics of aluminosilicate garnet solid solution: new experimental data, an optimized model, and thermometric applications. Contrib Mineral Petrol 126:137–151

Ghiorso MS, Gualda GAR (2015) An H_2O-CO_2 mixed fluid saturation model compatible with rhyolite-MELTS. Contrib Mineral Petrol 169:53–83

Ghose S (1982) Mg-Fe order-disorder in ferromagnesian silicates. I. Crystal chemistry. In: Saxena SK (ed) Advances in physical geochemistry, vol 2. Springer, Berlin, pp 4–57

Green EJ (1970) Predictive thermodynamic models of mineral systems. Am Mineral 55:1692–1713

Guggenheim EA (1937) Theoretical basis of Raoult's law. Trans Faraday Soc 33:151–159

Guggenheim EA (1952) Mixtures. Clarendon Press, Oxford, p 270

Guggenheim EA (1967a) Theoretical basis of Raoult's law. Trans Faraday Soc 33:151–159

Guggenheim EA (1967b) Thermodynamics. North Holland Publishing Co., Amsterdam

Helffrich G, Wood BJ (1989) Subregular models for multicomponent solutions. Am Mineral 74:1016–1022

Hildebrand JH (1929) Solubility XII. Regular solutions. J Am Chem Soc 51:66–80

Hildebrand JH, Scott RL (1964) The solubility of nonelectrolytes. Dover, New York, p 488

Hillert M (1998) Phase equilibria, phase diagrams and phase transformations: Their thermodynamic basis. Cambridge University Press, Cambridge

Hillert M (2001) The compound energy formalism. J Alloys Compd 320:161–176

Hillert M, Sundman B (2001) Predicting miscibility gaps in reciprocal liquids. Calphad 25:599–605

Holland TJB, Powell R (2003) Activity-composition relations for phases in petrological calculations: an asymmetric multicomponent formulation. Contrib Mineral Petrol 145:492–501

Huggins ML (1941) Solutions of long chain compounds. J Phys Chem 9:440

Jacob KT, Fitzner K (1977) The estimation of the thermodynamic properties of ternary alloys from binary data using the shortest distance composition path. Thermochim Acta 18:197–206

Jordan D, Gerster JA, Colburn AP, Wohl K (1950) Vapor-liquid equilibrium of C_4 hydrocarbon-furfural-water mixtures. Chem Eng Prog 46:601–613

Kikuchi R (1951) A theory of cooperative phenomena. Phys Rev 81:988–1003

Kohler F (1960) Zur Berechnung der Thermodynamischen Daten eines ternären Systems aus dem zugehörigen binären System. Monatsch Chem 91:738–740

Kress V (2000) Thermochemistry of sulfide liquids. II. Associated solution model for sulfide in the system O-S-Fe. Contrib Mineral Petrol 139:316–325

Liermann H-P, Ganguly J (2003) Fe^{2+}-Mg fractionation between orthopyroxene and spinel: experimental calibration in the system FeO-MgO-Al_2O_3-Cr_2O_3-SiO_2, and applications. Contrib Mineral Petrol 145:217–227

Meyer KH, van der Wyk A (1944) Properties of polymer solutions. XVIII. Thermodynamic analysis of binary systems with a chain-shaped component. Helv Chim Acta 27:845–858 (in French)

Muggianu YM, Gambino M, Bros JP (1975) Enthalpies of formation of liquid alloy bismuth-gallium-tin at 723 K. Choice of an analytical representation of integral and partial excess functions of mixing. J Chimie Phys 72:83–88

Mukhopadhyay B, Basu S, Holdaway M (1993) A discussion of Margules-type formulations for multicomponent solutions with a generalized approach. Geochim Cosmochim Acta 57:277–283

Nagamori M, Hatakeyama T, Kameda M (1970) Thermodynamics of Fe-S melts between 1100 and 1300 °C. Trans Japan Inst Met 11:190–194

Neuhoff PS, Hovis GL, Balassone G, Stebbins JF (2004) Thermodynamic properties of analcime solid solutions. Am J Sci 304:21–66

Newton RC, Charlu TV, Kleppa OJ (1977a) Thermochemistry of high pressure garnets and clinopyroxenes in the system CaO-MgO-Al_2O_3-SiO_2. Geochim Cosmochim Acta 41:369–377

Newton RC, Thompson AB, Krupka KM (1977b) Heat capacity of synthetic $Mg_3Al_2Si_3O_{12}$ from 350–1000 K and the entropy of pyrope. Trans Am Geophys Union EOS 58:523

Newton RC, Charlu TV, Kleppa OJ (1980) Thermochemistry of high structural state plagioclases. Geochim Cosmochim Acta 44:933–941

Orye RV, Prausnitz JM (1965) Thermodynamic properties of binary solutions containing hydrocarbons and polar organic solvents. Trans Farady Soc 61:1338–1346

Ottonello G (1992) Interactions and mixing properties in the (C2/c) clinopyroxene quadrilateral. Contrib Mineral Petrol 111:53–60

Ottonello G (2001) Thermodynamic constraints arising from the polymeric approach to silicate slags: the system CaO-FeO-SiO$_2$ as an example. J Non-Crystalline Solids 282:72–85

Pelton AD, Blander M (1986) Thermodynamic analysis of ordered liquid solutions by a modified quasichemical approach – applications to silicate slags. Met Trans B 17B:805–815

Powell R (1987) Darken's quadratic formulation and the thermodynamics of minerals. Amer Min 72:1–11

Prausnitz JM, Lichtenthaler RN, de Azevedo EG (1986) Molecular thermodynamics of fluid phase equilibria. Prentice-Hall, New Jersey, p 600

Redlich O, Kister AT (1948) Thermodynamics of non-electrolyte solutions, x-y-t relations in a binary system. Ind Eng Chem 40:341–345

Renon H, Prausnitz JM (1968) Local compositions in thermodynamic excess functions for liquid mixtures. J Amer Inst Chem Eng 14:135–144

Saxena SK, Fei Y (1988) Fluid mixtures in the carbon-hydrogen-oxygen system at high pressure and temperature. Geochim Cosmochim Acta 52:505–512

Shi P, Saxena SK (1992) Thermodynamic modeling of the carbon-hydrogen-oxygen-sulfur fluid system. Am Mineral 77:1038–1049

Shibue Y (1999) Calculations of fluid-ternary solid solutions equilibria: an application of the Wilson equation to fluid-(Fe, Mn, Mg)TiO$_3$ equilibria at 600°C and 1 kbar. Am Mineral 84:1375–1384

Stimpfl M, Ganguly J, Molin G (1999) Fe^{2+}-Mg order-disorder in orthopyroxene: equilibrium fractionation between the octahedral sites and thermodynamic analysis. Contrib Mineral Petrol 136:297–309

Sundman B, Ågren J (1981) A regular solution model for phases with several components and sublattices suitable for computer applications. J Phys Chem Solids 42:297–301

Temkin M (1945) Mixtures of fused salts as ionic solutions. Acta Physicochem URSS 20:411–420

Thompson JB Jr (1967) Thermodynamic properties of simple solutions. In: Abelson PH (ed) Researches in geochemistry, vol 2. Wiley, New York, pp 340–361

Toop GW (1965) Predicting ternary activities using binary data. Trans Am Inst Min Eng 233:850–855

Van Laar JJ (1910) Über Dampfspannungen von binaren Gemischen. Z Phys Vhem 72:723–751

Vinograd VL (2001) Configurational entropy of binary silicate solid solutions. In: Geiger CA (ed) Solid solutions in silicate and oxide systems. EMU Notes Mineral, vol 3:303–346. Eötvös University Press, pp 303–346

Wilson GM (1964) Vapor-liquid equilibrium. XI. A new expression for the excess free energy of mixing. J Amer Chem Soc 86:127–130

Wohl K (1946) Thermodynamic evaluation of binary and ternary liquid systems. Trans Amer Inst Chem Eng 42:215–249

Wohl K (1953) Thermodynamic evaluation of binary and ternary liquid systems. Chem Eng Prog 49:218–221

Wood BJ, Nicholls J (1978) The thermodynamic properties of reciprocal solid solutions. Contrib Mineral Petrol 66:389–400

Wood BJ, Holland TJB, Newton RC, Kleppa OJ (1980) Thermochemistry of jadeite-diopside pyroxenes. Geochim Cosmochim Acta 44:1363–1371

Zeng Q, Nekvasil H (1996) An associated solution model for albite-water melts. Geochim Cosmochim Acta 60:59–73

Equilibria Involving Solutions and Gaseous Mixtures

10

We have discussed in Chap. 6 the thermodynamic treatment of equilibrium relations among phases of fixed compositions. In this chapter, we expand the scope of equilibrium calculations to include phases of variable of compositions. It should be emphasized at the outset, and should also be obvious, that the equations derived for the calculation of equilibrium P-T-X relation among phases do not depend on the nature of the phases that have been chosen as convenient examples for the derivation of the equations. The formal P-T-X relations are generally valid.

Equilibrium relations are often illustrated by means P-T-X phase diagrams that show the fields of stability of various phases. A section is included in this chapter to present some of the basic concepts of these types of phase diagrams using petrologically important but simple systems. The concepts that are needed to interpret phase diagrams have been extensively discussed in the context of geological problems in a number of books (e.g. Philpotts 1990; Winter 2001; Ernst 1976; Ehlers 1987). Overall, there are a large number of books devoted to phase equilibrium and phase diagrams because of the importance of this subject in geology, materials science and ceramics. Thus, the section on phase diagrams has been kept short with the objective of exposing some of the essential thermodynamic, mass balance and geometric concepts.

10.1 Extent and Equilibrium Condition of a Reaction

Let v_i stand for the stoichiometric coefficient of a chemical species in a reaction, with a positive value for a product and a negative value for a reactant species, and Δn_i denote the change in the number of moles of a species at a given stage during the progress of the reaction. The ratio $\Delta n_i / v_i$ has the interesting property that its value is independent of the choice of the species. To appreciate this point, consider a simple reaction $H_2 + Cl_2 = 2HCl$. If at a given stage of the reaction, 5 mol of H_2 was consumed, in which case $\Delta n_{H2} = -5$, then from the stoichiometric relation of the reaction, $\Delta n_{Cl2} = -5$ and $\Delta n_{HCl} = 10$. However, since following the sign

© Springer Nature Switzerland AG 2020
J. Ganguly, *Thermodynamics in Earth and Planetary Sciences*,
Springer Textbooks in Earth Sciences, Geography and Environment,
https://doi.org/10.1007/978-3-030-20879-0_10

convention for stoichiometric coefficient (negative for a reactant and positive for a product) $v_{H2} = v_{Cl2} = -1$ and $v_{HCl} = 2$, we have

$$\frac{\Delta n_{H_2}}{v_{H_2}} = \frac{\Delta n_{Cl_2}}{v_{Cl_2}} = \frac{\Delta n_{HCl}}{v_{HCl}} = 5$$

For infinitesimal progress of a reaction, Δn_i. is replaced by dn_i.

The founder of the famous Belgian school of thermodynamics, De Donder (1872–1957), defined the extent of a reaction, ξ, in terms of the **species independent** ratio as

$$\frac{dn_i}{v_i} = d\xi \qquad (10.1.1)$$

where $d\xi$ is the infinitesimal change of the **extent of reaction** ξ. The latter is also referred to as the reaction progress variable or simply as progress variable. Noting the sign convention for the stoichiometric coefficient ($v_I > 0$ for product and < 0 for reactant), it should be easy to see that $d\xi > 0$ implies that the reaction proceeds to the right, and vice versa.

Let us now consider an arbitrary reaction, which may be represented symbolically as

$$\sum_i v_i A_i = 0 \qquad (10.1.2)$$

with the usual sign convention for v_i (see above). When Gibbs introduced the concept of chemical potential, he had in mind only the changes of moles of chemical species within a system due to exchange with the surrounding. However, the change in the number of moles of a species has the same effect on G whether it is due to mass exchange with the surrounding or chemical reaction within the system. Thus, using Eq. (8.1.7), the change in the Gibbs energy of a system due to internal chemical reaction at a constant P-T condition is given by

$$(\partial G)_{P,T} = \sum_i \mu_i dn_i,$$

which, on substitution of $v_i\, d\xi$ for dn_i (Eq 10.1.1) yields

$$\left(\frac{\partial G}{\partial \xi}\right)_{P,T} = \sum_i v_i \mu_i, \qquad (10.1.3)$$

Now, since for any spontaneous change at constant P-T condition, G of a system must decrease until equilibrium is achieved (Fig. 10.1), it follows that for a reaction

$$\left(\frac{\partial G}{\partial \xi}\right)_{P,T} = \sum_i v_i \mu_i \leq 0 \qquad (10.1.4)$$

Fig. 10.1 Schematic illustration of the change of Gibbs free energy, G, of a system as a function of the extent of reaction, ξ, at constant P-T condition. The equilibrium condition is given by the minimum of G

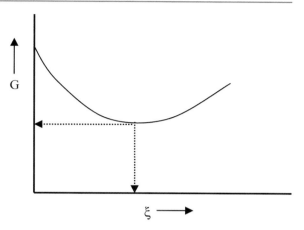

10.2 Gibbs Free Energy Change and Affinity of a Reaction

Following the classic and highly influential text by Lewis and Randall on thermodynamics (Lewis and Randall 1923), the quantity $\sum v_i \mu_i$ is commonly denoted by the symbol $\Delta_r G$ and is called the **Gibbs free energy change** or simply the **Gibbs energy change** of a reaction. Thus, according to Eq. (10.1.4), at constant P-T condition, a reaction proceeds in the direction of lower Gibbs energy.

In an alternative development of the direction of reaction progress in terms of entropy production, De Donder (1927) developed the concept of **Affinity**, A, which led to the result that a reaction progresses in the direction of increasing Affinity. From Eq. (10.1.4), one would therefore anticipate that

$$A = -\sum v_i \mu_i = -\Delta_r G,$$
$$v_i > 0 \text{ for products}$$
$$< 0 \text{ for reactants} \tag{10.2.1}$$

It is instructive here to see how De Donder developed the criterion for the direction of a chemical reaction from the principle of entropy production. From Eq. (8.1.4), we have

$$dS = \frac{dU + PdV}{T} - \frac{\sum\limits_{i} \mu_i dn_i}{T} \tag{10.2.2}$$

From the perspective of Gibbs' development, the first and second right hand terms in this equation represent entropy change of the system due to reversible exchange of heat (since $dU + PdV = TdS = \delta q(rev)$) and matter with the surrounding, respectively. De Donder recognized that in a closed system undergoing irreversible chemical reactions, the **internal** entropy production can be expressed according to the form of the second right hand term as (Kondepudi and Prigogine 2002),

$$dS_{int} = -\frac{\sum_i \mu_i d_i n_i}{T} \tag{10.2.3}$$

where $d_i n_i$ stands for the change of mole numbers of i due to irreversible chemical reaction. Using the relation (10.1.1) for $d_i n_i$, we then have

$$dS_{int} = -\frac{\sum_k \mu_i (\nu_i d\xi)}{T} \tag{10.2.4}$$

or

$$\boxed{dS_{int} = \left(\frac{A}{T}\right) d\xi} \tag{10.2.5}$$

where $A = -\sum \nu_i \mu_i$.

If we write a reaction as

$$r_1 A_1 + r_2 A_2 + \ldots = p_1 B_1 + p_2 B_2 + \ldots, \tag{10.2.a}$$

then

$$\boxed{A = \sum_i r_i \mu_{A_i} - \sum_j p_j \mu_{B_j}} \tag{10.2.6}$$

Now since, according to the second law (Eq. 2.4.9), $dS_{int} \geq 0$, it follows from Eq. (10.2.5) that $Ad\xi \geq 0$. Consequently, A and $d\xi$ must have the same sign. In other words, if $A > 0$, then $d\xi > 0$ (that is the reaction proceeds from left to right), and vice versa.

The two terms, Affinity and Gibbs free energy change of a reaction came into use in two independent developments of chemical thermodynamics, and their use in the literature often reflects a matter of personal preference. However, since the development of the concept of Affinity is linked to entropy production, it is more commonly used in irreversible thermodynamics that deals with entropy production. We return to the problem of entropy production in Appendix A.

10.3   Gibbs Phase Rule and Duhem's Theorem

The general idea of Gibbs Phase Rule was introduced in Sect. 6.1, where we also formally defined the terms phase, component and degree of freedom. Here we present the derivation of the Phase Rule, and along with that state and prove another important theorem known as the Duhem's theorem. Phase Rule addresses the problem of total number of independent **intensive** variables in a system at equilibrium, regardless of whether it is open or close, as long as it is not subjected to any variation of an external force field such as gravity. The phase rule is not concerned with the quantities of the different phases. On the other hand, the Duhem's theorem allows one to determine the number of variables that need to be fixed (whether these variables are intensive or extensive or a combination of both) so that the equilibrium properties of a **closed** system become completely characterized.

If the number of variables that we happen to be dealing with is V, and there are R independent relations among these variables, then the degree of freedom, that is the number of variables that can be varied independently, is given by (V − R). For example, if we have two variables y and x, and these are related by an equation such as y = mx + c, then there is one degree of freedom. That is, if we choose the value of one of the variables, the value of the other one is defined. However, if there is an additional relation between the two variables, such as of the form y = m′x + c′, then there is zero degree of freedom. That is the values of both variables are fixed by the requirement that these must simultaneously satisfy the two independent relations between them. In this example, the x and y values are the co-ordinates of the point of intersection of the two straight lines defined by the two relations.

10.3.1   Phase Rule

10.3.1.1   General Derivation

In order to derive the Phase Rule, let us consider a system in which the number of phases is P and that of the components is C. If all C components are present in each phase, then the number of intensive compositional variables (atomic fractions) is PC. In addition, pressure and temperature are two other intensive variables, assuming that the pressure on certain phase is not buffered from outside the system. (The latter situation could arise in certain geological problems, such as a solid-gas system in which the gas pressure could be buffered by communication of the gas with an external reservoir while the solid pressure is defined by the overburden or lithostatic pressure on the rock). Thus, the total number of intensive variables in the system subjected to uniform P-T condition is **PC + 2**.

Now we need to find out the number of independent relations among these intensive variables. It is easy to see that there is a set of such relations defined by the stoichiometric constraints. That is $\sum_i (X_i)^\alpha = 1$ for each phase where $(X_i)^\alpha$ is the atomic fraction of the component i in the phase α. The total number of such relations is **P**.

Additional relations among the intensive variables are obtained from the thermodynamic requirement of chemical equilibrium, that is, the chemical potential of each component must be the same in each phase in which it can be present. Thus, assuming that each component is present in all the phases, we have

$$(\mu_1)^1 = (\mu_1)^2 = (\mu_1)^3 = \ldots\ldots\ldots = (\mu_1)^P$$
$$(\mu_2)^1 = (\mu_2)^2 = (\mu_2)^3 = \ldots\ldots\ldots = (\mu_2)^P$$

C relations $\ldots\ldots\ldots\ldots\ldots\ldots\ldots\ldots\ldots\ldots\ldots\ldots\ldots$ (10.3.1)

$$(\mu_c)^1 = (\mu_c)^2 = (\mu_c)^3 = \ldots\ldots\ldots = (\mu_c)^P$$

$\longleftarrow$ (P-1) relations $\longrightarrow$

where the subscript refers to a component and the superscript refers to a phase. Each row in the above set of equations defines (P-1) relations (since there are P-1 equalities). Hence, the system of equations in (10.3.1) yield **C(P-1)** independent relations. Now assuming that there is no additional restriction on the chemical potential of any component through communication with an external reservoir, the degree of freedom of intensive variables, F, is given by

$$F = [PC + 2] - [P + C(P - 1)] = C - P + 2.$$

Here the terms within the first pair of square brackets indicate the total number of intensive variables, whereas those within the second pair indicate the number of independent relations among these variables.

In summary, the Phase Rule states that in a system consisting of P phases and C components, which (a) is in chemical equilibrium, (b) has a homogeneous pressure throughout, (c) is not subject to buffering of chemical potential of any component by mass exchange with an external reservoir, and (d) is not subject to variation of an external force such as gravity, the number of degrees of freedom of the **intensive variables** is given by

$$\boxed{F = C - P + 2}$$ (10.3.2)

10.3.1.2 Special Case: Externally Buffered Systems

It is now easy to modify the phase rule by removing the simplifying assumptions that have been used to derive it. Let us assume, as illustrated in Fig. 10.2, that the chemical potentials of a certain number of components, κ, are controlled from an external reservoir by exchange through a semi-permeable membrane, which are open only to those components. The chemical potential of any of these components in any phase is the same as that in the reservoir, but the chemical potentials of the components in the reservoir remain fixed since any exchange of a component between the reservoir and the system does not have any significant effect on its content in the reservoir. Thus, there are now $(C-\kappa)$ rows in the system of equations of the type (10.3.1) for the equality of the chemical potentials of the components among the phases, and consequently $(P-1)(C-\kappa)$ independent equations. However,

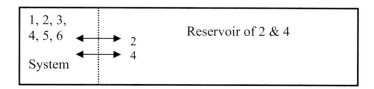

Fig. 10.2 Schematic illustration of the buffering of the chemical potentials of two components, 2 and 4, in a system by an external reservoir. The system and reservoir are separated by a membrane that is permeable only to the components 2 and 4

since the chemical potentials of κ number of components in each phase are controlled from outside, there are κP additional equations. Also, as before, we have P equations for the stoichiometric restrictions. Thus,

$$F = [PC + 2] - [(P-1)(C-\kappa) + \kappa P + P]$$

or

$$\boxed{F = (C-\kappa) - P + 2} \tag{10.3.3}$$

The maximum number of phases that can coexist in equilibrium in a system, P_{max}, is obtained by setting $F = 0$. If P and T are held constant (which means that we have exercised two degrees of freedom), then we subtract 2 from the right hand side of the above equation. Thus, at constant P-T condition

$$P_{max} = C - \kappa$$

The above equation explains the development of rocks with a small number of minerals and the apparently strange situation of monomineralic rocks even though there are many components in a natural environment (note that the number P increases with the number C). Equation (10.3.3) was first derived by Korzhinskii (1959) in a different way. (Korzhinskii's derivation of the modified form of Phase Rule was hotly debated but the final result is unquestionably correct, and he should be credited for the recognition that the standard form of phase rule needed modification to account for the development of rocks with **monomineralic** bands). Rumble (1982) discussed the petrographic, chemical (including stable isotopic compositions) and geological considerations that may be used to decide if some components within a rock under investigation had been buffered from outside.

It should now be obvious that if the pressure on the gas phase in a system, P_g, is buffered from outside and is different from that on the solid phase, P_s, then the Phase Rule should be further modified as (Ganguly and Saxena 1987).

$$\boxed{F = (C - \kappa) - P + 2 + I} \tag{10.3.4}$$

where $I = 0$ for $P_g = P_s$, and $I = 1$ for $P_g \neq P_s$.

10.3.2 Duhem's Theorem

Consider now a **closed system** with fixed masses of all components. The question is: **how many variables must be specified so that all equilibrium properties of the system, intensive and extensive, are defined?** The answer to this question is provided by the Duhem's theorem, which states that in a closed system, there are **only two** independent variables under equilibrium condition. Once the values of two variables, whether intensive or extensive or one of each type, are specified, all equilibrium properties of the system, intensive plus extensive, are defined. The number of intensive variables that can be specified is, of course, restricted by the Phase Rule. As an example, if we have a divariant closed system, and the pressure and temperature conditions are specified, then the composition and mass of each phase are defined. All other properties of the system derive from these three properties. Duhem's theorem, of course, does not tell us how to derive the equilibrium properties of the system, but knowing that all properties of a closed system have unique values when two variables are specified is an important step. Now let us see how Duhem's theorem is derived.

Consider a system consisting of C chemical species and P phases, and assume for simplicity that each phase contains all chemical species. In that case, there are C mole numbers in each phase, so that there are a total of PC mole numbers in the system. With pressure and temperature as the two other additional variables, we have a total of PC + 2 variables in the system. The requirement of chemical equilibrium among the phases, that is the system of equations in Eq. (10.3.1), yields $C(P - 1)$ independent relations, as discussed above. In addition, since the system is closed, the total number of moles of each component is fixed. Thus, for each component, j, there is a mass conservation relation of the type

$$(n_j^1 + n_j^2 + \ldots n_j^P) = N_j, \qquad (10.3.5)$$

where n stands for the mole numbers of j in the superscripted phase (p being the p th phase), and N_j is the total mole numbers of the chemical species j in the system. Obviously, there are a total of C relations of this type. Thus, the number of independent variables, υ, is given by

$$\boxed{\upsilon = [PC + 2] - [C(P - 1) + C] = 2} \qquad (10.3.6)$$

The restricted total variance of a closed system, as expressed by the Duhem's theorem, forms the basis of a variety of calculations in geological and planetary problems. For example, Spear (1988) used it to develop quantitative relationships among the changes of phase chemistry and modes as function of changes of P-T conditions in igneous and metamorphic systems. (This work is an extension of an earlier study by Spear et al. (1982) that dealt only with the relationship among changes of the intensive variables that has become known as the "Gibbs method"). Duhem's theorem also forms the basis of calculations of modal abundance and compositions of phases in a closed system at equilibrium (Sect. 10.13). Since the total variance of the closed system is 2, one can find out if in a given situation all independent relations among the variables (PC, as given by the terms within the second square bracket in Eq. 10.3.6), which are needed to perform the desired calculations, have been accounted for.

Problem 10.1 Show that the Phase Rule, as expressed by Eq. (10.3.2) holds even when all C components are not present in all phases.

Hint: Proceed by choosing a specific case in which the c th component is present only in φ phases where $\varphi < P$, and setting up a set of equations as in Eq. (10.3.1) for the equality of chemical potential of components in different phases.

10.4 Equilibrium Constant of a Chemical Reaction

10.4.1 Definition and Relation with Activity Product

Consider a balanced chemical reaction written in the form

$$\sum_i v_i C_i = 0, \tag{10.4.a}$$

($v_i > 0$ for products and <0 for reactants) where C_i is a component in a specific phase. As a specific example, let us consider the reaction

$$\underset{\text{Plag}}{3CaAl_2Si_2O_8} = \underset{\text{Grt}}{Ca_3Al_2Si_3O_{12}} + \underset{\text{Ky}}{2Al_2SiO_5} + \underset{\text{Qtz}}{SiO_2} \tag{10.4.b}$$

where $CaAl_2Si_2O_8$ is a component in the mineral plagioclase, and so on (Plag: plagioclase, Grt: garnet, Ky: kyanite, Qtz: quartz). Using the form of Eq. (10.4.a), this reaction is written as

$$(Grs)^{Grt} + 2(Al_2SiO_5)^{Ky} + (SiO_2)^{Qtz} - 3(An)^{Plag} = 0$$

where Grs and An stand for $Ca_3Al_2Si_3O_{12}$ (grossular component) and $CaAl_2Si_2O_8$ (anorthite component), respectively.

At any specified P, T, X condition, the Gibbs energy change of the reaction (10.4.a), can be expressed as (Eq. 10.2.1)

$$\Delta_r G(P, T, X) = \sum_i v_i \mu_i(P, T, X) \tag{10.4.1}$$

where μ_i is the chemical potential of component i in the phase φ_j. Upon substitution of the expressions of chemical potentials in terms of activities (i.e. $\mu_i = \mu_i^* + RT\ln a_i$: Eq. 8.4.6) and rearrangement of terms, we then have

$$\Delta_r G(P, T, X) = \left[\sum_i v_i (\mu_i)^*\right] + RT \ln \prod_i (a_i)^{v_i}, \qquad (10.4.2)$$

where the symbol $\prod_i$ stands for the product of all activity terms raised to the appropriate powers that are positive for the products and negative for the reactants. The quantity within the square brackets denotes the change of standard state chemical potentials in the reaction, and is denoted by $\Delta_r G^*(T)$.

Application of the last equation to the reaction (10.4.b) yields

$$\Delta_r G(P, T, X) = \Delta_r G^*(P, T)$$
$$+ RT\left[\ln\left(a_{Grs}^{Grt}\right) + 2\ln\left(a_{Al_2 SiO_5}^{Ky}\right) + \ln\left(a_{SiO_2}^{Qtz}\right) - 3\ln\left(a_{An}^{Plag}\right)\right] \qquad (10.4.3)$$

which can also be written as

$$\Delta_r G(P, T, X) = \Delta_r G^*(P, T) + RT \ln \frac{\left(a_{Grs}^{Grt}\right)\left(a_{Al_2 SiO_5}^{Ky}\right)^2\left(a_{SiO_2}^{Qtz}\right)}{\left(a_{An}^{Plag}\right)^3} \qquad (10.4.4)$$

From this analysis, it should be easy to see that for any reaction we can write

$$\Delta_r G(P, T, X) = \Delta_r G^*(T) + RT \ln \frac{\left(a_{C_1}^{P_1}\right)^{v_1''}\left(a_{C_2}^{P_2}\right)^{v_2''}\cdots}{\left(a_{C_1}^{R_1}\right)^{v_1'}\left(a_{C_2}^{R_2}\right)^{v_2'}\cdots} \qquad (10.4.5)$$

where, for the sake of clarity and distinctiveness, we have used in the logarithmic term P_i for the product phase and v_i'' for the **magnitude** of the stoichiometric coefficient of the product component C_i', with analogous labeling applying to the reactants. We call the ratio of activities in the above expression as a **reaction quotient**, and for the sake of brevity represent it by the symbol Q. Thus, in general, the Gibbs energy change of a reaction at any condition can be written as

$$\Delta_r G(P, T, X) = \Delta_r G^*(T) + RT \ln Q \qquad (10.4.6)$$

When **equilibrium** is achieved, $\Delta_r G(P, T, X) = 0$, so that

$$RT \ln Q_{eq} = -\Delta_r G^*(T) \qquad (10.4.7)$$

The above equation shows that at equilibrium, the reaction quotient assumes a specific value, Q_{eq}, which depends only on the choice of standard states at a given temperature. This specific value of Q is known as the **equilibrium constant, K**. Thus, we write,

$$RT \ln Q_{eq} \equiv RT \ln K = -\Delta_r G^*(T) \qquad (10.4.8)$$

Note that it is not required to have the standard states of all phases referred to the same pressure. We discuss in a later section (Sect. 10.5.1) the use of mixed standard states. Although it should be obvious, it is re-emphasized that because of the last relation, the equilibrium constant, K, is **not a function of composition** of the phases. It has a defined value at a fixed P-T condition, regardless of the composition of the phases.

10.4.2 Pressure and Temperature Dependencies of Equilibrium Constant

The formal expressions for the pressure dependence of equilibrium constant follow from Eq. (10.4.8), depending on the choice of standard state. We may choose the standard state of a component in a phase to be the state of the pure end-member phase at 1 bar and the temperature of interest. In that case, $\Delta_r G^*(T) = \Delta_r G^o(1 \text{ bar}, T)$, and consequently the equilibrium constant is independent of pressure. On the other hand, if the standard state is chosen to be the state of pure end-member phase at the P-T condition of interest, so that $\Delta_r G^*(T) = \Delta_r G^o(P, T)$, then obviously the equilibrium constant depends on pressure according to $RT(\partial \ln K / \partial P) = -\partial(\Delta_r G^o / \partial P) = -\Delta_r V^o$.

Summarizing,

(a) for standard state tied to a fixed pressure,

$$\left(\frac{\partial \ln K}{\partial P} \right)_T = 0; \text{ and} \qquad (10.4.9a)$$

(b) for standard state tied to the pressure of interest, $\Delta_r G^* = f(P, T)$:

$$\left(\frac{\partial \ln K}{\partial P} \right)_T = -\frac{\Delta V^*}{RT} \qquad (10.4.9b)$$

From Eq. (10.4.8), the temperature dependence of equilibrium constant is given by

$$\left(\frac{\partial \ln K}{\partial T}\right)_P = -\left(\frac{\partial(\Delta_r G^*/RT)}{\partial T}\right)$$

Using the rule for differentiating a ratio, we obtain

$$\left(\frac{\partial \ln K}{\partial T}\right)_P = -\frac{1}{R}\left(\frac{T\dfrac{\partial \Delta_r G^*}{\partial T} - \Delta_r G^*}{T^2}\right) = -\frac{1}{R}\left(\frac{-T\Delta_r S^* - \Delta_r G^*}{T^2}\right),$$

which, on substitution of the relation H = G + TS yields

$$\boxed{\left(\frac{\partial \ln K}{\partial T}\right)_P = \frac{\Delta_r H^*}{RT^2}} \qquad (10.4.10a)$$

or, since $dT = -T^2 d(1/T)$

$$\boxed{\left(\frac{\partial \ln K}{\partial(1/T)}\right)_P = -\frac{\Delta_r H^*}{R}} \qquad (10.4.10b)$$

These equations provide **general relations** for calculating change of equilibrium temperature at a fixed pressure as function of changing compositions of the phases.

The second relation, Eq. (10.4.10b), is a more convenient form for many practical applications. It may be remembered by rewriting Eq. (10.4.8) as

$$\ln K = -\frac{\Delta_r H^*}{RT} + \frac{\Delta_r S^*}{R}, \qquad (10.4.11)$$

and differentiating it with respect to (1/T) at constant values of ΔH^* and ΔS^*. It should be obvious, however, that Eq. (10.4.10b) is valid regardless of the nature of temperature dependence of ΔH^* and ΔS^* since no assumption was introduced in its derivation.

Problem 10.2 Beginning with Eq. (10.4.11), derive Eq. (10.4.10b) by differentiating both sides with respect to 1/T, but treating ΔH^* and ΔS^* as functions of temperature.

Problem 10.3 Using the phase diagram presented in Fig. 10.23, but without using any data from the literature for G, H and S, calculate the equilibrium constant of the GASP reaction at 800 °C, 23 kb, using pure component (P, T) standard state for each phase. Get any data that you need from the literature.

Hint: First determine the value of lnK at 800 °C on the equilibrium boundary of the reaction involving pure phases.

10.5   Solid-Gas and Homogeneous Gas Speciation Reactions

Solid-gas reactions and homogeneous reactions in a gas phase play important roles in a variety of geological and planetary problems. In this section we illustrate some of the strategies for calculating and using these reactions, using examples from both types of problems.

10.5.1   Condensation of Solar Nebula

Figure 10.3 shows the equilibrium condensation temperatures of different minerals as function of pressure from a gas of the composition of the solar nebula (Lewis and Prinn 1984). The pressure within the solar nebula was very low, less than 10^{-2} bar, and Lewis (1974) argued that the temperature gradient within the solar nebula must have been essentially adiabatic (isentropic), as originally suggested and calculated by Cameron and Pine (1973). The different symbols beside the adiabat in Fig. 10.3 indicate different planets at their inferred formation conditions. These conditions are constrained by the fact that a planet had to form at a specific place in the condensation sequence in order to satisfy its density, bulk composition and other properties (Lewis 1974). For example, Mars has too low a density (3.9 gm/cm^3 compared to 5.4 gm/cm^3 of Earth) to have formed with large amount of metallic Fe; major fraction of all Fe must have been oxidized to FeO and incorporated into ferromagnesian minerals before it formed. On the other hand, the density of Mars is too large for it to contain any appreciable amount of the hydrous mineral serpentine. Thus, the formation condition of Mars is placed between the condensation conditions of FeO and serpentine from the solar nebula. For further discussion of this fascinating topic, the reader is referred to the original work of Lewis (1974). A comprehensive set of calculations for the equilibrium condensation of different minerals as a function of temperature in the solar nebula is presented in two pioneering papers by Grossman (1972) and Grossman and Larimer (1974). The process of equilibrium condensation of minerals was subsequently explored by Saxena and Eriksson (1986) by minimization of Gibbs energy (see Sect. 10.13) and in a series of papers by Grossman and co-workers. The purpose of this section is to show how the equilibrium condensation temperature is calculated as a function of the total gas pressure, P_T, using equilibrium constants of reactions, and discuss some important insights about the nebular process that may be gained by comparing the results of such calculations with the observations in meteorites.

Consider the formation of forsterite, Mg_2SiO_4. Taking into account the gas species in the solar nebula, the appropriate reaction is

$$SiO + 2Mg + 3H_2O = Mg_2SiO_4 + 3H_2 \qquad (10.5.a)$$

(After its condensation, forsterite is replaced by enstatite, $MgSiO_3$, at slightly lower temperature   according   to $SiO + Mg_2SiO_4 + H_2O = 2MgSiO_3 + H_2$.   Thus,

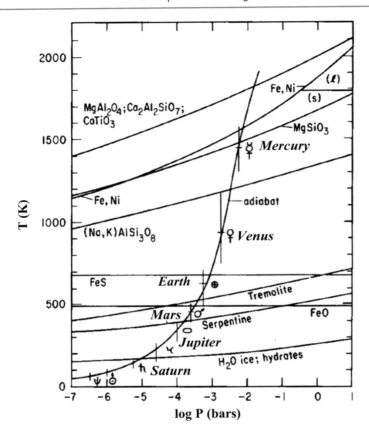

Fig. 10.3 Equilibrium condensation temperatures of minerals in solar gas, as calculated by Lewis and Prinn (1984). The first magnesian silicate to condense is forsterite (Mg_2SiO_4), but is replaced by enstatite ($MgSiO_3$) at slightly lower temperature. Adiabat represents the isentropic temperature profile as function of pressure in the solar nebula with the temperature decreasing away from the heliocenter. The locations of the different planets (Mercury to Neptune) on the adiabat are shown by conventional symbols. From Lewis and Prinn (1984). With permission from Academic Press-Elsevier

forsterite condensation condition is not shown in Fig. 10.3). At equilibrium, we have for the forsterite forming reaction (Eq. 10.4.8)

$$K_{(9.5.a)}(P, T) \equiv e^{-\Delta G^*/RT} = \frac{\left(a^{ol}_{Mg_2SiO_4}\right)\left(a^{g}_{H_2}\right)^3}{\left(a^{g}_{SiO}\right)\left(a^{g}_{Mg}\right)^2\left(a^{g}_{H_2O}\right)^3} \qquad (10.5.1)$$

We now specify the following **standard states**:

Solid: pure state at the P, T of interest so that $G^{*(s)}(T) = G^{o(s)}(P, T)$

Gas: pure gas at unit fugacity, T so that $G_i^{*(g)}(T) = G_i^{o(g)}[P(f = 1), T]$,

where $P(f = 1)$ implies the pressure at T at which the fugacity is 1 bar (Fig. 8.15). With this choice of standard state for gas, the activity of a gaseous species numerically equals its fugacity, $a_i^g = f_i^g$ (recall that by definition $a_i(P, T, X) = f_i(P, T, X)/f_i^*(T)$). It is further assumed that $f \approx 1$ when $P = 1$ (since all gases behave ideally as $P \to 0$). Thus, we have

$$\Delta G^*(P', T) \approx G_{fors}^o(P', T) + \Delta G_i^{o(g)}(1; T)$$

(Note that the two Gibbs energy terms on the right are associated with two different pressures). The Gibbs energy of pure forsterite is related to its value at 1 bar according to

$$G_{fors}^o(P', T) = G_{fors}^o(1 \text{ bar}, T) + \int_1^{P'} V_{fors}^o dP$$

Since the nebular pressure is much less than 1 bar, the integral in the above equation can be written as $V_{fors}^o(P' - 1) = -V_{fors}^o$. (Note that the last term is multiplied by 1 bar so that it has the unit of cm^3-bar). Thus, to a very good approximation,

$$\Delta G^*(P', T) \approx G_{fors}^o(1 \text{ bar}, T) - V_{fors}^o \times 1(\text{bar}) + \Delta G_i^{o(g)}(1; T) \qquad (10.5.2)$$

with the unit of volume being in J/bar, if the G values are in Joules.

The gas phase may be assumed to be behave ideally because of very low pressure in the solar nebula (Fig. 10.3) so that $f_i^g = P_i^g = P_T X_i^g$. Thus, we have

$$a_i^g(P, T, X) \approx P_T X_i^g, \qquad (10.5.3)$$

which, on substitution in Eq. (10.5.1), yields

$$e^{-\Delta G^*/RT} \approx \frac{\left(a_{Mg_2SiO}^{Ol}\right)\left(X_{H_2}^g\right)^3}{P_T^3 \left(X_{SiO}^g\right)\left(X_{Mg}^g\right)^2 \left(X_{H_2O}^g\right)^3} \qquad (10.5.4)$$

Here ΔG^* is to be calculated from Eq. (10.5.2) using the available thermochemical data.

Given the composition of the solar gas, and specific values of P_T and $a_{Mg_2SiO_4}^{Ol}$, the last equation can be solved for T (the only unknown in the equation) to yield the equilibrium condensation temperature of olivine of specific Mg content. However, the gas composition also changes as a function temperature by homogeneous reaction. Thus, the condensation temperature needs to be calculated by an iterative procedure, until the temperature at which the gas composition is computed becomes the same as that obtained from the solution of the above equation. After the condensation of a phase, the solar gas composition is recalculated by removing the components that condensed into the phase, and used as an initial gas composition for calculation of condensation of a lower temperature phase.

In chondritic meteorites, olivine and pyroxene grains within the chodrules show FeO/(FeO + MgO) ratios greater than 0.15. On the other hand, equilibrium condensation calculation carried out by Grossman (1972) using a solar gas composition shows only trace amounts of FeO in these silicate minerals because iron is more stable in co-condensing Fe-Ni alloy (Fig. 10.3). At equilibrium, oxidation of iron metal by reaction with gaseous H_2O is delayed until the temperature falls below 550 K. However, at this temperature, solid-gas reaction and diffusion of FeO into silicate are too slow to be able to cause any significant FeO enrichment of the silicates. Thus, to explain the observed FeO content of the silicate minerals, one needs to find a mechanism of increasing the $f(O_2)$ of the environment in which the chondrules formed so that Fe can be oxidized at sufficiently high temperatures. In the nebular setting, this may be achieved by increasing dust/gas ratio in local domains. Because the dust component is relatively enriched in oxygen, vaporization of the dust produces a gas enriched in oxygen compared to solar composition. This example shows how important insights about nebular process may be gained by comparing the results of equilibrium condensation with observations in meteorites. The question of equilibrium condensation in the dust-enrichment environment was addressed by Yoneda and Grossman (1995) and Ebel and Grossman (2000).

10.5.2 Surface-Atmosphere Interaction in Venus

On the basis of data gathered from several missions to Venus, the surface temperature of Venus has been inferred to be ~ 750 K. In addition, it was concluded that Venus has an atmosphere that is very rich ($\sim 95\%$) in CO_2. The P-T profile of the Venusian atmosphere roughly follows the adiabatic (isentropic) temperature gradient (Eq. 7.2.3) of CO_2 (Lewis and Prinn 1984).

The high surface temperature of Venus prompted Mueller (1963) to suggest that unlike Earth, the surface rocks and atmosphere of Venus should be at least in partial equilibrium since at such temperatures equilibrium is known to have been attained in many terrestrial metamorphic processes. Thus, he deduced the mineralogical character of crustal rocks from the atmospheric compositions. This idea of crust-atmosphere equilibrium in Venus has served as the basis of subsequent work on the nature of Venusian crust and atmosphere. A modern and detailed discussion of the subject may be found in Lewis and Prinn (1984). Here we show that the inferred CO_2 partial pressure in the atmosphere of Venus is very similar to that expected by equilibration with a crust that contains calcite, orthopyroxene and clinopyroxene.

In terrestrial metamorphic process, two of the important CO_2 producing reactions are

$$\underset{CaCO_3}{Cal} + \underset{siO_2}{Qtz} = \underset{CaSiO_3}{Wo} + CO_2 \qquad (10.5.b)$$

$$\underset{MgSiO_3}{Opx} + \underset{CaCO_3}{Cal} + \underset{SiO_2}{Qtz} = \underset{CaMgSi_2O_6}{Di} + CO_2 \qquad (10.5.c)$$

Since the mass and density of Venus are similar to those of the Earth, Mueller (1963) suggested that Venus has similar bulk composition as that of Earth and, consequently, the planetary evolution should have led to surface rocks in Venus that have terrestrial counterparts. Thus, the above reactions should also prevail in the Venusian crustal rocks.

Assuming that all minerals are in their respective pure states, and setting the standard states of minerals and CO_2 to be their respective pure states at 1 bar, T, so that the activities of all condensed components are unity at 1 bar, T, and $a(CO_2) \approx f(CO_2)$, the equilibrium constants of the above reactions are

$$K_{(10.5.b)} \equiv \exp\left(-\frac{\Delta_r G^0_{(10.5.b)}}{RT}\right) \approx f_{CO_2} \qquad (10.5.5)$$

$$K_{(10.5.c)} \equiv \exp\left(-\frac{\Delta_r G^0_{(10.5.c)}}{RT}\right) \approx f_{CO_2} \qquad (10.5.6)$$

where $\Delta_r G^°$ represents Gibbs energy change at 1 bar, T, when all phases are in their respective pure states. Owing to the very low atmospheric pressure, $f(CO_2) = P(CO_2)$.

Figure 10.4, which is taken from Mueller and Saxena (1977), shows the relationship between T and $P(CO_2)$, as calculated from the thermochemical data for the reactions at 1 bar, T. The range of inferred $P(CO_2)$ and T for Venus, shown as a large cross, agree with the calculated relations. The data for Earth's atmosphere are shown for comparison.

Mueller's "interaction model" was further explored by Lewis (1970) to derive useful information about the crustal mineralogy of Venus from the knowledge of atmospheric composition, as inferred from spectroscopic data. The minerals deduced to be stable in the crust, but not all of them together, include pyroxene, quartz, magnetite, calcite, halite, fluorite, tremolite, akermanite and andalusite. From the inferred mineralogy of the crust, Lewis (1970) suggested that the surface of Venus is a silica-rich differentiate rather than representing the average cosmic Mg/Si ratio of ~ 10.

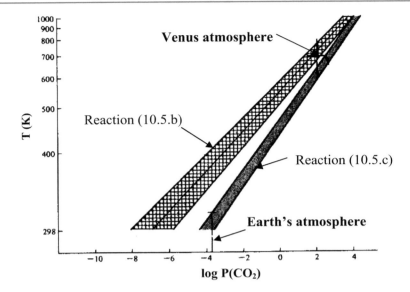

Fig. 10.4 Comparison of the observed P(CO$_2$)-T data in the atmosphere of Venus with the equilibrium curves of important CO$_2$ producing reactions. The bandwidths indicate uncertainties of the calculations arising from those in the thermochemical data. The P(CO$_2$)-T combination of the Earth's atmosphere is shown for comparison. From Mueller and Saxena (1977)

10.5.3 Metal-Silicate Reaction in Meteorite Mediated by Dry Gas Phase

An apparently puzzling observation in the Allende meteorite[1] is the iron enrichment of olivine crystals within narrow zones around metallic inclusions. If the iron enrichment is due to a simple exchange process, Mg$_2$SiO$_4$(Ol) + 2Fe(metal) = Fe$_2$SiO$_4$(Ol) + 2Mg(metal), then the metallic inclusions must show Mg enrichment, which is neither possible (because of extremely small solubility Mg in Fe) nor observed. This problem of iron enrichment of olivine was investigated by Dohmen et al. (1998) by controlled experiments in a Knudsen cell mass spectrometer and thermodynamic analysis of the experimental data. (In a Knudsen cell, a sample is uniformly heated within a container and the vapor phase that is generated by the sublimation of a solid is allowed to escape through a small hole at the top of the container, and analyzed by a quadrupole mass spectrometer). By putting physically separated blocks of forsterite and Fe metal in a Knudsen cell mass spectrometer at high temperature and controlled f(O$_2$) condition, and monitoring the composition of the effusing gas and the residual solid, Dohmen et al. (1998) found that Mg$_2$SiO$_4$ evaporates **stoichiometrically** according to

---

[1]The Allende meteorite fell in Pueblito de Allende, Chihuahua state, Mexico, on February 8, 1969. It is a carbonaceous chondrite and consists of the earliest formed minerals in the solar system.

$$Mg_2SiO_4(\text{solid }) \rightarrow 2Mg(g) + SiO(g) \text{ and } 3/2\ O_2(g) \qquad (10.5.d)$$

In addition, Fe(metal) evaporates to Fe(g)

$$Fe(\text{metal}) \rightarrow Fe(g) \qquad (10.5.e)$$

Homogeneous reaction within the gas phase then leads to the precipitation of fayalite, Fe_2SiO_4 according to

$$2Fe(g) + SiO(g) + 3/2\ O_2(g) \rightarrow Fe_2SiO_4(s) \qquad (10.5.f)$$

The net reaction leading to the formation of fayalite by a vapor mediated reaction between forsterite and Fe(m) is obtained by adding the last three reactions.

$$Mg_2SiO_4(s) + 2Fe(m) = Fe_2SiO_4(s) + 2Mg(g) \qquad (10.5.g)$$

for which the equilibrium constant is

$$K_{(10.5.g)}(P, T) = \left[\left(\frac{X_{Fe}}{X_{Mg}} \right)^2 \right]^{ol} \left[\frac{\left(P^g_{Mg} \right)^2}{\left(a^m_{Fe} \right)^2} \right] \qquad (10.5.7)$$

so that

$$\left[\frac{X_{Fe}}{X_{Mg}} \right]^{Ol} \propto \left[\frac{a^m_{Fe}}{P^g_{Mg}} \right]^{gas} \qquad (10.5.8)$$

In the experiments of Dohmen et al. (1998) $a^m_{Fe} = 1$, since Fe is in the pure (standard) state, so that the Fe content of olivine varies inversely as P_{Mg}. This work is an excellent demonstration of how physically separated condensed phases can react with each other through a dry vapor phase formed by the sublimation of the phases.

Dohmen et al. (1998) also found that in an experiment with a polycrystalline aggregate of olivine that was kept physically separated from metallic iron, the olivine grains near the surface showed a scatter in their Fe content between 2 and 4 mol%, but the individual grains were homogeneous (the temperature was high enough to homogenize the composition within the duration of the experiment). This variation of composition was most likely due to fluctuations of P_{Mg}. Similar effects may be expected in meteorites. In the latter case, it is the (a^m_{Fe}/P^g_{Mg}) ratio that affects the silicate mineral composition.

10.5.4 Effect of Vapor Composition on Equilibrium Temperature: T Versus X_v Sections

Let us consider a reaction that involves one or two volatile species that may be written as

$$A \leftrightarrow B + v_1 V_1 + v_2 V_2, \tag{10.5.h}$$

were A and B are solid phases, V_1 and V_2 are two volatile species and v_1 and v_2 are their respective stoichiometric coefficients. One of these coefficients may be zero, greater than zero or less than zero. Geologically significant examples of such reactions involving CO_2 and H_2O as vapor species are

$$Mg(OH)_2(Brucite) = MgO(Periclase) + H_2O \quad [v(CO_2) = 0] \tag{10.5.i'}$$

$$CaCO_3 + SiO_2 = CaSiO_3(Wollastonite) + CO_2 \quad [v(H_2O) = 0] \tag{10.5.i''}$$

$$Tremolite + 3CaCO_3 + 2SiO_2 = 5CaMgSi_2O_6(Diopside) + 3CO_2 + H_2O$$
$$[v(CO_2) > 0, v(H_2O) > 0] \tag{10.5.j}$$

$$3CaMg(CO_3)_2 + 4SiO_2 + H_2O = Talc + 3CaCO_3 + 3CO_2$$
$$[v(CO_2) > 0, v(H_2O) < 0] \tag{10.5.k}$$

(tremolite: $Ca_2Mg_5Si_8O_{22}(OH)_2$; Talc: $Mg_3Si_4O_{10}(OH)_2$).

Any change in the composition of the gas phase causes a change in the equilibrium temperature of a univariant solid-gas reaction at a constant pressure, or of equilibrium pressure at a constant temperature. However, because of the differences in the manner of appearance of the vapor species in the reactions, the isobaric equilibrium temperatures of the reactions vary as a function of the composition of the vapor phase, X^v, in qualitatively different ways for the three types of reactions (10.5.i' and 10.5.i'' are of the same type in that both these reactions directly involve only one volatile species). The T-X^v topologies of these types of reaction were first derived and discussed by Greenwood (1967).

Using pure component (P, T) standard states and rearranging terms, Eq. (10.4.10a) can be written as

$$(\partial T)_P = (\partial \ln K)_P \frac{RT^2}{\Delta_r H^\circ} \tag{10.5.9}$$

Differentiating both sides with respect to X_1^v (i.e. the mole fractions of the species 1 in the vapor phase)

$$\left(\frac{\partial T}{\partial X_1^v}\right)_P = \frac{RT^2}{\Delta_r H^\circ} \left(\frac{\partial \ln K}{\partial X_1^v}\right)_P \tag{10.5.10}$$

Assuming that the solid phases are in their respective pure states, we have at equilibrium for the reaction (10.5.h),

$$K = [(X_1\gamma_1)^{\nu_1}(X_2\gamma_2)^{\nu_2}]^g_{eq} \qquad (10.5.11)$$

where X and γ stand for the mole fraction and activity coefficient, respectively, of the indicated volatile or gaseous species ($1 \equiv V_1$, $2 \equiv V_2$), and the superscript g stands for the gas phase.

Combining the last two equations, and omitting the superscript V for the sake of brevity, we obtain

$$\left(\frac{\partial T}{\partial X_1}\right)_P = \frac{RT^2}{\Delta_r H^o}\left(\frac{\nu_1\partial\ln(X_1\gamma_1)}{\partial X_1} + \frac{\nu_2\partial\ln(X_2\gamma_2)}{\partial X_1}\right)_P$$

or

$$\left(\frac{\partial T}{\partial X_1}\right)_P = \frac{RT^2}{\Delta_r H^o}\left(\frac{\nu_1}{X_1} + \frac{\nu_2}{X_2}\left(\frac{\partial X_2}{\partial X_1}\right) + \frac{\nu_1\partial\ln\gamma_1}{\partial X_1} + \frac{\nu_2\partial\ln\gamma_2}{\partial X_1}\right)_P \qquad (10.5.12)$$

This equation is generally valid for the change of equilibrium temperature as a function of gas composition for reactions in which stoichiometric solids react with one or two volatile species, as in the examples above. There is, however, no restriction on the number of volatile species that could be present in the system.

10.5.4.1 Binary Vapor Phase

Let us now consider the special case of an ideal binary vapor phase. In that case, $\gamma_1 = \gamma_2 = 1$, and $dX_1 = -dX_2$ so that the last equation reduces to

$$\left(\frac{\partial T}{\partial X_1}\right)_P^{ideal} = \frac{RT^2}{\Delta_r H^o}\left(\frac{\nu_1}{X_1} - \frac{\nu_2}{X_2}\right)_P \qquad (10.5.13)$$

Using this equation, let us explore the topological properties of the three types of solid-gas reactions discussed above in the T-X^V space, if the mixing in the gas phase is **ideal**. We first want to find out if there is an extremum (maximum or minimum) on the univariant reaction line. Imposing the condition of extremum, $\partial T/\partial X_1 = 0$, in Eq. (10.5.13), we get

$$[\nu_2 X_1 = \nu_1 X_2 = \nu_1(1 - X_1]_{T(extm)} \qquad (10.5.14)$$

so that $X_1(\nu_1 + \nu_2) = \nu_1$, or

$$\left[X_1 = \frac{\nu_1}{\nu_1 + \nu_2}\right]_{T(extm)} \qquad (10.5.15)$$

where T(extm) stands for a temperature extremum. If the stoichiometric coefficients are of opposite signs, then X_1 is either greater than 1 or is a negative quantity. However, since $0 \leq X_1 \leq 1$, the above equation yields an acceptable result only when the stoichiometric coefficients are of the same sign, or in other words, both volatile species are in the same side of the reaction, as in reaction (10.5.j). Thus, reactions of this type will have an extremum of equilibrium temperature in the T-X^v space. Is this a maximum or minimum?

To answer this question, we take the second derivative of Eq. (10.5.13) with respect to X_1, which yields

$$\left(\frac{\partial^2 T}{\partial X_1^2}\right)_P^{ideal} = \frac{RT^2}{\Delta_r H^\circ} \left\{ -\left[\frac{v_1}{X_1^2} + \frac{v_2}{X_2^2}\right] + \frac{2RT}{\Delta_r H^\circ} \left(\frac{v_1}{X_1} - \frac{v_2}{X_2}\right)^2 \right\} \qquad (10.5.16)$$

We know from elementary calculus that if $\partial^2 T/\partial X_1^2 > 0$, the extremum is a minimum, whereas if it is less than 0, the extremum is a maximum. Since, Eq. (10.5.14) must be satisfied at the extremum, the last term in the above equation vanishes at the extremum condition. Thus, we have

$$\left(\frac{\partial^2 T}{\partial X_1^2}\right)_P^{ideal} = -\frac{RT^2}{\Delta_r H^\circ} \left[\frac{v_1}{X_1^2} + \frac{v_2}{X_2^2}\right] \qquad (10.5.17)$$

at the extremum of T versus X_1 relation.

Both v_1 and v_2 are positive quantities for reactions that have two gaseous species in the product side. Also $\Delta_r H^\circ > 0$. Thus, for this type of reactions (e.g. 10.5.j) $\partial^2 T/\partial X_1^2 < 0$ at the extremum and this proves the existence of a **thermal maximum**. Also, from Eq. (10.5.15), this thermal maximum appears at the vapor composition that is the same as that evolved by the reaction, if the gas phase behaves as an ideal mixture. As an example, for the reaction (10.5.j), the thermal maximum appears at $X(H_2O) = 1/(1 + 3) = 0.25$, which is the composition of the evolved fluid phase.

When the two volatile species appear on two opposite sides of a solid-gas reaction, as in reaction (10.5.k), the stoichiometric coefficients have opposite signs. It should be easy to see that in such cases, Eq. (10.5.15) yields either $X_1 < 0$ or $X_1 > 1$. Both these solutions violate the physical restriction that $0 \leq X_1 \leq 1$. Thus, reactions that involve two volatile species on two sides do not have an extremum. Instead, this type of reaction has an inflection point in the T-X space.

The T-X topologies of the three classes of reactions between stoichiometric solids and binary volatile phase, namely (a) $v_i > 0$, $v_j = 0$ (10.5.i' and 10.5.i''), (b) $v_i > 0$, $v_j > 0$ (10.5.j) and $v_i > 0$, $v_j < 0$ (10.5.k), are illustrated in Fig. 10.5. Appreciation of these topological properties has led to the understanding of the reason for intersection of mapped traces of solid-gas reactions in the field (Carmichael 1969), and to the retrieval of T-X^v conditions during metamorphic processes (e.g. Ghent et al. 1979).

Fig. 10.5 Schematic illustration of the effect of fluid composition on the equilibrium temperature of different types of solid-gas reactions (see text, reactions (10.5.i)–(10.5.k)) involving change in the mole fraction of two volatile species. Arrows indicate qualitative effect of an additional fluid species, with opposing arrow indicating no effect, in an ideal fluid mixture. From Ganguly and Saxena (1987)

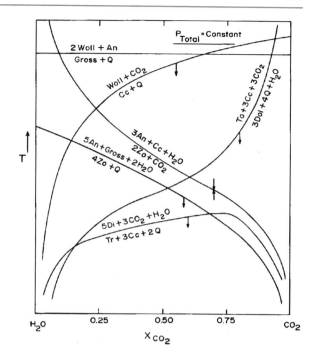

10.5.4.2 Ternary Vapor Phase

When there is significant content of a third volatile species, the topological relation shown in Fig. 10.5 may be viewed as those pertaining to a section with constant mole fraction of the third component. The effect of the third component (component 3) can be calculated by considering the displacement of equilibrium along a fixed X_1/X_2 ratio. Ganguly (1977) showed that if the ratio X_1/X_2 is kept fixed, and the gas phase behaves as an ideal solution, then

$$\left(\frac{\partial T}{\partial X_3}\right)^{ideal}_{P,X_1/X_2} = -\frac{RT^2}{\Delta_r H^o}\left(\frac{\nu_1 + \nu_2}{X_1 + X_2}\right) \tag{10.5.18}$$

Problem 10.4 Assuming $\Delta_r H^o$ to be independent of temperature, show that for a solid-gas reaction directly involving a single volatile species that is mixed in a multicomponent gas phase, and pure solid phases, the equilibrium temperature changes as a function of vapor composition at a constant pressure according to

$$\frac{1}{T(X_1)} = \frac{1}{T_o} - \frac{\nu_1 R}{\Delta_r H^o}\left(\ln a_1^v\right)_{eq} \tag{10.5.19}$$

where T_o and $T(X_1)$ are, respectively, the equilibrium temperature when the vapor phase consists only of the species 1 and has a mole fraction X_1 of the species 1. (Hint: See Eq. 10.4.10).

10.5.5 Volatile Compositions and Oxidation States of Natural Systems

The oxidation states and volatile compositions of the Earth's crust and mantle and other planetary objects are topics of major interest in the Earth and Planetary Sciences. A limited coverage of this topic is given below.

10.5.5.1 Oxidation States of Planetary Systems

It is customary in the Earth and Planetary Sciences communities to refer to the oxidation states of natural samples in terms of $f(O_2)$ values of solid oxygen fugacity buffers, often using representations such as IW-1, which is meant to imply that the $f(O_2)$ is one log unit below that of iron-wüstite buffer (Fe + ½ O_2 = FeO). The $\log f(O_2)$ versus T relation of some of the common reference buffers at 2 kb pressure are illustrated in Fig. 10.6.

The oxidation states of Earth, Mars and Moon are very different, the Moon being the most reduced and the Earth the most oxidized. The $f(O_2)$ of the Earth's mantle is found to be similar to that of QFM buffer ($3Fe_2SiO_4$(fayalite) + O_2 = $2Fe_3O_4$ (magnetite) + SiO_2(quartz)) and those of Moon and the asteroid 4 Vest around IW-1.

It is instructive now to calculate the relative abundance of H_2 and H_2O in the O-H vapor phase as a function of $f(O_2)$ according to the homogeneous equilibrium

$$H_2O = H_2 + 1/2\, O_2 \qquad (10.5.\mathit{l})$$

for which

$$K\,(P,T) = e^{\left(-\Delta G_i^*(T)/RT\right)} = \frac{a(H_2)a(O_2)^{1/2}}{a(H_2O)} \qquad (10.5.20)$$

where, as elsewhere in the book, $\Delta G_i^*(T)$ stands for the standard state free energy change of the reaction (l) at T. We now choose the standard states of the vapor species to be those of the respective pure states at 1 bar, T so that $\Delta G_i^*(T) = \Delta G_l^o(P = 1\,\text{bar},T)$. Also we make the reasonable assumption that at 1 bar pressure, $f_i^o \approx 1\,bar$, and consequently, according to Eq. (8.4.8), $a_i(P, T) \approx f_i(P, T)$. The last equation then reduces to

$$\log f(O_2) - \frac{2\Delta G_l^o(P=1,T)}{2.303\,RT} - 2\log\frac{f(H_2)}{f(H_2O)} \qquad (10.5.21)$$

Fig. 10.6 logf(O2) versus T
diagram for solid oxygen
buffers at 2 kb pressure. (a, b)
graphite buffer in the C-O and
C-O-H systems, respectively,
using a H/O ratio of 2/1 for
the latter. WI: wüstite-iron;
MI: magnetite-iron; MW:
magnetite-wüstite; QFM:
quartz-fayalite-magnetite;
HM: hematite-magnetite.
From Ganguly (1977)

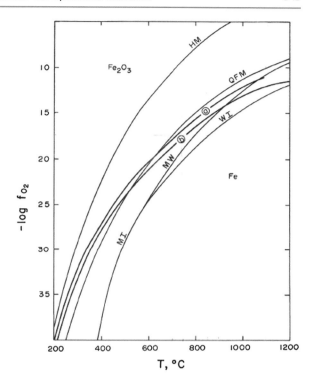

Using now the Eq. (8.4.9) that relates fugacity to mole fraction, the above relation
can be expressed as

$$\log f(O_2) - \frac{2\Delta G_1^o(P=1,T)}{2.303\ RT} - 2\log \frac{f_{H_2}^o(P,T)X_{H_2}(\gamma_{H_2})}{f_{H_2O}^o(P,T)X_{H_2O}(\gamma_{H_2O})} \qquad (10.5.22)$$

Since $f(O_2)$ in natural systems is very low (Fig. 10.6), we can write

$$X_{H_2O} + X_{H_2} \approx 1 \qquad (10.5.23)$$

if the vapor phase composition is essentially restricted to the O-H system. Additionally, if the vapor phase behaves as an ideal mixture of the volatile species (i.e.
$\gamma_i = 1$), the last two equations can be solved to determine the composition of the
vapor phase at specified P, T, $f(O_2)$ conditions. This exercise was carried out by
Sharp et al. (2013) for P-T condition of 1 bar, 1273 K to roughly estimate the
volatile composition of Earth's mantle, Mars and Asteroid. Their results are illustrated in Fig. 10.7, which is a modified version of the corresponding Figure in their
paper (Sharp: pers.com.).

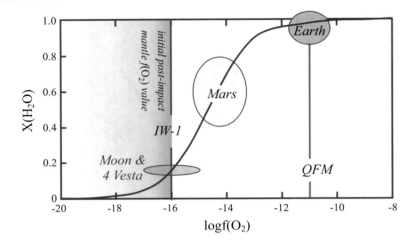

Fig. 10.7 Relationship between the mole fraction of water and $\log f(O_2)$ in the O-H system at 1 bar, 1273 K, and comparison with inferred the $f(O_2)$ conditions of crustal and mantle samples of Earth, Mars, Moon and asteroid 4 Vesta. From Sharp et al. (2013), as modified by Zachary Sharp (pers.com.)

It is now widely accepted that the Moon formed from the gravitational capture and collapse of material ejected from the Earth following a giant impact of a Martian size object. The $f(O_2)$ of the Earth and Moon was below IW buffer during their early history as a result of mixing of the iron core with the oxide mantle of the Earth. Sharp et al. (2013) argued that the $f(O_2)$ of the Earth and Moon increased subsequently as a result of degassing of hydrogen, but that of the Earth increased much more, and effectively to the limit possible by hydrogen degassing as illustrated in Fig. 10.7, possibly because of exhaustion of the $f(O_2)$ buffering capacity of Earth's mantle. They estimated that the entire Earth's mantle could have been oxidized to QFM buffer condition by degassing of 1.4 ocean equivalent of hydrogen (4.7×10^3 g) or only ¼ th of that amount if only the upper mantle was oxidized. This seems to be a modest amount of hydrogen loss over the course of Earth's accretion and atmospheric evolution.

Sharp et al. (2013) showed that hydrogen degassing would have raised the D/H ratio of the Earth significantly above the primordial nebular condition as a result of kinetic fractionation of the isotopes since the diffusion coefficient of $\mathcal{D}(D)/\mathcal{D}(H) \sim (m(H)/m(D))^{1/2} = 1.4$, where $\mathcal{D}$ and m stand, respectively, for the diffusion coefficient and mass. This model may at least partly account for the observed enrichment of deuterium in the Earth relative to that of the primordial nebula. (The D/H ratio of the Earth's ocean is around a factor of seven higher than that of the solar nebula and this has led to several suggestions about the origin of terrestrial water from exogenous sources such as carbonaceous chondrites and comets). Calculations at higher pressure and allowing for non-ideal mixing in the gas phase may be carried out using the last two equations, but the qualitative conclusions of Sharp et al. (2013) should be expected to hold.

10.5.5.2   Volatile Compositions: Metamorphic and Magmatic Systems

In both metamorphic and magmatic systems, volatile compositions are often established by equilibration with graphite. These volatile compositions play important roles in the mineralogical evolution of metamorphic rocks and in many volcanic and plutonic phenomena. The important volatile species in the magmatic and metamorphic systems can be represented within the four component system C-O-H-S. In the presence of graphite, we have two phases (graphite and volatile). From Phase Rule (Eq. 10.3.2), the degrees of freedom for this system is given by $F = C - P + 2 = 4$. Thus, in order to determine the volatile composition, four variables need to be fixed. Two of these are P and T. The other two variables may be the fugacities of any two volatile species (or two ratios of fugacities, or a combination of both). The choice of which volatile fugacities are to be kept fixed is a matter of convenience and dictated by practical considerations. It is obviously useful to choose those species for which we can obtain an independent estimate of fugacity values in natural assemblages. From this point of view, the calculations are often carried out at fixed values of $f(O_2)$ and $f(S_2)$.

After fixing P-T condition and two fugacity values, the composition of the vapor phase may be calculated from the equilibrium constants of independent homogeneous reactions within this phase and of reactions between graphite and volatile species, plus the constraint that the sum of the partial pressures of all volatile species must equal the imposed total pressure of the system.

The important volatile species in the metamorphic and magmatic systems are CO, CO_2, H_2O, H_2, O_2 and S_2. Thus, we write the following reactions and their corresponding equilibrium constants.

$$C + 1/2\, O_2 = CO \qquad\qquad (10.5.l)$$

$$C + O_2 = CO_2 \qquad\qquad (10.5.m)$$

$$C + 2H_2 = CH_4 \qquad\qquad (10.5.n)$$

$$H_2 + 1/2\, O_2 = H_2O \qquad\qquad (10.5.o)$$

$$H_2 + 1/2\, S_2 = H_2S \qquad\qquad (10.5.p)$$

$$1/2\, S_2 + O_2 = SO_2 \qquad\qquad (10.5.q)$$

In addition, we write

$$P(\text{total}) = \sum_i P_i$$

or

$$P(\text{total}) = \sum_i \frac{f_i}{\Phi_i} \qquad (10.5.24)$$

where $\Phi(i)$ is the fugacity coefficient of the species i. Since $f(O_2)$ and $f(S_2)$ are very small, of the order of 10^{-15} to 10^{-20} bars, these fugacity terms may be dropped in the above summation.

At fixed P, T condition, there are 10 intensive variables in the system consisting of nine different fugacities and the activity of carbon, a_C. One, thus, needs 10 equations to uniquely solve for all the unknowns. In addition to the six equations defined by the equilibrium constants of the reactions (10.5.l) through (10.5.q), plus (10.5.24), we have decided to fix the values of $f(O_2)$ and $f(S_2)$. Furthermore, the presence of graphite (or diamond) uniquely fixes a_C to unity. Thus, we again reach the same conclusion as derived by the application of Phase Rule, that the fugacities (and hence the mole fractions) of all volatile species are uniquely determined at constant P-T condition if two fugacity values are fixed in the presence of graphite or diamond. However, the advantage of using the Phase Rule lies in the fact that it tells us in a straightforward way how many intensive variables need to be fixed in order to obtain a unique solution for the problem. If there is no free graphite phase, then $F = 5$, and, thus, one more intensive variable needs to be fixed. This may be a fixed value of a_C or fixed value of the ratio of two fugacities.

Following the pioneering work of French (1966), Holloway (1981) solved the system of equations defined by the equilibrium constants of the reactions (10.5.l) through (10.5.q) plus Eq. (10.5.24) to simultaneously obtain the fugacity value of each volatile species at fixed P, T $f(O_2)$ and $f(S_2)$ conditions. These results provide a framework for understanding the evolution of magmatic volatile composition, in equilibrium with graphite, during the ascent of magma from the interior of the Earth. Holloway's results are illustrated in Fig. 10.8. The lower temperature approximates the solidus of granite melting in the presence of CO_2-H_2O fluids, while the higher temperature approximates the solidus for peridotite. It is evident that the mole fractions of the fluid species, especially those of H_2S and CO_2, are very sensitive to $f(O_2)$ (Fig. 10.8a). The $f(O_2)$ conditions of magma lie in the range from slightly above that defined by the $f(O_2)$ buffering reaction $Fe_2SiO_4 + O_2 = Fe_3O_4 + SiO_2$ (commonly referred to as QFM buffer) to two to three orders of magnitude below QFM buffer (Haggerty 1976). In the presence of graphite, the volatile compositions vary from H_2O rich at high pressure to CO_2 rich at low pressure. Other significant volatile species in equilibrium with graphite are H_2S and CH_4.

Solved Problem: Assuming ideal mixing and ideal gas behavior, calculate $P(O_2)$, $P(H_2)$ and $P(\text{total})$ developed by the stoichiometric dissociation of water (reverse of the reaction 10.5.o referred to below as o$'$) within a closed system at 800 K at $P(H_2O) = 2$ kb, using the following value of the free energy of formation of water from the elements at 1 bar: $\Delta G_{f,e}$ (water: 1 bar, 800 K) $= -203{,}496$ J/mol.

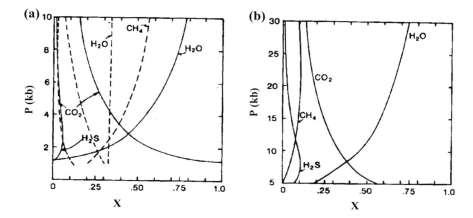

Fig. 10.8 Fluid composition in the C-O-H-S system in equilibrium with graphite as function of pressure, temperature and $f(O_2)$. X stands for mole fraction. (**a**) T = 750 °C, solid lines: $f(O_2)$ = QFM-1, dashed lines: $f(O_2)$ = QFM-2. (**b**) T = 1100 °C, $f(O_2)$ = QFM-2. QFM: Quartz-Fayalite-Magnetite buffer: $3Fe_2SiO_4$ (Fay) + O_2 = $2Fe_3O_4$ (Mag) + $3SiO_2$ (Qtz). From Holloway (1981)

We choose as standard states for the species in the O-H system the states of respective pure substance at 1 bar, T. Thus, since according to the assumptions, $f_i^o = P_i^o$ and $\gamma_i = 1$, the expression of the equilibrium constant for the dissociation of water, given by Eq. (10.5.20), reduces to

$$K\,(P, T) \approx e^{\left(-\Delta G_{o'}^o (1\text{bar,T})/RT\right)} = \frac{P(H_2)(P(O_2))^{1/2}}{P(H_2O)} \qquad (10.5.25)$$

Also, from the stoichiometry of the reaction (o'), $n(H_2) = 2n(O_2)$, where n stands for the mole numbers; thus, because of the ideal gas behavior,

$$P(H_2) = 2P(O_2) \qquad (10.5.26)$$

so that

$$P(O_2) \approx \left(\frac{K_{o'}P(H_2O)}{2}\right)^{2/3} \qquad (10.5.27)$$

The Gibbs free energy of the equilibrium (o') is given by $\Delta_r G(1 \text{ bar}, 800 \text{ K}) = -G_{f,\,e}(H_2O: 1 \text{ bar}, 800 \text{ K}) = 203{,}496$ J. Thus, $K_{o'}(1 \text{ bar}, 800 \text{ K}) = 5.16(10^{-4})$, and hence, from the last two relations, $P(O_2) = 1.4(10^{-7})$ bars and $P(H_2) = 2.8(10^{-7})$ bars at $P(H_2O) = 2$ kb, so that P(total) (= $P(H_2O) + P(H_2) + P(O_2)$) is effectively 2 Kb. (Similar calculation at 900 K yields $P(O_2) = 2.2(10^{-6})$ bars).

Comparing the above results with the $\log f(O_2)$ versus T relation illustrated in Fig. 10.6, we find that the oxidation states of regional metamorphic rocks are well within the stability field of hematite (F_2O_3), if it were established by the stoichiometric dissociation of water that is ubiquitous in low-medium grade rocks. Such high $f(O_2)$ condition would preclude the formation of the variety of ferromagnesian silicates (in which iron is in the divalent state) that characterize the regionally metamorphosed rocks and show systematic variation of mineralogy in response to increasing temperature. This point was first recognized by Miyashiro (1964) who drew attention to the common presence of graphite in metamorphic rocks and its effect on buffering the $f(O_2)$ condition. The $f(O_2)$ versus T conditions prevailing in presence of graphite within the systems C-O and C-O-H (H/O = 2/1) have been illustrated Fig. 10.6. (Note that according to the Phase Rule, the second system has 3 degrees of freedom (F = C − P + 2 = 3 − 2 (vapor + graphite) + 2 = 3) so that it is necessary to fix one more intensive variable, in addition to P and T, to define all intensive properties of the system; the H/O ratio was fixed to serve that purpose). These calculations were carried out from the system of equations in Eq. (10.5.l) through Eq. (10.5.o), but excluding those involving S species, plus the Eq. (10.5.24). For further discussion of buffering of $f(O_2)$ condition in regionally metamorphosed rocks, the interested reader is referred to Greenwood (1975) and Ganguly (1977).

*Problem 10.5 Assuming ideality of mixing (i.e. $\gamma_i = 1$), but non-ideal behavior of the end member species H_2 and H_2O, show that in the O-H system,

$$X(H_2) = \frac{1}{\psi + 1}$$

where $\psi = (f(O_2)^{1/2}/K)(\Phi_{H2}/\Phi_{H2O})$, with K being the equilibrium constant for the reaction $H_2O = H_2 + \frac{1}{2} O_2$, referred to pure gas at 1 bar, T standard state and Φ_i the fugacity coefficient of the specified species. To appreciate the effect of $f(O_2)$ on the vapor composition in the O-H system, and hence on the temperature of a dehydration equilibrium, calculate $X(H_2)$ at 2 kb, 800 K at $f(O_2)$ defined by the QFM and IW buffers ($\sim 10^{-22}$ and 10^{-27} bars, respectively); the fugacity coefficients of H_2 and water are 1.6 and 0.405 bars, respectively. Also to appreciate the effect of non-ideal behavior of the end members, calculate X_{H2} assuming $\Phi_i = 1$.

10.6 Equilibrium Temperature Between Solid and Melt

10.6.1 Eutectic and Peritectic Systems

The melting of a solid may be described by a reaction of the general form

$$A(\text{solid}) \leftrightarrow A(\text{liquid}) \tag{10.6.a}$$

If the composition of the melt changes, but that of the solid remains fixed, and the heat of melting is effectively independent of temperature, then the equilibrium temperature will change as a function of melt composition at a fixed pressure according to Eq. (10.4.10a). Since we are dealing with a melting process, we write $\Delta_r H°$ as ΔH_m^o, which stands for the heat of melting. In the above reaction, $K = (a_A^l)_{eq}$, and consequently we obtain from Eq. (10.4.10b), assuming ΔH_m^o to be independent of T,

$$\frac{1}{T_m} = \frac{1}{T_o} - \frac{R}{\Delta H_m^o} \left(\ln a_A^l \right)_{eq} \tag{10.6.1}$$

where T_m stands for the melting temperature at a specified melt composition, T_o is the melting temperature of pure A(solid) in the absence of any other component, and a_A^l is the activity of the component A in the liquid. The standard state is taken to be the pure state for both solid and liquid at the P-T condition of interest.

A mineralogical example of the reaction of the type (10.6.a) is the melting relation in the system $CaMgSi_2O_6$–$CaAl_2Si_2O_8$. The end member compositions represent the minerals diopside and anorthite, respectively. The melting reactions are

$$CaMgSi_2O_6(Di) \leftrightarrow CaMgSi_2O_6(liquid)$$
$$CaAl_2Si_2O_8(An) \leftrightarrow CaAl_2Si_2O_8(liquid)$$

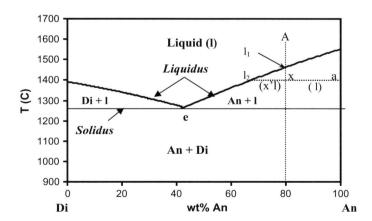

Fig. 10.9 Melting temperature versus composition in the binary system Diopside-Anorthite as calculated from thermochemical data at 1 bar, assuming ideal solution behavior of the melt. The eutectic point, which is given by the intersection of the two liquidus curves, is located at 1270 °C (T_e) and wt% An = 42. The fields of liquid and different phase assemblages are labeled. Below the solidus temperature (T_e), the stable assemblage is An + Di. For a melt of initial composition X (80 wt% An), crystallization begins with the precipitation of anorthite at temperature l_1. The mass ratio of liquid (l) to anorthite crystal (x'l) at 1400 °C is given by the ratio of the length segments ax to l_2x (l/x'l = ax/l_2x)

in which the melt composition varies between the two limiting compositions of the system. The end-member melting properties in the system at 1 bar are as follows: $T_o(Di) = 1665$ K, $T_o(An) = 1826$ K; $\Delta H_m^o(Di) = 77{,}404$ J/mol, $\Delta H_m^o(An) = 81{,}000$ J/mol. T_o and ΔH_m^o values are from Bowen (1915) and Robie et al. (1978), respectively. These values are substituted in Eq. (10.6.1) to calculate the melting behavior of diopside and anorthite at 1 bar, assuming ideal mixing of components in the liquid phase. The results are illustrated in Fig. 10.9. The calculated diagram is in good agreement with the phase diagram determined experimentally by Bowen (1915).

The melting temperatures of both diopside and anorthite decrease as the melt compositions change from the respective limiting compositions. The intersection of the two melting or liquidus curves generates an invariant point, according to the Phase Rule (Eq. 10.3.2). This invariant point is known as the **eutectic point** (e), and the corresponding temperature is often referred to as solidus temperature. (In general, **liquidus curve** refers to a curve in a phase diagram that defines the lowest temperature limit of the field of complete liquid, whereas the **solidus curve** to that defining the highest temperature limit for the field of complete solid). Note that the restriction to constant pressure leads to a reduction of the degree of freedom by one so that the Phase Rule becomes $F = C - P + (2 - 1) = C - P + 1$. Consequently, since there are two components and three phases at equilibrium (diopside, anorthite and melt) at the eutectic point, $F = 0$.

Figure 10.10 shows the melting behavior in a portion of the system $MgO\text{-}SiO_2$. At low pressures enstatite (En: $MgSiO_3$), which undergoes a polymorphic transformation to protoenstatite (PEn) at high temperature, does not melt to a liquid of its own composition, but yields at 1557 °C a slightly silica rich liquid according to the reaction $PEn = Fo + l(SiO_2\text{-rich})$. At 1 bar pressure, the composition of the liquid is given by the point P, which is known as a **peritectic** point. The low pressure melting behavior exhibited by PEn is known as **incongruent** melting. (At P > 3 kb, PEn melts to a liquid of its own composition so that the vertical line at $MgSiO_3$ up to the congruent melting temperature of PEn acts as a thermal divide in the phase diagram with an eutectic point on each side).

10.6.2 Systems Involving Solid Solution

If the solid phase in the reaction (10.6.a) represents a solid solution, then $K = \left(a_A^l/a_A^s\right)_{eq}$, and consequently we have from Eq. (10.4.10b), assuming again that ΔH_m^o is insensitive to temperature,

$$\frac{1}{T_m} = \frac{1}{T_{o(A)}} - \frac{R}{\Delta H_{m(A)}^o}\left(\ln\frac{a_A^l}{a_A^s}\right)_{eq} \tag{10.6.2}$$

where the subscript (A) indicates the properties of melting of pure A to a liquid of the same composition A. An analogous relation, namely

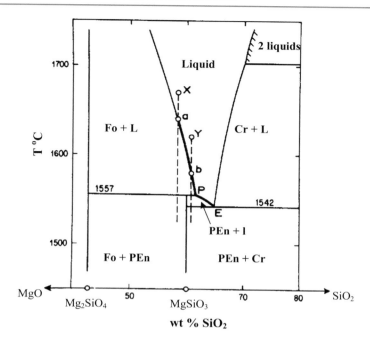

Fig. 10.10 Phase relations in a portion of the system MgO-SiO_2 (Bowen and Anderson 1914). P: Peritectic point; E: Eutectic point; Fo: Forsterite (Mg_2SiO_4); PEn: Proto enstatite ($MgSiO_3$); L: Liquid; Cr: Cristoballite (SiO_2); l: Liquid

$$\frac{1}{T_m} = \frac{1}{T_{o(B)}} - \frac{R}{\Delta H^o_{m(B)}} \left(\ln \frac{a^l_B}{a^s_B} \right)_{eq} \tag{10.6.3}$$

holds for the second component, B, in a binary system. After substituting appropriate expressions for the activities in terms of compositions, and considering the fact that in a binary solution there is only one independent compositional variable in each phase, the last two equations can be solved simultaneously to determine the compositions of the solid and melt phases at a specified value of T_m, if the T_o and ΔH^o_m values are known.

As an example, let us consider the melting of a **binary** solid solution $(A, B)_\nu F$. Substituting the activity expressions according to the ionic solution model (Eq. 9.1.1) in the last two equations along with the stoichiometric relation $X_A = 1 - X_B$ in a given phase, and re-arranging the terms, we obtain

$$\ln \left(\frac{X^l_A}{X^s_A} \right) + \ln \left(\frac{\gamma^l_A}{\gamma^s_A} \right) = -\frac{\Delta H^o_{m(A)}}{\nu R} \left(\frac{1}{T_m} - \frac{1}{T_{o(A)}} \right) \tag{10.6.4}$$

and

$$\ln\left(\frac{1 - X_A^l}{1 - X_A^s}\right) + \ln\left(\frac{\gamma_B^l}{\gamma_B^s}\right) = -\frac{\Delta H_{m(B)}^o}{R}\left(\frac{1}{T_m} - \frac{1}{T_{o(B)}}\right) \qquad (10.6.5)$$

In each phase, both γ_A and γ_B can be expressed in terms of one set of binary interaction parameters and one compositional variable. For example, if the solid phase behaves as a Regular Solution, then $\ln\gamma_A^s = (W^s/RT)(1 - X_A^s)^2$ and $\ln\gamma_B^s = (W^s/RT)(X_A^s)^2$ (Eq. 9.2.5), where W^s is the interaction parameter between A and B. Thus, if the binary interaction parameters in the solid and liquid, and the melting properties of the pure end-members are known, then there are three unknowns in the last two equations, viz. X_A^l, X_A^s and T_m. If any of these variables are specified, the other two can be determined by the simultaneous solution of these equations. An example of melting relation in a binary solid solution, that of plagioclase, $NaAlSi_3O_8$ (albite: Ab) $-$ $CaAl_2Si_2O_8$ (anorthite: An), as determined experimentally by Bowen (1913), is shown in Fig. 10.11.

For the special case of **ideal mixing** behavior of both melt and solid, we obtain from Eq. (10.6.4)

$$X_A^l = X_A^s e^{\varphi_1} \qquad (10.6.6)$$

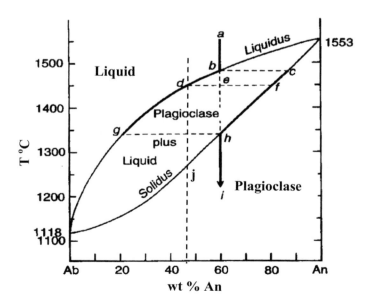

Fig. 10.11 Liquidus and solidus relations in the system Albite(Ab)-Anorthite(An) at 1 bar, after Bowen (1913). From Winter (2001)

and from Eq. (10.6.5)

$$(1 - X_A^l) = e^{\varphi_2} - X_A^s e^{\varphi_2}, \qquad (10.6.7)$$

where φ_1 and φ_2 are the quantities on the right hand side of Eqs. (10.6.4) and (10.6.5), respectively. Combining the last two equations, we then have

$$X_A^s = \frac{e^{\varphi_2} - 1}{e^{\varphi_2} - e^{\varphi_1}} \qquad (10.6.8)$$

Substitution of the value of X_A^s obtained from this equation into Eq. (10.6.7) yields the equilibrium composition of the coexisting liquid.

It should be obvious that the above equation is not restricted to the construction of melting relation of a solid solution, but is also applicable to subsolidus equilibrium between two solid solutions, such $(Mg, Fe)_2SiO_4$ (Ol) + SiO_2(Qtz) = (Mg, Fe) SiO_3(Opx) (Ol: olivine, Opx: orthopyroxene) and similar equation can be easily developed to construct P-X section at constant temperature. When the melting temperature (or sobsolidus equilibrium temperature) for both end-member phases of a solid solution are not available, one can make use of the data for the exchange equilibrium between the two solid solution phases (Sect. 11.1.1) that is, in fact, a linear combination of the end-member reactions. For example, the two end-member reactions that form the basis of the Eqs. (10.6.2) and (10.6.3), namely A(solid) = A(melt) and B(solid) = B(melt), where the both the solid and melt phases are solutions of the components A and B, may be subtracted from one another to write the exchange equilibrium A(solid) + B(melt) = B(solid) + A(melt). This approach of combining the data for one end-member reaction and exchange equilibrium between the reactant and product phases was first developed by Olsen and Mueller (1966) to construct P-X section and isopleths in the P-T space of olivine-orthopyroxene equilibrium. (In the problems considered here, there are three reactions between Solid-Melt/Solid systems, of which only two are independent. The choice of the independent reactions is dictated by the availability of data).

10.7   Azeotropic Systems

In some systems, the coexistence curves (the solidus and liquidus or boiling and condensation curves) show **coincident** extrema. Such systems are known as azeotropic systems. Two of the geologically important solid solutions that show coincident minima of solidus and liquidus (**negative** azeotropy) are the alkali feldspar ($NaAlSi_3O_8$–$KAlSi_3O_8$) and melilite ($Ca_2Al_2SiO_7$–$Ca_2MgSi_2O_7$). The phase diagram for the alkali feldspar binary is illustrated in Fig. 10.12. Azeotropic systems may also have coincident maximum (**positive** azeotropy) of the coexistence curves. Several liquid solutions (e.g. H_2O-HNO_3) show positive azeotropy. This

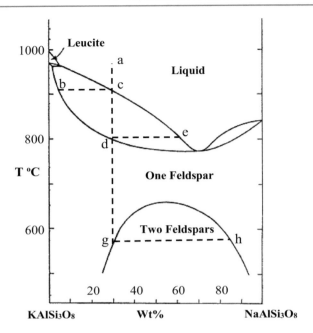

Fig. 10.12 Azeotropic behavior of the alkali feldspar system, $KAlSi_3O_8$ (orthoclase: Or)–$NaAlSi_3O_8$ (albite: Ab) at $P(H_2O = 2$ kb), according to Bowen and Tuttle (1950). (Note that the slight difference between the minima of the solidus and liquidus is due to drafting error of the solidus curve, since these two minima must coincide). The horizontal tie lines represent the equilibrium compositions of the coexisting phases. With permission from The University of Chicago Press

type of behavior is also known to exist between garnet and staurolite solid solutions that fractionate Fe and Mg (Ganguly and Saxena 1987).

Since the two phases have the same composition at the azeotrope, Eqs. (10.6.4) and (10.6.5) yield two explicit relations between the activity coefficients of the components in solid and liquid when $T_m = T_{az}$, if the heat of melting and end-member melting temperatures are known. Thus, if the mixing properties of both solid and liquid phases can be represented by Regular Solution models (Eq. 9.2.5), then the use of azeotropic condition leads to the explicit solution of the mixing properties of both phases.

The negative azeotropic behavior in the alkali feldspar and melilite binary systems is due to the existence of large solvus in both systems. The latter implies strong positive deviation from ideality of mixing in the solid solution that tends to make the solid solution less stable relative to the liquid in intermediate compositional range. However, existence of a solvus does not always lead to negative azeotropy.

Problem 10.6 Assuming the heat of melting to be insensitive to temperature, and ideal mixing in the melt, derive an expression (similar to Eq. 10.6.1) for the calculation of liquidus temperature between P and E as function of $X^l_{SiO_2}$, in Fig. 10.10. The melt composition at P and the enthalpy of melting of the end member phases are known. No other data are required for this formulation.

Hint: Begin with Eq. 10.4.10b.

Problem 10.7 Assuming ideal mixing properties of liquid and solid, calculate the T versus composition curves defining the equilibrium compositions of coexisting solid and melt in the system Mg_2SiO_4(forsterite)–Fe_2SiO_4(fayalite) at 1 bar pressure, with the composition given in wt%, using the following data. $T_{o(Fo)} = 2163$ K, $T_{o(Fa)} = 1490$ K, $\Delta H^\circ_{m(Fo)} = 114$ kJ/mol and $\Delta H^\circ_{m(Fa)} = 92.173$ kJ/mol (Robie et al. 1978; Navrotsky et al. 1989). Label the solidus and liquidus curves. Compare the symmetry of the melting loop bounded by the liquidus and solidus in your calculated diagram with that in Fig. 10.11, and provide a qualitative explanation for the difference.

Problem 10.8 At a constant P-T condition, the qualitative property of the Gibbs energy versus composition relation of phases that show negative azeotropic relation (Fig. 10.12) is illustrated in Fig. 10.13. Following the logic for the construction of this diagram (see also Sect. 8.11.2) draw (a) schematic Gibbs energy versus composition at the azeotropic (minimum) temperature, and (b) the counterpart of Fig. 10.13 for a system showing positive azeotropic behavior.

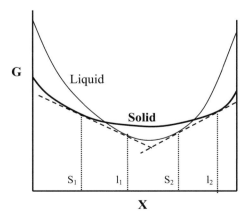

Fig. 10.13 Qualitative G versus X relation at a constant P-T condition somewhat above the minimum temperature of an azeotropic system, such as that of the line d-e in Fig. 10.10. The stable solid-liquid compositional pairs are defined by the points of common tangency

10.8   Reading Solid-Liquid Phase Diagrams

Phase diagrams involving melting of solids play critical role in the development of concepts of magmatic processes in nature. The purpose of this section is not an in-depth and extensive discussion of how to understand this type of phase diagrams, but to provide an exposure to some of the elementary but fundamental concepts that govern the interpretation of phase diagrams. As would be appreciated from the following discussions, interpretation of phase diagrams is governed by two fundamental requirements, namely, (a) adherence to the Phase Rule and (b) preservation of bulk composition in a closed system.

10.8.1   Eutectic and Peritectic Systems

The crystallization of a liquid in a binary eutectic system can be followed by considering a composition and temperature defined by the point A in Fig. 10.9. As the liquid cools to slightly below the liquidus temperature (point l_1), anorthite begins to precipitate, and the liquid composition evolves along the liquidus line towards e. Note that, in principle, there can be no equilibrium crystallization from a liquid that has cooled *exactly* to the liquidus temperature at constant composition. However, since a point like l_1 represents the theoretical limit of the existence of *only* liquid that has cooled at constant composition, we would say, following common practice, that crystallization begins at l_1. At any temperature, the mass (m) ratio of liquid (l) to anorthite crystals (x′l) is given by the requirement of conservation of total mass that leads to the so called **lever rule**. As an example, at 1400 °C, the lever rule yields $m(l)/m(x′l) = ax/l_2x$, where ax and l_2x are the lengths of the horizontal segments on two sides of x that defines the bulk composition. When the liquid composition and temperature descends to the eutectic point, diopside begins to crystallize from the liquid, and as a consequence the temperature remains fixed (because F = 0 at constant pressure), if equilibrium is maintained, even when heat is withdrawn from the system. After the liquid completely crystallizes to an assemblage of Di + An, the system gains a degree of freedom (F = 1 since at this stage C = P = 2), enabling further cooling with the withdrawal of heat. The crystallization behavior of a liquid on the other side of the eutectic point follows the same general principle, beginning with the crystallization of diopside.

Upon heating, a two phase mixture of Di and An of any proportion begins to melt at the eutectic temperature, T_e, which is indicated by a solid horizontal line in Fig. 10.9. The composition of this initial melt is always given by that of the eutectic point. At the eutectic temperature, one of the solid phases disappears, depending on the proportion of Di to An relative to that defined by the eutectic point, e. The melt composition then evolves along an appropriate liquidus curve upon further heating, as a result of progressive melting of the residual solid phase, until the point where the liquid composition becomes the same as that of the initial solid bulk composition. For example, if the solid bulk composition is given by that of x, then diopside disappears at the eutectic point and the liquid composition progressively evolves from e to l_1 upon progressive heating.

The general principle governing the crystallization behavior of a liquid with composition to the right of the peritectic point P in Fig. 10.10 is the same as that discussed above with reference to Fig. 10.9. For liquid composition to the left of the point P, crystallization begins at the liquidus temperature with the precipitation of Fo, followed by progressive crystallization of Fo and change of liquid composition along the liquidus curve until it reaches the peritectic point. At this condition, the liquid reacts with Fo to form PEn, leading to an isobaric invariant condition. Except for the special case that the initial liquid composition corresponds exactly to that of En, either liquid or Fo must be completely consumed (so that the system gains a degree of freedom) before the liquid can be cooled any further. Which of these two phases would be consumed depends on the initial liquid composition. If it is to the left of $MgSiO_3$ composition, such as given by X in Fig. 10.10, then liquid must be completely consumed leading to a final equilibrium assemblage of Fo plus PEn, the latter transforming to En upon subsolidus cooling. This assemblage with appropriate proportion of phases, as dictated by Lever Rule, conserves the bulk composition of the system. On the other hand, if the initial liquid composition lies between En and P, such as corresponding to the point Y in Fig. 10.10, Fo is completely consumed by the peritectic reaction, and the liquid finally descends to the eutectic point, E, yielding a final equilibrium product of PEn + Cr (cristoballite) (with these two phases transforming to En and quartz, respectively, at low temperature, if equilibrium prevails). *It is left to the reader to figure out the crystallization behavior of a liquid of composition $MgSiO_3$.*

Upon heating, an assemblage of Fo plus En or PEn always melts at the peritectic temperature (1557 °C at 1 bar), whereas that of En or PEn plus SiO_2 at the eutectic temperature (1542 °C at 1 bar), with the initial liquid composition given by the points P and E, respectively. For the Fo + PEn assemblage, PEn must melt completely before the temperature can be raised beyond the peritectic temperature. After complete melting of PEn, the T-X (temperature-composition) condition of liquid evolves along the liquidus line to the left with progressive melting of Fo until it reaches the bulk composition defined by the initial mixture of the solid phases. For example, if the initial bulk composition of the Fo plus PEn or En mixture is defined by the vertical dashed line connecting to the point X, then complete melting of the mixture takes place when the liquid composition reaches the point **a** on the liquidus line, since otherwise the bulk composition is not conserved.

If the initial mixture of PEn plus Cr has a bulk composition given by the vertical line connecting to the point Y, then Cr must be completely consumed at E, followed by the evolution of T-X condition of the liquid towards P. At P, En melts incongruently to Fo + l until it is consumed. The T-X condition of the liquid then evolves along the liquidus line towards left with progressive melting of Fo, until the latter melts completely at the point **b**.

Problem 10.9 Prove "Lever Rule".

Hint: Begin with the mass conservation relation: $m_i^l + m_i^s = m_i^{bulk}$ where m stands for mass. Then show that

$$X_m^l X_i^l + X_m^s X_i^s = X_i^{bulk}, \qquad (10.8.1)$$

where X_m^l and X_m^s represent the mass fractions of liquid and solid, respectively, thereby leading to the Lever rule.

Note that the last equation can be generalized to mass conservation relation in an n-phase system as

$$\sum_{j=1}^{n} X_m^j X_i^j = X_i^{bulk} \qquad (10.8.2)$$

10.8.2 Crystallization and Melting of a Binary Solid Solution

As an illustration of the crystallization and melting behavior in a binary system involving a solid solution, let us consider the melting of plagioclase in the system $NaAlS_3O_8$ (An: Albite) – $CaAl_2Si_2O_8$ (An: Anorthite). The experimentally determined melting diagram at 1 bar pressure is illustrated in Fig. 10.12. The crystallization behavior of a liquid may be illustrated by considering the cooling of a liquid with temperature and composition given by the point **a**. Crystallization begins when the liquid cools to the point **b** (strictly slightly below this point; see above), and precipitates a solid of composition **c**. Upon further cooling the liquid and solid compositions evolve along the liquidus and solidus curves, respectively. At any temperature, the equilibrium compositions of the coexisting liquid and solid are given by the horizontal tie lines such as d-f. Under equilibrium, complete crystallization takes place when the system cools to the temperature corresponding to the point **g**. At this temperature the solid has the same composition as the initial liquid and, thus, there can be no liquid present in the system since the bulk composition must always lie between the solid and liquid compositions. As the system cools at a fixed bulk composition, both the liquid and solid compositions become albite rich, but the relative abundance of the two phases keeps on changing such that the bulk composition of the system is preserved (Eq. 10.8.1). At any temperature, the mass ratio of the solid to liquid is given by the **lever rule**. For example, when the liquid cools to the point d, m(l)/m(plag) = ef/ed.

For equilibrium crystallization, a plagioclase solid solution that crystallizes at any point on the solidus, such as the point **f**, must react completely with the liquid to yield the equilibrium composition at a lower temperature. However, initially this adjustment takes place at the solid-liquid interface, and the equilibrium composition of the crystal is established in an outer segment. If the diffusive exchange of components between the crystal and liquid is too slow for the attainment of complete equilibrium during cooling, as in the case of plagioclase in a geological environment, then the interior composition of a crystal fails to adjust to the equilibrium composition at a given temperature, leading to the development of compositional zoning.

Consider now the removal of the crystals that formed at the point **f** from the system. In geological processes such removal of early formed crystals is a common phenomenon and is caused by the settling or flotation of crystals during growth or squeezing of liquid out of the crystal-liquid mush. Crystallization of a liquid accompanied by removal of early formed crystals is known as **fractional** crystallization. (The process of development of compositional zoning discussed above also represents fractional crystallization since the interior of a crystal is effectively removed from the interacting system). As the crystals are removed, the effective bulk composition of the crystallizing system now shifts to that of the point **d**. Consequently, if equilibrium is maintained, the final crystallization takes place at a temperature and composition corresponding to the point **j**. Thus, fractional crystallization extends the "descent" of liquid along the liquidus.

Unlike a eutectic system, the equilibrium melting behavior of a solid solution is exactly the reverse of its crystallization behavior. Thus, a plagioclase of composition 60 wt% An, corresponding to the bold vertical line in Fig. 10.10, begins to melt as its temperature increases to the point **h** (strictly slightly above this point). Progressive melting of the solid takes place upon further heating, with the solid and liquid compositions changing along the solidus and liquidus curves, respectively. Complete melting takes place when the system is heated to the temperature corresponding to **c**.

For the alkali feldspar system showing azeotropic behavior and a solvus (Fig. 10.12), complete crystallization of a liquid follows the same principle as the crystallization of plagioclase solid solution discussed above. For example, complete crystallization of the liquid of composition **a** takes place as the temperature decreases to **c**, with the compositions of solid and liquid evolving along the solidus and liquidus, respectively, during crystallization. Upon complete crystallization and further cooling, the solid solution of composition **d** begins to exsolve when the temperature decreases to the solvus. The limiting composition of the first albite rich exsolved phase is given by the point **h**. Continued cooling under equilibrium condition leads to the evolution of the compositions of the two alkali feldspars, one orthoclase-rich, and the other albite-rich, along the two limbs of the solvus curve.

10.8.3   Intersection of Melting Loop and a Solvus

In many systems, the melting loop and solvus interpenetrate to produce complex binary phase relations. Two such cases are illustrated in Fig. 10.14. The phase diagram for the system MnO-FeO at 1 bar is similar to the right panel of Fig. 10.14a, whereas the binary alkali feldspar system, $NaAlSi_3O_8$–$KAlSi_3O_8$,

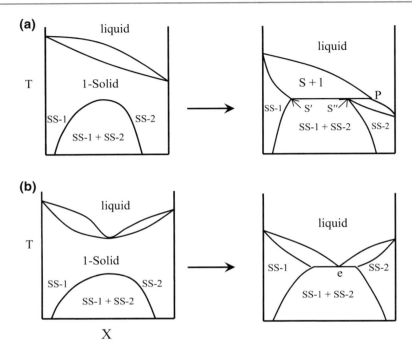

Fig. 10.14 Isobaric phase diagrams generated by the intersection of melting loop and solvus in a binary system. Right hand panels show the qualitative features of phase diagrams that are produced by the collapse of the melting loop on to the solvus relations shown on the left. The non-collapsed state may not always exist. SS: solid solution; P: peritectic point; e: eutectic point

shows a phase relation similar to the right panel of Fig. 10.14b, at $P(H_2O) \geq 3$ kb. The crystallization behavior of liquid in these two types systems are discussed in this section, from which the reader should be able to construct the melting behavior following the principles discussed above. In the alkali feldspar system, the interpenetration of the solvus and melting loop is due to the lowering of melting temperature as a result of progressive dissolution of O-H with increasing $P(H_2O)$ beyond 3 kb (Winkler 1979). and increase of solvus temperature with increasing pressure (the lowering of melting temperature being the dominant effect in the convergence of melting loop and solvus).

In the right panel of Fig. 10.14a, crystallization of a liquid follows the same principles discussed for crystallization of a binary solid solution in the last section (Fig. 10.11) until the liquid composition reaches the point P on the liquidus line. At this point the liquid and the solid solution-1 of composition S' react to produce another immiscible solid solution of composition S'', thus leading to an isobaric invariant condition. As in Fig. 10.10, the point P represents a peritectic point. Further equilibrium cooling of the system is only possible after the complete consumption of either S' or l, depending on the initial composition (l_o) of the liquid, whether it is to the right or left of the point S''. If the liquid gets completely consumed

(l_o to the left of S″), then further cooling leads to compositional evolution of the two unmixed solid solutions along the two solvus limbs as long as kinetics of solid state readjustment of composition is favorable. On the other hand, if the solid S′ gets consumed (l_o between S″ and P), the solid and liquid compositions evolve, upon further equilibrium cooling, along the solidus and liquidus curves, respectively, below the peritectic temperature, limited by the bulk compositional constraint.

In Fig. 10.14b, interpenetration of the melting loop and the solvus leads to the development of a eutectic point, e. Liquid of any composition descends to e upon equilibrium cooling, with the composition of coexisting solid evolving along one of the two solidus curves, depending on the initial liquid composition (l_o), up to the eutectic temperature, T_e. At this temperature, two immiscible solids of compositions defined by the terminals of the horizontal line at T_e crystallize simultaneously from the liquid until the liquid is completely consumed. Further cooling of the system leads to the evolution of the compositions of two solids along the two limbs of the solvus as long as kinetics remains favorable.

Problem 10.10 In Fig. 10.10, describe the melting behavior of a mixture of Fo + En showing the temperatures for the beginning and end of melting (or solidus and liquidus temperature, respectively), initial melt composition, invariant situation, and the solid phase/phases that melt as the temperature of the system is progressively raised.

Problem 10.11 Schematically illustrate the isobaric phase diagram when the solvus intersects the melting loop of an azeotropic system on the left side of the temperature minimum.

10.8.4 Ternary Systems

In an isobaric ternary system, an invariant point (eutectic or peritectic) in a bounding binary join transforms to a univariant line because of the gain of a degree of freedom resulting from the addition of a component. This univariant line is called a **cotectic** or a **reaction line** depending on whether it is connected to an eutectic point or a peritectic point, respectively, in a terminal binary. A ternary phase diagram involving both eutectic and peritectic relations in the bounding binaries, but not involving any solid solution, is illustrated in Fig. 10.15. The cotectic and reaction lines represent projections on to a basal ternary plane of lines marked by the intersections of adjacent liquidus surfaces, and slope toward the invariant points within the ternary system. The point E denotes a ternary eutectic point that is defined by the intersection of three cotectic lines. As the liquid cools along a cotectic line, two phases whose fields are separated by this line crystallize simultaneously from the liquid. The line P-P' is a reaction line (i.e. a translation of binary peritectic or reaction point into the ternary system) along which the phase B reacts with the liquid to form the incongruently melting phase B′. The point p′ is a ternary peritectic point and is at a higher temperature than the ternary eutectic point E.

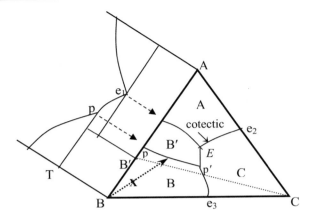

Fig. 10.15 A schematic isobaric ternary melting diagram with eutectic (e) and peritectic (p) relations in the A-B binary and only eutectic relations in the B-C and A-C binaries. The phase relations in the A-B-C triangle represent projections from the liquidus relations in the ternary space. The binary phase relations in only the A-B join are shown schematically, with the arrows indicating projections of the eutectic and peritectic points on to the base. E denotes the projection of a ternary eutectic point that is formed by the intersection of three cotectic lines. p-p' represents the projection of a univariant reaction line: $B + l \rightarrow B'$. The stability fields of the phases in the ternary space are labeled. Crystallization of a liquid of initial composition (x) is discussed in the text

The principles governing crystallization of melt in the ternary system may be discussed with reference to the crystallization of a liquid of composition X in Fig. 10.15. The liquid begins to crystallize the phase B as it cools to the liquidus surface. With progressive cooling and continued crystallization of B from the liquid, the composition of the liquid changes away from the initial composition X along the straight line joining the points B and X. After the liquid cools to the temperature of the reaction line P-P', it reacts with B that has crystallized earlier to form the phase B'. The composition of the liquid now evolves along the reaction line until it reaches the point P'. At this point, the phase C begins to crystallize, thus leading to an isobaric invariant condition. Consequently, either B or liquid must be completely consumed by reaction between one another before the system can be cooled any further under equilibrium condition. Now, in order to preserve the bulk composition in a closed system, the initial composition must always be contained within the polygon formed by joining the compositions of the phases in the system. Consequently, for the bulk composition marked by the point X, liquid must be completely consumed at P', leaving a final equilibrium assemblage of B + B' + C. If the initial liquid composition falls between *P-P'* and the dotted line B'-C, then at P', the phase B is completely consumed. In that case, the liquid composition evolves from P' to E along the cotectic line P'-E with simultaneous crystallization of B' and C. Finally, at the ternary eutectic point the phase A joins the crystallization process that continues to take place at a fixed temperature until the remaining liquid has completely crystallized. Obviously, the assemblage A + B + C preserves the initial composition of the system if it is located between P-P' and B'-C.

Upon heating, beginning of melting of an assemblage of solids depends on its bulk composition. If it is given by a point below the dotted line B'-C, then melting begins at the ternary peritectic temperature, with the initial melt composition given by the point P', whereas if the bulk composition lies above B'-C, then the melting begins at the ternary eutectic temperature, with the initial melt composition given by the point E. Let us now consider the melting behavior of an assemblage B + B' + C, with the bulk composition being given by the point X. In this case, all C is consumed at the ternary peritectic temperature. The T-X condition of the liquid then evolves along the reaction line toward P with progressive incongruent melting of B' to B + l, until it reaches the tip of the arrow. At this point all B' is completely consumed, and the T-X condition of the liquid evolves along the liquidus surface towards the point X, with the compositional variation being restricted to the dotted line through this point. Complete melting of the phase B takes place at X.

From the above discussion of melting behavior in a ternary system, we note a geologically important point that regardless of the relative proportion of the initial solid phases (e.g. proportion of minerals in the Earth's upper mantle that is the site of primary magmas), there are only a restricted number of initial magma compositions that are defined by the invariant points within the system. Since magma escapes from the source region after only a small amount of partial melting, with the melt forming no more than a few percent of the system, the primary magmas derived from the Earth's mantle are of very limited compositional types.

Problem 10.12 Describe the melting behavior (initial melting, evolution of melt composition, change of the solid assemblage, and final melt composition) of two assemblages of A + B' + C, with the bulk compositions in the field of (a) B' and (b) C in Fig. 10.15.

10.9 Natural Systems: Granites and Lunar Basalts

10.9.1 Granites

As illustrated in Fig. 10.16, the azeotropic behavior of the alkali feldspar system is responsible for the development of a minimum on the melting temperature surface in the system $NaAlSi_3O_8$-$KAlSi_3O_8$-SiO_2 system (Bowen and Tuttle 1964). The minimum shifts somewhat with changing $P(H_2O)$ and addition of anorthite ($CaAl_2Si_2O_8$) component to the system, and transforms to a ternary eutectic when $P(H_2O) \geq 3$ kb, corresponding to the appearance of an eutectic point in the alkali feldspar binary, similar to Fig. 10.14b. With increasing $P(H_2O)$, the alkali feldspar melting loop drops to lower temperature because of the dissolution of H_2O in the liquid and consequent decrease in the Gibbs energy of the liquid phase, ultimately leading to the intersection of the solvus and the melting loop (the solvus also rises

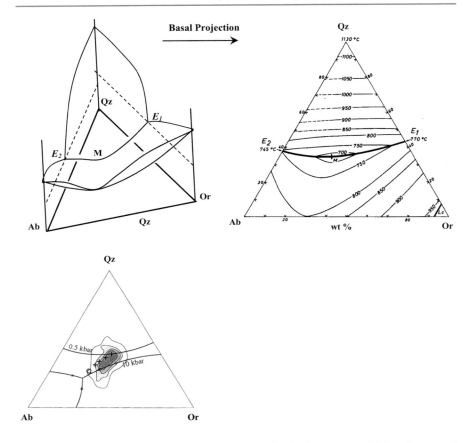

Fig. 10.16 (**a**) Schematic illustration of the melting relation in the ternary model "granite system" Ab-Or-Qz at $P(H_2O) < 3$ kb. (**b**) Projection of the liquidus relation at $P(H_2O) = 2$ kb on to the basal plane. From Winkler (1976), drawn on the basis of data from Bowen and Tuttle (1964). (**c**) Compositions of natural granites normalized to the system Ab-Or-Qz along with the location of minimum temperatures in the ternary system at different $P(H_2O)$. Crosses: azeotropic minima at $P(H_2O)$ of 0.5, 2, 3 and 4 kb; circle: ternary eutectic point at $P(H_2O) = 5$ kb. From Anderson (2005). With permission from Cambridge University Press

in temperature slightly with increasing $P(H_2O)$). The locus of the minimum, including the eutectic, as a function of pressure is commonly referred to as the **granite melting minimum**. Thus, fractional crystallization of a "granite system", or partial melting of a pelitic rock, which has "granitic components", in the presence of water, is expected to produce liquids of composition clustering within a limited domain corresponding to the collection of minima in the natural systems. Indeed, it is found that the compositions of large number of granites and rhyolites (more than two thousand), normalized to the system Ab-Or-Qz, cluster around the composition of thermal minima in the system Ab-Or-Qz. A compilation of granite composition, normalized to the Ab-Or-Qz system, is shown in Fig. 10.16c along with the location of the azeotropic and eutectic minima as a function of pressure (Anderson 2005). Such coincidence of compositions could not be fortuitous, and cannot be explained by any non-magmatic hypothesis of the origin of granite. These observations completely resolved the long standing controversy among petrologists about the origin of granite in favor of magmatic origin.

10.9.2   Lunar Basalts

Samples returned by several lunar missions reveal plagioclase-rich basalts to be important components of lunar highlands. These basalts are commonly referred to as Fra Mauro basalt after the landing site, known as the Fra Mauro Hills, of the Apollo 14 mission that returned the most abundant samples. The natural question that arises in the study of these samples is whether or not the compositionally different basalt samples are genetically related and represent samples at different stages of crystallization in a low pressure environment. The compositions of the Fra Mauro basalts can be described fairly well in terms of the pseudo-ternary system $(Mg, Fe)_2SiO_4$ (Ol)–$CaAl_2Si_2O_8$ (An)–SiO_2 (Si), as illustrated in Fig. 10.17. The line separating the olivine and pyroxene fields is a reaction line along which olivine and liquid react to form pyroxene. Also shown in this figure are the projections of the experimentally determined cotectic and reaction lines in the system at 1 bar pressure, dry and low $f(O_2)$ condition. There is no qualitative change of the phase diagram for (dry) pressures up to a 5 kb. Below this pressure the only changes involve small shifts of the liquidus surface and the univariant lines.

The phase diagram is qualitatively the same as Fig. 10.15, except for the presence of two binary eutectics in the Ol-An join and the resultant cotectic lines that intersect at an invariant point in the ternary space. It is noteworthy that the projections of the Fra Mauro basalt compositions cluster near the trough or the univariant lines in the ternary system that has the minimum temperature at the eutectic

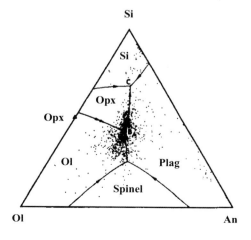

Fig. 10.17 Low-pressure liquidus relations in the system olivine-anorthite-silica along with the lunar (Fra Mauro) basaltic compositions normalized to the system composition. From Walker et al. (1981)

point labeled as C. The observed distribution of the compositions of the Fra Mauro basalt in relation to the low-pressure liquidus relations in the Ol-An-Si ternary system is only possible if the basalt samples represent crystallization products from liquid that had repeatedly erupted from a magma chamber undergoing crystallization in a low pressure environment below ~ 5 kb that corresponds to the outer 100 km of Moon.

10.10 Pressure Dependence of Eutectic Temperature and Composition

The pressure dependence of eutectic temperature is governed by the general Clapeyron-Clausius relation (Eq. 6.4.4) and repeated below in the appropriate form for the eutectic melting

$$\left(\frac{\partial T}{\partial P}\right)_e = \left(\frac{\Delta_m V}{\Delta_m S}\right)_e = \left(\frac{T\Delta_m V}{\Delta_m H}\right)_e \qquad (10.10.1)$$

where $\Delta_m V$ and $\Delta_m S$ denote, respectively, volume and entropy change upon eutectic melting and are given for a binary system by

$$\Delta_m V = X_1 \Delta_m V_1^o + X_2 \Delta_m V_2^o + \Delta V^{mix(m)} \qquad (10.10.2a)$$

and

$$\Delta_m S = X_1 \Delta_m S_1^o + X_2 \Delta_m S_2^o + \Delta S^{mix(m)} \qquad (10.10.2b)$$

Here the mole fractions refer to those at the eutectic point and the superscripted mix(m) terms refer only to the melt phase since the solids at the eutectic point are assumed to be stoichiometric. For the special case of an ideally mixing binary liquid, $\Delta V^{mix} = 0$, and $\Delta S^{mix} = -R(X_1 \ln X_1 + X_2 \ln X_2) > 0$. Thus, while $\Delta_m V$ is the weighted average of the volume changes of melting of the two end members, $\Delta_m S$ is greater than the corresponding weighted average property. Thus, in an ideally mixing system, the positive contribution from the $\Delta S^{mix(m)}$ would cause smaller pressure dependence of the eutectic temperature compared to those of the end members. The melting behavior of CsCl-NaCl system, illustrated in Fig. 10.18, shows this behavior.

A particularly interesting case arises when the eutectic solution exhibits a strong negative deviation from ideality ($\Delta V^{mix} < 0$), thereby leading to $\Delta V_m \sim 0$, and consequently very little pressure effect on the eutectic temperature. This behavior is exhibited by the Fe-FeS eutectic temperature up to ~ 60 kb, as determined by Brett and Bell (1969), Ryzhenko and Kennedy (1973) and Usselman (1975). The data from the first two sources are illustrated in Fig. 10.19. This behavior of the Fe-FeS eutectic system is very likely responsible for the core formation in the early history

Fig. 10.18 Pressure
dependence of melting
temperature in the system
CsCl-NaCl. From Kim et al.
(1972). See the latter for
references of Vaidya and
Kennedy (1971) and Akella
et al. (1969)

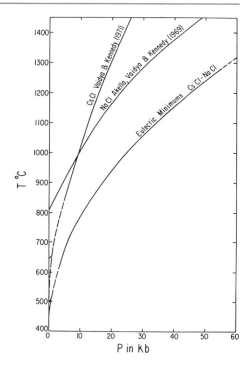

of the Earth, enabling formation of heavy iron rich melt up to quite high pressure
without completely melting the silicates as the Earth heated up after accretion, and
segregation of the melt into a core. (The release of the gravitational potential energy
caused further heating of the Earth, thereby contributing to the melting of the
silicates to from an early magma ocean).

The qualitative nature of pressure dependence of the eutectic composition in a
system in which the liquidus curves on either side of the eutectic point remain
subparallel at different pressures, as in a system with nearly ideal mixing behavior
of the melt phase, may be understood by simple geometric considerations. It shows
that an eutectic composition shifts towards the phase with smaller (or smallest, if
there are more than two phases) pressure dependence, as illustrated for the melting
behavior in the Fe-FeO system (Fig. 10.20) calculated by Komabayashi (2014).[2] In
a formal way we may state that in a system in which the melt phase shows nearly
ideal mixing behavior, the sign of $(\partial X_2/\partial P)_e$ is the same as that of the term

---

[2]The available experimental data in this system up to 50 GPa requires a non-ideal mixing model
for the melt phase, but the experimental data at higher pressure up to ~ 330 GPa is incompatible
with the adopted nonideal mixing model, but match the calculated relation with an ideal mixing
model quite well.

(a)

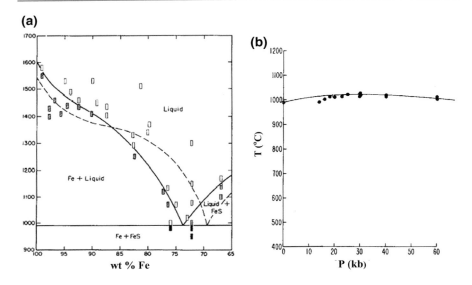

wt % Fe

P (kb)

Fig. 10.19 (**a**) Melting relation in the Fe-FeS system (**a**) at 1 bar (dashed lines) and 30 kb (solid lines); symbols: experimental data at 30 kb; open rectangles: liquid; hatched rectangles: solid + liquid; filled rectangles: solid. From Brett and Bell (1969). With permission from Elsevier. (**b**) Pressure dependence of the eutectic melting temperature of the Fe-FeS system; symbols: experimental data. From Ryzhenko and Kennedy (1973). With permission from American Journal of Science

Fig. 10.20 Eutectic phase diagram of the system Fe-FeO as a function of pressure up to the interface of the molten outer core and solid inner core of the Earth that are made primarily of Fe. From Komabayashi (2014). With permission from American Geophysical Union

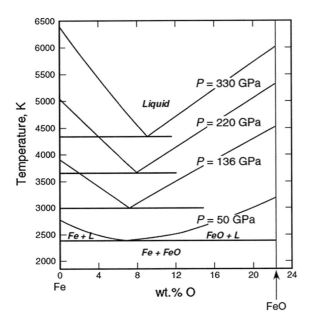

$$\left(\frac{\Delta_m V^\circ}{\Delta_m S^\circ}\right)_1 - \left(\frac{\Delta_m V^\circ}{\Delta_m S^\circ}\right)_2 \qquad (10.10.3a)$$

or, using the relation that at equilibrium, $\Delta S = \Delta H/T$, the sign of the term $[(\Delta_m V^\circ/\Delta_m H^\circ)_1 - (\Delta_m V^\circ/\Delta_m H^\circ)_2]$. Here the superscript zero indicates, as usual, pure phase. When the melt phase shows nonideal mixing property, the above statement remains valid if $\Delta_m V^\circ$, $\Delta_m S^\circ$ and $\Delta_m S^\circ$ are replaced respectively by the partial molar volume, entropy and enthalpy of melting of the components 1 and 2 at the eutectic composition. The full equation for a binary system can be found in Prigogine and Defay (1954, Eq. 22.27), but in the modern era of computer aided computing, it is advantageous to calculate the eutectic composition and temperature at different pressures (see Sect. 10.6.1), as illustrated in Fig. 10.20. Comparing the eutectic compositions in Fig. 10.20 with the data at 1 bar pressure, one finds that the oxygen content of the eutectic melt in the Fe-FeO system increases from ~ 1 wt% at 1 bar to ~ 9 wt% at ~ 330 GPa, which is the pressure at the boundary of the inner core of the Earth. As discussed in Sect. 7.3.2, oxygen and sulfur have been proposed as potential light alloying elements in the outer core.

In Fig. 10.18, we find that the eutectic temperature in the system CsCl-NaCl has an initial dT/dP slope closer to that of the melting curve of pure CsCl, but with increasing pressure, the slope becomes similar to that of the melting curve of pure NaCl. This is because of the fact that at low pressures, T_m of CsCl is lower than that of NaCl, making the eutectic melt relatively enriched in CsCl component. However, because of lower pressure dependence of T_m of NaCl, the eutectic melt becomes progressively enriched in NaCl, according to Eq. (10.10.2), and thus the pressure dependence of the eutectic temperature becomes similar to that of the melting temperature of NaCl.

10.11   Reactions in Impure Systems

Calculations of the displacement of mineral reactions in natural assemblages due to the effect of compositional change of phases relative to the compositions (usually pure end-members) used in laboratory calibrations play very important roles in the re-construction of the P-T-fluid composition and geodynamic history of rocks. The mineral reactions that a rock had experienced are deciphered by careful petrographic studies. The methods of calculation of these compositional effects are discussed in this section. A specific example of reconstruction of P-T path and geodynamic history of a section of a tectonically active regime, namely the Himalayas, building on the result of a solved problem has been discussed in the Sect. 10.11.1.

10.11.1   Reactions Involving Solid Solutions

To develop the thermodynamic formulation for calculating the effect of changing solid solution compositions on the equilibrium P-T condition of a reaction, let us consider a specific mineral reaction, as follows, that has been widely used in the petrological literature.

$$3CaAl_2Si_2O_8(An) = Ca_3Al_2Si_3O_{12}(Grs) + 2Al_2SiO_5 + SiO_2 \qquad (10.11.a)$$

This is a univariant reaction in a system that involves only pure end-members (P = 4, C = 3, thus, F = 1), and has been calibrated experimentally by several workers because of its importance in determining P-T condition of a rock; the most thorough experimental calibration is by Koziol and Newton (1988), which is illustrated in Fig. 10.21. At the P-T conditions of experimental studies, the stable aluminosilicate polymorph is kyanite. However, in natural assemblages, it is often sillimanite because of the displacement of the equilibrium condition into the sillimanite field due to compositional effects on the equilibrium conditions.

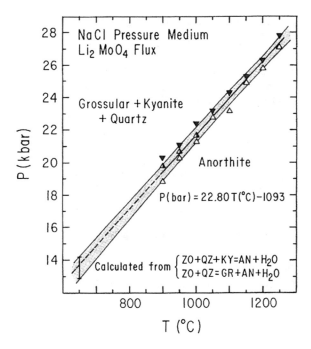

Fig. 10.21 Experimentally determined equilibrium boundary of the reaction 3 Anorthite (An) = Grossular (Gr) + 2 Kyanite (Ky) + Quartz (Qtz). The starting mixture consisted of all four crystalline phases. Filled symbols: Growth of Gross + Ky + Qtz; Open symbols: The reverse; Half-filled symbols: No detectable reaction. The equilibrium condition at 650 °C was calculated from two other reactions, as indicated in the figure. From Koziol and Newton (1988). With permission from Mineralogical Society of America

In geological assemblages, An (anorthite) and Grs (grossular) are dissolved components in plagioclase and garnet solid solutions, respectively. The consequent changes in the chemical potentials of the An and Grs components, as dictated by Eq. (8.4.7), result in a change of the equilibrium P-T conditions of the above reaction. In general, a major geological interest in the calculation of the displacement of P-T condition of a reaction as function of the solid solution compositions lies in the fact that it enables one to reconstruct the condition at which an observed assemblage of phases related by a reaction had formed.

We now develop a general formulation to calculate the displacement of a univariant reaction in the P-T space due to compositional changes of the phases. One may calculate either the change of temperature at a fixed pressure or the reverse. It should be easy to see that Eq. (10.6.1) serves the purpose of calculation of isobaric temperature change. Here the activity ratio represents Q_{eq} or K so that this expression transforms to

$$\frac{1}{T} = \frac{1}{T_o} - \frac{R}{\Delta_r H^o}(\ln K) \tag{10.11.1}$$

where T_o is the equilibrium temperature of the pure end-member reaction at a fixed pressure, P, and $\Delta_r H^o$ is the enthalpy change of that reaction at P, T_o. It is again assumed that $\Delta_r H^o$ does not change significantly between T_o and T. (If this assumption is not valid, then we need to integrate the "root equation", Eq. (10.4.10b) expressing $\Delta_r H^o$ as function of temperature).

It is, however, often more convenient to calculate the change of pressure at a fixed temperature since it is related to $\Delta_r V$ for which more complete data are available, at least for natural systems. To derive the appropriate expression, we begin by re-stating the expression for the Gibbs energy change of a reaction (Eq. 10.4.5), viz.

$$\Delta_r G(P, T, X) = \Delta_r G^*(T) + RT \ln Q(P, T, X) \tag{10.11.2}$$

Here, as usual, $\Delta G^*(T)$ is the standard state free energy change of the reaction at T, and

$$RT \ln Q(P, T, X) = RT \ln \underbrace{\left[\prod_i (X_i)^{\nu_i}\right]}_{Q_x} + RT \ln \underbrace{\left[\prod_i (\gamma_i)^{\nu_i}\right]}_{Q_\gamma} \tag{10.11.3}$$

where X_i and γ_i are, respectively, the mole fraction and activity coefficient of a component i in a specific phase (see Sect. 10.4). Denoting the terms within the two square brackets by Q_x and $Q_\gamma(P, T, X)$, respectively, we write

$$RT \ln Q(P, T, X) = RT \ln Q_x + RT \ln Q_\gamma(P, T, X) \tag{10.11.3'}$$

The equilibrium pressure, P_e, for the impure phase reaction at T is obtained by transforming the right hand term of Eq. (10.11.2) to a pressure dependent expression, and imposing the condition of equilibrium, $\Delta_r G(P_e, T, X) = 0$. The logical set-up of the problem is illustrated in Fig. 10.22.

Let us now choose pure component (P, T) standard states, as we have done earlier, so that $\Delta_r G^*(T) = \Delta_r G^\circ(P, T)$. Instead of calculating the value of $\Delta_r G^\circ(P, T)$ from thermochemical tables, it is sometimes better to calculate it from well constrained experimental data on the equilibrium condition of the reaction involving only pure phases (e.g. Figure 10.21). For the second approach, we write

$$\Delta_r G^*(T') \equiv \Delta_r G^\circ(P, T') = \underbrace{\Delta_r G^\circ(P_0, T')}_{0} + \int_{P_0}^{P} (\Delta_r V^\circ)_{T'} \, dP \qquad (10.11.4)$$

where P_0 is the equilibrium pressure of the pure end member reaction at T', and $(\Delta_r V^\circ)_{T'}$ is the volume change of the reaction at T' as a function of pressure. The first term on the right of the second equality is zero since it represents the Gibbs energy change of the pure end-member reaction at an equilibrium P-T condition. Thus, Eq. (10.11.2) reduces to

$$\Delta_r G(P, T', X) = \int_{P_0}^{P} (\Delta_r V^\circ)_{T'} \, dP + RT' \ln Q(P, T', X) \qquad (10.11.5)$$

Now setting $P = P_e$, where P_e is the equilibrium pressure of the impure phase reaction at T' (Fig. 10.22), we can write

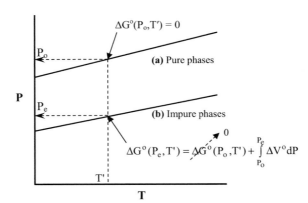

Fig. 10.22 Schematic set-up of the derivation of equation to calculate the displacement of equilibrium boundary of a reaction due to compositional effect. (**a**) Location of the equilibrium boundary when the reaction involves only pure phases; (**b**) displacement of the equilibrium boundary due to compositional effect. ΔG°: Gibbs energy change of the reaction when all phases are in their pure states

$$\Delta_r G(P_e, T', X) = 0 = \int_{P_o}^{P_e} (\Delta_r V^o)_{T'} dP + RT' \ln K(P_e, T') \qquad (10.11.6)$$

where we have now used K for Q_{eq}, following usual convention. We now decompose K into collections of compositional (K_x) and activity coefficient (K_γ) terms in the same form as in Eq. (10.11.3) so that $K = K_x K_\gamma$. To reiterate, K is not a function of composition at fixed P-T condition, but that does not imply that K_x and K_γ are independent of compositions.

For an ideal solution with respect to the chosen standard state, $\gamma = 1$ for all components, and consequently the solution for P_e at T' can be easily obtained from the last equation with the value of Q_e calculated simply from the equilibrium compositions of the phases according to Eq. (10.11.3). For reactions involving only solid phases, $\Delta_r V^o$ varies quite weakly as a function of pressure since the volumes of individual phases tend to have similar pressure dependence. Thus, for this type of reactions, little error is usually introduced in the calculation of P_e if $\Delta_r V^o$ is assumed to be constant.

For the general case of a reaction involving non-ideal solutions, we need to further develop Eq. (10.11.6) by expressing RTlnQ as a function of pressure. Since by our choice of standard state, $\mu_i^*(T) = \mu_i^o(P, T)$, we have from Eq. (8.4.11)

$$RT \left(\frac{\partial \ln \gamma_i}{\partial P} \right)_{T,X} = \left(\frac{\partial \mu_i}{\partial P} \right)_{T,X} - \left(\frac{\partial \mu_i^o}{\partial P} \right)_{T,X} \qquad (10.11.7)$$
$$= \overline{V}_i - V_i^o$$

so that

$$RT \ln \gamma_i(P, T, X) = RT \ln \gamma_i(1, T, X) + \int_1^P \overline{V}_i dP - \int_1^P V_i^o dP \qquad (10.11.8)$$

where $\overline{V}_i$ and V_i^o are, respectively, the partial molar volume and molar volume of the component i in a phase. (Throughout this and the following sections of the chapter, the symbol $\overline{V}_i$, has been used for partial molar volume of a component, instead of the symbol v_i used elsewhere in the book, so that it is not confused with the stoichiometric symbol v).

Using the last expression, the ratio of activity coefficients, K_γ can be expressed as

$$RT \ln K_\gamma(P, T, X) = RT \ln K_\gamma(1, T, X) + \int_1^P (\Delta_r \overline{V})_T dP - \int_1^P (\Delta_r V^o)_T dP$$

$$(10.11.9)$$

Substituting this relation in Eq. (10.11.6), we obtain the **general equation** for calculating the equilibrium pressure, P_e, for impure phase reaction as

$$
RT \ln\left[K_x K_\gamma (1, T, X)\right] + \int\limits_{1}^{P_e} (\Delta_r \overline{V})_T dP - \int\limits_{1}^{P_o} (\Delta_r V^o)_T dP = 0 \qquad (10.11.10)
$$

where P_o and P_e are, respectively, the equilibrium pressures at T for the pure phase and impure phase reactions with a specified value of K_x. To follow the derivation of the last equation, note that

$$
\int\limits_{P^o}^{P_e} (\Delta_r V^o) dP - \int\limits_{1}^{P_e} (\Delta_r V^o) dP = - \int\limits_{1}^{P_o} (\Delta_r V^o) dP
$$

Again, to a very good approximation, we may assume both $\Delta \overline{V}$ and ΔV^o to be insensitive to pressure for solid-solid reactions, at least within the range of crustal pressures. Furthermore, the deviations of $\Delta_r \overline{V}$ and $\Delta_r V^o$ from constant values with increasing pressure should be similar and, thus, substantially self-compensating since the two terms have opposite signs. In addition, for geological problems we typically have $P_o \gg 1$ and $P_e \gg 1$ so that the last two integrals in Eq. (10.11.10) may be simplified as $P_e(\Delta_r \overline{V})$ and $P_e \Delta V^o$, respectively. The P-T effects on $\Delta_r V$ of solid-solid reactions, however, need to be taken into account when the calculations are extended to the conditions in the Earth's mantle.

Problem 10.13 Derive Eq. (10.11.10) choosing pure component (1 bar, T) standard state.

Solved Problem: As an example for the calculation of the displacement of the equilibrium P-T condition of reaction (10.11.a) due to compositional changes of the phases, let us calculate the equilibrium P-T curve using the mineral compositions given for an assemblage (sample No. 64/86) of garnet, plagioclase, sillimanite and quartz from the Sikkim, Himalayas, by Ganguly et al. (2000). The garnet in the above sample is compositionally zoned in that the grains have uniform core compositions, but there is compositional zoning within ~ 100 μm of the grain boundary that is due to the readjustment of garnet composition to the changing P-T conditions during the exhumation of the rocks. (This is typical of garnet compositions from metamorphic rocks that were exhumed from granulite facies conditions where the temperature is high enough, >650 °C, to lead to homogeneity of garnet compositions during metamorphism). It was argued by Ganguly et al. (2000) that at the peak metamorphic condition that we want to reconstruct, the observed plagioclase composition was in equilibrium with the observed composition of the garnet core. (Dasgupta et al. 2004, presented a detailed discussion of how to choose equilibrium compositions of compositionally zoned garnets and other minerals to retrieve peak metamorphic conditions). The average plagioclase and garnet-core compositions of the sample, as obtained from several spot analyses by an electron microprobe, are as follows.

Garnet: $X_{Fe} = 0.64, X_{Mg} = 0.27, X_{Mn} = 0.02, X_{Ca} = 0.07$
Plagioclase: $X_{Ca} = 0.35, X_{Na} = 0.63, X_K = 0.02$

We calculate the equilibrium P-T condition of reaction (10.11.a), which is often referred to by the acronym GASP (**G**arnet, **A**luminosilicate, **P**lagioclase), using the above mineral compositions across the kyanite-sillimanite transition boundary. First, we treat the aluminosilcate to be kyanite, and then calculate the effect of kyanite-sillimanite transition when the P-T condition falls in the sillimanite field (Fig. 10.23).

Since quartz and aluminosilcate (AS) are essentially pure, we have $a(Al_2SiO_5)^{AS} = a(SiO_2)^{Qz} = 1$, according to our choice of standard states (i.e. pure components at P,T). Using Ionic Solution model (Sect. 9.1.1) for both garnet and plagioclase, $a(Grs)^{Grt} = [(X_{Ca}\gamma_{Ca})^3]^{Grt}$ and $a(An)^{Plag} = (X_{Ca}\gamma_{Ca})^{Plag}$. Thus,

$$K = \frac{a_{Grs}^{Grt}}{(a_{An}^{Plag})^3} = \frac{[(X_{Ca}\gamma_{Ca})^3]^{Grt}}{[(X_{Ca}\gamma_{Ca})^3]^{plag}} \qquad (10.11.11)$$

If we assume ideal solution behavior for both garnet and plagioclase, then $Q_\gamma = 1$ and $\Delta_r\overline{V} = \Delta_r V^o$. Thus, assuming that $\Delta_r V$ to be insensitive to pressure, Eq. (10.11.10) reduces to

$$P_e(T) = P_o(T) - \frac{RT \ln Q_x}{(\Delta_r V^o)_T}$$

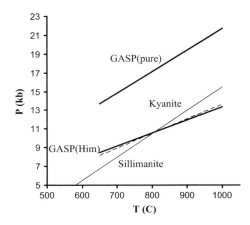

Fig. 10.23 Calculation of displacement of the equilibrium boundary of the reaction 3 Anorthite (An) = Grossular (Gr) + 2 Kyanite (Ky) + Quartz (Qtz) (acronym: GASP) due to compositional effects. GASP(pure): Reaction involving only pure phases, as determined by Koziol and Newton (1988), Fig. 10.21. GASP(Him): Displacement of the equilibrium according to the composition of natural sample from the Himalayas, as discussed in the text. Short dashed line: ideal solution approximation; bold solid line: non-ideal treatment. Light solid line: Kyanite/sillimanite boundary (Holdaway 1971)

At 1 bar, 298 K, $\Delta_r V^o = -66.2$ cm^3. From the given compositional data, $Q_x = (0.07/0.35)^3 = 0.01$. Assuming $\Delta_r V^o$ to be insensitive to temperature, we now solve for $(P_e - P_o)$ as a function of T, and then for P_e, using the expression for P_o as a function of T given in Fig. 10.21. In the field of sillimanite, the calculated equilibrium boundary involving kyanite as the aluminum silicate phase is metastable. The stable boundary involving sillimanite instead of kyanite has been calculated following the procedure for accounting for polymorphic transition discussed in Sect. 6.7.2. However, the difference between these two boundaries is only ~ 400 bars at 1000 °C due to small free energy change associated with the transformation of 2 Ky = 2 Sill that relates the GASP equilibria in the kyanite and sillimanite fields. The calculated equilibrium boundary is illustrated by a dashed curve in Fig. 10.23.

Both garnet and plagioclase are, however, known to exhibit non-ideal mixing behaviors. The activity coefficient terms needed to calculate $RTlnK_\gamma$ (1, T, X) (Eq. (10.11.10)) can be obtained from the available data on the mixing properties of the garnet and plagioclase at 1 bar (Ganguly et al. 1996; Fuhrman and Lindsley 1988, respectively). Both garnet and plagioclase show effectively linear variation of molar volume versus composition, in which case one can use $\Delta \overline{V} \approx \Delta V^o$. The solid line for GASP in Fig. 10.23 illustrates the results for the solution of P_e versus T relation, according to Eq. (10.11.10), incorporating the effects of non-ideal mixing in garnet and plagioclase and treating ΔV as function of pressure and temperature. (These calculations were carried out using a computer program that is available from the author on request). The results are very similar to those obtained using the assumption of ideal mixing for both garnet and plagioclase solid solutions. This indicates substantial self-cancellation between the effects of deviation from ideal mixing in garnet and plagioclase such that $(\gamma_{Ca}^{Grt}/\gamma_{Ca}^{Plag}) \sim 1$ for the compositions of the Himalayan sample, and also justifies the assumption of insensitivity of $\Delta_r V$ to changes of pressure and temperature.

Note that the reaction defining the coexistence of the four phases. Grt, Plag, aluminosilcate and quartz remains univariant even though we have increased the number of components. This is because we have fixed the value of each component in each phase, thereby eliminating the additional degrees of freedom that are introduced by adding the new components.

To constrain the peak P-T condition of the investigated Himalayan rock, we need at least one more relation between P and T to be defined by another independent reaction that has a P-T slope much different from that of GASP. The Fe–Mg exchange reaction between Grt and Bt is well suited for this purpose (the thermodynamics of element exchange equilibrium has been discussed in the next chapter). Simultaneous solution of the GASP and Grt-Bt exchange equilibrium using the observed compositions yields 10.4 kb, 800 °C as the peak P-T condition of the assemblage.

Petrographic observations show that the rock had experienced breakdown reactions of Grt plus Sillimanite to Cordierite ($2Fe_3Al_2Si_3O_{12} + 4Al_2SiO_5 + 5SiO_2 = 3Fe_2Al_4Si_5O_{18}$) and of Spinel to Cordierite ($2FeAl_2O_4 + 5SiO_2 = Fe_2Al_4Si_5O_{18}$) during its exhumation. Following similar procedure as above for the

GASP equilibrium (Fig. 10.23), calculations of the displaced P-T conditions of these equilibria relative to laboratory calibrations define another small region in the P-T space, as illustrated in Fig. 10.24, that the rock must have crossed through during its exhumation. Consequently, exhumation of the rock from ~ 30 km (corresponding to the peak pressure) to 12 km had taken place at nearly isothermal condition. Solution of heat transfer equation, subject to the constraint on the P-T path, yields an exhumation velocity of at least 15 km/year within this depth interval. However, garnet crystals adjacent to biotite in the same sample shows diffusion induced compositional zoning that could not have developed in an isothermal P-T path. Thus, the exhumation velocity must have greatly slowed down at shallower depths. The reconstructed P-T path that yields the observed compositional zoning of garnet is illustrated in Fig. 10.24 (see Ganguly et al. 2000 for further discussions; also it may be noted incidentally that the second stage of slow exhumation has been confirmed by subsequent studies using cosmogenic and radiogenic isotopic systems by Vance et al. 2003 and Harris et al. 2004, respectively). This illustrative example should convey an idea of how thermodynamic calculations may be integrated with petrographic observations and modeling of heat and chemical diffusion to develop insights into the geodynamic history of rocks.

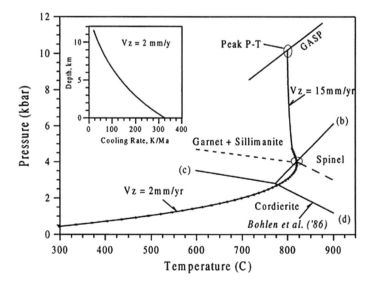

Fig. 10.24 Two stage exhumation paths of the Himalayan sample, as reconstructed from the phase equilibrium calculations and retrograde compositional zoning in garnet. The inset shows the cooling rate as a function of depth for $V_z = 2$ mm/year, which is required to produce the observed retrograde Fe–Mg zoning in garnet. The solid lines labelled (**b**), (**c**) and (**d**) denote the equilibrium reaction boundaries in the FeO-Al$_2$O$_3$-SiO$_2$ system

10.11.2   Reactions Involving Solid Solutions and Gaseous Mixture

10.11.2.1   Thermodynamic Formulations

It is sometimes convenient to treat reactions of this type in terms of fugacities of the gaseous species and activities for the condensed components. With that objective, we decompose the first term on the left of Eq. (10.11.10), which provides the general solution for equilibrium pressure at a fixed temperature, into solid (s) and gas (g) terms as follows.

$$RT \ln \left[K_x K_\gamma (1, T, X) \right] = RT \ln \underbrace{\left[K_x^s K_\gamma^s (1, T, X_s \right]}_{K'(s)} + RT \ln \underbrace{\left[K_x^g K_\gamma^g (1, T, X_g \right]}_{K'(g)},$$

$$(10.11.12)$$

where X_s and X_g stand for the compositions of the solid and gaseous phases, respectively. For brevity, we would henceforth write $K'(s)$ and $K'(g)$ for the terms within the first and second square brackets, respectively, as indicated above.

Let us now consider a simple solid-gas reaction involving only one gaseous species i, such as $A + B = C + D + \nu H_2O$. In that case,

$$RT \ln K'(g) = \nu_i RT \ln a_i^g = \nu_i RT \ln \frac{f_i(1, T, X_g)}{f_i^*(T)} \qquad (10.11.13)$$

In deriving Eq. (10.11.10), we have chosen pure component at P, T of interest as the standard states for each component. In the above equation, the P, T of interest is 1 bar, T. Thus, $f_i^*(T) = f_i^o(1, T)$, and hence

$$RT \ln K'(g) = \nu_i RT \ln \frac{f_i(1, T, X_g)}{f_i^o(1, T)} \qquad (10.11.14)$$

The two $\int \Delta V dP$ integral terms in Eq. (10.11.10) can be decomposed into solid and gaseous components as $\int (\Delta_r V)_s dP + \int (\Delta_r V)_g dP$. Now, from the relationship between fugacity and chemical potential, i.e. $d\mu_i = RT d\ln f_i$, we have

$$\int_1^{P_e} \overline{V}_i dP = RT \ln f_i(P_e, T, X_g) - RT \ln f_i(1, T, X_g) \qquad (10.11.15a)$$

and

$$\int_1^{P_o} V_i^o dP = RT \ln f_i^o(P^o, T) - RT \ln f_i^o(1, T) \qquad (10.11.15b)$$

Here $\overline{V}_i$ and V_i^o are, respectively, the partial molar volume and molar volume of the component i; P_o is the equilibrium pressure at T in the pure end-member system and P_e is that in the impure system.

Using the last three equations and rearranging terms, Eq. (10.11.10) can be written as

$$
\begin{aligned}
& RT \ln K'(s)(1, T, X) + RTv_i \ln \frac{f_i(P, T, X_g)}{f_i^o(P_o, T)} \\
& + \int_1^{P_e} (\Delta_r \overline{V}_s) dP - \int_1^{P_o} (\Delta_r V_s^o) dP = 0
\end{aligned}
\tag{10.11.16}
$$

If the reaction involves more than one gas species, then the second term on the left is replaced by $RT \sum v_i \ln\left(f_i(P, T, Xg)/f_i^o(P_o, T)\right)$.

Solved problem: As an application of Eq. (10.11.6), let us consider an assemblage of staurolite, garnet, kyanite and quartz, as reported by Ghent et al. (1979) from Mica Creek, British Columbia. The equilibrium P-T condition of this assemblage, as deduced by Ghent et al. (1979), is 8.2 kb, 640 °C. The compositions of the mineral phases are as follows. $X_{Fe}(staur) = 0.77$, $X_{Fe}(Grt) = 0.68$, while kyanite and quartz are effectively pure so that $a(Al_2SiO_5)^{ky} = a(SiO_2)^{Qz} = 1$. The garnet is compositionally zoned, and the above composition represents the rim composition of garnet that was in equilibrium with staurolite, which is homogeneous. There is no significant substitution in any site except the divalent cation site in either mineral. We want to estimate the composition of the vapor phase at the P-T condition of formation (equilibration) of the assemblage.

The following reaction relation applies to the observed assemblage of Mica Creek.

$$
\begin{array}{cc}
\text{Staur} & \text{Garnet} \\
6Fe_2Al_9Si_{3.75}O_{22}(OH)_2 + 12.5SiO_2 = & 4Fe_3Al_2Si_3O_{12} + 23Al_2SiO_5 + 6H_2O
\end{array}
$$

$$\tag{10.11.a}$$

The equilibrium reaction boundary is illustrated in Fig. 10.25.

Using ionic solution model, we now write

$$
a(Fe{-}\text{end member})^{Staur} = \left[(X_{Fe}\gamma_{Fe})^2\right]^{staur}
$$

$$
a(Fe{-}\text{end member})^{Grt} = \left[(X_{Fe}\gamma_{Fe})^3\right]^{Grt}
$$

Fig. 10.25 Equilibrium
dehydration boundary of
Fe-staurolite plus quartz, as
calculated by Pigage and
Greenwood (1982) using the
experimental data of Ganguly
(1972) and Rao and Johannes
(1979). With permission from
American Journal of Science

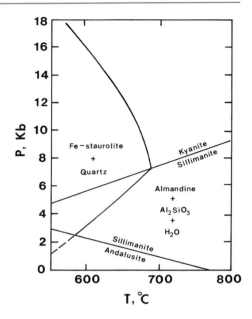

Thus,

$$K'(s) = \left(\frac{X_{Fe}^{Grt}}{X_{Fe}^{Staur}}\right)^{12} \left(\frac{\gamma_{Fe}^{Grt}}{\gamma_{Fe}^{Staur}}\right)^{12}$$

Assuming the ratio of activity coefficients to be approximately unity, which is a very reasonable assumption given the available mixing property data, we have $\ln K'_s \approx 12 \ln(0.68/0.77) = -1.492$. Also, from the available data, $\Delta_r \overline{V} = \Delta_r V^\circ = -148.26\,cm^3$. The first equality implies that the molar volumes of the minerals change linearly with composition. The inferred equilibrium temperature of the Mica Creek assemblage is $\sim\!640\,°C$, at which $P_o = 13{,}200$ bars (Fig. 10.25). Burnham et al. (1969) have experimentally determined the fugacity of pure water as function of pressure and temperature to 10.0 kbar and 1000 °C. Extrapolation of these data yields $f^\circ(H_2O: P^\circ, T) = 27.98$ kbar at the above P_o-T condition. Substitution of the values of K'_s, ΔV and $f^\circ(H_2O)$ in Eq. (10.11.16) then yields

$$P_e(bars) \approx -18{,}620 + 3032 \ln f_{H_2O}(P_e, T, X_v) \tag{10.11.17}$$

As noted above, the inferred pressure (P_e) of this assemblage is 8.2 kbar. Using this result, the fugacity of water and composition of the fluid phase that was in equilibrium with the Mica Creek assemblage can be calculated as follows. The fugacity of water in the vapor phase can be expressed as (Eq. 8.4.8).

$$f_{H_2O}(P_e, T, X_g) = f^o_{H_2O}(P_e, T)a_{H_2O} \qquad (10.11.18)$$

Combining the last two equations and rearranging terms, we then have

$$\ln(a_{H_2O}) = \frac{P_e + 18,620 - 3032\,\ln f^o_{H_2O}(P_e, T)}{3032} \qquad (10.11.19)$$

From the data of Burnham et al. (1969), the fugacity of pure water at the inferred P_e-T condition (8.2 kbar, 640 °C) is 8163 bars. Thus, we obtain $a(H_2O) = 0.85$. If the vapor phase behaves as an ideal mixture, then $a(H_2O) = X(H_2O) = 0.85$.

Occasionally, we are also interested in knowing the displacement of the equilibrium condition of solid-gas reaction due **only** to the effect of a change of **solid compositions**, holding the gas to be in the pure state. As an illustration, let us calculate P_e at 640 °C when the solid phases have the above compositions while the gas phase is pure water. Equation (10.11.17) then transforms to

$$P_e(\text{bars}) \approx -18,620 + 3032\,\ln f^o_{H_2O}(P_e, T), \qquad (10.11.20)$$

The equation is solved by successive approximations, changing the value of P_e until the difference between the right and left hand terms become negligible. This procedure yields $P_e(640\ °C) = 10.2$ kbar.

10.12   Retrieval of Activity Coefficient from Phase Equilibria

As discussed above, calculation of displacement of an equilibrium boundary among a set of phases due to compositional effects requires data on the activity coefficients of the components in solution. It should, therefore, be evident that if the displacement of equilibrium boundary due compositional effects is known, one can retrieve the activity coefficient of a component in a solution if data for the activity coefficients for other components are available. If there are several components with unknown activity coefficients, then one would need data on compositional effects of additional equilibria so that there as many relations as the number of unknowns.

As an example of this procedure, the experimentally determined effect of variation of garnet composition on the equilibrium pressure of the GASP reaction (10.11.a) at 1000 °C, as determined by Ganguly et al. (1996), is illustrated in Fig. 10.26. The garnet composition lies in the ternary system Mg-Mn-Ca, but the Mg/(Mg + Mn) value was kept fixed at 0.68. The experiments were carried out at different pressures at 1000 °C, using starting mixtures of garnet of known composition along with anorthite, kyanite and quartz of effectively pure end-member compositions. Each starting mixture was held for a certain length of time at a fixed pressure, quenched and then analyzed to determine the change of the garnet composition in an electron microprobe. Since at each P-T condition, the initial garnet composition in the starting mixture is different from what it should have been

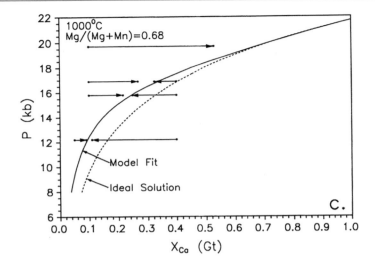

Fig. 10.26 Experimentally determined displacement of the equilibrium pressure of the reaction Grossular (Grs) + 2 Kyanite (Ky) + Quartz (Qtz) = 3 Anorthite (Ann) (GASP) at 1000 °C as a function of Ca content of garnet (Gt) in the ternary system Mg-Mn-Ca with the Mg/(Mg + Mn) value kept fixed at 0.68. The ideal solution line does not adequately fit the experimental constraints on the equilibrium garnet compositions that are contained between the opposing arrow heads at fixed pressure. A model non-ideal fit matching the experimental constraints is shown by a solid line. From Ganguly et al. (1996)

to be in equilibrium with anorthite, kyanite and quartz, the garnet composition evolved towards the equilibrium value through reaction with or breakdown to the other phases in the starting mixture. The initial and final garnet compositions in each experiment are connected by a horizontal line with arrow, with the latter showing the direction of evolution of composition. The equilibrium compositions of garnet at each pressure at 1000 °C were bracketed by approaching these compositions from two opposing sides.

The displacement of equilibrium pressure of a reaction due to the effect of compositional variation under isothermal condition is given by Eq. (10.11.10). Since all phases except garnet are in their respective pure states, $K_x K_\gamma = [(X_{Ca} \gamma_{Ca})^{Grt}]^3$. As K_x, P_o and P_e (i.e. the equilibrium pressure at specified values of K_x and T) are known, one can solve Eq. (10.11.10) to obtain γ_{Ca}^{Grt} at 1000 °C as a function of garnet composition and pressure (P_e). These data can then be cast into an analytical expression of γ according to an appropriate solution model. The 'ideal solution' line in Fig. 10.26 shows the solution of Eq. (10.11.10) using $K_\gamma = 1$. The disagreement between this line and the experimentally constrained equilibrium compositions of garnet implies that the garnet is a non-ideal solution in the Mg-Mn-Ca ternary system. It is left to the reader to calculate the 'ideal solution' line making the reasonable assumption, for the sake of simplicity, that $\Delta \overline{V} = \Delta V° \sim$ constant, and also to retrieve the values of γ_{Ca}^{Grt} at the experimentally constrained P-X conditions at 1000 °C.

In the study of Ganguly et al. (1996), the garnet solution model was retrieved from optimization of a large set of experimental data on the displacement of GASP equilibrium as function of garnet composition within the Fe-Mg-Mn-Ca quaternary system, calorimetric data on the enthalpy of mixing in the Ca-Mg join (Fig. 9.4), effect of garnet composition on K_D(Fe-Mg) (Eq. 11.1.2) in the garnet-olivine and garnet-biotite (Fig. 11.3). The activity coefficient of Ca in garnet, γ_{Ca}^{Grt}, was expressed by a Sub-Regular quaternary model (Sect. 9.3.1), and the interaction parameters were solved numerically using the governing equations for equilibrium (e.g. Eq. 10.11.10) and an optimization program to obtain the statistical best fit to all data. The 'model fit' in Fig. 10.26 describes the calculated equilibrium pressure as function of garnet composition using the retrieved subregular mixing parameters.

As an additional example of the retrieval of activity coefficient from phase equilibria data, note that Eq. (10.6.3) or its complement, Eq. (10.6.4) enables one to constrain the thermodynamic mixing behavior of the system at the liquidus temperatures from the experimentally determined phase equilibrium data. If both solid and liquid phases behave as ideal solutions, $\log(X_i^l/X_i^s)$ is proportional to $(1/T_m - 1/T_o)$. Deviation from this proportional relation implies non-ideality of either or both solid and liquid solutions. If the mixing behavior of one of the phases is known, then the other can be uniquely retrieved as function of temperature within the bounded interval of the experimental data. Figure 10.27 shows the relationship between $\log(X_{Di}^m/X_{Di}^s)$ versus $(1/T_m - 1/T_o)$ derived from the melting relation of diopside-jadeite solid solution at 1400–1650 °C (Ganguly 1973). One finds a proportionality relation between the two sets of variables, implying ideal mixing behavior of the clinopyroxene solid solution at the near solidus conditions.

10.13   Equilibrium Abundance and Compositions of Phases

10.13.1   Closed System at Constant P-T

According to Duhem's theorem (Sect. 10.3), all properties of a closed system are, in principle, completely determined if any two variables are fixed. Thus, if P-T condition of a closed system is defined, then the compositions, abundance of the phases, as well as other intensive properties, such as $f(O_2)$, become fixed. Once the nature, composition and abundance of the phases are known, one can derive other properties of the system from combining the properties of the phases.

Several methods of calculating the abundance of phases in a closed system at fixed P-T condition are discussed by Smith and Misen (1991). The most widely used method involves minimization of Gibbs free energy, G, subject to the bulk compositional constraint. This is a problem of **constrained optimization**, and commonly this type of problems is solved by using the method of **Lagrangian multipliers**. A brief exposition of the theoretical foundation of the method of constrained G minimization using Langrangian multipliers is given below.

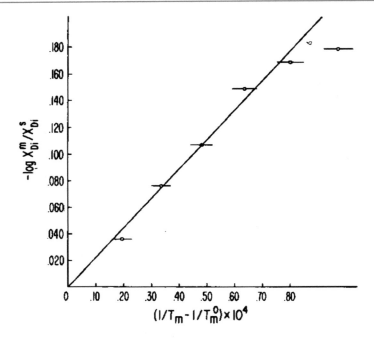

Fig. 10.27 Relationship between $\log(X^m_{Di}/X^s_{Di})$ versus $(1/T_m - 1/T_o)$ derived from the equilibrium compositions of coexisting melt and diopside-jadeite solid solution at 1400–1650 °C (Ganguly 1973). One finds a proportionality relation between the two sets of variables, implying ideal mixing behavior of both the melt and the clinopyroxene solid solution at the near solidus conditions. With permission from Elsevier

Consider a closed system consisting of the components CaO, FeO, MgO, Al_2O_3 and SiO_2 (CMAS), in which the molar abundance of each of these components is specified. We would refer to these oxide components as **basic** components. For example, let us say that there are 95 mol of MgO in the system. Thus, whatever be the nature, composition and abundance of different phases that form at a fixed P-T condition, and regardless of the number of phases (or end-member components of phases) that contain MgO (e.g. garnet: $(Mg, Fe, Ca)_3Al_2Si_3O_{12}$; ortho-pyroxene: $(Mg, Fe, Ca)SiO_3$ etc.), the total number of moles of MgO ($n_{MgO(T)}$) must always be 95. Thus, we write

$$n_{MgO(Pyr)}n_{Pyr} + n_{MgO(Enst)}n_{Enst} + \ldots = 95$$

where $n_{MgO(Pyr)}$ is the number of moles of MgO in a mole of pyrope component $(Mg_3Al_2Si_3O_{12})$ in the mineral garnet, $n_{MgO(Enst)}$ is the number of moles of MgO in a mole of enstatite component $(MgSiO_3)$ in the mineral orthopyroxene, and so on, and n_{Pyr}, n_{Enst} etc., denote the number of moles of the specified components (Pyr, Enst etc.). Now, if N^α stands for the number of moles of a phase α (Grt, Opx etc.), then we can re-write the above mass conservation relation as

$$n_{MgO(Pyr)} \left(X_{Pyr}^{Grt} N^{Grt} \right) + n_{MgO(Enst)} \left(X_{Enst}^{Opx} N^{Opx} \right) + \ldots = 95 \qquad (10.13.1)$$

A similar relation holds for every other basic component in the system. The choice of a set of **basic** components in a system is arbitrary, but must be such that any end-member component of a phase can be expressed as a linear combination of the basic components.

From the general principle of constrained optimization using Lagrangian multipliers, it can be shown that at the minimum of G, subject to the mass conservation constraints, the following relation must be satisfied (e.g. Ganguly and Saxena 1987).

$$\boxed{\mu_i^\alpha - \sum_k \lambda_k n_{k(i)} = 0} \qquad (10.13.2)$$

where $n_{k(i)}$ is the number of moles of the basic component k in the end-member formula unit i (e.g. $n_{MgO(Pyr)}$ or $n_{MgO(Enst)}$ in Eq. (10.13.1)), and λ_k is a constant that is known as the Lagrangian multiplier of the basic component k. The constant multipliers are, however, of no practical interest in our problem, but constitute an essential part of the mathematical method.

To illustrate the procedure of constrained minimization using Lagrangian multipliers, let us consider the end-member components in garnet, namely, grossular (Grs: $Ca_3Al_2Si_3O_{12}$), almandine (Alm: $Fe_3Al_2Si_3O_{12}$) and pyrope (Prp: $Mg_3Al_2Si_3O_{12}$). According to the last expression, we have

$$\left. \begin{array}{l} \text{Gros(Grt):} \ \mu_{Gros}^{Grt} - \lambda_{CaO}(3) \ - \lambda_{Al_2O_3}(1) - \lambda_{SiO_2}(3) = 0 \\[4pt] \text{Alm(Grt):} \ \mu_{Alm}^{Grt} \ - \lambda_{FeO}(3) \ - \lambda_{Al_2O_3}(1) - \lambda_{SiO_2}(3) = 0 \\[4pt] \text{Pyr(Grt):} \ \mu_{Pyr}^{Grt} \ - \lambda_{MgO}(3) - \lambda_{Al_2O_3}(1) - \lambda_{SiO_2}(3) = 0 \end{array} \right\} \qquad (10.13.3)$$

where, using the "ionic solution" model (Sect. 9.1)

$$\mu_{Gros}^{Grt} = \mu_{Gros}^o(P, T) + 3RT \ln X_{Ca}^{Grt} + 3RT \ln \gamma_{Ca}^{Grt}, \qquad (10.13.4)$$

and so on. Similar relations can be written for every end member component of other phases. For every phase α, there is also a stoichiometric relation of the type

$$\sum_i X_i^\alpha = 1 \qquad (10.13.5)$$

The system of Eqs. (10.13.1), (10.13.2) and (10.13.5) can be solved numerically to obtain the compositions and abundances of the various phases and the values of the Lagrangian multipliers. There are exactly as many independent equations as the number of unknowns in these relations. However, the non-linearity of the system of equations for each component analogous to Eq. (10.13.4) poses major technical problems when there are many phases and many components in the system. In addition, the free energy surface of systems consisting of phases with highly non-ideal mixing properties has local minima. One needs to introduce suitable algorithm strategies to find the global minimum (e.g. Ghiorso 1994). In general, the problem of G minimization is solved by successive approximations in a system consisting of many phases and many components. From the possible set of phases, one first chooses a suitable subset and finds the combination of phases and their compositions within the subset that yields a minimum of G. The subset is then varied and by successive approximations one finds the combination of phases and their compositions that yields the minimum G for the entire system. For further details about the method of G minimization using Lagrangian multipliers, the reader is referred to Eriksson (1974) and Eriksson and Rosen (1973).

White et al. (1958) proposed a method that avoids the problem of nonlinearity in the Langrangian method by approximating the continuous G-X surface of each (solution) phase by stepwise variation of G among a set of arbitrarily defined points or "pseudocompounds" (Fig. 10.28). Because there are no compositional degrees of freedom associated with the pseudocompounds, the approximated problem is linear, as discussed below, and can be solved by identifying the pseudocompounds that minimize the Gibbs free energy of the system. The method was essentially ignored as the number of pseudocompounds that are needed to deal with a multicomponent and multiphase system is extremely large, exceeding the capacity of commonly available computers at the time it was proposed. It was later successfully adopted by Connolly (1990, 2005) to develop a program known as perple_x (available on-line) for constrained G minimization problems along with linked data bases for a large set of rock-forming minerals. The program has also been found to be useful for thermomechanical modeling as it computes the bulk material properties from the equilibrium assemblages obtained from G-minimization. The general idea of this "pseudocompound method" is as follows.

In thermodynamics, the molar Gibbs free energy-composition space of a solution is described by the properties of the pure end member phases and logarithmic terms that express the mixing properties among the end member phases (Eq. 8.6.3). As discussed above, one can approximate the G-X relation of a solution by defining pseudocompounds, ps(i), at discrete steps (Fig. 10.28). The molar free energy of

Fig. 10.28 Approximation of the molar Gibbs free energy, G_m, of a solution by the free energies of pseudocomponents spaced at $X = 0.2$, where X is the mole fraction of a component

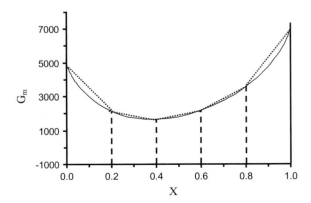

each pseudocomponent, $G_{m,ps(i)}$, is calculated from the true end member free energies and their mixing properties according to Eq. (8.6.3).

The total Gibbs free energy of a system consisting of k components of fixed composition is given by the relation

$$G = \sum_{i=1}^{k} n_i G_{m(i)} \tag{10.13.6}$$

where n_i is the number of moles of the component i that has a molar Gibbs free energy of $G_{m(i)}$. Both conventional stoichiometric phases and pseudocompounds are included in this expression. This equation is linear and can be solved to find the minimum of G subject to the bulk compositional constraints by linear programming technique, such as Simplex (Press et al. 1990). The solution yields the proportion of the stoichiometric phases and the pseudocompounds in the system. The data for the pseudocompounds can be combined to find the composition and abundance of the solution phases. The quality of solution using this approach depends on the choice of pseudocompounds. Using an initially crude spacing of compositions defining the pseudocompounds, one may find an approximate solution, and refine it by choosing more narrow compositional spacing of the pseudocompounds around the initial solution.

Constrained G minimization programs with linked data bases for the rock forming systems (e.g. Holland and Powell 1998) are available on-line. Some of these are Thermo-calc by Holland and Powell, perple_x by Connolly (1990), Theriak-Domino by de Capitani (de Capitani and Brown 1987; de Capitani 1994) and MELT family of programs by Ghiorso and co-workers in the OFM Research web page, the latter being specifically geared to problems of partial melting in the Earth's mantle and planetary systems. Some examples of G minimization calculations in problems of Earth and Planetary Sciences are given below.

Figure 10.29 shows an application of the constrained G-minimization approach to calculate the adiabatic density profile in the Earth's mantle between ∼100 and 800 km depths, corresponding to pressures of 6.42 and 29.38 GPa, respectively (Ganguly et al. 2009). The bulk composition of the mantle is assumed to be given by the pyrolite model of Ringwood (1982) and restricted to the system CaO-FeO-MgO-Al_2O_3-SiO_2 that constitutes ∼98 wt% of the pyrolite composition. The abundance and compositions of the phases were calculated by the method of Lagrangian multipliers (using the program FACTSAGE of Eriksson and Pelton), and these are converted to densities using appropriate P-V-T relations. The calculated pyrolite density profile is compared with the density profile given by the Preliminary Reference Earth Model (PREM) of Dziewonski and Anderson (1981) that is inverted from the geophysical properties. There is good match between the PREM and the thermodynamically calculated density of the mantle. It is also worth noticing that the calculated density profile shows density jumps at 400 and 660 km depths which also show rapid changes (or discontinuities) of seismic velocities. However, unlike the PREM, the calculated profile shows a small density jump at

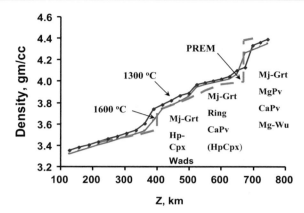

Fig. 10.29 Calculated density versus depth along the adiabatic temperature profile in a pyrolite bulk composition at 1300 and 1600 °C, and comparison with the data in PREM. The vertical lines indicate depths of major mineralogical transformations. Note that with decreasing temperature, the depth of the density jump decreases near 400 km (top of the transition zone) and increases near 670 km (bottom of transition zone). Some of the critical minerals with major influence on the density jumps are indicated within the different fields. Ol: olivine, Grt: Garnet, Cpx: clino-pyroxene, Opx: orthopyroxene, HP-Cpx: high pressure clinopyroxene, Wads: wadsleyite (β-Mg$_2$SiO$_4$), Rng: Ringwoodite (γ-Mg$_2$SiO$_4$), Ilm: Ilmenite, Mg-Pv: Mg-perovskite, Ca-Pv: Ca-perovskite, MgWu: magnesiowüstite; Mj-Grt: majoritic garnet

500 km depth. There is a seismic discontinuity at this depth, but it is not reported to be a global discontinuity. The density discontinuity should, however, be global. The non-global nature of the 500 km seismic discontinuity may be due to failure to resolve a small discontinuity everywhere.

Figure 10.30 illustrates phase relations as function of temperature and water content at a fixed pressure of 6 kb for the bulk composition of a sample from Sikkim in eastern Himalaya that was calculated using the perple_x software (Sorcar et al. 2014). These calculations show that the rock must have had at least ~ 2 wt% water for the observed mineral assemblage, shown within a rectangle in the Figure, to be stable. This is a very important piece of information since the water content of a rock needs to be specified for calculation of P-T relations.

Figure 10.31 shows the results of calculation using the MELTS software for equilibrium and fractional crystallization of primitive mid-ocean ridge basalt or MORB (Ghiorso 1997) (also see the next section for another example of MELTS application). The upper panels show the mineral abundances as function of temperature at 1 kb for equilibrium and fractional crystallization of MORB. The abundances for fractional crystallization are calculated by extracting the crystals at 2–5 °C steps, recalculating the residual melt composition at each step, and allowing these to crystallize as closed systems. The lower panel shows the evolution of melt composition during equilibrium (left panel) and fractional (right panel) crystallizations.

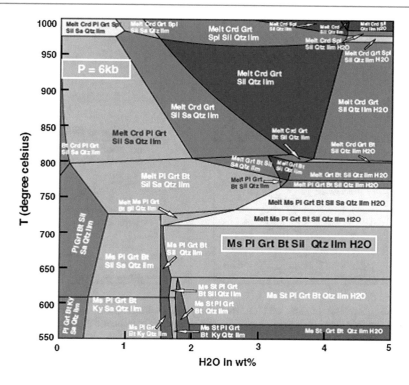

Fig. 10.30 Calculated phase relations for the bulk composition of a sample from the Sikkim Himalaya as function of temperature and water content at a fixed pressure of 6 kb that was inferred from geobarometric calculation. The observed assemblage is shown in the right hand side. Mineral abbreviations: Ms: muscovite, Pl: plagioclase, Grt: garnet, Bt: biotite, Sill: sillimanite, Qtz: quartz, Ilm: ilmenite, Crd: cordierite. From Sorcar et al. (2014)

As an example of the application of G-minimization calculation to planetary problem, we show in Fig. 10.32 the calculations by Saxena and Hrubiak (2014) on the condensation of the solar nebula as function of the P-T condition along the nebular adiabat. The calculations were carried out in the multicomponent system that represents the bulk composition of the solar nebula, and allowing for the effect of solid solutions. The bulk elemental composition and uncompressed or zero pressure density (ρ_o) of the condensates at different P-T conditions along the adiabat have been calculated from their modal abundance and composition, as obtained from the G-minimization calculations. The side bar illustrates the elemental composition of the Earth, as estimated by Allegre et al. (1995). Comparison of the estimated ρ_o (3.96 g/cm^3) by Stacey (2005) and elemental composition of the Earth with the calculations presented in the Fig. 10.31 suggests $\sim$750 K, 10^{-3} bars as the condensation condition of the Earth from the solar nebula. Similar exercise suggests $\sim$1500 K and pressure of slightly greater than 10^{-2} bars as the condition of formation of Mercury ($\rho_o = 5.0$ g/cm^3, 64 wt% Fe) and 600 K, $10^{-3.5}$ bars as

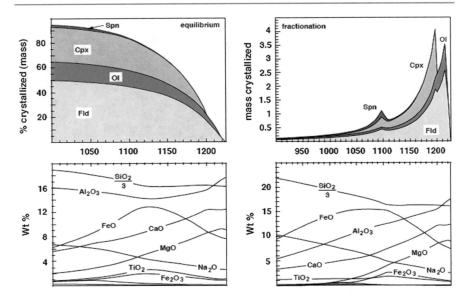

Fig. 10.31 Equilibrium (left) and fractional (right) crystallizations of mid-ocean ridge basalt (MORB), as calculated by minimizing Gibbs free energy of the system using the MELT software (Ghiorso and Sack 1995). The upper panels show the modal abundances of the phases as function of temperature at 1 kb, while the lower panels show the corresponding evolution of the melt compositions. From Ghiorso (1997)

that for Mars (ρ_o = 3.70 g/cm^3) (ρ_o values are also taken from Stacey 2005). Considering the uncertainties of the ρ_o data (Stacey 2005), these conditions are comparable to those illustrated in Fig. 10.3 that is based on simpler calculations relating to the appearance of key minerals of end-member compositions on the nebular adiabat.

10.13.2   Closed System at Constant V-T

While computation of equilibrium assemblages of closed systems at constant P-T condition is of greatest interest in dealing with natural processes, in geological problems, we also encounter cases that in which the computation needs to be carried out for closed systems for which V-T, instead of constant P-T, are held constant. An example of such a problem is the evolution of a magmatic system confined by rigid walls. Obviously, in such cases, it is the Helmholtz Free Energy, F, instead of Gibbs Free Energy, G, that is to be minimized to determine the equilibrium phase assemblages and track change of pressure of the system as the magma cools and undergoes crystallization. An interesting example of this is the study carried out by Fowler and Spera (2008) for isochoric (constant V) cooling of

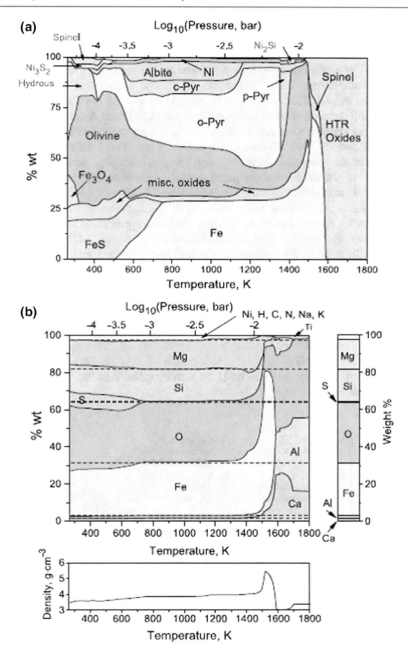

Fig. 10.32 (a) Calculated modal abundance (wt%) of minerals that would form under equilibrium condition from a gas phase of the composition of the solar nebula as a function of P-T condition along the nebular adiabat. (b), (c) are the corresponding elemental abundances and zero pressure density of the condensates. The side bar shows the estimated bulk composition of the Earth according to Allegre et al. (1995). c-Pyr, o-Pyr, p-Pyr: clino-, ortho- and proto-pyroxene, respectively. From Saxena and Hrubiak (2014)

several acid igneous systems confined within rigid walls to gain insight into the mechanisms of explosive volcanisms.

The calculated change of pressure in the magma chambers resulting from cooling and crystallization processes in four acid igneous systems, namely the Bandelier Tuff, Bishop Tuff, Yellowstone Tuffs and Campanian Ignimbrites, are illustrated in Fig. 10.33. Fowler and Spera (2008) carried out these calculations using the MELTS algorithm of Ghiorso and Sack (1995) and Asimow and Ghiorso (1998). There are two scenarios explored for each system by fractionating (a) only the crystals and (b) both crystals and exsolved fluid. The exsolved phases were isolated from the melt and only the melt composition was treated as the bulk composition of the system for the purpose of minimization of F at constant V-T.

It is found that initially isochoric crystallization leads to a decrease of pressure exerted by the magma on the wall rock. A likely consequence of this phenomenon, as suggested by Fowler and Spera (2008), could be episodes of implosive assimilation during the early stages of solidification. However, as the temperature approaches the solidus, the pressure on the wall rock increases rapidly (see Fig. 7.4 and the associated discussion to understand the physical reason) to values that are sufficient to cause magma fracturing, propagation of magma filled cracks or formation of dykes, and ultimately explosive eruptions. In reality, evolution of a magma-hydrothermal system upon cooling lies between the extremes of isobaric and isochoric crystallization. Fowler and Spera (2008) explored both extremes and found near solidus dynamic instability leading to violent eruptions to be a natural consequence of crystallization and fractionation of crystals plus/minus vapor in both cases; however, the pressure build up is more extreme in the case of isochoric crystallization.

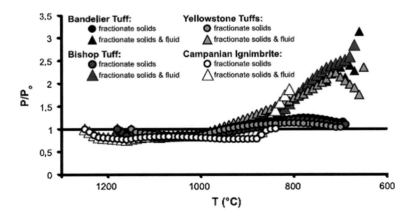

Fig. 10.33 Pressure versus temperature relations at constant volume in four natural systems, as calculated by Fowler and Spera (2008) by minimizing Helmholtz free energy (F) at discrete temperature steps. P/P_o represents the pressure exerted on the wall rock normalized by initial pressure, P_o, before the onset of cooling

10.13.3 Minimization of Korzhinskii Potential

Aside from the closed system cases discussed above that require minimization of either the Gibbs free energy (G) or Helmholtz free energy (F) for computation of equilibrium assemblage of phases, in geological problems we sometimes encounter situations that require minimization of new types of potentials since the stipulated conditions are different from those that we have considered above. Two such cases are (a) system in which chemical potentials of one or more components are held constant through communication with an external reservoir, and (b) magmatic assimilation process at condition of constant enthalpy, pressure and mole numbers of all components. These problems were addressed by Ghiorso and Kelemen (1987). As argued by these authors, isenthalpic calculations provide a close approximation to the magmatic process in which the assimilates are less refractory than the minerals with which the magma is saturated. The appropriate potentials to minimize in these processes are to be found by Legendre transformation of suitable functions.

The general principle of Lagendre transformation has been discussed in Sect. 3.1, and applied to derive the thermodynamic potentials H, F and G. To recapitulate, the partial Legendre transform of a function $Y = Y(x_1, x_2, x_3, \ldots)$ is given by

$$(I_{x_i})_{x_j \neq x_i} = Y - \left(\frac{\partial Y}{\partial x_i}\right)_{x_j \neq x_i} x_i \qquad (10.13.7)$$

where $(I_{x_i})_{x_j \neq x_i}$ is the partial Legendre transform of Y with respect to x_i when all other variables $(x_j \neq x_i)$ are held constant.

It was shown in Sect. 3.1 that if $Y = U$ (internal energy) and $m = \pm(\partial Y/\partial x_i)$, then the equilibrium condition at constant m and $x_j \neq x_i$ is given by the minimization of the partial Legendre transform I_{x_i} Following this lead, we can now construct the appropriate potentials to minimize for the cases (a) and (b).

For the case (a), the chemical potential of a component k along with P and T are to be held constant. Thus, the derivative term in Eq. (11.13.7) must equal $\pm\mu_k$ at constant P, T and $n_j \neq n_k$ (that is at constant mole numbers of all components other than n_k). Now, since $\mu_k = (\partial G/\partial n_k)_{P,T,n_{jk}}$, the function Y corresponds to G, and the Legendre transform of the function is given by

$$(I_{n_k})_{P,T,n_j \neq n_i} = G - \left(\frac{\partial G}{\partial n_k}\right)_{P,T,n_j \neq n_i} n_k \qquad (10.13.8)$$
$$= G - \mu_k n_k$$

It can be easily shown that the transformed function $(G - \mu_k n_k) < 0$ for any spontaneous process and reaches a minimum when equilibrium is achieved, that is $(G - \mu_k n_k) \leq 0$. It is left to the reader as an exercise to show this property. These types of potentials that are to be minimized to compute equilibrium assemblages in an open system with fixed chemical potentials of the mobile components are often referred to as **Korzhinskii potentials** in the geochemical and petrological literature, in recognition of the fact that these potentials were first introduced by the Russian petrologist, Korzhinskii (1959) (Korzhinskii's ideas were later refined and more clearly formalized by Thompson (1970)). The components whose chemical potentials are fixed by an external reservoir were called "perfectly mobile components" by Korzhinskii and K-components by Thompson (1970). If there are several such components, then the last term in the above equation is given by $-\Sigma \mu_k n_k$.

For the case (b), we need to find a function that can be differentiated with respect to an appropriate variable at constant P and n (number of moles of all components) to yield H. The function is G/T. It is left to the reader to show that $H = ((\partial G/T)/\partial(1/T))_{p, n}$. Thus,

$$
\begin{aligned}
\left(I_{1/T}\right)_{p,n} &= \frac{G}{T} - \left(\frac{\partial(G/T)}{\partial(1/T)}\right)_{p,n} \frac{1}{T} \\
&= \frac{G}{T} - \frac{H}{T} = -S
\end{aligned}
\tag{10.13.9}
$$

The magma and assimilates are taken to constitute an isolated system. Now since for such a system, $dS \geq 0$ (second law: Eq. 2.4.8), S is maximized and consequently the transformed function $(I_{1/T})_{p, n}$ is minimized. Thus, the equilibrium phase assemblage and composition for isenthalpic assimilation is obtained by maximizing the entropy of the system at constant P and n.

Figure 10.34 shows an example of the consequence of isenthalpic assimilation on the modal abundance of phases crystallized from a MORB (Mid-Ocean Ridge Basalt). Here the assimilates are taken to be pelites at an initial temperature of 500 °C. The bulk compositions of the assimilates A and B are identical except that B includes 1.35 wt% of H_2O, whereas A is anhydrous. Mineralogically A is composed of quartz-ilmenite-K-feldspar-orthopyroxene-spinel-plagioclase, whereas B is composed of quartz-ilmenite-muscovite-biotite-garnet-plagioclase. The calculations presented in Fig. 10.33 show that assimilation of a hydrous assemblage results in lower final temperature (lower T-axis in the figure) and a smaller mass of crystals compared to those for an equivalent anydrous assemblage of the same mass. As discussed by Ghiorso and Kelemen (1987) assimilation of wall-rock in magma may fundamentally alter the modal abundance and nature of the crystallizing minerals and the consequent evolution of magma composition.

*Problem 10.14 Show that at constant P, T condition, the Korzhinskii potential $(G - n_k \mu_k) \leq 0$ if the chemical potential of the mobile component k is fixed through communication with an external reservoir while the mole numbers of all other components are held constant. (Hint: see Sect. 3.2, and consider U to depend on the mole numbers of components).

Fig. 10.34 Cumulative mass of solid crystallized for isenthalpic assimilation of pelites with initial temperature at 500 °C in a magnesian mid-ocean ridge basalt (MORB: FAMOUS 527-1-1) at 3 kbar. Both A and B have the same bulk composition except that B includes 1.35 wt% H$_2$O). Ma/Mc refers to the ratio of mass assimilated to mass crystallized. See text for further details. From Ghiorso and Kelemen (1987)

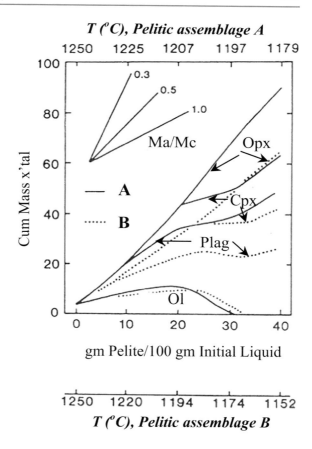

gm Pelite/100 gm Initial Liquid

References

Akella J, Ganguly J, Grover R, Kennedy GC (1972) Melting of lead and zinc to 60 kbar. J Phys Chem Solids 34:631–636

Allegre CJ, Poirier J-P, Humler E, Hofmann A (1995) The chemical composition of the Earth. Earth Planet Sci Lett 134:515–526

Anderson GM (2005) Thermodynamics of natural systems. Cambridge University Press, Cambridge

Asimow PD, Ghiorso MS (1998) Algorithmic modifications extending MELTS to calculate subsolidus phase relations. Amer Min 83:1127–1131

Bowen NL (1913) Melting phenomena in the plagioclase feldspars. Am J Sci 35:577–599

Bowen NL (1915) The crystallization of haplodioritic, and related magmas. Am J Sci 40:161–185

Bowen NL, Anderson O (1914) The system MgO-SiO$_2$. Am J Sci 37:487–500

Bowen NL, Tuttle OF (1950) The system NaAlSSi3O8 – KalSi3O8 – H2O. J Geol 58:489–511

Bowen NL, Tuttle OF (1964) Origin of granite in the light of experimental studies in the system NaAlSi$_3$O$_8$-KAlSi$_3$O$_8$-SiO$_2$-H$_2$O. Geol Soc Am Mem 74:153

Brett R, Bell PM (1969) Melting relations in the system Fe-rich portion of the system Fe-FeS at 30 kb pressure. Earth Planet Sci Lett 6:479–482

Burnham CW, Hollaway JR, Davis NF (1969) Thermodynamic properties of water at 1,000 °C and 10,000 bars. Geol Soc Am Special Paper 132:96

Cameron AGW, Pine MR (1973) Numerical models of primitive solar nebula. Icarus 18:377–406

Carmichael DM (1969) On the mechanism of prograde metamorphic reactions in quartz-bearing pelitic rocks. Contrib Miner Petrol 20:244–267

Connolly JAD (1990) Multivariate phase diagrams: an algorithm based on generalized thermodynamics. Am J Sci 290:666–718

Connolly JAD (2005) Computation of phase equilibria by linear programming: a tool for geodynamic modeling and an application to subduction zone decarbonation. Earth Planet Sci Lett 236:524–541

Dasgupta S, Ganguly J, Neogi S (2004) Inverted metamorphic sequence in the Sikkim Himalayas: crystallization history, P-T gradient and implications. J Met Geol 22:395–412

de Capitani C (1994) Gleichgewichts-Phasendiagramme: Theorie und Software. Beihefte zum European J Mineral 72. Jahrestagung der Deutschen Mineralogischen Gesellschaft 6:48

de Capitani C, Brown TH (1987) The computation of chemical equilibrium in complex systems containing non-ideal solutions. Geochim Cosmochim Acta 51:2639–2652

De Donder T (1927) L'Affinite'. Gauthiers-Villars, Paris

Dohem R, Chakraborty S, Palme H, Rammansee W (1998) Solid-solid reactions mediated by a gas phase: an experimental study of reaction progress and the role of surfaces in the system olivine + iron metal. Am Mineral 83:970–984

Dziewonski AM, Anderson DL (1981) Preliminary reference earth model. Phys Earth Planet Inter 25:297–356

Ebel DS, Grossman L (2000) Condensation in dust-enriched systems. Geochim Cosmochim Acta 64:339–366

Ehlers EG (1987) Interpretation of geologic phase diagrams. Dover

Eriksson G (1974) Thermodynamic studies of high temperature equilibria XII. SOLAGASMIX, a computer program for calculation of equilibrium compositions in multiphase systems. Chem Scr 8:100–103

Eriksson G, Rosen E (1973) Thermodynamic studies of high temperature equilibria XIII. General equations for the calculation of equilibria in multiphase systems. Chemica Scripta 4:193–194

Ernst WG (1976) Petrologic phase equilibria. W.H. Freeman and company, San Francisco

Fowler SJ, Spera FJ (2008) Phase equilibria trigger for explosive volcanic eruptions. Geophy Res Lett 35(8). doi:10.1029/2008GL033665

French BM (1966) Some geological implications of equilibrium between graphite and a C-H-O gas phase at high temperatures and pressures. Rev Geophys 4:223–253

Fuhrman ML, Lindsley DH (1988) Ternary feldspar modeling and thermometry. Am Mineral 73:201–206

Ganguly J (1972) Staurolite stability and related parageneses: theory, experiments and appliations. J Petrol 13:335–365

Ganguly J (1973) Activity-composition relation of jadeite in omphacite pyroxene: theoretical deductions. Earth Planet Sci Lett 19:145–153

Ganguly J (1977) Compositional variables and chemical equilibrium in metamorphism. In: Saxena SK, Bhattacharya S (eds) Energetics of geological processes. Springer, Berlin, pp 250–284

Ganguly J, Saxena SK (1987) Mixtures and mineral reactions. Springer, Berlin, Heidelberg, New York, Paris, Tokyo, p 291

Ganguly J, Cheng W, Tirone M (1996) Thermodynamics of aluminosilicate garnet solid solution: new experimental data, an optimized model, and thermometric applications. Contrib Miner Petrol 126:137–151

Ganguly J, Dasgupta S, Cheng W, Neogi S (2000) Exhumation history of a section of the Sikkim Himalayas, India: records in the metamorphic mineral equilibria and compositional zoning of garnet. Earth Planet Sci Lett 183:471–486

Ganguly J, Freed AM, Saxena SK (2009) Density profiles of oceanic slabs and surrounding mantle: Integrated thermodynamic and thermal modeling, and implication for the fate of slabs at the 660 km discontinuity. Phys Earth Planet Int 172:257–267

Ghent ED, Robins DB, Stout MZ (1979) Geothermometry, geobarometry and fluid compositions of metamorphosed calcsilicates and pelites, Mica Creek, British Columbia. Am Mineral 64:874–885

Ghiorso MS (1994) Algorithms for the estimation of phase stability in heterogeneous thermodynamic systems. Geochim Cosmochim Acta 58:5489–5501

Ghiorso MS (1997) Thermodynamic models of igneous processes. Ann Rev Earth Planet Sci 25:221–241

Ghiorso MS, Kelemen PB (1987) Evaluating reaction stoichiometry in magmatic systems evolving under generalized thermodynamic constraints: examples comparing isothermal and isenthalpic assimilation. In: Mysen B (ed) Magmatic processes: physicochemical principles, Geochemical Society Special Publication No. 1

Ghiorso MS, Sack RO (1995) Chemical mass transfer in magmatic processes. IV. A Revised and internally consistent thermodynamic model for the interpolation and extrapolation of liquid-solid equilibria in magmatic systems at elevated temperatures and pressures. Contrib Miner Petrol 119:197–212

Greenwood HJ (1967) Mineral equilibria in the system MgO-SiO_2-H_2O-CO_2. In: Abelson PH (ed) Researches in geochemistry, vol 2. Wiley, New York, pp 542–547

Greenwood HJ (1975) Buffering capacity of pore fluids by metamorphic reactions. Am J Sci 275:573–593

Grossman L (1972) Condensation in the primitive solar nebula. Geochim Cosmochim Acta 36:597–619

Grossman L, Larimer JW (1974) Early chemical history of the solar nebula. Rev Geophys Space Phys 12:71–101

Haggerty SE (1976) Opaque mineral oxides in terrestrial igneous rocks. In: Rumble D III (ed) Oxide minerals, Mineralogical Society of America short course notes, vol 3, chap 8

Harris NBW, Caddick M, Kosler S, Goswami MI, Vance D, Tindle AG (2004) The pressure-temperature-time path of migmatites from the Sikkim Himalaya. J Metamorph Geol 22:249–264

Holdaway MJ (1971) Stability of andalusite and aluminosilicate phase diagram. Am J Sci 271:97–131

Holland TJB, Powell R (1998) An internally consistent data set for phases of petrological interest. J Metamorph Geol 16:309–343

Holloway JR (1981) Volatile interactions in magmas. In: Newton RC, Navrotsky A, Wood BJ (eds) Thermodynamics of minerals and melts. Adv Phys Geochem 1. Springer, pp 273–293

Kim K, Vaidya SN, Kennedy GC (1972) The effect of pressure on the melting temperature of the eutectic minimums in two binary systems: NaF-NaCl and CsCl-NaCl. J Geophys Res 77:6984–6989

Komabayashi T (2014) Thermodynamics of melting relations in the system Fe-FeO at high pressure: Implications for oxygen in the Earth's core. J Geophys Res Solid Earth 119:4164–4177

Kondepudi D, Prigogine I (2002) Modern Thermodynamics: From Heat Engines to Dissipative Structures. John Wiley, New York, p 486

Korzhinskii DS (1959) Physicochemical basis of the analysis of the paragenesis of minerals. Consultants Bureau, New York, p 142

Koziol A, Newton RC (1988) Redetermination of anorthite breakdown reaction and improvement of the plagioclase-garnet-Al_2SiO_5-quartz geobarometer. Am Mineral 73:216–223

Lewis J (1970) Venus: atmospheric and lithospheric composition. Earth Planet Sci Lett 10:73–80

Lewis JS (1974) Temperature gradient in solar nebula. Science 186:440–443

Lewis JS, Prinn RG (1984) Planets and their atmospheres: origin and evolution. Academic Press, p 470

Lewis GN, Randall M (1923) Thermodynamics and free energy of chemical substances. McGraw-Hill, New York

Miyashiro A (1964) Oxidation and reduction in the Earth's crust with special reference to the role of graphite. Geochim Cosmochim Acta 28:717–729

Mueller RF (1963) Chemistry and petrology of venus: preliminary deductions. Science 141:1046–1047

Mueller RF, Saxena SK (1977) Chemical petrology. Springer, Berlin, p 394

Navrotsky A, Ziegler D, Oestrine R, Maniar P (1989) Calorimetry of silicate melts at 1773 K: measurement of enthalpies of fusion and of mixing in the systems diopside-anorthite-albite and anorthite-forsterite. Contrib Min Petrol 101:122–130

Olsen E, Mueller RF (1966) Stability of orthopyroxenes with respect to pressure, temperature, and composition. J Geol 74:420–625

Philpotts AR (1990) Principles of igneous and metamorphic petrology. Prentice Hall, New Jersey, p 498

Pigage LG, Greenwood HJ (1982) Internally consistent estimates of pressure and temperature: the staurolite problem. Am J Sci 282:943–968

Press WH, Flannery BP, Teukolsky SA, Vetterling WT (1990) Numerical recipes. Cambridge, p 702

Prigogine I, Defay R (1954) Chemical thermodynamics. Longmans, p 543

Rao BB, Johannes W (1979) Further data on the stability of staurolite + quartz and related assemblages. Neues Jarb Mineral Monatsh 1979:437–447

Ringwood AE (1982) Phase transformations and differentiation in subducted lithosphere: implications for mantle dynamics, basalt petrogenesis, and crustal evolution. J Geol 90:611–643

Robie RA, Hemingway BS, Fisher JR (1978) Thermodynamic properties of minerals and related substances at 298.15 K and 1 bar (10^5 pascals) pressure and at higher temperatures. US Geol Surv Bull 1452:456

Rumble D III (1982) The role of perfectly mobile components in metamorphism. Ann Rev Earth Planet Sci 10:221–233

Ryzhenko B, Kennedy GC (1973) Effect of pressure on eutectic in system Fe-FeS. Am J Sci 273:803–810

Saxena SK, Eriksson G (1986) Chemistry of the formation of the terrestrial planets. In: Saxena SK (ed) Chemistry and physics of terrestrial planets, Adv Phys Geochem, vol 6. Springer, Berlin, pp 30–105

Saxena SK, Hrubiak R (2014) Mapping the nebular condensates and chemical composition of the terrestrial planets. Earth Planet Sci Lett 393–119

Sharp ZD, McCubbin FM, Shearer CK (2013) A hydrogen-based oxidation mechanism relevant to planetary formation. Earth Planet Sci Lett 380:88–97

Smith WR, Missen R (1991) Chemical reaction equilibrium analysis: theory and algorithms. Krieger Pub Co, Florida

Sorcar N, Hoppe U, Dasgupta S, Chakraborty S (2014) High-temperature cooling histories of migmatites from the High Himalayan Cystallines in Sikkim, India: rapid cooling unrelated to exhumation? Contrib Mineral Petrol 167:957–980

Spear FS (1988) The Gibbs method and Duhem's theorem: the quantitative relationships among P, T, chemical potential, phase composition and reaction progress in igneous and metamorphic systems. Contrib Min Petrol 99:249–256

Spear FS, Ferry JM, Rumble D (1982) Analytical formulation of phase equilibria: the Gibbs method. In: Ferry JM (ed) Characterization of metamorphism through mineral equilibria, Rev Mineral Mineral Soc Am, pp 105–145

Stacey FD (2005) High pressure equations of state and planetary interiors. Rep Prog Phys 68:341–383

Thompson JB Jr (1970) Geochemical reactions in open systems. Geochim Cosmochim Acta 34:529–551

Usselman TM (1975) Experimental approach to the state of the core: part I. the liquidus relations of the Fe-rich portion of the Fe-Ni-S system from 30 to 100 kb. Am J Sci 275:278–290

Vance D, Bickle M, Ivy-Ochs S, Kubik PW (2003) Erosion and exhumation in the Himalaya from cosmogenic isotope inventories of river sediments. Earth Planet Sci Lett 206:273–288

Walker D, Stolper EM, Hays J (1981) Basaltic volcanism: the importance of planet size. Lunar Planet Sci Conf Proc 10:1996–2015

White WB, Johnson SM, Dantzig GB (1958) Chemical equilibrium in complex mixtures. J Chem Phys 28:751–755

Winkler HGF (1976) Petrogenesis of metamorphic rocks. Springer, Berlin

Winter JD (2001) An introduction to igneous and metamorphic petrology. Prentice Hall, p 697

Yoneda S, Grossman L (1995) Condensation of CaO-MgO-Al_2O_3-SiO_2 liquids from cosmic gases. Geochim Cosmochim Acta 59:3413–3444

Element Fractionation in Geological Systems

11

Element fractionation between coexisting phases, such as those between two minerals, mineral and melt, mineral and a vapor phase, and molten metal and silicate liquid plays a variety of important roles in geological and planetary problems. In this chapter, we discuss the general thermodynamic formalisms of element fractionation, with illustrative applications to (a) geothermometry, (b) interpretation of rare earth element pattern of basaltic magma, and (c) that of siderophile element abundance in the Earth's mantle that bears on the general problem of magma ocean in the early history of the Earth (siderophile elements are those that preferentially fractionate into a metal phase relative to a silicate).

11.1 Fractionation of Major Elements

11.1.1 Exchange Equilibrium and Distribution Coefficient

A detailed thermodynamic treatment of element fractionation and its extension to fractionation of isotopes of an element, such as ^{18}O and ^{16}O, between coexisting minerals is given by Ganguly and Saxena (1987) that may be consulted for additional details. In general, equilibrium fractionation of two species of the same valence state, i and j (e.g. Fe^{2+} and Mg), between two phases α and β can be treated by an exchange reaction of the form

$$i - \alpha + j - \beta = i - \beta + j - \alpha \tag{11.1.a}$$

As an example, we write the following exchange equilibrium to treat the fractionation of Fe^{2+} and Mg between coexisting garnet and biotite.

$$1/3(\mathbf{Fe_3})Al_2Si_3O_{12}(Grt) + K(\mathbf{Mg})Al_3Si_3O_{10}(OH)_2(Bt)$$
$$= 1/3(\mathbf{Mg_3})Al_2Si_3O_{12}(Grt) + K(\mathbf{Fe})Al_3Si_3O_{10}(OH)_2(Bt) \tag{11.1.b}$$

© Springer Nature Switzerland AG 2020
J. Ganguly, *Thermodynamics in Earth and Planetary Sciences*,
Springer Textbooks in Earth Sciences, Geography and Environment,
https://doi.org/10.1007/978-3-030-20879-0_11

Note that the reaction is balanced in such a way that there is one mole of an exchangeable species (Fe^{2+} or Mg) in each side of the reaction, conforming to the form of the reaction (11.1.a). This type of reaction balancing is not a thermodynamic requirement, but is a matter of practical convenience as it reduces the exponents of all terms in the expression of equilibrium constant to unity. To appreciate it, consider the activity expressions of end-member components in garnet in which the substitutions are restricted to the eight (VIII)-and six (VI)-coordinated sites. According to the ionic solution model (Sect. 9.1 and Eq. 9.1.7b)

$$a_{Alm}^{Grt} = \left[{}^{VIII}(x_{Fe}\gamma_{Fe})^3 \right] \left[{}^{VI}(x_{Al}\gamma_{Al})^2 \right] \gamma_{Alm(rec)} \qquad (11.1.1a)$$

$$a_{Pyr}^{Grt} = \left[{}^{VIII}(x_{Mg}\gamma_{Mg})^3 \right] \left[{}^{VI}(x_{Al}\gamma_{Al})^2 \right] \gamma_{Pyr(rec)} \qquad (11.1.1b)$$

where Alm and Pyr represent the iron and magnesium end-member components of garnet, as written in the reaction (11.1.b), the terms within the square brackets indicate site-mole fraction (x) and site-activity coefficients (γ) and the γ_{rec} terms indicate parts of the activity coefficient terms that are due to reciprocal interactions. The expression of equilibrium constant, K, for reaction (11.1.b) contains the ratio $(a_{Pyr}^{Grt}/a_{Alm}^{Grt})^{1/3}$ that reduces, upon substitution of the above activity expressions, to $[{}^{VIII}(X_{Mg}/X_{Fe})\,{}^{VIII}(\gamma_{Mg}/\gamma_{Fe})]^{Grt}$ (γ'_{rec}), where γ'_{rec} stands for the ratio of the two reciprocal activity coefficient terms, $[\gamma_{Pyr(rec)}/\gamma_{Alm(rec)}]$.

Following the above procedure, the equilibrium constant for an exchange reaction of the type (11.1.a) can be written as

$$K = \underbrace{\left[\frac{(X_i/X_j)^\beta}{(X_i/X_j)^\alpha}\right]}_{K_D} \underbrace{\left[\frac{(\gamma_i/\gamma_j)^\beta}{(\gamma_i/\gamma_j)^\alpha}\right]}_{K_{\gamma(site)}} K_{\gamma(rec)} \qquad (11.1.2)$$

where $K_{\gamma(rec)}$ stands for the collection of all reciprocal activity coefficient terms. From Eq. (9.1.7a), it is easy to see that $K_{\gamma(rec)} = 1$ if the sites that do not participate directly in the exchange reaction are filled by only one type of ions. Conventionally, the collection of mole fraction terms within the first set of square brackets in Eq. (11.1.2) is indicated by the symbol $K_D(i\text{-}j)$ or simply K_D and referred to as a **distribution coefficient**.

Using the last relation, we can write

$$\ln K_D = \ln K(P,T) - \ln K_{\gamma(site)}(P,T,X) - \ln K_{\gamma(rec)}(P,T,X) \qquad (11.1.3)$$

Consequently, in general, K_D is a function of P, T and X. Note, however, $K_{\gamma(site)}$ in the above equation not only reflects the nonideal interactions of i and j, but also of those of i and j with other ions that substitute in the same site. In the special case in which both phases are ideal solutions ($K_\gamma = 1$), K_D is a function only of P and T.

11.1.2   Temperature and Pressure Dependence of K_D

The temperature dependence of $\ln K_D$ can be adequately expressed according to the form of the temperature dependence of $\ln K$. Thus, using Eq. (10.4.11), we write

$$\ln K_D = A + \frac{B}{T} \qquad (11.1.4)$$

It is easy to see that the terms A and B be are proportional to the entropy and enthalpy change of the reaction, respectively. For exchange equilibria involving two mineral solid solutions, the temperature dependence of ΔH and ΔS terms are weak so that $\ln K_D$ varies as a linear function of reciprocal temperature within a temperature interval of at least a few hundred degrees. The reason for the weak temperature dependence lies in the fact that the temperature dependence of ΔH and ΔS is given by ΔC_p (Eq. 3.7.5), which is very small for an exchange equilibrium since both sides of the reaction involve the same phases.

The distribution coefficients are usually calibrated experimentally, and fitted according to the form of the last equation, as illustrated in Fig. 11.1. In this figure, K_D refers to the distribution coefficient of Fe^{2+} and Mg between orthopyroxene and

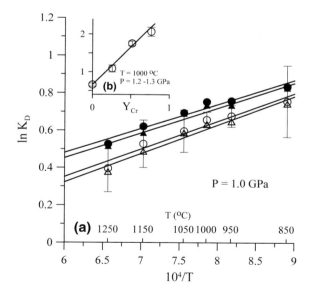

Fig. 11.1 Experimentally determined calibration of Fe^{2+}-Mg distribution coefficient, K_D, between orthopyroxene and spinel as a function of temperature. In the *main figure*, the polybaric experimental data have been reduced to a constant pressure of 1.0 GPa according to Eq. (11.1.6), and fitted using the linear form of Eq. (11.1.4). Vertical bars represent $\pm 1\sigma$. Triangles and circles represent different methods of estimation of Fe^{3+} in spinel, with (filled) and without (open) corrections for the effect of Al in Opx. The *inset* shows the dependence of $\ln K_D$ on the mole fraction of Cr, Y_{Cr}, in the octahedral site of spinel. From Liermann and Ganguly (2003)

spinel, and is defined as $K_D = (X_{Fe}/X_{Mg})^{Spnl}/(X_{Fe}/X_{Mg})^{Opx}$ (Liermann and Ganguly 2003), corresponding to the Fe^{2+}-Mg exchange reaction

$$MgAl_2O_4(Spnl) + FeSiO_3(Opx) \leftrightarrow FeAl_2O_4(Spnl) + MgSiO_3(Opx) \quad (11.1.c)$$

The pressure dependence of $\ln K_D$ is obtained by differentiating both sides of Eq. (11.1.3) with respect to pressure. This procedure yields

$$\left(\frac{\partial \ln K_D}{\partial P}\right)_T = -\frac{\Delta \overline{V}}{RT} \quad (11.1.5)$$

where $\Delta \overline{V}$ is the partial molar volume change of the exchange reaction. Derivation of this relation is left to the reader as a problem (see below). However, mineral solid solutions that are of common interest in geothermometry show nearly linear volume versus composition relation. Thus, to a very good approximation, $\Delta \overline{V} \approx \Delta V°$. Also, since the same minerals are involved in both sides of a reaction, the pressure dependence of $\Delta V°$ is quite small even though the volume of an individual standard state component depends significantly on pressure. Thus, we write, for the purpose of geothermometry based on solid state exchange reaction

$$\ln K_D(P) \approx \ln K_D(P^*) - \frac{\Delta V°(P - P^*)}{RT} \quad (11.1.6)$$

This equation was used to normalize the polybaric experimental data for K_D to a constant pressure of 10 kb that are illustrated in Fig. 11.1.

11.1.3   Compositional Dependence of K_D

For non-ideally behaving systems, one adds compositionally dependent terms to the expression of $\ln K_D$ in Eq. (11.1.3) by expanding the K_γ terms, using appropriate solution models (Chap. 9), and thus expresses $\ln K_D$ as function of P, T, and X. Compositional dependence of $K_D(i\text{-}j)$ within a binary or quasibinary system (that is a system that has more than two components, but the contents of all but two components are kept fixed) is usually illustrated by **Roozeboom plots** in which $\overline{X}_i^\alpha$ is plotted against $\overline{X}_i^\beta$ where $\overline{X}_i$ represents the binary mole fraction of i, i.e. i/(i + j). In experimental studies, the system is typically restricted to the i-j binary, so that the binary mole fraction is the same as total mole fraction. Two examples of such plots are shown in Fig. 11.2. These represent data for experimentally determined Fe/Mg fractionation between (a) spinel, $(Fe, Mg)Al_2O_4$, and orthopyroxene, $(Fe, Mg)SiO_3$, (Liermann and Ganguly 2003) and (b) olivine, $(Fe, Mg)_2SiO_4$, and orthopyroxene, $(Fe, Mg)SiO_3$ (von Seckendorff and O'Neill 1993). Each data set is at a constant P-T condition, as shown in the figures. In these experiments, a starting material consisting of two phases of known compositions is held at fixed P-T condition so that the minerals exchange Fe and Mg and evolve towards compositions that represent an equilibrium pair.

For constant value of K_D, or in other words when $K_D(i\text{-}j)$ is independent of the i/j ratio, the equilibrium distribution curve is symmetrical with respect to the diagonal line connecting X_i^α versus X_i^β, as in Fig. 11.2a. The independence of $K_D(i\text{-}j)$ on i/j ratio implies that either both solid solutions behave ideally (or nearly so) in the binary system ($K_{\gamma(\text{site})} = 1$), or that there is essentially the same departure from ideality of mixing of i and j in both phases so that the effects of nonideal behavior on K_D cancel out. Figure 11.2b, on the other hand, shows asymmetric distribution curve, implying that K_D is a function of Fe/Mg ratio of the phases, and thus, the Fe-Mg mixing property of at least one phase is nonideal. The simplest explanation of the distribution data illustrated in Fig. 11.2 is that the mixing of Fe^{2+} and Mg in both spinel and orthopyroxene is essentially ideal, whereas in olivine these cations mix nonideally. One can retrieve the values of the non-ideal mixing parameters by modeling the compositional dependence of the distribution data.

As an example of the effect of non-ideal mixing of components that substitute in the same site as that occupied by the exchangeable species, we illustrate the effect of Mn on $K_D(\text{Fe-Mg})$ between garnet and biotite. The Fe-Mg exchange reaction is given by (11.1.b), and the effect of Mn of $K_D(\text{Fe-Mg})$ is illustrated in Fig. 11.3. The data illustrated in this figure are for natural samples from Pecos Baldy, New Mexico (Williams and Grambling 1990) that formed at essentially constant P-T condition. Substitution of Mn in biotite relative to that in garnet is negligible so that the observed compositional dependence is due to the non-ideal mixing of Mn in garnet

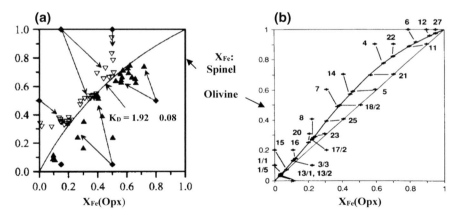

Fig. 11.2 Equilibrium fractionation of Fe^{2+} and Mg between (**a**) spinel and orthopyroxene (T = 1000 °C, P = 9 kbar) and (**b**) olivine and orthopyroxene (T = 1000 °C, P = 16 Kbar) as determined experimentally by Liermann and Ganguly (2003) and von Seckendorff and O'Neill (1993), respectively. In (**a**), the initial compositions are shown by filled diamond symbols and the evolved compositions by triangles. The most evolved compositions can be fitted well by a constant distribution coefficient, K_D, which describes a symmetric curve. In (**b**), the initial compositions are numbered and connected to the evolved compositions by lines. The equilibrium distribution curve is asymmetric which implies that K_D depends on Fe^{2+}/Mg ratio. In both sets of experiments, the equilibrium distribution is constrained by reversal experiments, that is by approaching K_D from two sides of the equilibrium distribution curves. (**a**) From Liermann and Ganguly (2003) and (**b**) from von Seckendorff and O'Neill (1993)

solid solution. According to Ganguly et al. (1996), the mixing of Mn with Fe and Mg can be adequately described by a Regular Solution model. The mixing of Fe and Mg in biotite is known to be almost ideal. Thus, expanding the $RT\ln K_\gamma$ term in Eq. (11.1.3) in terms of a ternary Regular Solution model (Eq. 9.3.3) for garnet, and rearranging terms, we obtain

$$RT \ln K_D = RT \ln K(P,T) + W^G_{FeMg}(X_{Mg} - X_{Fe}) + (W^G_{Mg-Mn} - W^G_{Fe-Mn})X^{Grt}_{Mn}$$

(11.1.7)

where W^G_{i-k} denotes the Regular Solution free-energy interaction parameter between i and k. Substitution of the W parameters from Ganguly et al. (1996) into the above expression leads to the relationship between $\ln K_D$ versus $X_{Mn}(Gt)$ that is illustrated in Fig. 11.3 by a solid line (Ganguly et al. 1996).

To illustrate the effect of cation substitution in a site other than that participating in the exchange reaction, let us consider the experimental data of Liermann and Ganguly (2003) on the effect of Cr on Fe-Mg exchange between spinel and orthopyroxene, as illustrated within the inset of Fig. 11.1. The spinels used in this study can be represented as $^{IV}(Fe^{2+}, Mg)$ $^{VI}(Cr, Al)_2O_4$ where a left-hand roman superscript indicates the co-ordination number of the site. The important point to note that even though Fe^{2+} and Mg mix almost ideally within the tetrahedral site (Fig. 11.2a), variation of Cr/Al ratio affects $K_D(Fe\text{-}Mg)$. This is the effect of reciprocal activity coefficient term in Eq. (11.1.3). Writing the reciprocal parts of the activity coefficients of $FeAl_2O_4$ and $MgAl_2O_4$ according to Eq. (9.1.7), it can be shown from Eq. (11.1.3) that

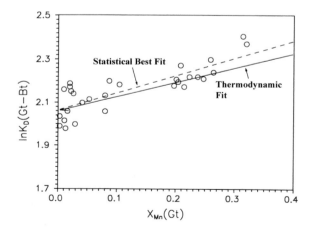

Fig. 11.3 Effect of Mn substitution in garnet on the Fe-Mg distribution coefficient between garnet and biotite in a suite of rocks from Pecos Baldy, New Mexico. The circles represent the measured data by Williams and Grambling (1990), whereas the thermodynamic fit represents the fit according to Eq. (11.1.7) with the interaction parameters from Ganguly et al. (1996). The effect of W_{FeMg} term is negligible

$$\ln K_D(Fe - Mg) \approx \ln K - \ln K\gamma(\text{site}) + \frac{\Delta G^o_{rec}}{RT} Y^{Sp}_{Cr} \qquad (11.1.8)$$

where Y^{spnl}_{Cr} is the atomic fraction of Cr in the octahedral site of spinel (we use the symbol Y to emphasize that the substitution is in a site that is not occupied by the exchangeable ions), and ΔG^o_{rec} is the standard state Gibbs energy change of the homogeneous (reciprocal) exchange reaction in spinel

$$FeCr_2O_4 + MgAl_2O_4 \leftrightarrow MgCr_2O_4 + FeAl_2O_4 \qquad (11.1.d)$$

Thus, if $K_{\gamma(\text{site})} \approx 1$, $\ln K_D$ is expected to vary linearly as function of the atomic fraction of Cr given by $Cr/(Cr + Al)$, conforming to the experimental data.

It can be argued that, in general, ΔG^o_{rec} depends weakly on temperature (Ganguly and Saxena, 1987) so that the value of ΔG^o_{rec} extracted from one temperature may be used to calculate compositional dependence of $\ln K_D(Fe^{2+}-Mg)$ on X_{Cr} at other temperatures. Indeed calculations of $\ln K_D(Fe^{2+}-Mg)$ between spinel and orthopyroxene versus $X_{Cr}(sp)$ at different temperatures using the ΔG^o_{rec} value extracted from the data in the inset of Fig. 11.1 show good agreement with those calculated from more elaborate theory (Liermann and Ganguly 2003).

11.1.4   Thermometric Formulation

If the activity coefficient terms are relatively simple, such as given by a Regular Solution model, then it is possible to derive a simple thermometric expression that can be solved with the aid of a calculator or a computer spreadsheet. For relatively complex solution model, a computer program may be needed. As an illustration of the development of a relatively simple thermometric formulation, let us consider the one formulated by Liermann and Ganguly (2003) using the experimental data on Fe^{2+}-Mg fractionation between spinel and orthopyroxene. They first normalized the polybaric experimental data to a constant pressure of 1 bar, using Eq. (11.1.6), and volumetric data for the end members ($\Delta V^o = -0.628$ cm^3), and fitted the data using the linear form of Eq. (11.1.4). This procedure yields A = $-0.351(\pm 0.102)$ and B = 1217($\pm$ 120) for 1 bar pressure. (This is one of four sets of coupled A and B parameters derived by Liermann and Ganguly (2003) using different methods of estimation of Fe^{3+} iron in their experiments, with and without corrections for the effect of Al substitution in Opx, that yield different values of $K_D(Fe^{2+}-Mg)$. However, this set seems to yield somewhat better temperature estimates than others.) Using Eq. (11.1.6), we then have

$$\ln K_D = -0.351 + \frac{1217}{T} + \frac{7.626(10^{-3})(P-1)}{T} \qquad (11.1.9)$$

where P is in bars. For $P \gg 1$ that represents typical geological situation, $P - 1 \approx P$. Incorporation of the effects of mixing of Al with Fe and Mg in orthopyroxene according to Eq. (11.1.7), with Mn replaced by Al, and of Cr substitution in the octahedral site of spinel, according to Eq. (11.1.8), yields

$$\ln K_D \approx -0.351 + \frac{1217}{T} + \frac{76.26\,P(GPa)}{T}$$
$$- \frac{\Delta W_{Al}(X_{Al})^{Opx}}{RT} + \frac{\Delta G^o_{rec}(Y_{Cr})^{spnl}}{RT} \quad (11.1.10)$$

where the pressure is in GPa, ΔW_{Al} stands for the term $(W_{MgAl} - W_{FeAl})$ in orthopyroxene, and x and y stand, respectively, for the atomic fraction in the site occupied and not occupied by the exchange-cations. This equation is rearranged to yield a thermometric expression

$$\boxed{T \approx \frac{1217 + 76.26\,P(GPa) - C(X_{Al})^{Opx} + D(Y_{Cr})^{spnl}}{\ln K_D - A}} \quad (11.1.11)$$

where $C = \Delta W_{Al}/R = 1863$ K and $D = \Delta G^o_{rec}/R = 2345$ K.

Problem 11.1 Derive Eq. (11.1.5).

Hint: Start with Eq. (11.1.3). Then derive an expression for $\partial \ln\gamma_i/\partial P$ and from that an expression for $\partial \ln K_\gamma/\partial P$.

Problem 11.2 Expand Eq. (11.1.7) to include the effects of non-ideal interaction of Ca with other cations (Fe, Mg and Mn) in garnet, using Regular Solution model (Eq. 11.1.7). There will be an additional term similar to the last term in Eq. (11.1.7) involving X_{Ca} in Grt.

11.2   Trace Element Fractionation Between Mineral and Melt

11.2.1   Thermodynamic Formulations

Trace element patterns of basaltic magma offer significant information on their genetic relations, as first shown in a pioneering paper by Gast (1968). Since then experimental determination and theoretical estimation of trace element fractionation between mineral and melt and modeling of the observed trace element patterns of melts have become a very active field. The thermodynamic treatment of trace element fractionation between mineral and melt relies on the expected Henry's law behavior of solute at high dilution (trace element), and the principle of conservation of mass. To treat the equilibrium fractionation of an element or an ion between liquid and solid, we write a fusion reaction of the form

$$i(\text{liquid}) \leftrightarrow i(\text{solid}) \quad (11.2.a)$$

If i is a trace element, then it is convenient to express its content in a phase in terms of ppm, that is $gm/10^6$ gm instead of its mole fraction which is an extremely small quantity. As discussed earlier (Sect. 8.4), the activity expression can be cast in terms of any conveniently chosen measure of the content of a species. Thus, we write for the above reaction

$$K_a(P, T) = \frac{a_i^s}{a_i^l} = \underbrace{\left[\frac{C_i^s}{C_i^l}\right]}_{D_i^{s/l}} \left[\frac{\gamma_i^s}{\gamma_i^l}\right] \tag{11.2.1}$$

where C_i^l stands for the content of i **in terms of ppm**. Conventionally, the ratio of terms within the first square brackets on the right is denoted by the symbol $D_i^{s/l}$, and referred to as a mineral-melt **partition coefficient** of the species i. For the sake of brevity, we will henceforth drop the subscript i from the symbolic representation of partition coefficient and concentration.

Now, since i is a very dilute component, it is expected to satisfy the Henry's law (Eq. 8.8.1), in which case the ratio of the two activity coefficients becomes a constant at fixed P-T-X_{solv} condition (note that adherence to Henry's law means that the $a_i \propto$ [i] in a solvent (solv) of fixed composition.) Consequently, $D_i = f(P, T, X_{solv})$, but independent of [i] within the domain of validity of Henry's law. Thus, we write

$$D^{s/l} = \frac{C^s}{C^l} = f(P, T, X_{solv}) \tag{11.2.2}$$

This is known as the **Nernst distribution law**. In this and following equations, the subscript i for D, C^S, C^l and C^b has been omitted

The principle of conservation of mass requires that

$$C^l X_m^l + C^{s(1)} X_m^{s(1)} + C^{s(2)} X_m^{s(2)} + \ldots + C^{s(n)} X_m^{s(n)} = C^b \tag{11.2.3}$$

where s(j) is a solid phase j, X_m^l and $X_m^{s(j)}$ stand, respectively, for the mass fractions of liquid and solid j in the **total** system, and C^b is the bulk content of i (in ppm, or in whatever unit C^l and C^s have been expressed), that is the content of i in the total system. In the last equation, $C^{s(j)}$ can be expressed in terms of C^l according to

$$C^{s(j)} = D^{s(j)/l}(C^l) \tag{11.2.4}$$

where $D^{s(j)/l}$ stands for the distribution coefficient of i between solid j and liquid at fixed P-T-X_{sol} condition. Equation (11.2.3) then reduces to

$$C^l \left(X_m^l + \sum_{j=1}^n X_m^{s(j)} D^{s(j)/l} \right) = C^b \tag{11.2.5}$$

Denoting the terms within the summation sign collectively as $\langle D \rangle^{s/l}$, and writing, following common practice, X_m^l as F, the above expression can be re-written as

$$\boxed{\frac{C^l}{C^b} = \frac{1}{F + \langle D \rangle^{s/l}}}$$
(11.2.6)

The term $\langle D \rangle^{s/l}$ is a function of F and mass fractions of all solid phases in the total system. A limiting case, commonly known as **modal melting**, is that in which the relative proportions of the solids do not change during melting. In other words, the mass of each solid undergoes the same fractional change during melting, i.e. $X_m^{s(j)}(F) = \alpha X_m^{s(j)}(F = 0)$, where α is a constant fractional quantity. Even in this case, $\langle D \rangle^{s/l}$ changes because the mass fraction of each solid within the total system changes during melting.

It is convenient to recast the last expression in terms of a weighted average distribution coefficient, $\bar{D}^{s/l}$, in which a weighting factor for a distribution coefficient is not the mass fraction of the associated solid in the total system, as for $\langle D \rangle^{s/l}$, but its mass fraction in the **solid part** of the system, $\bar{X}_m^j$, that is

$$\bar{D}^{s/l} = \sum_j \bar{X}_m^j D^{s(j)/l}$$
(11.2.7)

If the individual distribution coefficients remain effectively constant during the melting, then the modal melting process can be treated in terms of a constant weighted average distribution coefficient. Since $\bar{X}_m^j = w^{s(j)}/W_T^s$, where $w^{s(j)}$ and W_T^s are, respectively, the weight of solid j and total weight of all solid phases, we have, from the definitions of $\langle D^{s/l} \rangle$ and $\bar{D}^{s/l}$ in Eqs. (11.2.5) and (11.2.7)

$$\langle D \rangle^{s/l} - \bar{D}^{s/l} = \sum_j D^{s(j)/l}\left(X_m^{s(j)} - \bar{X}_m^j\right)$$
(11.2.8)

Also,

$$X_m^{s(j)} - \bar{X}_m^j = \frac{w^{s(j)}}{W_T^s + w_l} - \frac{w^{s(j)}}{W_T^s} = -\frac{w_l w^{s(j)}}{\left(W_T^s + w_l\right)W_T^s}$$

where w_l is the weight fraction of liquid. Thus,

$$X_m^{s(j)} - \bar{X}_m^j = -F\bar{X}_m^j$$
(11.2.9)

Substituting the last expression into Eq. (11.2.8), we obtain

$$\langle D \rangle^{s/l} = \bar{D}^{s/l} - F\sum_j D^{s(j)/l}\left(\bar{X}_m^j\right) = \bar{D}^{s/l} - F\bar{D}^{s/l}$$
(11.2.10)

Equation (11.2.5) then reduces to

$$\boxed{\frac{C^l}{C^b} = \frac{1}{\bar{D}^{s/l} + F(1 - \bar{D}^{s/l})}} \quad (11.2.11)$$

This expression, which was first derived by Shaw (1970) in a different way, can be used to model the change in the normalized trace element content of melt as function of melt fraction during both melting and crystallization process, using a bulk partition coefficient that remains constant in the case of modal melting if the individual $D^{s(j)/l}$ remains constant.

When the total melt fraction F is maintained in equilibrium with the solid until it is extracted and isolated, as above, the melting process is called **batch melting**. If, however, the melt is extracted continuously, and collected in one place to form a cumulative melt fraction F, the process is called **fractional melting**. The aggregate liquid composition, $\bar{C}^l$, is given by (Shaw 1970)

$$\frac{\bar{C}^l}{C^b} = \frac{1}{F}\left[1 - (1 - F)^{1/\bar{D}^{s/l}}\right] \quad (11.2.12)$$

For **non-modal batch** melting, it can be shown by combining Eq. (11.2.11) and mass balance restrictions (Shaw 1970) that

$$\frac{\bar{C}^l}{C^b} = \frac{1}{\bar{D}_o^{s/l} + F(1 - P)} \quad (11.2.13)$$

where $\bar{D}_o^{s/l}$ is the initial average distribution coefficient and $P = \sum_j p_j D^{s(j)/l}$ with p_j being the mass fraction of the mineral j the melt.

Significant dependence of $D^{s(j)/l}$ on solvent composition in both melt and solid phases has been demonstrated in some cases that are of importance to the melting process beneath mid-ocean ridges (e.g. Salters and Longhi 1999). The above equations may still be used in these cases by changing $\bar{D}^{s/l}$ in a stepwise fashion during the melting or crystallization process. In addition, as discussed by Ottonello (1997), the Henry's law proportionality constant may change at extreme dilution, thereby affecting D, as a consequence of changes of solubility mechanism of the trace elements in solids.

Since the ratio of activity coefficients in Eq. (11.2.1) is constant in the Henry's law limit of solute content, the dependences of $D^{s/l}$ on changes of pressure and temperature conditions are given by those of the equilibrium constant, K. Thus,

$$\frac{\partial \ln D^{s/l}}{\partial(1/T)} = -\frac{\Delta H^o}{R} \quad (11.2.14)$$

and

$$\frac{\partial \ln D^{s/l}}{\partial P} = -\frac{\Delta V^{\circ}}{RT} \tag{11.2.15}$$

where ΔH° and ΔV° are, respectively, the enthalpy and volume change of the fusion reaction (11.2.a) involving pure (or standard state) phases. Since ΔH° is negative, $D^{s/l}$ decreases with increasing temperature at constant pressure. At low pressure, the fusion reaction (11.2.a) is associated with negative volume change. Consequently, $D^{s/l}$ has positive pressure dependence at low pressure. However, because of greater compressibility of liquid relative to the solid, ΔV° would be progressively less negative with increasing pressure, and ultimately change sign at sufficiently high pressure. As a consequence, with increasing pressure, $D^{s/l}$ would initially increase, reach a maximum at a critical pressure ($\Delta V^{\circ} = 0$) and then decrease. (Asahara et al. (2007) found that K_D for the reaction FeO (magnesiowüstite: mw) = Fe (metallic liquid) + O (metallic liquid), which is defined as $K_D = (X_O^{Fe-1})(X_{Fe}^{Fe-1})/X_{FeO}^{mw}$, has minima at ~ 10 GPa at temperatures between 2373 K and 3073 K. These results conform to the above analyses of the pressure dependence of $D^{s/l}$. Note that Ashara et al. (2007) have written the liquid component in the right side of the reaction, which is opposite to what we have done above; hence K_D goes through a minimum, instead of a maximum, at a constant temperature.)

The behavior of $D^{s/l}$ as function of P and T below the critical pressure where $D^{s/l}$ versus P attains a maximum value is illustrated in Fig. 11.4, using the Na and REE (3+) partition coefficient between clinopyroxene and melt, as compiled by McDade et al. (2003). The broken grey line denotes the mantle solidus. As noted by these authors, the partition coefficient for Na increases whereas that for REE decreases

Fig. 11.4 Effect of pressure and temperature on the partition coefficients of Na (dashed lines) and of a trivalent REE (solid line) between clinopyroxene and melt. The thick grey line denotes the mantle solidus. From McDade (2003). With permission from Elsevier

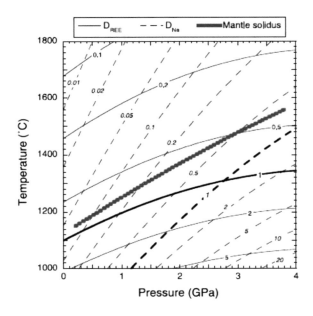

with increasing pressure along mantle solidus. Adiabatic upwelling and melting would be associated with decreasing partition coefficient since the process is nearly isothermal.

11.2.2 Illustrative Applications

11.2.2.1 Trace Element Pattern of Basalt Derived from Garnet-Peridotite

In addition to the problems of dependence of $D^{s/l}$ on solvent composition, pressure and temperature, there may also be significant disequilibrium effects due to slow diffusivity of some elements, e.g. REE in garnet, relative to the time scale of melt extraction. This would prevent complete equilibration between the mineral and melt, and thus affect the relationship between normalized trace element content of melt and the melt fraction or the extent of partial melting of the source rock (Tirone et al. 2005). Despite these limitations, Eqs. (11.2.11), (11.2.12) and (11.2.13) have been proved to be useful in understanding the mineralogy of the source region and the extent of partial melting that gave rise to a certain type of basaltic rock. As an illustration, we calculate the normalized REE content of melt as a function of the extent of partial melting of a garnet-peridotite consisting of the phases olivine (60%), orthopyroxene (20%), clinopyroxene (10%), and garnet (10%). As melting proceeds, $\bar{D}^{s/l}$ changes because of the change in the relative abundances of the minerals, in addition to the effects of changing P-T conditions that lead to the increase of melt fraction. We assume, however, that for small amount of partial melting, $\bar{D}^{s/l}$ remains approximately constant. The results of calculation for 2, 4 and 8% partial melting of garnet peridotite are illustrated in Fig. 11.5. In these

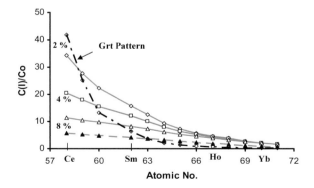

Fig. 11.5 Rare Earth Element patterns of melt, normalized to that of the source region, derived by 2, 4 and 8% partial melting of a garnet peridotite (40% olivine, 20% orthopyroxene, 10% garnet, 10% clinopyroxene). The short-dashed line labeled "Grt Pattern" shows the pattern that would develop if garnet were the only mineral in the source region. The dashed-line connecting filled triangles shows the effect of lowering the individual element distribution coefficient by a factor of two for 8% partial melting

calculations, $D^{s/l}$ values for individual elements are taken from the available data in the literature. We find that for small amount of partial melting the liquid is enriched, relative to the source rock, in light REE (or LREE). The normalized REE pattern of the liquid fans out from the heavy REE (HREE), with a counter-clockwise rotation with increasing partial melting.

The LREE enrichment pattern of liquid shown in Fig. 11.5 represents a signature of the presence of garnet in the source region, and thus constrains a minimum depth within the Earth's mantle for the generation of the melt, since at pressures below 20 kb ($\sim$ 60 km), garnet-peridotite transforms to spinel-peridotite. The effect of garnet on the trace element pattern of the melt is illustrated by showing the pattern (dotted line connecting diamond symbols) that would be generated for 2% partial melting if garnet were effectively the only mineral in the rock so that $\bar{D}^{s/l} = D^{Grt/l}$ for each REE.

The bulk distribution coefficient may change during partial melting due to change in the modal abundances of the minerals and changing P-T conditions as the body of rock undergoing partial melting ascends upwards in the Earth's interior. The effect of lowering of $D^{s(j)/l}$ by a factor of 2 for 8% partial melting is shown in Fig. 11.5 by a dashed line connecting solid triangles. The effect of adiabatic decompression is expected to cause a decrease of $D^{s(j)/l}$, as discussed above.

11.2.2.2   Highly Incompatible Trace Element as Indicator of Source Region of Melt

Trace element with values of $D^{s/l} \ll 1$ are referred to as highly incompatible trace elements since these strongly partition to liquid, and are hence highly incompatible in the crystalline sites ($D^{s/l} > 1$ are referred to as compatible trace element). It is easy to see from Eqs. (11.2.11) and (11.2.13) that the ratio of two highly incompatible trace elements, i and k, in a melt is essentially the same as that in the source rock prior to melting, i.e.

$$\left(\frac{C_i}{C_k}\right)^l = \left(\frac{C_i}{C_k}\right)^o \qquad (11.2.16)$$

Thus, the ratio of highly incompatible trace element provides important constraint about the source region.

11.2.3   Estimation of Partition Coefficient

Because of lack of sufficient data on the dependence of $D^{s/l}$ on pressure, temperature and solvent composition, geochemists often use a constant average value of $D^{s/l}$ to model trace element evolution as a function of melt fraction F, even though $D^{s/l}$ may change substantially as the P, T and solvent (melt and solid) compositions change during the course of melting. Blundy and Wood (1994) developed a method of estimation of trace element partition coefficient by expressing ΔG^o of the fusion reaction (11.2.a) in terms of the lattice strain energy associated with the substitution of a trace element i in the crystal lattice for a host element j. With some modification, but without affecting the final result, the Blundy-Wood model can be developed as follows. Limitations of the model and further improvements using computer simulations have been discussed by Allen et al. (2001).

The standard state Gibbs energy, G_i^o, of either solid or liquid may be represented as a sum of two terms as

$$G_i^o = G_j^o + \Delta G_{j \to i}^o$$

where the first right hand term is the standard Gibbs energy of the phase with a host cation j, and the second right hand term is the free energy change associated with replacing j by i. Thus, we can write for the reaction (11.2.a)

$$
\begin{aligned}
\Delta G_a^o &= G_i^o(s) - G_i^o(l) \\
&= \left(G_j^o(s) + \Delta G_{j \to i}^o(s) \right) - \left(G_j^o(l) + \Delta G_{j \to i}^o(l) \right) \\
&= -\underbrace{\left(G_j^o(l) - G_j^o(s) \right)}_{\Delta G_j^o(f)} + \left(\Delta G_{j \to i}^o(s) - \Delta G_{j \to i}^o(l) \right)
\end{aligned}
\tag{11.2.17}
$$

The first parenthetical term after the last equality is simply the free energy of fusion of the solid with the host cation j, $\Delta G_j^o(f)$.

In the absence of significant chemical effects such as those due to crystal field effect (Sect. 1.7.2) and change of bonding energy, the term $\Delta H_{j \to i}^o$ should be essentially the same as the strain energy, $\Delta E_{strain(j \to i)}$, associated with the complete replacement of the host cation j by the cation i. Also, since the liquid structure is open and flexible, the substitutional strain energy for the liquid should be negligible compared to that for the solid. Using these ideas, and decomposing each free energy term in the last equation into an enthalpic and an entropic term (according to the relation G = H − TS), the equilibrium constant for the fusion reaction (11.2.a) can be expressed as

$$K_a = e^{-\Delta G_a^o / RT} = K_a^o e^{\left(-\Delta E_{strain(j \to i)}^s \right) / RT} \tag{11.2.18}$$

where K_a^o is a function of only P and T, and is given by

$$K_a^o = \left[e^{(\Delta S_j^{o,f} + \Delta S_{j \to i}^o(l) - \Delta S_{j \to i}^o(s)) / R} \right] \left[e^{-\Delta H_j^{o,f} / RT} \right] \tag{11.2.19}$$

Combining Eqs. (11.2.18) and (11.2.1), and rearranging terms, we obtain

$$D_i^{s/l} = \left[K_a^o e^{\left(-\Delta E_{strain(j \to i)}^s \right) / RT} \right] \left[\frac{\gamma_i^s}{\gamma_i^l} \right] \tag{11.2.20}$$

In the Henry's law limit, the activity coefficients become constant at constant P-T and solvent composition. Thus, if the solute i obeys Henry's law, then the γ_i terms can be combined with K_b^o to define a new parameter $D_o^{s/l}$ that is a constant at constant P-T condition and constant solvent composition. Thus, we finally have

$$D_i^{s/l} = D_o^{s/l} e^{-(\Delta E_{strain}(s))/RT} \qquad (11.2.21)$$

Blundy and Wood (1994) expressed the strain energy term according to a relation derived by Brice (1975) for the mechanical strain energy around a homovalent cation defect in an elastically isotropic medium. This leads to an expression of $D_i^{s/l}$ in terms of the radius of the host cation site, r_o, that of the substituting trace element, r_i, and an effective Young's modulus, E, for the crystallographic site in which the substitution takes place. The expression is

$$D_i^{s/l} \approx D_o^{s/l} \exp\left[-4\pi LE\left(\frac{r_o}{2}(r_i - r_o)^2 + \frac{1}{3}(r_i - r_o)^3\right)/RT\right] \qquad (11.2.22)$$

where L is the Avogadro's number and R is the gas constant. According to this relation, $D_i^{s/l}$ versus r_i has a parabola like relation, attaining a maximum value, $D_o^{s/l}$, at $r_i = r_o$, that is when the strain energy associated with a substitution vanishes. Hence, $D_o^{s/l}$ is referred to as **strain-free** partition coefficient.

Figure 11.6 shows good agreement between the experimental data and the form of the relation between $D_i^{s/l}$ and r_i that is predicted by the above expression. Thus, if

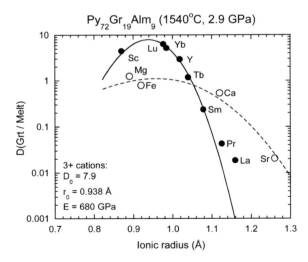

Fig. 11.6 Effect of ionic radius on trace element partitioning between garnet, $Prp_{73}Alm_9Grs_{19}$, and melt. The symbols denote experimental data at 1540 °C, 29 kbar whereas the fitted curves are according to the Blundy-Wood model, Eq. (11.2.22). From Van Westrenen et al. (2000). With permission from Elsevier

sufficient data are available for mineral-melt partition coefficients of cations of the same charge in a given mineral, $D_o^{s/l}$, E and r_o can be retrieved by statistical fitting of the data, and then used to predict the partition coefficient of other cations of the same charge and chemical properties for which no experimental data are available. One can also a priori fix r_o from the size of the best-fit cation in the particular site in which substitutions of the trace elements take place, such as the eight-coordinated radius of Ca if the substitutions take place in the eight coordinated M2 site of clinopyroxene. The magnitude of E relates to the stiffness of the crystallographic site, and determines the tightness of the parabola-like $D_i^{s/l}$ versus r_i curve, with tighter or more compressed form being related to larger value of E.

Using the last expression, one can easily derive a relationship between the partition coefficients of two trace elements, eliminating the strain free partition coefficient, $D_o^{s/l}$. This permits evaluation of the unknown partition coefficient of an element from knowledge of the known partition coefficient of another element in the same mineral/melt system. For heterovalent substitution, i^{m+} substituting for j^{n+} (m $\neq$ n), Blundy and Wood (1994) suggested scaling of the pre-exponential term from the available data on partition coefficients for cations with charges m+ and n+. For example, they find that D_o^{3+}/D_o^{2+} is 0.14 $\pm$ 0.06 for clinopyroxene. They used this value to successfully predict the partition coefficient of trivalent REE between clinopyroxene and melt with Ca^{2+} being the host cation in clinopyroxene. A comprehensive understanding of the REE and Y partitioning behavior between clinopyroxene and basaltic melt is of importance to the understanding of the process of melting of upwelling mantle such as in the mid-ocean ridges.

Sun and Liang (2012) carried out a detailed study of the predictive success of the Blundy-Wood model by comparing the results with a large number of experimental data at high P-T conditions. On the basis of this comparison, they developed empirical expressions for $D_o^{s/l}$ as a linear function of clinopyroxene (X_{Al}^T, X_{Mg}^{M2}) and

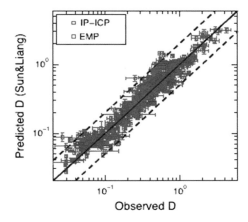

Fig. 11.7 Comparison between experimental (344 data) and predicted partition coefficients of trivalent REE and Y between clinopyroxene and basaltic melt. The dashed lines represent $\pm 2\sigma$ error envelope. From Sun and Liang (2012)

melt $(X_{H_2O}^{melt})$ compositions and reciprocal temperature $(1/T(K))$, r_o as a linear function of clinopyroxene composition and E as a linear function of r_o. These adjustments significantly improved the agreement between the predictions from the Blundy-Wood model and the experimental data set (344 data). The results of Sun and Liang are illustrated in Fig. 11.7.

11.3   Metal-Silicate Fractionation: Magma Ocean and Core Formation

It is commonly accepted that in the early period of the Earth's history, a significant portion of the mantle underwent melting to form what is referred to as the "terrestrial magma ocean" as a result of heat produced by giant impacts. The segregation of iron rich metal to form the Earth's core has taken place by settling of metal droplets through the magma ocean that led to the scavenging of the siderophile elements into the core. This is also true for Moon, Mars and the asteroid Vesta. The metal droplets were likely to have ponded at the base of the terrestrial magma ocean and finally descended as diapirs to form the Earth's core (Fig. 11.8).

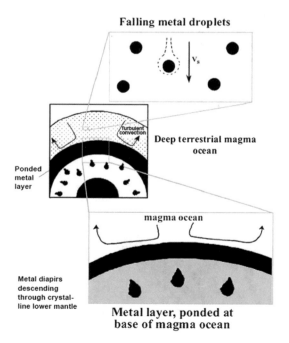

Fig. 11.8 Schematic illustration of the process of metal-silicate separation through the formation of a silicate magma ocean in the early history of Earth. The central panel shows a deep magma ocean overlying the lower mantle (white) and ponding of falling metal droplets at the base of the magma ocean. The metal in the ponded layer periodically descends as large diapirs to form the Earth's core. From Rubie et al. (2003). With permission from Elsevier

The abundance of **siderophile** elements, such as Ni, Co, Mo, W etc. in the Earth's mantle is found to be considerably larger than what one would predict from the equilibrium fractionation of these elements between molten iron metal and silicate melt at 1 bar pressure and moderate temperatures (1200–1600 °C). This is known as the **"excess siderophile"** problem of the Earth's mantle, and has been a topic of considerable interest and research in recent years (see reviews by Righter and Drake 2003; Wood et al. 2006 and Rubie et al. 2007). The general consensus that seems to have emerged is that the abundances of the siderophile elements appear to be in excess in the Earth's mantle because the expected abundances are calculated on the basis of low pressure partitioning data while the equilibration of these elements between liquid metal and silicate melt had likely taken place at high pressure condition in a deep magma ocean. (The "excess siderophile" problem is also true for the mantles of Mars, Moon and Vesta.)

Although the Earth's mantle has apparent "excess" problems for the moderately siderophile elements Ni and Co, these elements have almost the same relative abundance as in the chondritic meteorites (Ni/Co: Earth's mantle = 18.2; Chondrites = 21.2), which are thought to constitute the precursor materials that accreted to form the Earth. Thibault and Walter (1995) and Li and Agee (1996) experimentally determined the pressure dependence of the partition coefficients of Ni and Co between liquid metal and silicate melt as a function of P at 2123–2750 K, and found the logD versus P trend for the two metals to intersect at ~ 28 GPa. In addition, the D values at this pressure are found to be reduced to a level that is compatible with the inferred concentrations of Co and Ni in the Earth's mantle (Fig. 11.9a). Thus, these experimental results seem to simultaneously resolve two problems (i.e. apparent overabundance and chondritic relative abundance of Co and Ni in the mantle), and also suggest a deep magma ocean with a pressure of $\sim$ 28 GPa near the bottom. However, subsequent experimental study by Kegler et al. (2005) show that the distribution coefficients, K_D, of both Co and Ni between liquid metal and silicate melt have similar pressure dependencies (Fig. 11.9b). The break in the slope of $\ln K_D$ versus P at ~ 3 GPa has been ascribed to a change of coordination number of Co^{2+} and Ni^{2+} in the silicate melts (Keppler and Rubie 1993). In addition to pressure, oxygen fugacity has also been found to have significant influence on the metal/silicate partition coefficients (Fig. 11.10). Thus, the effect of changing $f(O_2)$ also needs to be accounted for in evaluating metal-silicate partitioning in a magma ocean. In this section, we consider the thermodynamics of metal/silicate equilibrium, and address the problem of pressure dependencies of the partition (D) and distribution (K_D) coefficients of Co and Ni between metallic liquid and silicate melt (We return again to the problem of core formation in the Earth and Mars in Sect. 13.7).

Fig. 11.9 Pressure
dependencies of the
(**a**) partition and
(**b**) distribution coefficients of
Ni and Co between Fe-liquid
and silicate melt. The data in
(**a**) are from Thibault and
Walter (1995) and Li and
Agee (1996) and in (**b**) are
from Kegler et al. (2005). The
horizontal patterned box in
(**a**) shows the range of D
values that explain the
inferred abundances of Ni and
Co in the terrestrial mantle
and core, assuming
equilibrium partitioning.
From (**a**) Wood et al. (2006)
(with permission from
Nature) and (**b**) Keggler et al.
(2005)

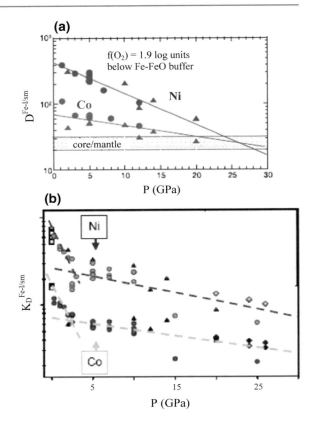

11.3.1 Pressure Dependence of Metal-Silicate Partition Coefficients

As pointed out by Capobianco et al. (1993), the partitioning of an element, such as
Ni, between liquid metallic Fe and silicate melt involves a change of the oxidation
state of the element, and therefore cannot be treated in terms of a transfer equilibrium analogous to reaction (11.2.a); instead the problem must be treated on the
basis of a redox reaction that conserves both mass and charge. Thus, the appropriate
reaction governing the partitioning of Co between Fe-liquid (Fe-l) and silicate
melt (sm) is given by

$$Co(Fe - liquid) + 1/2\,O_2(g) = CoO(silicate\ melt) \qquad (11.3.a)$$

An analogous reaction describes the partitioning of Ni. Choosing the pure state of
the condensed phases at P and T to be their respective standard states, as we have
often done before, and that of O_2 gas at 1 bar (which effectively equals unit
fugacity) and T to be the standard state of oxygen, we have at equilibrium

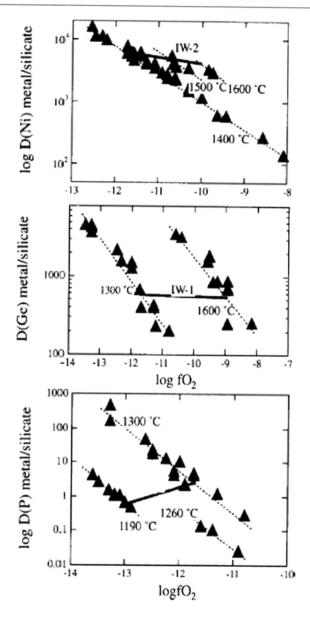

Fig. 11.10 Effect of oxygen fugacity (fO₂) on the metal/silicate partition coefficient Ni, Ge and P. IW-1: one log unit below the f(O₂) of iron-wüstite buffer. From Righter (2003)

$$\Delta_r G(P,T,X) = 0 = \underbrace{\Delta_r G^o_{con}(P,T) - G^o_{O_2}(1bar,T)}_{\Delta_r G^*(T)} + RT \ln \frac{a^{sm}_{CoO}}{\left(a^{Fe-1}_{Co}\right)\left(f^g_{O_2}\right)^{1/2}}$$

$$(11.3.1)$$

where the superscript o indicates pure phase, and $\Delta_r G^o_{con} = G^o_{CoO(sm)} - G^o_{Co(Fe-1)}$. The first two terms on the right of the above equation constitute the standard state free energy change of the reaction (11.3.a) at T, while the appearance of fugacity instead of activity of oxygen in the denominator of the equation is a consequence of the specific choice of standard state for $O_2(g)$ $(a(O_2) = f(O_2)/f^*(O_2)$, while at $P = 1$ bar, $f^*(O_2) \sim 1$). By convention, $G^o_{O_2}(1 \text{bar}, T) = 0$. Now, defining the **partition coefficient** of Co as

$$D^{Fe-1/sm}_{Co} = \frac{X^{Fe-1}_{Co}}{X^{sm}_{CoO}} \qquad (11.3.2)$$

Eq. (11.3.1) can be written as

$$RT \ln D^{Fe-1/sm}_{Co} = \Delta_r G^o_{con} + RT \ln \gamma^{sm}_{CoO} - RT \ln \gamma^{Fe-1}_{Co} - \frac{1}{2} RT \ln f^g_{O_2} \qquad (11.3.3)$$

The pressure dependencies of the first and last terms on the right are evaluated according to the relations $(\partial G/\partial P)_T = V$ and $RT(\partial \ln f_i/\partial P)_T = v_i$ (Eq. 10.11.15a), respectively, where v_i is the partial molar volume of i. In order to evaluate the pressure dependence of the activity coefficient terms at constant temperature, we first write $\ln \gamma = f(P, T, X)$. Thus, using the total derivative of $\ln \gamma$ at constant temperature, we have (Eq. B.1.4)

$$\left(\frac{\partial \ln \gamma_i}{\partial P}\right)_T = \left(\frac{\partial \ln \gamma_i}{\partial P}\right)_{T,X}\left(\frac{\partial P}{\partial P}\right)_T + \left(\frac{\partial \ln \gamma_i}{\partial X_i}\right)_{P,T}\left(\frac{\partial X_i}{\partial P}\right)_T$$

Assuming that a dilute siderophile element obeys Henry's law, the last term in the above expression drops out since γ_i is independent of X_i within the Henry's law limit at constant P-T condition and solvent composition (Sect. 8.8.1), in which case, using Eq. (10.11.7)

$$\left(\frac{\partial \ln \gamma_i}{\partial P}\right)_T \approx \left(\frac{\partial \ln \gamma_i}{\partial P}\right)_{T,X} = v_i - V^o_i \qquad (11.3.4)$$

where V^o_i and v_i are, respectively, the end-member molar volume and partial molar volume of i. Thus, we obtain

$$\left(\frac{\partial \ln D_{Co}^{Fe-1/sm}}{\partial P}\right)_T = \frac{1}{RT}\left(\Delta_r v_{Co} - \frac{1}{2}v_{O_2}\right) \tag{11.3.5}$$

where $\Delta_r v_{Co} = v_{CoO(sm)} - v_{Co(Fe-1)}$.

For the general case of an arbitrary oxidation state of the partitioning metallic ion in silicate melt, one writes (Capobianco et al. 1993)

$$M(Fe-1) + n/4 O_2 = M^{n+}O_{n/2}(\text{silicate melt}) \tag{11.3.b}$$

where n is the charge on the metal ion. Following the derivations of the Eqs. (11.3.3) and (11.3.5), we then have the general expressions

$$RT \ln D_M^{Fe-1/sm} = \Delta_r G_{con}^o + RT \ln \gamma_{MO_{n/2}}^{sm} - RT \ln \gamma_M^{Fe-1} - \frac{n}{4}RT \ln f_{O_2}^g \tag{11.3.6}$$

and

$$RT\left(\frac{\partial \ln D_M^{Fe-1/sm}}{\partial P}\right)_T = \left(\Delta_r v_M - \frac{n}{4}v_{O_2}\right) \tag{11.3.7}$$

where the partition coefficient of M is defined as

$$D_M^{Fe-1/sm} = \frac{X_{MO_{n/2}}^{Fe-1}}{X_M^{sm}} \tag{11.3.8}$$

and

$$\Delta_r v_M = v_{M^{n+}O_{n/2}(sm)} - v_{M(Fe-1)} \tag{11.3.9}$$

The second term on the right of Eq. (11.3.7) (i.e. v_{O2}) contributes to negative pressure dependence of $D_M^{Fe-1/sm}$, but the net pressure dependence of D may be either positive or negative depending on the values of $\Delta_r v_M$ and v_{O2}.

11.3.2   Pressure Dependence of Metal-Silicate Distribution Coefficients

The distribution of a species between Fe-liquid (Fe-l) and silicate melt (sm) can also be treated in terms of an exchange reaction as discussed in Sect. 11.1.1. Thus, we write

$$Fe(Fe-1) + CoO(\text{silicate melt}) = Co(Fe-1) + FeO(\text{silicate melt}) \tag{11.3.c}$$

for which the **distribution coefficient**, K_D, is (see Eq. 11.1.2)

$$K_D(Co - Fe)^{Fe-1/sm} = \frac{(X_{Co}/X_{Fe})^{Fe-1}}{(X_{CoO}/X_{FeO})^{sm}} \qquad (11.3.10)$$

According to Eq. (11.1.3)

$$\ln K_D(Co - Fe)^{Fe-1/sm} = \ln K_{3.c} - \ln K_{\gamma(3.c)} \qquad (11.3.11)$$

where $K_{\gamma(3.7)}$ is the ratio of the activity coefficients defined as

$$K_{\gamma(3.c)} = \frac{(\gamma_{Co}/\gamma_{Fe})^{Fe-1}}{(\gamma_{CoO}/\gamma_{FeO})^{sm}} \qquad (11.3.12)$$

Differentiating both sides of Eq. (11.3.11) with respect to pressure, and rearranging terms, we obtain

$$\left(\frac{\partial \ln K_D(Co - Fe)^{Fe-1/sm}}{\partial P}\right)_T = -\frac{\Delta_r v_{Co}}{RT}$$

$$= -\frac{1}{RT}\left[\left(v_{Co(Fe-1)} - v_{Fe(Fe-1)}\right) - \left(v_{CoO(sm)} - v_{FeO(sm)}\right)\right] \qquad (11.3.13)$$

Derivation of this expression, which follows similar manipulations as those used to derive Eq. (11.3.5) including assumption of Henry's law behavior of Co and Fe, is left to the reader as an exercise.

For a metal that oxidizes to a valence state n, we can write the exchange reaction in the general form

$$n/2\, Fe(Fe - 1) + M^{n+}O_{n/2}(\text{silicate melt}) = M(Fe - 1) + n/2\, FeO(\text{silicate melt}) \qquad (11.3.d)$$

for which

$$\left(\frac{\partial \ln K_D(M - Fe)^{Fe-1/sm}}{\partial P}\right)_T = -\frac{\Delta_r v_M}{RT}$$

$$= -\frac{1}{RT}\left[\left(v_{M(Fe-1)} - \frac{n}{2}v_{Fe(Fe-1)}\right) - \left(v_{MO_{n/2}(sm)} - \frac{n}{2}v_{Feo(sm)}\right)\right] \qquad (11.3.14)$$

11.3.3 Pressure Dependencies of Ni Versus Co Partition- and Distribution-Coefficients: Depth of Terrestrial Magma Ocean

From Eq. (11.3.5) and analogous relation for $D_{Ni}^{Fe-1/sm}$, we have at constant temperature

$$\frac{\partial}{\partial P}\left(\Delta \ln D_{Ni-Co}^{Fe-1/sm}\right) = \frac{1}{RT}\left(\Delta_r v_{Ni} - \Delta_r v_{Co}\right)$$

$$= \frac{1}{RT}\left[\left(V_{Co(Fe-1)} - V_{Ni(Fe-1)}\right) - \left(V_{CoO(sm)} - V_{NiO(sm)}\right)\right]$$

(11.3.15)

where $\Delta \ln D_{Ni-Co}^{Fe-1/sm} = \ln D_{Ni}^{Fe-1/sm} - \ln D_{Co}^{Fe-1/sm}$. Similarly, from Eq. (11.3.13) and analogous expression for the distribution coefficient of Ni, we have

$$\frac{\partial}{\partial P}\left(\Delta \ln K_D(Ni - Co)^{Fe-1/sm}\right) = \frac{1}{RT}\left[\left(V_{Co(Fe-1)} - V_{Ni(Fe-1)}\right) - \left(V_{CoO(sm)} - V_{NiO(sm)}\right)\right]$$

(11.3.16)

where $\Delta \ln K_D(Ni\text{-}Co)^{Fe\text{-}l/sm} = \ln K_D(Ni\text{-}Fe)^{Fe\text{-}l/sm} - \ln K_D(Co\text{-}Fe)^{Fe\text{-}l/sm}$. The right hand terms of the last two equations are exactly the same. Consequently, the data on the pressure dependence of lnD and lnK$_D$, as illustrated in Fig. 11.9, are *mutually incompatible*. Thus, the issue of the relative pressure dependence of the partition or distribution coefficient of Ni and Co between Fe-liquid and silicate melt remains unresolved. However, we seek reasonable resolution of the problem as follows.

Since Fe $(3d^64s^2)$, Co $(3d^74s^2)$ and Ni $(3d^84s^2)$ occupy three successive positions in a row in the periodic table, it is reasonable to assume that $v_{Ni/NiO} - v_{Co/CoO} \approx v_{Co/CoO} - v_{Fe/FeO}$, where $v_{Ni/NiO}$ means the partial molar volume of Ni or NiO, and so on, with the partial molar volumes of metals taken in Fe-liquid, and those of metal-oxides taken in silicate melt. Thus, we write, using Eq. (11.3.16)

$$\frac{\partial}{\partial P}\left(\Delta \ln K_D(Ni - Co)^{Fe-1/sm}\right) = -\frac{1}{RT}\left[\left(v_{Ni(Fe-1)} - v_{Co(Fe-1)}\right) - \left(v_{NiO(sm)} - v_{CoO(sm)}\right)\right]$$

$$\approx -\frac{1}{RT}\left[\left(v_{Co(Fe-1)} - v_{Fe(Fe-1)}\right) - \left(v_{CoO(sm)} - v_{FeO(sm)}\right)\right]$$

(11.3.17)

The quantity after the approximation sign is the same as that describing the pressure dependence of lnK$_D$(Co-Fe)$^{Fe\text{-}l/sm}$ (Eq. 11.3.13). Thus, using the experimental data illustrated in Fig. 11.9b, the quantity within the square brackets after the approximation sign is evaluated to be ~ 8 cm^3/mol. For brevity, we now denote the volumetric quantities within the first and second square brackets in the last equation as

$$\Delta_r V_{Ni-Co} = \left(V_{Ni(Fe-l)} - V_{Co(Fe-l)}\right) - \left(V_{NiO(sm)} - V_{CoO(sm)}\right) \qquad (11.3.18)$$

$$\Delta_r v_{Co-Fe} = \left(v_{Co(Fe-l)} - V_{Fe(Fe-l)}\right) - \left(v_{CoO(sm)} - V_{FeO(sm)}\right) \qquad (11.3.19)$$

Assuming $\Delta_r v_{Co-Fe}$ to be insensitive to pressure, as suggested by the linear trend of $\ln K_D$ versus P data at P > 3 GPa in Fig. 11.9b, we now integrate Eq. (11.3.17) between two pressures P* and P^{ref} to obtain

$$[\Delta \log K_D(Ni - Co)]_{P*} - [\Delta \log K_D(Ni - Co)]_{pref} \approx -\frac{8\left(P^* - P^{ref}\right)}{2.303RT} \qquad (11.3.20)$$

where

$$\Delta \log K_D(Ni - Co) = \log K_D(Ni - Fe)^{Fe-l/sm} - \log K_D(Co - Fe)^{Fe-l/sm}$$

at the specified pressure shown as a subscript. This equation shows that

$$\Delta \log K_D(Ni - Co)_{P*} < \Delta \log K_D(Ni - Co)_{pref}$$

at $P^* > P^{ref}$, that is, the $\log K_D$ versus P trends of Ni and Co converge with increasing pressure. Now, if the $\log K_D$ versus P trends for Ni and Co intersect at the pressure P*, then the first term on the left of Eq. (11.3.20) is zero, so that

$$P^* \approx P^{ref} + \frac{2.303RT\left(\Delta \log K_D(Ni - Co)_{pref}\right)}{8} \qquad (11.3.21)$$

Choosing P^{ref} as 5 GPa, and retrieving $\Delta \log K_D(Ni-Co)$ at 5 GPa from Fig. 11.9b, we finally obtain $P^* \approx 32.3$ GPa.

   Thus, given the reasonable premise used in deriving the last expression, that is $\Delta_r v_{Ni-Co} \approx \Delta_r v_{Co-Fe}$, we conclude that the dual problem of "excess abundance" of Ni and Co in the mantle and the chondritic relative abundance of these elements may be resolved if equilibrium was achieved between Fe-liquid and **magma ocean** at a depth corresponding to ~ 32 GPa, which is similar to the depth of intersection of the $\log D$ versus P trends of Ni and Co in Fig. 11.9a. A 20% error in the inferred value of $\Delta_r v_{Ni-Co}$ yields a pressure range of 27–41 GPa for the intersection of the $\log D$ versus P trends of Ni and Co.

11.4   Effect of Temperature and f(O₂) on Metal-Silicate Partition Coefficient

Various estimates about the temperature of core formation in the Earth vary between 2000 and 3750 K at 25–50 GPa (Rubie et al. 2007). Despite major advancements in the high pressure-temperature experimental studies, it is still necessary to extrapolate the laboratory experimental partitioning data to the high

P-T conditions relevant to core formation. These extrapolations, however, need to be guided by thermodynamic principles.

The expression for the temperature dependence of the distribution coefficient at a constant pressure is readily obtained from Eq. (11.3.6) as

$$\left(\frac{\partial \ln D_M^{Fe-1/sm}}{\partial T}\right)_P = \frac{1}{R}\left(\frac{\partial (\Delta_r G_{con}^o / T)}{\partial T}\right)_P + \left(\frac{\partial \ln \left(\gamma_{MO_{n/2}}^{sm}/\gamma_M^{Fe-1}\right)}{\partial T}\right)_P - \frac{n}{4}\left(\frac{\partial \ln f_{O_2}}{\partial T}\right)_T$$

$$(11.4.1)$$

To evaluate the first two terms on the right, we note that

$$\frac{\partial (G/T)}{\partial T} = -\frac{H}{RT^2} \tag{11.4.2}$$

And

$$\left(\frac{\partial \ln \gamma_i}{\partial T}\right)_{P,X} = \frac{1}{R}\left(\frac{\partial (\mu_i - \mu_i^o)}{\partial T}\right)_{P,X} = -\frac{h_i - H_i^o}{RT^2} \tag{11.4.3}$$

where h_i is the partial molar enthalpy of the component i. (The first equality in the last equation follows from the relation $\mu_i = \mu_i^o + RT\ln X_i + RT\ln\gamma_i$.) Assuming Henry's law behavior of the siderophile elements, we obtain $(\partial \ln\gamma_i/\partial T)_P = (\partial \ln\gamma_i/\partial T)_{P,X}$. The derivation of the last relation is analogous to that of the first equality of Eq. (11.3.4). Thus, substitution of the last two relations in Eq. (11.4.1), and rearrangement of terms yield

$$\frac{\partial \ln D_M^{Fe-1/sm}}{\partial T} = -\frac{\Delta_r h}{RT^2} - \frac{n}{4}\frac{\partial \ln f_{O_2}}{\partial T} \tag{11.4.4}$$

where $\Delta_r h = h(MO_{n/2})^{sm} - h(M)^{Fe-l}$, from which we obtain

$$\ln D_M^{Fe-1/sm}(T_2) = \ln D_M^{Fe-1/sm}(T_1) - \frac{1}{R}\int_{T_1}^{T_2}\frac{\Delta_r h}{T^2}dT$$
$$- \frac{n}{4}(\ln f_{O_2}(T_2) - \ln f_{O_2}(T_1)) \tag{11.4.5}$$

It is sometimes convenient to replace $-dT/T^2$ by $d(1/T)$ so that, if $\Delta_r h$ is **constant** over a chosen temperature interval, then

$$\ln D_M^{Fe-1/sm}(T_2) = \ln D_M^{Fe-1/sm}(T_1) + \frac{\Delta_r h}{R}\left(\frac{1}{T_2} - \frac{1}{T_1}\right)$$
$$- \frac{n}{4}(\ln f_{O_2}(T_2) - \ln f_{O_2}(T_1)) \tag{11.4.6}$$

There are no data for the partial molar enthalpies in the systems of interest, which makes the extrapolation of lnD versus T very difficult. Capobianco et al. (1993) made the extrapolations assuming $\Delta_r h = \Delta_r H°$.

According to Eq. (11.4.6), logD versus $\log f(O_2)$ relation is expected to be linear with a slope of $-n/4$, if the effect of the variation of the activity coefficient terms are negligible. The data summarized in Fig. 11.10 show that slopes of logD versus $\log f(O_2)$ relations of the different cations are almost exactly $-n/4$ if n is the number of charges in the conventional valence state of the ions (2 for Ni, 4 for Ge and 5 for P). Thus, it seems quite reasonable to predict $-n/4$ for the slopes of the logD versus $\log f(O_2)$ relations of other trace elements, with n representing the number of charges in their conventional valences.

References

Allan NL, Blundy JD, Purton JA, Yu Lavrentiev M, Wood BJ (2001) Trace element incorporation in minerals and melts. In: Geiger CA (ed) Solid solutions in silicate and oxide systems. Eötvös University Press, Budapest, Hungary, pp 251–302

Asahara Y, Frost DJ, Rubie DC (2007) Partitioning of FeO between magnesiowüstite and liquid iron at high pressures and temperatures: implications for the composition of the Earth's core. Earth Planet Sci Let 257:435–449

Blundy J, Wood B (1994) Prediction of crystal-melt partition coefficients from elastic modulii. Nature 372:452–454

Brice JC (1975) Some thermodynamic aspects of growth of strained crystals. J Cryst Growth 28:249–253

Capobianco CJ, Jones JH, Drake MJ (1993) Meta-silicate thermochemistry at high temperature: magma oceans and the "excess siderophile element" problem of the Earth's upper mantle. J Geophys Res 98:5433–5443

Ganguly J, Saxena SK (1987) Mixtures and mineral reactions. Springer, Berlin, Heidelberg, New York, Paris, Tokyo, p 291

Ganguly J, Cheng W, Tirone M (1996) Thermodynamics of aluminosilicate garnet solid solution: new experimental data, an optimized model, and thermometric applications. Contrib Miner Petrol 126:137–151

Gast P (1968) Trace element fractionation and the origin of tholeiitic and alkaline magma types. Geochim Cosmochim Acta 32:1057–1086

Kegler P, Holzheid A, Rubie DC, Frost DJ, Palme H (2005) New results on metal/silicate partitioning of Ni and Co at elevated pressures and temperatures. Lunar and planetary science conference XXXVI, Abstr # 2030

Keppler H, Rubie DC (1993) Pressure-induced coordination changes of transition-metal ions in silicate melts. Nature 364:54–55

Li J, Agee CB (1996) Geochemistry of mantle-core differentiation at high pressure. Nature 381:686–689

Liermann H-P, Ganguly J (2003) Fe^{2+}-Mg fractionation between orthopyroxene and spinel: experimental calibration in the system $FeO-MgO-Al_2O_3-Cr_2O_3-SiO_2$, and applications. Contrib Miner Petrol 145:217–227

McDade P, Blundy JD, Wood BJ (2003) Trace element partitioning on the Tinquillo Lherzolite solidus at 1.5 GPa. Phys Earth Planet Int 139:129–147

Ottonello G (1997) Principles of geochemistry. Columbia University Press, p 894

Righter K (2003) Metal-silicate partitioning of siderophile elements and core formation in the early Earth. Ann Rev Earth Planet Sci 31:135–174

Righter K, Drake MJ (2003) Partition coefficients at high pressure and temperature. In: Carlson RW (ed) The mantle and core: treatise on geochemistry, Elsevier, pp 425–450

Rubie DC, Melosh HJ, Reid JE, Liebske C, Righter K (2003) Mechanisms of metal-silicate equilibration in the terrestrial magma ocean. Earth Planet Sci Let 205:239–255

Rubie DC, Nimmo F, Melosh HJ (2007) Formation of Earth's core. In: Stevenson DJ (ed) Evolution of the Earth: treatise on geophysics, vol 9. Elsevier, Amsterdam, pp 51–90

Salters VJM, Longhi J (1999) Trace element partitioning during the initial stages of melting beneath mid-ocean ridges. Earth Planet Sci Lett 166:15–30

Shaw DM (1970) Trace element fractionation during anatexis. Geochim Cosmochim Acta 34:237–243

Sun C, Liang Y (2012) Distribution of REE between clinopyroxene and basaltic melt along a mantle adiabat: effects of major element composition, water and temperature. Contrib Mineral Petrol 163:807–823

Thibault Y, Walter MJ (1995) The influence of pressure and temperature on the metal- silicate partition coefficients of nickel and cobalt in a model C1 chondrite and implications for metal segregation in a deep magma ocean. Geochim Cosmochim Acta 59:991–1002

Tirone M, Ganguly J, Dohmen R, Langenhorst F, Hervig R, Becker H-W (2005) Rare earth diffusion kinetics in garnet: experimental studies and applications. Geochim Comsochim Acta 69:2385–2398

Van Westrenen W, Blundy J, Wood BJ (2000) Effect of Fe^{2+} on garnet-melt trace element partitioning: experiments in FCMAS and quantification of crystal-chemical controls in natural systems. Lithos 53:189–203

von Seckendorff V, O'Neill HStC (1993) An experimental study of Fe-Mg partitioning between olivine and orthopyroxene at 1173, 1273 and 1423 K and 1.6 GPa. Contrib Miner Petrol 113:196–207

Williams ML, Grambling JA (1990) Manganese, ferric iron and the equilibrium between garnet and biotite. Am Miner 75:886–908

Wood BJ, Walter MJ, Wade J (2006) Accretion of the Earth and segregation of its core. Nature 441:825–833

Electrolyte Solutions and Electrochemistry

12

An electrolyte is a compound which dissociates into charged species, either partially or completely, after dissolving in a solution. In general, we can write the following dissociation reaction for an electrolyte:

$$M_{v_+} A_{v_-} = v_+ (M^{z+}) + v_- (M^{z-}) \qquad (12.a)$$

where v_+ and v_- are the numbers of positive and negative ions (or ionic complexes), respectively, and z_+ and z_- are their respective formal charges. For example, H_2SO_4 dissociates in an aqueous solution according to $H_2SO_4 = 2H^+ + (SO_4)^{2-}$. In this case, $v_+ = 2$, $z_+ = 1+$, $v_- = 1$, $z_- = 2-$. Thermodynamics of electrolyte solutions play very important roles in the understanding of chemical equilibrium and redistribution of components in a variety of geochemical processes, such as those relating to ocean-atmosphere and fluid-rock interactions, solute transport in an aqueous solution, formation of sedimentary rocks, magma-hydrothermal systems etc.

Electrolytes are classified as **strong** or **weak** electrolytes according to whether these dissociate strongly or weakly in an aqueous solution. In dealing with electrolyte solutions, it is customary to use **molality** as a measure of content of the solute species. Molality is defined as the number of moles per kg of pure solvent, usually water.[1] The content of dilute electrolytes in a solution is commonly reported as **ppm** values, which indicate the mass of a solute per 10^6 g of solution. It should be easy to see that the values reported as ppm can be converted to molality according to the relation

$$\text{molality} = (\text{ppm/molecular weight of the solute}) \times 10^{-3}$$

---

[1]A similar sounding measure of content is **molarity**, M, which is defined as the moles of solute per liter of solution. The molarity measure is avoided since the solvent density changes as a function of P-T condition making M a function of P-T.

© Springer Nature Switzerland AG 2020
J. Ganguly, *Thermodynamics in Earth and Planetary Sciences*,
Springer Textbooks in Earth Sciences, Geography and Environment,
https://doi.org/10.1007/978-3-030-20879-0_12

A solution may be neutral or charged electrically. The properties of an ion in an electrically neutral solution are much different from those in a charged solution. In dealing with the thermodynamics of electrolyte solution, the solution, as a whole is treated as electrically neutral.

12.1   Chemical Potential

The chemical potential of a species in an electrolyte solution is defined in the usual way (Box 8.1.2). For example, the chemical potential of a positive ion, μ_+ is defined as

$$\mu_+ = \left(\frac{\partial G}{\partial n_+}\right)_{P,T,n_-,n_u,n_o} \tag{12.1.1}$$

where G is the total Gibbs free energy of the neutral solution, n_+ and n_- are the number of moles of positive and negative ions, respectively, n_u is the number of moles of the undissociated solute, and n_o is the number of moles of the solvent. Because of the electroneutrality condition, the chemical potential of a charged species cannot be measured directly by conventional methods as these require a macroscopic change in the number of moles of a charged species, while holding the number of moles of the species of opposite charge constant (along with the other variables). However, it is formally correct to use chemical potentials of charged species in the treatment of electrolyte solution, and a specific linear combination of μ_+ and μ_- can be measured, as discussed below, following the exposition of Denbigh (1981).

If μ_u is the chemical potential of the undissociated part of the electrolyte, then from reaction (12.a), we have at equilibrium

$$\mu_u = \nu_+\mu_+ + \nu_-\mu_-, \tag{12.1.2}$$

Let us now consider the rate of change of Gibbs free energy of the solution with respect to the addition of an electrolyte. If dm_e stands for an infinitesimal addition of the moles of electrolyte (e), then we define a chemical potential μ of the electrolyte as a whole as the rate of change of G with respect to m_e. Thus,

$$\mu = \left(\frac{\partial G}{\partial n_e}\right)_{P,T,n_o} \tag{12.1.3}$$

Note that we are not concerned here what happens to the electrolyte after it is dissolved. Combining the expression of the total derivative of G of the solution with Eq. (12.1.2) and the relationship among m and the various dissolved species, it can be shown that at equilibrium (e.g. Denbigh 1981).

$$\mu = \nu_+ \mu_+ + \nu_- \mu_- \qquad (12.1.4)$$

Thus, since μ is measurable, the linear combination of the chemical potentials of the ionic species represented on the right side of the above equation is also measurable. Comparing Eqs. (12.1.2) and (12.1.4).

$$\mu = \mu_u = \nu_+ \mu_+ + \nu_- \mu_- \qquad (12.1.5)$$

12.2   Activity and Activity Coefficients: Mean Ion Formulations

The activity expressions for the dissociated and undissociated (u) solute species are based on their molalities as

$$a_i = m_i \gamma_i \qquad (12.2.1)$$

where a_i and m_i stand respectively for the activity and molality of the species i. Because of the problem associated with the measurement of individual ion activities, it is customary in the treatment of electrolyte solution to introduce linear combination of the ion activities that can be measured, and this has led to the development of the concept of what is known as the **mean ion activity** and **mean ion activity coefficient**.[2]

Expressing the chemical potentials of the charged species in terms of their respective standard state properties and activities in the usual way (Eq. 8.4.5), we have from the last two expressions

$$\begin{aligned}
\mu = \mu_u &= \nu_+ \left[\mu_+^* + RT \ln\left(m_+ \gamma_+\right) \right] + \nu_- \left[\mu_-^* + RT \ln\left(m_- \gamma_-\right) \right] \\
&= \nu_+ \mu_+^* + \nu_- \mu_-^* + RT \ln\left[(m_+)^{\nu_+} (m_-)^{\nu_-}\right] + RT \ln\left[(\gamma_+)^{\nu_+} (\gamma_-)^{\nu_-} \right]
\end{aligned} \qquad (12.2.2)$$

where m_+ and m_- stand for the molalities of the positive and negative ions, respectively, and the superscript * stands, as usual, for the standard state at the temperature of interest. The activity coefficient product within the last square brackets is used to define a **mean ion activity coefficient**, $\gamma_\pm$, of the electrolyte as

$$(\gamma_\pm)^{\nu} = (\gamma_+)^{\nu_+} (\gamma_-)^{\nu_-} \qquad (12.2.3a)$$

or

---

[2]As we would see later, one often uses a quantity know as pH that is defined as pH $= -\log a_H^+$. The activity of H^+ ion may be determined by a combination of experimental and theoretical procedure, but as discussed by Pitzer (1995), "the basic uncertainty of single ion activity remains."

$$\gamma_\pm = \left[(\gamma_+)^{v+} (\gamma_-)^{v-} \right]^{1/v} \tag{12.2.3b}$$

where $v = v_+ + v_-$. Similarly, one can define a **mean ion activity** of the electrolyte as

$$a_\pm = m_\pm \gamma_\pm \tag{12.2.4}$$

where $m_\pm$, known as the **mean ion molality**, is defined in a similar manner as $\gamma_\pm$, using the individual ion molalities. Equation (12.2.3b) can be generalized as

$$\gamma_{\pm,k} = \left(\prod_i \gamma_i^{v_i} \right)^{1/v}, \tag{12.2.5}$$

where k stands for an electrolyte, i stands for a dissociated species, v_i for its stoichiometric coefficient, and $v = \sum v_i$.

12.3   Mass Balance Relation

An ion in a solution may exist both as a free ion and by complexing with another ion, or simply by complexing with other ions. For example, in an aqueous solution of NaCl and Na_2SO_4, sodium may be present as free Na^+ ion, NaCl, $NaSO_4^-$, Na(OH) and Na_2SO_4. In that case one can define a total molality of sodium as

$$m_{Na}(tot) = m_{Na^+} + m_{NaCl} + m_{NaSO_4^-} + m_{Na(OH)} + 2(m_{Na_2SO_4})$$

(the multiplier 2 in the last term is due to the fact that there are 2 mol of Na in a mole of Na_2SO_4). In general, we can then write

$$\boxed{m_i(tot) = \sum_k v_{i(k)} m_k} \tag{12.3.1}$$

where $v_{i(k)}$ is the number of moles of the ith ion in the solute k and m_k is the number of moles of that solute.

12.4   Standard State Convention and Properties

12.4.1   Solute Standard State

When dealing with a major component (j) in a solution, it is usually desirable to choose a standard state such that $a_j = X_j$ as $X_j \to 1$. As discussed in Sect. 8.8.2, this objective is realized by choosing a standard state of pure component at the P-T condition of interest. In dealing with a dilute solution, it is advantageous (to simplify life) to choose a standard state for the solute such that $a_i = m_i$ as $m_i \to 0$, since it is the property of the solute in dilute concentration that is of interest. Note that the manipulation of the theoretical framework so that the activity coefficient assumes a unit value (or any other value) does not affect the final outcome if the analysis is carried out in a self-consistent manner.

According to Henry's law for dilute solute (Eq. 8.8.1'), $f_i = K_H^* m_i$ as $m_i \to 0$, where i is an actual solute and K_H^* is a constant that is commonly referred to as the Henry's law constant (it is simply an activity coefficient that is independent of the concentration of the solute within a specified range). Let us now extrapolate f_i along the line $f_i = K_H^* m_i$, which we would refer to as 'Henry's law line', to $m_i = 1$ and choose the resultant hypothetical state of the solute as its standard state (Fig. 12.1a). This is often referred to as the **solute standard state** based on the properties at the infinite dilution. The slope of the Henry's law line, K_H^*, then becomes $f_i^*/1 = f_i^*$. Thus, the Henry's law transforms to

$$\text{limit } m_i \to 0, \quad f_i = f_i^* m_i$$

Since by definition, $a_i = f_i/f_i^*$ (Eq. 8.4.8), we then have

$$\text{limit } m_i \to 0 \quad a_i = m_i, \qquad (12.4.1)$$

as illustrated in Fig. 12.1b.

For any ionic species, the solute standard state defined according to the above procedure is a hypothetical state since an ionic species in a real solution does not

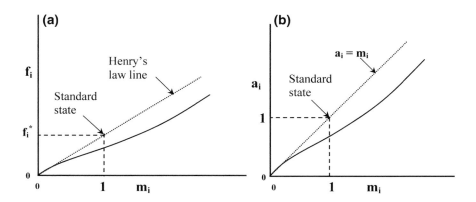

Fig. 12.1 Illustration of the choice of solute standard state by extrapolating the Henry's law line (dotted line) to unit molality in the (**a**) fugacity versus molality space and (**b**) activity versus molality space. The solid lines show schematic behaviors of a real solution with negative deviation from ideality. In (**a**), the Henry's law line follows relation $f = K_H(m_i)$ where K_H is a constant, whereas f^* is the fugacity of the hypothetical standard state at unit molality

follow Henry's law up to unit molality. However, for a neutral solute, the Henry's law behavior may be followed in the real solution up to unit molality, in which case, the standard state chosen according to the above procedure constitutes a real state. The solute standard state conforming to the property of **unit activity at the unit molality**, whether it is real or hypothetical, will henceforth be designated by the superscript symbol x.

Instead of the activity of an actual solute, we may want to deal with the activity of a strong electrolyte (e.g. NaCl), a_e, even though it has almost completely dissociated in a solution. In that case, as discussed in Sect. 8.8.1 (Eq. 8.8.12), we would find that

$$a_e \propto (m^\bullet)^\nu, \quad \text{as } m^\bullet \to 0,$$

where $\propto$ indicates proportionality and $m^\bullet$ is the molality of the electrolyte that one would calculate had there been no dissociation (simply by knowing the amount of the electrolyte added to the solution), and $\nu = \nu^+ + \nu^-$. It is easy to see that the above choice of "solute standard state" defined by the extrapolation of the Henry's law behavior to unit molality leads to the relation

$$a_e = (m^\bullet)^\nu \quad \text{as } m^\bullet \to 0 \tag{12.4.2}$$

Although it should be obvious, it is reiterated that both Eqs. (12.4.1) and (12.4.2) are statements of the Henry's law in which the proportionality constants have been made to assume the value of unity through a clever choice of standard state.

12.4.2 Standard State Properties of Ions

Because of the problem with the experimental determination of single ion properties in a solution that is effectively electrically neutral, one obtains Gibbs free energy of formation of combination of ions, such as of $H^+(aq)$ and $(OH)^-(aq)$, NH_4^+ (aq) and $(OH)^-(aq)$, and so on. However, it is cumbersome to have listing of ΔG_f of combination of ions. The convenient alternative to this practice is to have ΔG_f of each ion in combination with a common ion, and assume the ΔG_f of the latter to be zero. This is done by assuming that $\Delta_r G_f$ (1 bar, T) = 0 for the reaction

$$1/2\, H_2(g) \leftrightarrow H^+ (1 \text{ molal ideal aqueous solution}) + e^-, \tag{12.4.a}$$

This is equivalent to saying that the standard state free energy of formation of H^+ ion in an one molal ideal aqueous solution from a gaseous state is zero at any temperature at 1 bar pressure. It follows that since $\partial(\Delta_r G)/\partial T = -\Delta_r S$, and $\Delta_r G = \Delta_r H - T\Delta_r S$, both entropy and enthalpy of changes of the above reaction must also be zero.

To see how ΔG_f^x of other ions are calculated in a systematic way by assigning a zero value to $\Delta G_f^x(H^+)$, let us consider the following example, which is discussed by Denbigh (1981). For the ion pair H^+(aq) and $(OH)^-$(aq),

$$\Delta G_f^x(1\ bar, 298\ K) = -157,297.48\ J/mol,$$

whereas for the ion pair NH_4^+ (aq) and $(OH)^-$(aq),

$$\Delta G_f^x(1\ bar, 298\ K) = -236,751.64\ J/mol.$$

Now, since ΔG_f^x (H^+) (1 bar, 298 K) has been assigned a zero value, we have ΔG_f^x $(OH)^-$ (1 bar, 298 K) = −157,297.48 J/mol, and consequently, ΔG_f^x [NH_4^+ (aq)] (1 bar, 298 K) = −236,751.64 + 157,297.48 = −7954.16 J/mol.

12.5   Equilibrium Constant, Solubility Product and Ion Activity Product: Survival of Marine Carbonate Organisms

When the reaction (12.a) is at equilibrium, we can write, in the usual way

$$K \equiv \exp(-\Delta_r G^*/RT) = \frac{(m_+)^{\nu^+}(m_-)^{\nu^-}(\gamma_\pm)^\nu}{a_{M_{\nu+}A_{\nu-}}} \quad (12.5.1)$$

where $\gamma_\pm^\nu$ is used to replace the term $(\gamma_+^{\nu^+})(\gamma_-^{\nu^-})$ (Eq. 12.2.3a). Now if $M_{\nu+}A_{\nu-}$ is a solid electrolyte, and we have chosen pure state of that electrolyte at the P-T of interest as its standard state, then

$$RT \ln K(P,T) = \nu_+\left(\mu_+^\times\right) + \nu_-\left(\mu_-^\times\right) - \mu^o_{M_{\nu+}A_{\nu-}}$$
$$= \nu_+\left(\mu_+^\times\right) + \nu_-\left(\mu_-^\times\right) - G^o_{M_{\nu+}A_{\nu-}} \quad (12.5.2)$$

The equilibrium constant, K, defined by last relation, is called the **solubility product**, and is usually designated as K_{sp}. On the other hand, the product of the activities of the dissociated solute in the form $\Pi a_i^{\nu_i}$ (i.e. numerator of the right hand term in Eq. 12.5.1) is called the **ion activity product** (IAP).

It follows from Eq. (12.5.1) that a solution is in equilibrium with a pure solid electrolyte when IAP = K_{sp}, since for solid the standard state is pure solid (P,T) so that the activity of a pure solid component is unity. It is easy to see that if IAP > K_{sp}, and the undissociated solute does not enter into solid solution, then the solution must precipitate additional solute in order to reduce the IAP to its equilibrium value, and vice versa. But what happens if electrolyte $M_{z+}A_{z-}$ enters into a solid solution? In that case, of course, one needs to compare the reaction quotient Q (i.e. the entire right hand term of Eq. 12.5.1), not just the IAP, with K_{sp}. The solution would precipitate the solid electrolyte when $Q > K_{sp}$, and vice versa.

As a simple illustration of the application of the concept of solubility product and ion activity product, let us consider the question of survival of **marine organisms** with carbonate shells in the sea water, as discussed by Anderson (2005). The appropriate reaction to consider in this case is

$$CaCO_3 \leftrightarrow Ca^{2+} + (CO_3)^{2-} \tag{12.5.a}$$

From the calculated values of the molalities Ca and $(CO_3)^{2-}$ in the near surface sea water, IAP for the above reaction equals $10^{-7.87}$. The K_{sp} value of the reaction depends on the polymorphic from of $CaCO_3$, calcite (Cal) or aragonite (Arg), and is given by the relation $K_{sp} = \exp(-\Delta_r G^*/RT)$ where $\Delta_r G^* = (CO_3^{-2}) + G_f^x(Ca^{2+}) - G_f^o(Calc/Arag)$. Using the standard state data, we have

$$K_{sp}(Cal) = 10^{-8.304}$$

and

$$K_{sp}(Arg) = 10^{-8.122}$$

both of which are less than the IAP. Thus, the carbonate shells of marine organisms will not dissolve in the near surface sea water (marine organisms actually secrete aragonite, so $K_{sp}(Arg)$ is the more relevant solubility product to consider). However, at depths greater than ~ 5 km, the IAP falls below K_{sp}, and as a consequence the marine organisms with carbonate shells do not survive below this depth. This is known as the **carbonate compensation depth** of oceans.

The fate of marine organisms with carbonate shells is being challenged with increase of CO_2 content of the Earth's atmosphere resulting from fossil fuel burning. The problem may be appreciated by considering the fact that increasing $P(CO_2)$ leads to an increase in the acidity of the ocean water (see Sect. 12.10) and thus a decrease of the ratio of $IAP/K_{sp}(Arg)$ (Ω_A), as illustrated in Fig. 12.2, which is reproduced from Millero and DiTrolio (2010) (Although increasing $P(CO_2)$ leads to increase of both H^+ and CO_3^- in ocean water – see Sect. 12.10, the concentration of H^+ increases twice as fast as CO_3^- so that there is a net decrease in the concentration of CO_3^-).

The dependence of (Ω_A) on $P(CO_2)$ was calculated by Millero and DiTrolio (2010) and that of $P(CO_2)$ on time by Caldeira and Wickett (2003) (see Sect. 12.10). Experimental studies show that reduction of the aragonite saturation parameter, Ω_A, slows down the calcification rates of different coral species; this relationship is also borne out by historical records (Millero and DiTrolio 2010; Langdon and Atkinson 2005). The saturation parameter may even fall below unity in the late 21st century, resulting in dissolution of the aragonite skeletal structures of marine organism.

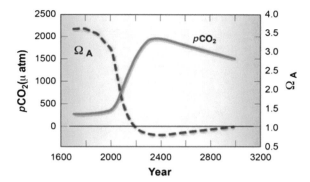

Fig. 12.2 Estimated change of P(CO$_2$) in the Earth's atmosphere and corresponding saturation state of aragonite in the ocean water with time. From Millero and DiTroilo (2010). With permission from Mineralogical Society of America

Problem 12.1 Consider the dissociation of a solid electrolyte in an aqueous solution, and assume that it is present in a solid solution, e.g. the dissociation reaction (12.5.a) with CaCO$_3$ dissolved in a solid solution (Ca,Mg)CO$_3$. Will the solid necessarily dissolve when IAP < K$_{sp}$?

Answer: No (but discuss the reason).

12.6   Ion Activity Coefficients and Ionic Strength

12.6.1   Debye-Hückel and Related Methods

A convenient starting point in the discussion about the ion activity coefficients in a dilute electrolyte solution is a theoretical expression known as the **Debye-Hückel limiting law**. In deriving this law, which yields the activity coefficient of an ion at very high dilution, it is assumed that (**a**) the deviation from ideality in a dilute ionic solution is entirely due to the electrical interactions between the ions, (**b**) the ions are point charges, (**c**) the solute is completely dissociated and (**d**) the repulsive forces between the ions are unimportant. Since the repulsive force drops off very rapidly with the distance of separation, the last assumption is justified in a very dilute ionic solution where the ions are far apart. Within the above framework, Debye and Hückel derived an expression for the individual ion activity coefficients, which is as follows.

$$\log \gamma_i = -AZ_i^2 \sqrt{I} \qquad (12.6.1)$$

where Z_i is the charge of the ionic species i, A is a constant, in unit of $(kg/mol)^{\frac{1}{2}}$, and depends on the dielectric constant and density of the solvent, and I is the ionic strength of the solution that is given by

$$I = 1/2 \sum m_i z_i^2 \quad \text{mol/kg} \tag{12.6.2}$$

As an example, the ionic strength of one molal solution of $La_2(SO_4)_3$, which dissociates in the solution according to $La_2(SO_4)_3 \rightarrow 2La^{3+} + 3(SO_4)^{2-}$, is $\frac{1}{2}[2(3)^2 + 3(2)^2] = 15$ mol/kg.

Debye and Hückel (1923) modified their earlier theory to account for the effects of finite sizes of the ions and short range interactions between them, assuming that the ions are nondeformable spheres of equal radii. A widely used extension (Robinson and Stokes 1959; Helgeson 1969) of the modified Debye-Hückel expression of activity coefficient is

$$\log \gamma_i = -\frac{Az_i^2 \sqrt{I}}{1 + Ba_i \sqrt{I}} + B^{\bullet}I, \tag{12.6.3}$$

where the last term $(B^{\bullet}I)$ represents the extension to the original Debye-Hückel (1923) form. This expression will be referred to as the **extended Debye-Hückel formulation**. Here B is another constant that is characteristic of the solvent, a_i is the "distance of closest approach" between ions of opposite charge, and $B^{\bullet}$ (often called B-dot, and also referred to as "deviation function") is an adjustable parameter to fit the experimental data. In fact the a_i term can also be treated as an adjustable parameter. The unit of Ba_i is $(kg/mol)^{\frac{1}{2}}$.

The above equation seems to work well up to around 1 molal solute concentration. Helgeson and Kirkham (1974a, b) have given values of A and B for water from 0 to 300 °C. At 0 °C, A = 0.4911 and B = 0.3244 whereas at 300 °C, these values are 1.2555 and 0.3965, respectively, with the unit of B in $kg^{\frac{1}{2}}/(mol^{\frac{1}{2}}\text{-Å})$. It is easy to see that when the solution becomes very dilute, Eq. (12.6.3) yields the Deby-Hückel limiting law (Eq. 12.6.1) since the denominator of the first equation approaches unity as $I \rightarrow 0$.

Because of the problem in measuring individual ion activity coefficients, γ_+ and γ_- are combined according to Eq. (12.2.3) to yield an expression for $\gamma_{\pm}$, which can be compared with experimentally determined values. Substituting the expression for the individual ion activities (Eq. 12.6.3), into Eq. (12.2.3), and using a single adjustable $\mathring{a}$ parameter, we obtain

$$\nu(\log \gamma_{\pm}) = -\frac{A\sqrt{I}}{I + B\mathring{a}\sqrt{I}} \left(\nu_+ z_+^2 + \nu_- z_-^2\right) + \nu B^{\bullet}I \tag{12.6.4}$$

However, since the electroneutrality condition requires that $|\nu_+ z_+| = |\nu_- z_-|$, the term within the parentheses can be written as $(\nu_- z_- z_+ + \nu_+ z_+ z_-) = z_- z_+(\nu_+ + \nu_-) = (z_- z_+)$. Thus, we finally obtain

$$\log \gamma_\pm = -\frac{A|z_+z_-|\sqrt{I}}{1 + B \mathring{a} \sqrt{I}} + B^\bullet I \qquad (12.6.5)$$

In a solution consisting of the dissociation products of a single electrolyte, the expressions for single ion and mean ion activity coefficients satisfy the relation $\gamma_i \to 1$ as $m_i \to 0$, which is imposed by the choice of solute standard state discussed in Sect. 12.4.1 (Fig. 12.1). This is because $I \to 0$ as $m_i \to 0$ (for the dissociation of a single electrolyte, the molalities of the dissociated species are proportional to each other). The numerator on the right hand side of Eq. (12.6.5) reflects the effect of long-range Coulomb forces. The modification of these forces by short-range interactions between the ions in the crudest approximation of hard-sphere model is given by the term in the denominator. The B-dot term partially takes care of the short-range interaction between ions and the solvent molecules and other types of short range interactions between ions that cannot be adequately accounted for by the hard sphere model.

Helgeson (1969) summarized the values of the different parameters in Eq. (12.6.3) for concentrated NaCl aqueous solutions that may be used to calculate the individual activity coefficients of dissolved ions up to $\sim 300\,^\circ\mathrm{C}$ at 1 bar pressure. Values of A, B and $\mathring{a}$ parameters for aqueous solutions up to $60\,^\circ\mathrm{C}$ can also be found in Garrels and Christ (1965). The latter workers have calculated individual ion activity coefficients in aqueous solutions using the classic Debye-Hückel formulation (i.e. Eq. 12.6.3, but without the B-dot term) for several ions in aqueous solution, and compared the results with those calculated from the experimental data on mean ion activity coefficients, $\gamma_\pm$, according to the "mean-salt method" that is discussed below. The results, which are illustrated in Fig. 12.3, show good agreement between the two methods up to ionic strength of 0.1. The departure of Debye-Hückel prediction from the experimentally constrained result of the mean-salt method at higher ionic strength may be accounted for by the B-dot term.

12.6.2 Mean-Salt Method

The "mean-salt method" was introduced by Garrels and Christ (1965) in the geochemical literature for the calculation of individual ion activity coefficients, γ_i, in aqueous solutions at ionic strengths where the classic Debye-Hückel formulation for γ_i fails. The method relates γ_i to the mean ion activity coefficients that can be measured experimentally. In developing this method, it was assumed that in an aqueous solution, $\gamma_{K^+} = \gamma_{Cl^-}$, as seems to be indicated by various lines of evidence (this relationship was first proposed by MacInnes (1919), and is sometimes referred to as the MacInnes' convention). Thus, using Eq. (12.2.3b),

$$\gamma_{\pm KCl} = [(\gamma_{K^+})(\gamma_{Cl^-})]^{1/2} = \gamma_{K^+} = \gamma_{Cl^-} \qquad (12.6.6)$$

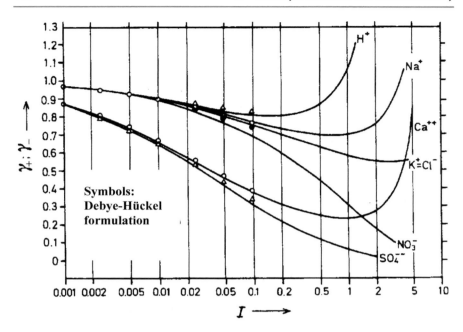

Fig. 12.3 Individual ion activity coefficients versus ionic strength. The symbols represent calculations using the modified Debye-Hückel formulation (Eq. 12.6.3, but without the B-dot term), whereas the lines represent calculations of individual ion activity coefficients from the measured values of mean ion activity coefficient according to the "mean salt method". From Garrels and Christ (1965). The Debye-Hückel model fails at I > 0.1. With permission from Harper and Row—Pearson Education

Using the $\gamma(K^+)$ and $\gamma(Cl^-)$ values calculated according to this relation from the measured values of $\gamma_\pm$ (KCl), one can systematically calculate the individual activity coefficients of many other ions from appropriate experimental data on the mean ion activity coefficients. For example, for a solution of a MCl_2 electrolyte, where M is a divalent cation, one can write, using Eqs. (12.2.3b) and (12.6.6)

$$\gamma_{\pm MCl_2} = \left[(\gamma_{M^+})(\gamma_{Cl^-})^2 \right]^{1/3} = \left[(\gamma_{M^+})(\gamma_{\pm KCl})^2 \right]^{1/3}$$

so that

$$\gamma_{M^+} = \frac{(\gamma_{\pm MCl_2})^3}{(\gamma_{\pm KCl})^2} \qquad (12.6.7)$$

A similar method may be developed to calculate the individual activity of an anion that combines with K to form an electrolyte, e.g. KF.

In case the electrolyte does not contain either K or Cl, one needs to employ a connection with these ions by introducing an intermediate step, which is called the "double bridge method" by Garrels and Christ (1965). For example, to calculate $\gamma(Cu^{2+})$ from the measured values of $\gamma_{\pm}(Cu_2SO_4)$, one needs to use $\gamma(SO_4)^{2-}$ that is determined from the experimental data for $\gamma_{\pm}(K_2SO_4)$.

12.7   Multicomponent High Ionic Strength and High P-T Systems

In modeling natural processes, we usually need to deal with solutions that have multiple electrolytes, high ionic strength, and are subjected to P-T conditions that are often far removed from the typical 1 bar, 298 K condition of the conventional domain of electrolyte thermodynamics. As an example, progressive evaporation of sea water to form brines would lead to multicomponent solution of high ionic strength that cannot be handled with simple extension of Debye-Hückel formulation of ion activity coefficients. In addition, modeling of such geological processes as fluid-rock interactions in the mid-ocean ridges or in the Earth's crust require an understanding of the behavior of electrolyte solutions at relatively high P-T conditions. There are two groups of major contributions to address these types of problems, one by Pitzer (1973, 1975, 1987) and the other by Helgeson and coworkers (Helgeson and Kirkham 1974a, b, 1976; Helgeson et al. 1981; Tanger and Helgeson 1981, 1988; Shock et al. 1992). These are often referred to as Pitzer equations/model and HKF model.

The Pitzer equations were originally developed to treat multicomponent electrolytes of relatively high ionic strengths at 1 bar, 25 °C. Some attempts were later made to extend it to higher P-T conditions (e.g. Pitzer 1987), but here we limit our discussion to see how the properties of multicomponent electrolytes are treated in the Pitzer equations since this is what the Pitzer equations are famous for. Very good exposition of the Pitzer formulation can be found in Harvie and Weare (1980) and Wolery (1992).

Pitzer introduced an expression of ΔG^{xs} of a multicomponent electrolyte solution as

$$\frac{\Delta G^{xs}}{RT} = n_w \left[f(I) + \sum_{ij} \lambda_{ij} m_i m_j + \sum_{ijk} \xi_{ijk} m_i m_j m_k \right] \qquad (12.7.1)$$

where n_w is number of kilograms of solvent, $f(I)$ is a "Debye-Hückel function" accounting for the long-range electrical interactions to first order, and λ_{ij} and ξ_{ijk} are binary and ternary interaction parameters that are, in general, functions of the ionic strength. The first and second summations are to be carried out over every binary and ternary sub-systems, respectively, of the multicomponent solution.

Except for the $f(I)$ term, the above expression is formally similar to that of ΔG^{xs} of non-electrolyte solution (Eq. 9.3.2). It can be extended to include quaternary interactions, but it was found that such higher order terms are almost never required to treat the properties of multicomponent electrolyte solutions (We recall that quaternary interactions are not needed for multicomponent non-electrolyte solutions if the bounding binaries conform to the regular or sub-regular models, as are often the case). The "Debye-Hückel function" used in the above expression is not the usual Debye-Hückel expression of activity coefficient, but a similar expression derived by Pitzer (1973). As in the case of non-electrolyte solution, the binary and ternary interaction parameters are to be determined by fitting experimental data in the binary and ternary sub-systems of the multicomponent solution. Once there is an expression for ΔG^{xs}, all other excess thermodynamic properties, including the ion activity coefficients, can be obtained by systematic thermodynamic operations on the expression of ΔG^{xs}, as discussed in Sect. 8.6. Figure 12.4 illustrates the success of Pitzer formulation to represent ion activity coefficients in aqueous solution to high ionic strength. A computer program, EQ3NR, was developed in the Lawrence Livermore National Laboratory (Wolrey 1992) to carry out calculations using the Pitzer model. The program has self-consistent values of the different parameters that are needed for this purpose.

The Pitzer model has been used successfully to calculate the development of mineral sequences in a number of geological environments at 1 bar, 25 °C. Figure 12.5 shows agreement between the phase diagram determined experimentally and calculated by Harvie and Weare (1980) using Pitzer model for the system Na-Mg-Cl-SO$_4$-H$_2$O at 1 bar, 25 °C. Harvie and Weare (1980) calculated the

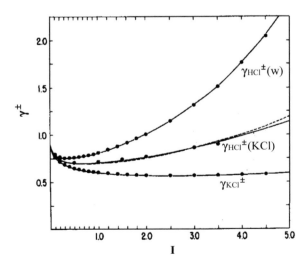

Fig. 12.4 The mean activity coefficient of HCl and KCl in different systems as function of ionic strength, I. $\gamma^{\pm}_{HCl}(w)$ and $\gamma^{\pm}_{HCl}(KCl)$: mean ion activity coefficient of HCl in HCl-H$_2$O and KCl-H$_2$O systems, respectively; $\gamma^{\pm}_{KCl}$: mean ion activity coefficient of KCl in KCl-H$_2$O. The solid curve represents calculation using Pitzer equation whereas the dashed curve is a simplified version of the Pitzer equation discussed in Harvie et al. (1984). Filled circles: experimental data. From Harvie et al. (1984). With permission from Elsevier

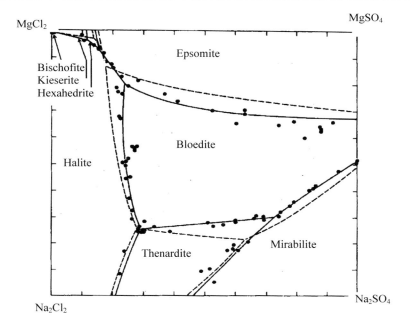

Fig. 12.5 Comparison of the calculated and experimental (symbols) phase diagram for the Na-Mg-Cl-SO₄-H₂O reciprocal system. Solid lines: calculations by Harvie and Weare (1980) using the Pitzer model. The dashed lines represent results of another model discussed by Harvie et al. (1984). From Harvie et al. (1984). With permission from Elsevier

development of mineralogical sequence due to progressive evaporation of sea water. The calculated sequence essentially matches the observed sequence of mineral zones in the classic Zechstein evaporite deposit in Germany. Simplified calculations carried out earlier did not have similar level of agreement with the observed sequence, thus leading to complex hypotheses about the formation of this deposit. The calculations of Harvie and Weare (1980) almost conclusively prove that the Zechstein deposit formed by the progressive evaporation of sea water.

The HKF group of papers developed internally consistent data set for the standard state properties of species in aqueous solutions that can be used to calculate the equilibrium constants of a wide variety of geochemical reactions from 1 to 5 kbar and 0 to 1000 °C. These calculations may be carried out using a software package, SUPCRT92 (Johnson et al. 1992) that has been used widely in geochemical and materials science literature. The standard state properties have been calculated through a combination of fundamental theory and available experimental data that help constrain the theoretical parameters. A comprehensive discussion of HKF papers is beyond the scope of this chapter, but it is noted that the HKF formulations account for the effects of both long- and short-range ionic interactions, local collapse of solvent structure around solvated ions, the concentration dependence of dielectric constant of electrolyte solutions, and the effect of ion association on ionic strength. An impression of the level of success of the HKF models may be conveyed by showing examples of the agreement between predicted and measured properties. In this spirit, Fig. 12.6 is reproduced from Helgeson et al. (1981) and Fig. 12.7 from Tanger and Helgeson (1988) showing, respectively, the variation of mean activity coefficient of NaCl and standard partial molal volume of NaCl as function of pressure and temperature.

12.8 Activity Diagrams of Mineral Stabilities

At constant P-T condition, variation of fluid compositions affects the mineral stabilities. It is, thus, useful to construct mineral stability diagrams as function of the activities of fluid species at constant P-T conditions. These types of diagrams are known as **activity diagrams**, since both axes in the plot represent activities (or combination of activities) of selected aqueous species. Bowers et al. (1984) have presented an entire book of such diagrams for rock-forming minerals for pressures and temperatures up to 5 kb and 600 °C using the standard state thermodynamic properties from Helgeson et al. (1981). We first present below the thermodynamic methodology for the construction of activity-activity diagrams, and then discuss illustrative applications. The standard states of the mineral components and H_2O

Fig. 12.6 Comparison of experimental mean ion activity coefficients (symbols) of NaCl with those calculated from HKF model as function of ionic strength at various temperatures (indicated in °C). The curve for 25 °C is at 1 bar pressure whereas the other curves are at pressures defined by the liquid-vapor equilibrium at the specified temperatures. From Helgeson et al. (1981). With permission from American Journal of Science

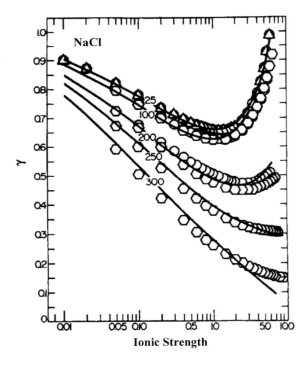

Fig. 12.7 Comparison of standard partial molal volume of NaCl as a function of pressure and temperature, as calculated by Tanger and Helgeson (1988: curves) with the experimental data (symbols). P_{SAT} refers to pressure along liquid-vapor saturation pressures for H_2O. From Tanger and Helgeson (1988). With permission from American Journal of Science

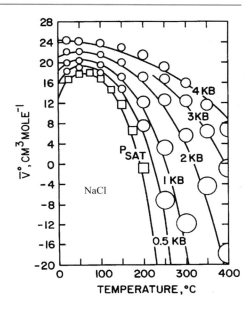

have been chosen to be pure components at P-T, while the dilute aqueous species are referred to the solute standard states at unit molality defined by the Henry's law line (Fig. 12.1).

12.8.1   Method of Calculation

To illustrate the calculation of activity diagrams, let us consider the problem of the stability of K-feldspar ($KAlSi_3O_8$), kaolinite ($Al_2Si_2O_5(OH)_4$) and muscovite ($KAl_3Si_3O_{10}(OH)_2$) in the presence of water at 1 bar, 298 K, which has been discussed earlier by Anderson (2005). The reaction relations among the three minerals and aqueous ions can be written as

$$KAlSi_3O_8 + 1/2\,H_2O + H^+\,(aq) = 1/2\,Al_2Si_2O_5(OH)_4 + 2SiO_2(aq) + K^+\,(aq)$$

$$(12.8.a)$$

$$3/2\,KAlSi_3O_8 + H^+\,(aq) = 1/2\,KAl_3Si_3O_{10}(OH)_2 + 3\,SiO_2(aq) + K^+\,(aq)$$

$$(12.8.b)$$

and

$$KAl_3Si_3O_{10}(OH)_2 + 3/2\,H_2O + H^+\,(aq) = 3/2\,Al_2Si_2O_5(OH)_4 + K^+\,(aq)$$

$$(12.8.c)$$

To illustrate the general procedure of calculation of equilibrium boundary in the activity space, let us now consider the reaction (12.8.a) and assume, for simplicity, that the minerals and H_2O are present essentially in their respective pure states so that $a(H_2O) = a(mineral) = 1$. Thus, at equilibrium at 1 bar, T, we have

$$K_a(1, T) = e^{-\Delta_r G_a^*/RT} = \frac{\left(a_{SiO_2}^{aq}\right)^2 \left(a_{K^+}^{aq}\right)}{\left(a_{H^+}^{aq}\right)} \tag{12.8.1}$$

with

$$\begin{aligned} \Delta_r G_a^* = &\left[1/2(\Delta G_{f(kaolinite)}^o) + 2\Delta G_{f(SiO_2:aq)}^\times + \Delta G_{f(k^+:aq)}^\times\right] \\ &- \left[\Delta G_{f(k-spar)}^o + 1/2(\Delta G_{f(H_2O)}^o) + \Delta G_{f(H^+:aq)}^\times\right] \end{aligned} \tag{12.8.2}$$

Using the data from Wagman et al. (1982), and noting that $\Delta G_{f(H^+:aq)}^\times = 0$, according to convention,

$$\begin{aligned} \Delta_r G_a^*(1 \text{ bar}, 298 \text{ K}) = &[1/2(-3799.7) + 2(-833.411) + (-283.27)] \\ &- [-3742.9 + 1/2(-237.129)] = 11.523 \text{ kJ} \end{aligned} \tag{12.8.3}$$

Since there are three activity terms in Eq. (12.8.1), two of these terms need to be combined for a two dimensional activity diagram. The manner in which the different activity terms should be combined is dictated by the requirement that all equilibria in the system are to be represented in a single diagram. Thus, we decide to plot the ratio $a(K^+)/a(H^+)$ on the y-axis, and $a(SiO_2)$ on the x-axis. In fact, for reasons that should be obvious from the derivation below, we plot the logarithms of the activity terms instead of the activity terms themselves.

The next step is to calculate the slopes of the equilibrium reaction boundaries in the activity diagram. This is easily done from the expression of K. Thus, for the equilibrium (12.8.a), we obtain by taking the logarithm of Eq. (12.8.1), and rearranging terms

$$\log \frac{a_{K^+}}{a_{H^+}} = \log K_a - 2 \log(a_{SiO_2}) \tag{12.8.4}$$

Similarly,

$$\log \frac{a_{K^+}}{a_{H^+}} = \log K_b - 3 \log(a_{SiO_2}) \tag{12.8.5}$$

and

$$\log \frac{a_{K^+}}{a_{H^+}} = \log K_c, \tag{12.8.6}$$

The last expression implies a zero slope of the equilibrium boundary of the reaction (12.8.c) in the log (a_{K^+}/a_{H^+}) versus log (a_{SiO2}) plot.

Equations (12.8.4) and (12.8.5) are linear equations in which the coefficients of log $[a(SiO_2)]$ define the slopes and the log K terms define the intercepts of the equilibrium boundaries in the activity plot. The log K terms can be calculated from the available data on the standard state free energies, using the relation ln K = $-\Delta_r G^*/RT$. Thus, using $\Delta_r G^*$ (1 bar, 298 K) for equilibrium (a) calculated above, log K_a (1 bar, 298 K) = -2.019. Similarly, log K_b (1 bar, 298 K) = -4.668 and log K_c = 3.281.

The reaction boundaries defining the equilibrium stability limits of muscovite, kaolinite and K-feldspar at 1 bar, 298 K, as calculated from the above values of equilibrium constants and the last three equations, are illustrated in Fig. 12.8. According to reactions (12.8.a) and (12.8.b), increasing $a(SiO_2(aq))$ stabilizes K-feldspar with respect to either muscovite or kaolinite. Thus, K-feldspar stability field is located on the higher $a(SiO_2)$ side. Also, the change of breakdown product of K-feldspar must cause a reduction of its field of stability relative to that defined by the extension of a single reaction past the transition between kaolinite and muscovite. This requirement leads to the topology shown in Fig. 12.8.

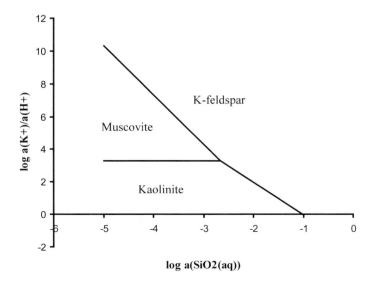

Fig. 12.8 Stability relations of muscovite, kaolinite and K-feldspar at 1 bar, 298 K, as function of activities of ions dissolved in an aqueous solution

12.8.2 Illustrative Applications

12.8.2.1 Spring Waters

As examples of applications of activity diagrams of aqueous species to geological problems, we consider here the studies of Norton and Panichi (1978) and Marini et al. (2000) of spring waters in Italy.

Norton and Panichi (1978) studied the chemistry of the spring waters in the Abano region in northern Italy and determined the source region and subsurface circulation path by comparing the water chemistry with those expected from equilibration with different minerals in the bed rock. Figure 12.8 shows the stabilities of the minerals kaolinite, Ca-montmorillonite and Mg-montmorillonite at 1 bar, 75 °C (348 K) in a log $(a_{Ca++}/(a_{H+})^2)$ versus log $(a_{Mg++}/(a_{H+})^2)$ plot (It is left to the reader as an exercise to figure out the reactions that relate the equilibrium boundaries to these activity ratios). It is assumed that the mineral phases and H_2O have unit activities.

The dashed lines in Fig. 12.9 represent the saturation conditions of the aqueous solution for calcite ($CaCO_3$) and dolomite ($CaMg(CO_3)_2$) at $P(CO_2) = 10^{-2}$ bars. The saturation condition is calculated as follows. For the dissolution of calcite in an aqueous solution, one can write

$$CaCO_3 + 2\,H^+\,(aq) = Ca^{2+}\,(aq) + CO_2(g) + H_2O \tag{12.8.d}$$

for which

$$K_d \equiv e^{-\Delta_r G_d^*} = \frac{\left(a_{Ca^{2+}}^{aq}\right)(a_{CO_2})(a_{H_2O})}{(a_{CaCO_3})\left(a_{H^+}^{aq}\right)^2} \tag{12.8.7}$$

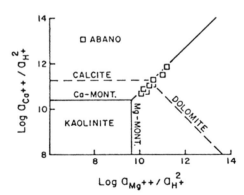

Fig. 12.9 Measured compositions (squares) of water samples from Abano region, Italy, plotted on calculated activity diagram depicting the stability of minerals in equilibrium with an aqueous phase at 1 bar, 75 °C and unit activity of H_2O. Dashed lines represent saturation surfaces of calcite and dolomite at $P_{CO_2} = 10^{-2}$ bars. From Norton and Panichi (1978). With permission from Elsevier

With the standard state of CO_2 as pure CO_2 at unit fugacity at T, so that $a(CO_2) =$ $f(CO_2) \approx P(CO_2)$ and $G^*(CO_2) \approx G^o(CO_2)$ at 1 bar, T, and assuming that $CaCO_3$ is in the pure state so that $a(CaCO_3) = 1$, the last equation reduces to

$$K_d \equiv e^{-\Delta_r G_d^*} \approx \frac{\left(a_{Ca^{2+}}^{aq}\right)(P_{CO_2})}{\left(a_{H^+}^{aq}\right)^2} \qquad (12.8.8)$$

with

$$\Delta_r G_d^* \approx \left[\Delta G_{f(Ca^{2+}:aq)}^{\times} + \Delta G_{f(CO_2)}^{o} + \Delta G_{f(H_2O)}^{o}\right]$$
$$- \left[\Delta G_{f(CaCO_3)}^{o} + 2\Delta G_{f(H^+:aq)}^{\times}\right]$$

with all G values at 1 bar, 348 K. Thus, for calcite saturation, we have

$$\log\left(\frac{a_{Ca^{2+}}^{aq}}{\left(a_{H^+}^{aq}\right)^2}\right) \approx \log K_d(1\ bar,\ 348\ K) - \log P_{CO_2} \qquad (12.8.9)$$

Consequently, in a plot of $\log\left(a_{Ca^{2+}}/a_{H^+}^2\right)$ versus $\log\left(a_{Mg^{2+}}/a_{H^+}^2\right)$, the calcite saturation curve is a horizontal line (i.e. the coefficient for the x-axis term is zero) with an intercept given by the value of $\log(K_d/P_{CO_2})$. It is now left to reader to develop the equation for dolomite saturation curve (Problem 12.2).

The square symbols in Fig. 12.9 represent the measured water compositions in the Abano hot spring region. The compositions plot on the equilibrium boundary between Ca-montmorillonite and Mg-montmorillonite, and thus imply that the fluid had passed through bed rocks containing these minerals and essentially achieved equilibrium with them. In addition, several fluid compositions (not shown in Fig. 12.9) fall above or close to the calcite and dolomite saturation curves. This implies that the fluid had also passed through bed rocks containing these carbonate minerals. The oxygen isotopic composition indicates that the hot spring fluid is derived from meteoric water infiltration into Permian and Mesozoic aquifers in the pre-Alps, which lie north of the Abano region. Thus, the combination of isotope geochemistry and thermodynamic calculations leads to an understanding of the source region and circulation path of fluid that finally found surface expression as hot springs.

Marini et al. (2000) studied the spring water chemistry in the Bisagno valley, Genoa, Italy, and developed model for the evolution of the water chemistry in terms of reaction kinetics between the bed rock minerals and aqueous fluid. The kinetic aspect of this work is beyond the scope of the present chapter, but the readers are encouraged to read this analysis for a good introduction to kinetic modeling.

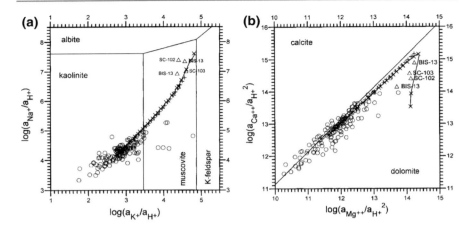

Fig. 12.10 Measured (open symbols) and calculated (crosses) compositions of the Bisagno valley spring waters, Genoa, Italy, plotted on activity diagrams depicting the stability of minerals. Reproduced from Marini et al. (2000). With permission from Elsevier

Figure 12.10 shows the measured fluid compositions (circles) in two activity diagrams, and also the computed kinetic evolution paths of the fluid compositions (crosses).

There is good agreement between the measured and computed fluid compositions. The measured fluid compositions fall mostly in the field of kaolinite and had clearly evolved within the stability field of this mineral (Fig. 12.10a). The fluid compositions also evolved along the equilibrium boundary between calcite and dolomite (Fig. 12.10b), which implies that the fluid achieved equilibrium with these minerals as it moved through the bed rocks. The measured compositions do not show any buffering by the kaolinite-muscovite boundary. This is in good agreement with computed fluid compositions that account for the reaction kinetics taking into consideration the reacting surface areas. As noted by Marini et al. (2000), the observed equilibration of the fluid with the carbonates (Fig. 12.10b) and lack of buffering by the muscovite-kaolinite boundary (Fig. 12.10a) are consistent with the relatively higher dissolution rates of the carbonates in aqueous solution.

One interesting aspect of the compositional trend of fluid of the Bisagno valley is that the fluid compositions seem to have evolved towards the albite, K-feldspar, muscovite invariant point, representing the situation of final stable equilibrium between the aqueous solution and the constituent minerals of local aquifer rocks However, probably due to kinetic reasons, the evolutionary path does not follow the muscovite, kaolinite phase boundary, as expected based on a purely thermodynamic ground. These calculations serve to illustrate the point that while activity plots of stability of minerals in equilibrium with aqueous solutions provide useful framework for the understanding of fluid-rock reactions, the fluid compositions do not necessarily evolve in thermodynamic equilibrium with the minerals.

Problem 12.2 Develop the equation for dolomite saturation of aqueous solution and determine the slope of the line in the $\log\left(a_{Ca^{2+}}/a_{H^+}^2\right)$ versus $\log\left(a_{Mg^{2+}}/a_{H^+}^2\right)$ activity diagram (Fig. 12.9).

12.8.2.2   Stability of Magnesium Silicates

As a final example of the use of activity diagram, we illustrate the stability of magnesium silicates (Fig. 12.11) that was calculated by Faure (1991), using $\log\left(a_{Mg^{2+}}/a_{H^+}^2\right)$ and $\log(a_{H_2SO_4})$ as the two axes. To illustrate the calculation of mineral stabilities in the activity plot, let us consider the dissolution reaction of forsterite in water, viz.,

$$1/2\,Mg_2SiO_4 + 2H^+\,(aq) \leftrightarrow Mg^{2+} + 1/2\,H_4SiO_4(aq) \qquad (12.8.e)$$

for which the expression of equilibrium constant and standard state thermodynamic properties yield

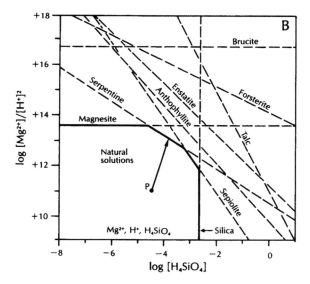

Fig. 12.11 Solubility limits of pure magnesium silicates in water at 1 bar, 25 °C. With increase in the activity ratio of $[Mg^{2+}/H^+]$ and/or activity of $[H_4SiO_4]$, natural water will precipitate one of the phases, magnesite, serpentine, sepiolite, or amorphous silica, depending on the manner in which these activities change. This diagram shows that at the surface of the earth, the only phases that are thermodynamically stable in contact with natural water are magnesite, serpentine, sepiolite and amorphous silica. From Faure (1991)

$$\log \frac{a_{Mg^{2+}}}{(a_{H^+})^2} = -0.5 \log(a_{H_2SiO_4}) + 14.2 \qquad (12.8.10)$$

This equation is represented by the dashed curve labeled "Forsterite" in Fig. 12.11. Forsterite is stable in equilibrium with an aqueous solution with composition above the dashed curve, but dissolves in aqueous solutions with composition below this curve.

If equilibrium is achieved, all magnesium silicate minerals would dissolve in an aqueous solution with composition below the heavy line in the lower left side of the activity diagram. The arrow indicates the equilibrium evolution of the composition of an aqueous solution with initial composition at the point P as a result of dissolution of enstatite. The trajectory of fluid composition intersects the stability limit of serpentine. Thus under equilibrium condition, an aqueous solution with initial composition marked by the point P would precipitate serpentine after sufficient dissolution of enstatite. Figure 12.11 shows that only magnesite, serpentine, sepiolite and amorphous silica are thermodynamically stable in contact with natural water on the surface of the Earth.

12.9 Electrochemical Cells, Nernst Equation and f(O₂) Measurement by Solid Electrolyte

12.9.1 Electrochemical Cell and Half-Cells

In an electrochemical cell, electrons released in one part the cell by oxidation of an electrode flow through a conducting wire to another electrode and are used for reduction reaction, as illustrated in Fig. 12.12. Following IUPAC (International

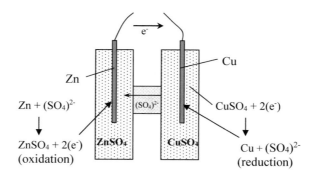

Fig. 12.12 Schematic illustration of an electrochemical cell consisting of two half-cells. Electrons are released in the left half-cell by the oxidation of a Zn electrode, and are transferred to the right half-cell, where these react with a CuSO₄ solution to cause deposition of Cu on to the Cu-electrode. The sulfate ions released due to the reduction of Cu²⁺ are transferred to the left half-cell through a semi-permeable membrane

Union of Pure and Applied Chemistry) convention, the part of the cell undergoing reduction reaction is drawn on the right and the oxidizing part on the left. In this figure, an electrode of Zn metal on the left undergoes oxidation by reaction with a solution according to

$$Zn(metal) \rightarrow Zn^{2+}(solution) + 2(e^-) \qquad (12.9.a)$$

The released electrons travel via a conducting wire to the right electrode of Cu that is immersed in a solution consisting of Cu^{2+} ions. The electrons react with the solution according to

$$Cu^{2+}(solution) + 2(e^-) \rightarrow Cu(metal) \qquad (12.9.b)$$

resulting in the deposition of metallic Cu on the electrode. The two parts of the cell are called **half-cells**, and the above reactions constitute half-cell reactions. The net reaction in the full-cell is

$$Zn(metal) + Cu^{2+}(solution) \rightarrow Zn^{2+}(solution) + Cu(metal) \qquad (12.9.c)$$

The net reaction can be observed by immersing a Zn metal rod in a $CuSO_4$ solution. One would observe dissolution of Zn metal in the solution to form $ZnSO_4$ (Zn + $CuSO_4 \rightarrow ZnSO_4$ + Cu) and complementary precipitation of Cu metal.

Let us suppose that the solution in the left half-cell is $ZnSO_4$ and that in the right half-cell is $CuSO_4$. Oxidation of a small amount of Zn starts the process of electron transfer to the Cu electrode where it reacts with $CuSO_4$ to form metallic copper ($CuSO_4 + 2(e^-) \rightarrow$ Cu (metal) + $(SO_4)^{2-}$). If the cells are completely isolated except with respect to electron transfer, then the process will stop because of charge build up in the half-cells. In order for the process to continue, the half-cells need to be connected by a semi-permeable membrane that permits transfer of $(SO_4)^{2-}$ ions from right to the left half-cell, and thus enables continued release of electrons as a result of the reaction Zn (metal) + $(SO_4)^{2-} \rightarrow ZnSO_4 + 2(e^-)$.

12.9.2 Emf of a Cell and Nernst Equation

The electrical potential difference between the two half-cells can be measured at any time during the process of electron transfer by connecting the electrodes to a voltmeter or a potentiometer using similar metal wire leads. Following the IUPAC convention, the electromotive force (*emf*), E, of the full cell is given by

$$E = E(reducing\ electrode : right) - E(oxidizing\ electrode : left)$$

If the potential on a charge χ is changed by E as it moves from one part to another in a uniform electrical field, then the electrical work done on the charge is χE. Thus, the electrical work done on the system when n moles of electrons (negative charge) are subjected to a potential change E is given by $-nF'E$, where F' is the Faraday constant, which is the electric charge in Coulombs (C) carried by one mole of electron ($F' = 96{,}484.56$ C/mol or J/V-mol). As discussed in Sect. 3.2, at equilibrium under constant P-T condition, the Gibbs free energy change of a reaction in a system subjected to non-PV work is given by the reversible non-PV work done **on** the system (Eq. 3.2.4; Box 3.2.1). Thus, when the electrical work is the only non-PV work done on a system, in which the electron transfer takes place from left to right in a reversible manner, we have at constant P-T condition

$$\Delta_r G = -nF'E \tag{12.9.1}$$

This is known as the **Nernst relation**. Now for a reaction $mA + nB = pC + qD$, we have from Eq. (10.4.5)

$$\Delta_r G = \Delta_r G^* + RT \ln \frac{(a_C)^p (a_D)^q}{(a_A)^m (a_B)^n} \tag{12.9.2}$$

Combining the last two equations, we then have, at equilibrium at constant P-T condition,

$$\boxed{E = E^* - \frac{RT}{nF'} \ln \frac{(a_C)^p (a_D)^q}{(a_A)^m (a_B)^n}} \tag{12.9.3}$$

where E^* ($=-\Delta_r G^*/nF'$) is the emf of the cell when the all ions are present in their respective standard states. (Here the numerator relates to the reduced assemblage and the denominator to the oxidized one.) This equation, which is of great fundamental importance in electrochemistry, is often referred to as the **Nernst equation**. The quantity R/F equals $8.617(10^{-5})$ V/K.

12.9.3 Oxygen Fugacity Measurement Using Solid Electrolyte Sensor

As an application of the Nernst relation to a problem in experimental petrology, let us consider determination of $f(O_2)$ in a gas mixing furnace by a solid electrolyte sensor (e.g. zirconia sensor). In experiments carried out in such furnaces, it is a common practice to flow a fixed ratio of CO_2 and CO through the furnace to control the $f(O_2)$ via the redox equilibrium

$$2CO + O_2 = 2CO_2 \tag{12.9.d}$$

for which

$$\Delta_r G_d = \Delta_r G_d^o + RT \left(2\ln \frac{f(CO_2)}{f(CO)} - \ln f(O_2) \right) \tag{12.9.4}$$

The way the above reaction proceeds from left to right via a solid electrolyte is as follows. Oxygen from a known source, say air, for which the $f(O_2)$ is defined, flows through a solid electrolyte as O^{2-} ions ($O_2(gas) + 4e^- = 2O^{2-}$) and reacts with CO in the gas mixture to form CO_2, thereby releasing $4e^-$ ($2O^{2-} + 2CO = 2CO_2 + 4e^-$). A schematic representation of the process is shown below.

Air side (reduction)   Gas side (oxidation)

$$O_2(gas) + 4e^- \rightarrow 2O^{2-} \rightarrow 2O^{2-} + 2CO \rightarrow 2CO_2 + 4e^-$$

←

Electron flow

Here the reduction is on left and oxidation is on the right (electron transfer takes place from right to left), which is opposite to the IUPAC convention.

At equilibrium, $\Delta_r G_d = 0$. Hence, from Eq. (12.9.4), the standard state Gibbs energy change of the reaction (12.9.d) is given by

$$\Delta_r G_d^o = RT \left(\ln f(O_2)_{eq} - 2\ln \frac{f(CO_2)}{f(CO)} \right) \tag{12.9.5}$$

where $f(O_2)_{eq}$ stands for the equilibrium oxygen fugacity in the imposed CO-CO_2 mixture in the furnace. When the $f(O_2)$ is that in the air or a in a reference source, the Eq. (12.9.4) becomes

$$\Delta_r G_d = \Delta_r G_d^o + RT \left(2\ln \frac{f(CO_2)}{f(CO)} - \ln f(O_2)_{air} \right) \tag{12.9.6}$$

where the subscript 'air' refers to air or any other reference source of oxygen. Combining the last two equations, we now have

$$\Delta_r G_d = RT \left(\ln \frac{\ln f(O_2)_{eq}}{\ln f(O_2)_{air}} \right) \tag{12.9.7}$$

Thus, at equilibrium in a CO-CO_2 mixture

$$\boxed{E = \frac{RT}{4F'} \left(\ln \frac{\ln f(O_2)_{eq}}{\ln f(O_2)_{air}} \right) = 0.0496T \left(\log \frac{\ln f(O_2)_{eq}}{\ln f(O_2)_{air}} \right) mV/K} \tag{12.9.8}$$

Application of this and analogous relationship requires that there is no significant leakage during voltage measurement and the solid electrolytes have negligible electronic conduction.

12.9.4   Standard Emf of Half-Cell and Full-Cell Reactions

If E* values of many half-cell reactions are available, then these can be paired in different combinations to yield the E* values of a variety of full-cell reactions. However, E of a half-cell reaction cannot be measured directly. Thus, in order that the E* of half-cells can be combined to yield that of full-cells in a self-consistent manner, a half-cell is combined with a standard hydrogen electrode (**SHE**) to determine the full-cell emf. The SHE is a gas electrode in which H_2 gas is bubbled over a specially treated piece of platinum. The dissociation of the gas (catalyzed by Pt) releases two moles of electrons according to

$$2H^+(\text{solution}) + 2(e^-) = H_2(g)$$

$$\xleftarrow{\hspace{2cm}} \text{dissociation} \tag{12.9.e}$$

(Following IUPAC convention, the reduced part is written on the right). The E* value of a full-cell, that is of SHE (or hydrogen-platinum electrode) plus the half-cell of interest, is taken to be the E* of the half-cell itself, assuming that E* (SHE) equals zero (This procedure is analogous to assigning a zero value to $\Delta G_f^x(H^+)$ in one molal ideal aqueous solution, as discussed in Sect. 12.4.2).

As an example of the above procedure, let us consider the half-cell reaction

$$Zn(\text{solid}) \rightarrow Zn^{2+} + 2e^- \tag{12.9.f}$$

This half-cell reaction is combined with the SHE half-cell reaction (12.9.e) to generate the full-cell reaction $Zn(\text{solid}) + 2H^+ \rightarrow Zn^{2+} + H_2(g)$. The measured value of E* of the full-cell reaction, which happens to be 0.763 V, is then assigned to be the E* of the half-cell reaction (12.9.f).

A compilation of E* values of half-cell reactions relative to SHE half-cell for which E* is taken to be zero is provided by Ottonello (1997; Table 8.14). As an example of combining these half-cell data to calculate E* of a full-cell reaction, let us consider the reaction (12.9.c). The E* values for the half-cell components represented by the reactions (12.9.a) and (12.9.b) are, respectively, 0.763 and 0.340. Since the addition of these two reactions lead to the overall reaction (12.9.c), the E* (12.9.c) = E*(12.9.a) + E*(12.9.b) = 0.763 + 0.340 = 1.103 V [the E* values in the tabulation are usually given for the reduction reactions, that is with the electrons on the left side of the reaction; thus the sign of E* value needs to be changed if the half-cell reaction being considered is an oxidation reaction, as in reaction (12.9.a)].

12.10   Hydrogen Ion Activity in Aqueous Solution: pH and Acidity

The activity of H^+ ion in aqueous solution is commonly expressed in terms of a quantity known as pH, and is defined as

$$pH = - \log(a_{H^+}) \qquad (12.10.1)$$

Consider now the dissociation of H_2O to H^+ and $(OH)^-$ ions:

$$H_2O(l) = H^+(aq) + (OH)^-(aq) \qquad (12.10.a)$$

for which the equilibrium constant at 1 bar, 298 K is $1.008(10^{-14})$. Thus,

$$K_a = \frac{\left(a_{H^+}^{aq}\right)\left(a_{(OH)^-}^{aq}\right)}{a_{H_2O}} = 1.008(10^{-14}) \qquad (12.10.2)$$

When the water is essentially pure, $a(H_2O) = 1$ at 1 bar, 298 K, in which case

$$\log\left(a_{H^+}^{aq}\right) + \log\left(a_{(OH)^-}^{aq}\right) = -14 \qquad (12.10.3)$$

In an acidic solution, $a_{H^+} > a_{(OH)^-}$, whereas in a basic or alkaline solution, the reverse is true. For a neutral solution, $a_{H^+} = a_{(OH)^-}$. Thus, we have

$$\begin{array}{l} \log\left(a_{H^+}^{aq}\right) > -7, \quad \text{or pH} < 7 : \text{Acid solution} \\ \log\left(a_{H^+}^{aq}\right) < -7, \quad \text{or pH} > 7 : \text{Basic solution} \end{array}$$

Cross section of pH variation in the Pacific and Atlantic oceans can be found in Millero and DiTrolio (2010). The pH of ocean water has been increasing with time as a result of the increase of anthropogenic CO_2 in the Earth's atmosphere that causes an increase of the H^+ ion concentration in the ocean water (CO_2(-gas) + H_2O = $2H^+$ + CO_3^-(aqueous)). Caldeira and Wickett (2003) calculated this variation of ocean pH as a function of depth and time on the basis of the measured and projected increase of $P(CO_2)$ in the Earth's atmosphere. Their results are illustrated in Fig. 12.13.

12.11   Eh-pH Stability Diagrams

The electrical potential of a cell, measured against a standard hydrogen electrode (SHE) is referred in the geochemical literature as Eh. Both Eh and pH are measurable quantities in natural environments by the use of appropriate electrodes (A good discussion of the topic from a practical standpoint can be found in Garrels and Christ (1965). Anderson (2005) discussed the problems attending the measurement of Eh in natural environments). It is, thus, useful to represent stabilities of minerals and metals at constant P-T condition by diagrams that use Eh and pH as two axes. These are known as Eh-pH diagrams. An additional appeal of such diagrams stems from the fact that H^+ is involved, or can be made to be involved with some manipulation, in the description of the stability relations of a large number of minerals. The Eh-pH diagrams owe their popularity to the seminal studies of the Belgian scientist (metallurgist) Pourbaix (1949, and 54 other papers between 1952

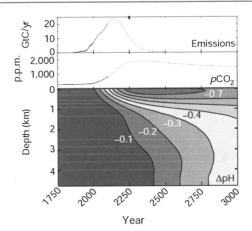

Fig. 12.13 Calculated increase in the acidity of ocean water due to the increase of CO_2 emissions from burning of fossil fuels. The calculations are based on the measured and projected change of CO_2 in the Earth's atmosphere. From Caldeira and Wickett (2003). ΔpH represents the changes in ocean pH based on horizontally averaged chemistry

and 1957 that are cited in Garrels and Christ 1965) The subsequent development and popularity of these diagrams in geochemical literature have been primarily due to R. M. Garrels and co-workers. A compilation of Eh-pH diagrams that are useful for geochemical problems can be found in Brookins (1988).

The construction of Eh-pH diagrams is based on the Nernst equation, Eq. (12.9.2), and is illustrated below by considering specific examples. First, let us formulate the stability of water at 1 bar, 298 K in terms of Eh and pH. For this purpose, we write a reaction between liquid water and gaseous oxygen as

$$O_2(g) + 4H^+(aq) + 4(e^-) = 2\,H_2O(l) \qquad (12.11.a)$$

The Nernst equation for this reaction is

$$Eh = E_a^* - \frac{RT}{4F'} \ln \frac{(a_{H_2O})^2}{a_{O_2}^g \left(a_{H^+}^{aq}\right)^4} \qquad (12.11.1)$$

With water in essentially pure state, $a(H_2O) = 1$. Also, with the standard state of gas as pure state at unit fugacity (~ 1 bar) and T, $a(O_2)$ effectively equals its partial pressure. Thus, after inserting the numerical value RT/F' with T = 298 K, the above expression reduces to

$$Eh = E_a^* + (0.0148) \log P_{O_2} - (0.0591)pH \qquad (12.11.2)$$

The E^* value is obtained from the Nernst relation (Eq. 12.9.1) and the tabulated data on standard Gibbs free energy of formations. Since $G_{f,e}$ (Gibbs free energy of formation from elements) for O_2 (g) and H^+ (aq) are zero by convention,

$$\Delta_r G_a^* = 2\Delta G_{f,e}^o(H_2O) = -474.26\,kJ/mol,$$

according to the data in Wagman et al. (1982), so that $E_a^* = (-\Delta_r G_a^*/4F') = 1.23$ V.

Since the total pressure is fixed at one bar, the **upper limit** of stability (i.e. the stability under the most oxidizing condition) is obtained by setting $P(O_2) = 1$ bar. The **lower limit** of stability of water in an Eh-pH diagram corresponds to the condition of $P(H_2) = 1$. The $P(O_2)$ corresponding to $P(H_2) = 1$ bar can be calculated from the equilibrium constant of the reaction

$$2H_2O(l) = 2H_2(g) + O_2(g), \tag{12.11.b}$$

for which log K(1 bar, 298 K) = −83.1. Thus, at $P(H_2) = 1$ bar, $P(O_2) = 10^{-83.1}$ bars. The absolute upper ($P(O_2) = 1$ bar) and lower ($P(O_2) = 10^{-83.1}$ bar) stability limits of water in an Eh-pH diagram, as calculated from Eq. (12.11.2), are illustrated in Fig. 12.14.

As an illustration of the representation of mineral stabilities in an Eh-pH diagram, let us consider, following Garrels and Christ (1965), a simple example showing the stability of iron oxides in the presence of H_2O. For the reaction between metallic Fe and magnetite (Fe_3O_4), we write the half-cell reaction involving water (again with the reduced assemblage on the right, according to IUPAC convention)

$$Fe_3O_4(s) + 8H^+(aq) + 8(e^-) = 3Fe(s) + 4H_2O \tag{12.11.c}$$

The half-cell reaction for the stability of magnetite with respect to hematite (Fe_2O_3) in the presence of water is given by

$$3Fe_2O_3(s) + 2H^+ + 2(e^-) = 2Fe_3O_4(s) + H_2O(l) \tag{12.11.d}$$

Assuming that the solid phases (s) are pure end members, so that the activity of each solid component, along with that of nearly pure water, is unity, we have

$$Eh_c = E_c^* - \frac{RT}{8F'}\ln\frac{1}{(a_{H^+})^8} \tag{12.11.3}$$

and

$$Eh_d = E_d^* - \frac{RT}{2F'}\ln\frac{1}{(a_{H^+})^2} \tag{12.11.4}$$

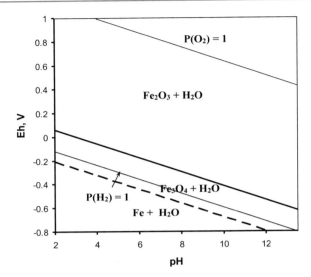

Fig. 12.14 Eh-pH diagram at 1 bar, 298 K showing the stability fields of water and iron oxides. The upper and lower stability limits of water, shown by thin lines, are defined by the conditions of $P(O_2) = 1$ bar and $P(H_2) = 1$ bar, respectively. The boundary between Fe_3O_4 and Fe in the presence of H_2O is shown by a dashed line, since it is a metastable boundary (below the equilibrium stability limit of H_2O)

Calculating E* values from the standard state free energy data of the phases from Wagman et al. (1982) according to the Nernst relation (Eq. 12.9.1), we then have

$$Eh_c = -0.087 - 0.591\,pH; \quad \Delta_r G_c^* = 66,884\,J \tag{12.11.5}$$

and

$$Eh_d = 0.214 - 0.591\,pH; \quad \Delta_r G_d^* = -41,329\,J \tag{12.11.6}$$

The upper and lower stability limits of magnetite in the Eh-pH space, as calculated according to the last two relations, are illustrated in Fig. 12.14 along with the stability limits of water. The Fe_3O_4/Fe boundary, as defined by the equilibrium (12.11.c), is metastable since it falls below the lower stability limit of water (Fig. 12.15).

Next we consider the solubility of iron oxides in water. To treat the equilibrium of the iron oxides with respect to Fe^{2+}(aq), we write the reactions

$$Fe_3O_4(s) + 8H^+\,(aq) + 2(e^-) = 3Fe^{2+}\,(aq) + 4H_2O \tag{12.11.e}$$

$$Fe_2O_3(s) + 6H^+\,(aq) + 2(e^-) = 2Fe^{2+}\,(aq) + 3H_2O \tag{12.11.f}$$

for which, at unit activities of H_2O, Fe_3O_4 and Fe_2O_3

Fig. 12.15 Composite Eh-pH diagram at 1 bar, 298 K showing the stability fields of hematite and magnetite in the presence of water, and contours of constant activities of Fe^{2+} in water. When $a(Fe^{2+}) < 10^{-6}$, the solids may be considered to be essentially insoluble in water

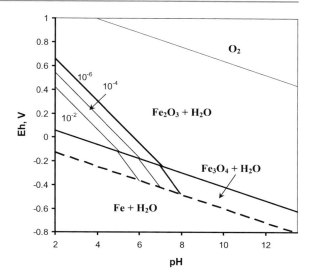

$$Eh_e = E_e^* - \frac{RT}{2F'} \ln \frac{\left(a_{Fe^{2+}}^{aq}\right)^3}{\left(a_{H^+}^{aq}\right)^8} \qquad (12.11.7)$$

and

$$Eh_f = E_f^* - \frac{RT}{2F'} \ln \frac{\left(a_{Fe^{2+}}^{aq}\right)^2}{\left(a_{H^+}^{aq}\right)^6} \qquad (12.11.8)$$

Thus, for the solubility equilibrium of Fe_3O_4

$$Eh_e = E_e^* - 0.089 \log\left(a_{Fe^{2+}}^{aq}\right) - 0.236\,pH \qquad (12.11.9)$$

and that for Fe_2O_3

$$Eh_f = E_f^* - 0.059 \log\left(a_{Fe^{2+}}^{aq}\right) - 0.177\,pH \qquad (12.11.10)$$

with $E_e^* = 0.88$ and $E_f^* = 0.66$ V (as before, the E* values are calculated from the standard free energy data tabulated in Wagman et al. (1982) and Eq. (12.9.1)).

Garrels and Christ (1965) suggested that when the activity of dissolved species in equilibrium with a solid is less than 10^{-6}, the solid may be treated as effectively not dissolved at all. Thus, a field in the Eh-pH diagram should be considered to be a domain where the solid is immobile in contact with water if dissolved species in equilibrium with a solid have a sum total activity of less than 10^{-6}. Figure 12.15

shows the fields of iron oxides with no significant dissolution in water at 1 bar, 298 K, along with the stability field of water at the same condition and contours of constant activity of Fe^{2+}(aq).

12.12 Chemical Model of Sea Water

The ionic strength of typical stream water is ~ 0.01 mol/kg, whereas that of sea water is ~ 0.7 mol/kg (Garrels and Christ 1965). These ionic strengths suggest negligible interaction among the dissolved ionic species in a stream water but significant interaction among them in sea water. The problem of complex formation due to these interactions in sea water was first addressed in a classic study by Garrels and Thompson (1962). These calculations can now be carried out using a public domain computer program EQ3NR from the Lawrence Livermore Laboratory that was written by Wolery (1992) for geochemical aqueous speciation-solubility calculations. It is, however, worth discussing the work of Garrels and Thompson (1962) in order to develop an understanding of how this type of calculations may be carried out with simplifications from insightful approximations. Thus, I first discuss the work of Garrels and Thompson (1962) and then compare the results with those in the EQ3NR test file.

The composition of average surface sea water with 19‰ (‰ implies parts per thousand) chlorinity at 25 °C that was used by Garrels and Thompson (1962) is shown below (Table 12.1). Because of the potential for complex formation, the species reported in this table are not necessarily the actual species present in the sea water. These simply represent the total content of different ions in the sea water.

When an ion is distributed over different ionic complexes and neutral species, one writes a mass balance constraint that must conserve the content of a given species in the analysis of sea water in Table 12.1. For example, the abundance of the various sulfate bearing species must satisfy the relation

$$m_{SO_4^{2-}}(\text{total}) = 0.028 = m_{NaSO_4^-} + m_{KSO_4^-} + m_{CaSO_4^\circ}$$
$$+ m_{MgSO_4^\circ} + m_{SO_4^{2-}}(\text{free}) \tag{12.12.1}$$

assuming that there is no other sulfate species in significant amount. There is also a dissociation reaction for each of the metal-sulfate species such as

Table 12.1 Composition of average surface sea water with 19‰ (parts per thousand) chlorinity at 25 °C (pH = 8.15)

Ion	Molality	Ion	Molality
Na^+	0.48	$(SO_4)^{2-}$	0.028
Mg^{2+}	0.054	$(HCO_3)^-$	0.0024
Ca^{2+}	0.010	$(CO_3)^{2-}$	0.00027
K^+	0.010		
Cl^-	0.56		

Source Garrels and Christ (1965)

$$NaSO_4^{2-} = Na^+ + SO_4^{2-}, \tag{12.12.a}$$

yielding a relation of the type

$$K = \frac{(m_{Na^+})(m_{SO_4^{2-}})}{m_{NaSO_4^-}} K_\gamma \tag{12.12.2}$$

where K_γ represents the usual ratio of the activity coefficients of the different species.

There are four equations of similar form for the four metal-sulfate species. Now, Na^+ may be complexed to ions besides SO_4^{2-}, such as NaCl, Na_2CO_3 etc. Thus, one can write mass balance relation for Na^+ similar to Eq. (12.12.1) and equilibrium constant relations similar to Eq. (12.12.2) for the dissociation of each Na-complex. This process is repeated for every ion, thus yielding as many independent mass balance plus equilibrium constant relations as the number of unknown species. Therefore, one can, in principle, solve for the abundance of each species (since the number of independent relations equal the number of unknowns in the system). The system of equations can be solved using a computer to obtain the molality of each species, if the activity coefficients are known. The activity coefficients of the various species at the ionic strength (0.7 mol/kg) of sea water, as estimated by Garrels and Thompson (1962), are shown in Table 12.2.

The ionic strength of 0.7 mol/kg was calculated from the data in Table 12.1 that do not include complexing with the metal cations. However, strictly speaking, the ionic strength cannot be calculated until the complexing is known. Thus, the ionic strength should be recalculated after the molalities of the different species have been determined using the initial value of the ionic strength of 0.7 mol/kg, and the process should be repeated until there is no more significant change of ionic strength. This is a simple task in computer calculation. However, it was found that the recalculated value of the ionic strength of the sea water after the first set of calculation is not sufficiently different from the initial value to warrant repetition of the calculation.

Table 12.2 Activity coefficients (γ) of various species in sea water at 25 °C (ionic strength: 0.7; chlorinity: 19‰), as calculated by Garrels and Thompson (1962)

Dissolved species	γ	Dissolved species	γ	Dissolved species	γ
$(NaHCO_3)^o$	1.13	$(HCO_3)^-$	0.68	Na^+	0.76
$(MgCO_3)^o$	1.13	$(NaCO_3)^-$	0.68	K^+	0.64
$(CaCO_3)^o$	1.13	$(NaSO_4)^-$	0.68	Mg^{2+}	0.36
$(MgSO_4)^o$	1.13	$(KSO_4)^-$	0.68	Ca^{2+}	0.28
$(CaSO_4)^o$	1.13	$(MgHCO_3)^+$	0.68	Cl^-	0.64
		$(CaHCO_3)^+$	0.68		
		$(CO_3)^{2-}$	0.20		
		$(SO_4)^{2-}$	0.12		

Garrels and Thompson (1962) made the initial simplifying assumption that the cations Na^+, K^+, Ca^{2+} and Mg^{2+} are present essentially as free ions in sea water so that m_{Na+}(total) $= 0.48 \approx m_{Na+}$(free), m_{K+}(total) $= 0.01 \approx m_{K+}$(free), and so on. This is a reasonable simplification since the molalities of the metal cations far exceed those of the complexing anions SO_4^{2-}, CO_3^- and HCO_3^-, and Cl^- is present almost completely as free ion. Iterative solutions of the equations show that the initial assumption is justified and only small portions of Ca^{2+} and Mg^{2+} are complexed with other cations.

The abundance of the major dissolved species in sea water, as derived by Garrels and Thompson (1962), are summarized in the Table 12.3. The results in the EQ3NR test file for sea water are shown as parenthetical numbers in the same table; the italicized numbers were calculated on the basis of B-dot (Eq. 12.6.3) and related activity coefficient expressions, whereas upright numbers were calculated using Pitzer equations, the general idea of which has been discussed in Sect. 12.7 (For some of the ions, there are slight differences between the total molalities used in the calculations of Garrels and Thompson (1962) and in the EQ3NR program file). For ions with total molalities greater than 0.002, there is good agreement

Table 12.3 Major dissolved species in sea water at 25 °C (chlorinity: 19‰, pH = 8.15), as calculated by Garrels and Thompson (1962)

Cation	Molality (total)	Free ion, %	Me-SO$_4$ pair, %	Me-HCO$_3$ pair, %	Me-CO$_3$ pair, %	Me-Cl
Na^+	0.48 (*0.47*)	99 (*95*)	1 (*1.4*)			(*3.5*)
		(100)				
K^+	0.010 (*0.010*)	99 (*98*)	1 (*1.6*)			
		(100)				
Mg^{2+}	0.054 (*0.053*)	87 (*77*)	11 (*14*)	1	0.3 (*0.16*)	(*8.7*)
		(99.8)				
Ca^{2+}	0.010 (*0.010*)	91 (*90*)	8 (*6.5*)	1	0.2 (*0.24*)	(*2.2*)
		(99.8)				
Anion	Molality (total)	Free ion, %	Ca-anion pair, %	Mg-anion pair, %	Na-anion pair, %	K-anion pair, %
SO_4^{2-}	0.028 (*0.028*)	54 (*47*)	3 (*2.4*)	22 (*27*)	21 (*23*)	0.5
		(100)				
HCO_3^-	0.0024 (*0.0019*)	69 (*69*)	4 (*1.76*)	19 (*9.6*)	8 (*19.7*)	
		(100)				
CO_3^{2-}	0.0003 (*0.0001*)	9 (*21.7*)	7 (*22.3*)	67 (*56*)	17	
		(30.2)	(16)	(54)		
Cl^-	0.56 (*0.54*)	100 (*96*)			(*3*)	
		(100)				

The parenthetical numbers are outputs from the test case file of EQ3NR computer program of Wolery (1992)

among the different results except for the abundance of Mg^{2+} and Ca^{2+} species. The calculations carried out with the Pitzer equations show the free ions to constitute nearly the total molalities of these species, whereas the other approaches show significantly less abundance of the free ions. It may be recalled, however, that Garrels and Thompson (1962) started their calculations assuming free ions to constitute 100% of the total molalities of Mg^{2+} and Ca^{2+}.

References

Anderson GM (2005) Thermodynamics of natural systems. Cambridge
Bowers TS, Jackson KJ, Helgeson HC (1984) Equilibrium activity diagrams for coexisting minerals and aqueous solutions at pressures and temperatures to 5 kb and 600 °C. Springer, Berlin, Heidelberg. New York, Tokyo, p 397
Brookins DG (1988) Eh-pH diagrams for geochemistry. Springer, Berlin, Heidelberg, New York
Caldeira K, Wickett ME (2003) Anthropogenic carbon and ocean pH. Nature 425:365
Debye P, Hückel E (1923) The theory of electrolytes. I. Lowering of freezing point and related phenomena. Z Phys 24:185–206
Denbigh K (1981) The principles of chemical equilibrium. Dover
Faure G (1991) Principles and applications of inorganic geochemistry. McMillan, New York, p 626
Garrels RM, Christ CL (1965) Solutions, minerals and equilibria. Harper and Row, New York, p 450
Garrels RM, Thompson ME (1962) A chemical model for sea water at 25 °C and one atmosphere total pressure. Am J Sci 260:57–66
Harvie CE, Weare JH (1980) The prediction of mineral solubilities in natural waters: the Na-K-Mg-Ca-Cl-SO$_4$-H$_2$O system from zero to high concentration at 25 °C. Geochim Cosmochim Acta 44:981–999
Harvie CE, Møller N, Weare JH (1984) The prediction of mineral stabilities in natural waters: the Na-K-Mg-Ca-H-Cl-SO$_4$-OH-HCO$_3$-CO$_3$-CO$_2$-H$_2$O system to high ionic strength at 25 °C. Geochim Cosmochim Acta 48:723–751
Helgeson HC (1969) Thermodynamics of hydrothermal systems at elevated temperatures and pressures. Am J Sci 267:729–804
Helgeson HC, Kirkham DH (1974a) Theoretical prediction of thermodynamic behavior of aqueous electrolytes at high pressures and temperatures: II. Debye-Hückel parameters for activity. Am J Sci 274:1199–1261
Helgeson HC, Kirkham DH (1974b) Theoretical prediction of thermodynamic properties of aqueous electrolytes at high pressures and temperatures: III. Equations of states for species at infinite dilution. Am J Sci 276:97–240
Helgeson HC, Kirkham DH (1976) Theoretical predictions of the thermodynamic properties of aqueous electrolytes at high pressures and temperatures. III. Equation of state of aqueous species at infinite dilution. Am J Sci 276:97–240
Helgeson HC, Kirkham DH, Flowers GC (1981) Theoretical prediction of the thermodynamic behavior of aqueous electrolytes at high pressures and temperatures: IV, calculation of activity coefficients, osmotic coefficients, and apparent molal and standard and relative partial molal properties to 600 °C and 5 kb. Am J Sci 281:1249–1516
Johnson JW, Oelkers EH, Helgeson HC (1992) SUPCRIT92: a software package for calculating the standard molal thermodynamic properties of minerals, gases, aqueous species, and reactions from 1 to 50000 bar and 0 to 1000 °C. Comput Geosci 18:889–947

Langdon C, Atkinson MJ (2005) Effect of elevated pCO$_2$ on photosynthesis and calcification of corals and interactions with seasonal change in temperature/irradiance and nutrient enrichment. J Geophys Res 110:1–16

MacInnes DA (1919) The activities of the ions of strong electrolytes. J Am Chem Soc 41:1086–1092

Marini L, Ottonello G, Canepa M, Cipoli F (2000) Water-rock interaction in the Bisagno valley (Beneoa, Italy): application of an inverse approach to model spring water chemistry. Geochim Cosmochim Acta 64:2617–2635

Millero FJ, DiTrolio BR (2010) Use of thermodynamics in examining the effects of ocean acidification. Elements 6:299–303

Norton DL, Panichi C (1978) Determination of the sources and circulation paths of thermal fluids: the Abano region, northern Italy. Geochim Cosmochim Acta 42:183–294

Ottonello G (1997) Principles of geochemistry. Columbia University Press, p 894

Pitzer KS (1973) Thermodynamics of electrolytes-I. Theoretical basis and general equations. J Phys Chem 77:268–277

Pitzer KS (1975) Thermodynamics of electrolytes V. Effects of higher order electrostatic terms. J Sol Chem 4:249–265

Pitzer KS (1987) Thermodynamic model of aqueous solutions of liquid-like density. In: Carmichael ISE, Eugster HP (eds) Thermodynamic modeling of geological materials: minerals, fluids and melts, Reviews in Mineralogy, vol 2. Mineralogical Society of America, Washington DC, pp 97–142

Pitzer KS (1995) Thermodynamics (International Edition). McGraw Hill Inc., p 626

Pourbaix MJN (1949) Thermodynamics of dilute solutions. E. Arnold, London

Robinson RA, Stokes RH (1959) Electrolyte solutions. Butterworths, London, p 571

Shock EL, Oelkers EH, Johnson JW, Sverjensky DA, Helgeson (1992) Calculation of the thermodynamic properties of aqueous species at high pressures and temperatures. J Chem Soc Faraday Trans 88(6):803–826

Tanger JV IV, Helgeson HC (1988) Calculation of the thermodynamic and transport properties of aqueous electrolytes at high pressures and temperatures: revised equations of state for the standard partial molal properties of ions and electrolytes. Am J Sci 288:19–98

Wagman DD, Evans WH, Parker VB, Schumm RH, Hakow I, Bailey SM (1982) The NBS tables of chemical thermodynamic properties. J Phys Chem Ref Data 2(Supplement No. 2) (American Chemical Society, Washington, DC)

Wolery TJ (1992) EQ3NR, a computer program for geochemical aqueous speciation—solubility calculations: theoretical manual, user's guide, and related documentation (version 7.0). Lawrence Livermore National Laboratory, California

Surface Effects

13

The atoms located at or within a few atomic layers of the surface of a phase are in a different energetic environment than those within the interior of the phase. Thus, the surface properties are different from the properties of the interior of a phase. An example of the difference between the surface and interior configurations of atoms is shown in Fig. 13.1, which represents a molecular dynamic simulation of the interior and near-surface structural configurations of ice at 1 bar, 270 K.

The surface properties play important roles in such problems as the propagation of cracks and brittle failure of rocks, development of crystal morphology and microstructure of rocks as observed in thin sections, coarsening of grains and exsolution lamella, solubility and melting temperature of a solid, nucleation, interconnectivity of fluid in a solid matrix and capillary rise of liquid. Usually the special properties of the surface have no significant influence on the stability of a phase, but when the surface to volume ratio of a solid exceeds a critical limit, that is the grain size becomes sufficiently small, the stability of the solid would be different from what one would calculate from its bulk thermodynamic properties. In this section we discuss the fundamental aspects of the surface properties along with illustrative applications to natural processes. The foundation of the thermodynamics of surfaces, as of many other aspects of thermodynamics, was laid by J. W. Gibbs (see Gibbs 1961: scientific papers).

13.1 Surface Tension and Energetic Consequences

In dealing with surface effects, it is customary to introduce a property known as the **surface tension**, σ, which is the force per unit length, and thus has the unit of J/m^2 (force/length $\equiv$ N/m; N: J/m). The surface tension opposes expansion of a surface. Thus, in order for a surface to expand, work must be done against this opposing force. This work is known as the **surface work**. To develop the expression of surface work, let us consider a simple example of reversible expansion of a

© Springer Nature Switzerland AG 2020

J. Ganguly, *Thermodynamics in Earth and Planetary Sciences*,

Springer Textbooks in Earth Sciences, Geography and Environment,

https://doi.org/10.1007/978-3-030-20879-0_13

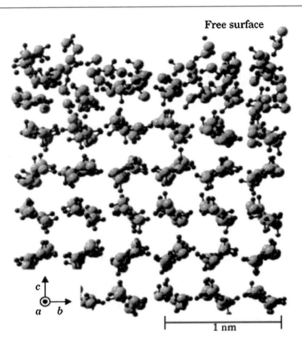

Fig. 13.1 Molecular-dynamics simulation of the near-surface and interior arrangements of water molecules in ice at 1 bar, 270 K. The large gray circles represent oxygen atoms and the small black ones, hydrogen atoms. The thin lines represent the covalent bonds connecting hydrogen and oxygen atoms. From Rosenberg (2005). With permission from American Institute of Physics

rectangular surface of an isotropic material by a length dx, and let l be the length of the side that is being displaced (Fig. 13.2). The work (force times displacement) done on the system is then given by $(\sigma l)dx = \sigma A_s$, where (σl) is the force due to the surface tension, and A_s is the enlargement of the surface area. In the case of solids σA_s is the work needed for the creation of a new surface of area A_s.

With the incorporation of surface work, the change in the internal energy of a closed system is given by (Eq. 2.1.3),

$$dU = \delta q - PdV + \sigma dA_s \qquad (13.1.1)$$

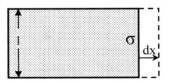

Fig. 13.2 Expansion of the rectangular surface of a material against the force due to surface tension (force per unit length), σ

Consequently, since $G = U + PV$ and $F = U - TS$, we obtain for a reversible process ($\delta q_{rev} = TdS$, according to the second law) in an open system

$$dG = -SdT + VdP + \sum_i \mu_i dn_i + \sigma dA_s \qquad (13.1.2)$$

and

$$dF = -SdT - PdV + \sum_i \mu_i dn_i + \sigma dA_s \qquad (13.1.3)$$

These equations are completely general for isotropic material of any shape. For anisotropic solids, the surface tension has directional properties.

From the last two expressions, we have

$$\sigma = \left(\frac{\partial G}{\partial A_s}\right)_{P,T,n_i} \qquad (13.1.4)$$

and

$$\sigma = \left(\frac{\partial F}{\partial A_s}\right)_{V,T,n_i} \qquad (13.1.5)$$

13.2 Surface Thermodynamic Functions and Adsorption

In order to deal with the problem of surface properties, one first needs to have an unambiguous definition of the surface or interface between two phases. If we consider two homogeneous phases, A and B, in mutual contact, then the different properties do not change discontinuously from one phase to another across a plane, but there is a small region of a few molecular layers through which the properties show a continuous change. For example, if we consider the change of the concentration of a component from one phase to another, then the concentration versus distance profile of the component would look something like the curve shown in Fig. 13.3.

Following Gibbs (see Gibbs 1961, scientific papers), one now defines an interface between the two phases as any geometrical surface within the zone of continuous variation of concentration that pass through "all points which are similarly situated with respect to the condition of adjacent matter." Thus, for the chosen example, a plane situated at x′ as indicated by a vertical dashed line, and any plane parallel to it within the zone of concentration variation can be defined as an interface. (If we consider a spherical grain of uniform composition immersed into a homogeneous liquid, then according to Gibbs' definition the interface is a surface within the zone of continuous variation of composition that passes through points of

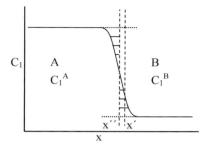

Fig. 13.3 Concentration (C_1) versus distance (x) curve of a major component (1) across the interface of two phases, A and B, each with homogeneous concentrations of the chosen component. The areas indicated by horizontal lines on two sides of the vertical dashed line at x'' are equal. By convention, the plane at x'', which satisfies the "equal area constraint" for the C_1-x curve of the solvent, is chosen as an interface between the phases A and B. At this interface, the surface concentration of the solvent is zero. An interface chosen at any other location between the zones of homogeneous compositions of the two phases, such as at x', has a non-zero concentration of the component 1

the same radial distance from the center of the sphere). The volume of the entire system is divided between the two phases. Thus, the volumes of the phases A and B are the entire volumes to the left and right, respectively, of the chosen interface, say the plane at x' in Fig. 13.3.

If V_A and V_B respectively the volumes of the phases A and B with respect to a chosen interface, C_i^A and C_i^B are the concentrations (mass/volume) of the component i within the interiors of the two phases, A_s is the surface of area of the interface and Γ_i is the concentration of the component i per unit surface area of the interface, then

$$\Gamma_i A_s = n_i - n_i^A - n_i^B$$
$$= n_i - C_i^A V_A - C_i^B V_B, \tag{13.2.1}$$

where n_i is the total number of moles of the component i in the entire system, and n_i^A and n_i^B are its mole numbers in the interior of the phases A and B, respectively. It should be easy to see that because of the non-uniqueness in the position of interface, there is also a non-uniqueness in the value of the surface concentration. When there are two or more components, it is conventional to choose an interface such that it has zero concentration of the solvent or a major component, which we call the component 1, i.e. $n_1 = C_1^A V_A + C_1^B V_B$. Geometrically it means that the interface should be chosen at a value of x-axis such that the areas between the extensions of C_1 (dotted lines) and the concentration variation curves on two sides of the interface are equal, as illustrated in Fig. 13.3. In general, with this definition of the interface, there would be non-zero interface concentration, which could be either positive or negative, of other components at the interface.

Once an interface is defined, the thermodynamic functions of the interface are defined in a formally similar way as the surface concentration. Thus, the Gibbs free energy per unit surface area, G_s, is related to the bulk phase properties according to $G_s A_s = G - G^A - G^B$, where G is the Gibbs free energy of the entire system and G^A and G^B are the Gibbs free energies of the phases A and B, respectively. Now since $G = \sum n_i \mu_i$, we have

$$G_s A_s = G - \sum_i n_i^A \mu_i^A - \sum_i n_i^B \mu_i^B \qquad (13.2.2)$$

In general, we write

$$Y_s A_s = Y - \sum_i n_i^A y_i^A - \sum_i n_i^B y_i^B \qquad (13.2.3)$$

where Y is any thermodynamic property with Y_s as its counterpart per unit surface area, and y_i is the related partial molar property of the component i in the indicated phases.

The chemical potentials of the component in either phase are unaffected by the change of surface area. Thus, combining Eqs. (13.1.2) and (13.2.2), we obtain for a change of surface area of a closed system (i.e. constant value of n_i) at constant T and P

$$G_s dA_s = \sigma dA_s - \sum_i \mu_i^A dn_i^A - \sum_i \mu_i^B dn_i^B \qquad (13.2.4)$$

Now since at equilibrium the chemical potential of a component must be the same throughout the system, the above equation reduces to

$$G_s dA_s = \sigma dA_s - \sum_i \mu_i \left(dn_i^A + dn_i^B \right)$$

where $\mu_i = \mu_i^A = \mu_i^B$. From the definition of surface concentration, (Eq. 13.2.1), the quantity within the last parentheses equals $-\Gamma_i dA_s$ for a closed system (for which $dn_i = 0$). Thus, dividing both sides by dA_s, we obtain

$$\boxed{G_s = \sigma + \sum_i \mu_i \Gamma_i} \qquad (13.2.5)$$

For a one-component system, $\Gamma_i = 0$ by the convention of the choice of the position of an interface, as discussed above. Thus, for a one component system, and **only** for a one component system, the surface tension is the same as the surface free energy per unit area.

Differentiating the last expression and substituting the relation $dG_s = -S_s dT + V_s dP + \sum \mu_i d\Gamma_i = -S_s dT + \sum \mu_i d\Gamma_i$ (since $V_s = 0$), we obtain, for an uncharged surface,

$$d\sigma = -S_s dT - \sum_i \Gamma_i d\mu_i \qquad (13.2.6)$$

This relation, which was derived by Gibbs, is known as the **Gibbs adsorption equation**. It plays very fundamental in the further development of the field of surface thermodynamics.

Instead of dealing with a surface within a small interfacial volume, one may deal with the interfacial volume itself. In that case one obtains (Cahn 1979)

$$d\sigma = -[S]dT + [V]dP - \sum_i [\Gamma_i]d\mu_i \qquad (13.2.7)$$

where [...] indicates properties of the interfacial volume.

13.3   Temperature, Pressure and Compositional Effects on Surface Tension

The temperature and pressure dependence of surface and interfacial tensions follow easily from Eq. (13.2.6). Upon differentiating it with respect to temperature, we have

$$\left(\frac{\partial\sigma}{\partial T}\right) = -S_s - \sum_i \Gamma_i \left(\frac{\partial\mu_i}{\partial T}\right)_{P,n_{j\neq i}}$$
$$= -S_s + \sum_i \Gamma_i s_i \qquad (13.3.1)$$

Similarly, differentiation with respect to pressure at constant temperature yields

$$\left(\frac{\partial\sigma}{\partial P}\right) = -\sum_i \Gamma_i v_i \qquad (13.3.2)$$

In these equations, v_i and s_i are, respectively, the partial molar volume and partial molar entropy of the component i in the **bulk phase**. The P-T dependence of interfacial tension when $d\sigma$ is expressed according to Eq. (13.2.7) should be obvious.

Let us now consider the surface of a phase consisting of two components, 1 and 2, in solution. At constant temperature, Eq. (13.2.6) reduces to

$$d\sigma = -\Gamma_1 d\mu_1 - \Gamma_2 d\mu_2 \qquad (13.3.3)$$

If the boundary is chosen to conform to the condition that $\Gamma_1 = 0$ (Fig. 13.3), then

$$\Gamma_{2(1)} = -\left(\frac{\partial \sigma}{\partial \mu_2}\right)_T = -\frac{1}{RT}\left(\frac{\partial \sigma}{\partial \ln a_2}\right)_T \qquad (13.3.4a)$$

or

$$\Gamma_{2(1)} = -\frac{1}{RT}\left(\frac{\partial \sigma}{\partial \ln f_2}\right)_T \qquad (13.3.4b)$$

where f stands for fugacity (Eq. 8.4.1) and $\Gamma_{2(1)}$ is the surface concentration of component 2 per unit area when that of component 1 is zero. For a non-dissociating solute obeying Henry's law or ideal solution behavior, $d\mu_2 = RTd\ln X_2$, so that

$$\boxed{\Gamma_{2(1)} = -\frac{1}{RT}\left(\frac{\partial \sigma}{\partial \ln X_2}\right)_T} \qquad (13.3.5)$$

(It should be obvious that if the solute dissociates in the solution, then the factor RT on the right needs to be multiplied by the number of dissociated species). Thus, if there is a decrease of surface tension with increasing concentration (more rigorously, activity) or fugacity of the solute 2, then $\Gamma_{2(1)} > 0$, that is there is an actual surface excess of the solute at the chosen interface where $\Gamma_1 = 0$. Because of the effect of impurity, the surface tensions of pure materials are very difficult to obtain experimentally. Using the last expression, the surface concentration of a solute may be determined from the change of surface tension as a function of the mole fraction of the solute.

The effects of addition of oxygen, sulfur and carbon on the surface tension of molten iron in contact with air, as determined by Halden and Kingery (1955), are shown in Fig. 13.4. As we would see later, the effect of addition of sulfur and oxygen on lowering the surface tension of molten iron has interesting consequences for the problem of **core formation** in Earth and Mars.

Fig. 13.4 Effect of addition of carbon, sulfur and oxygen on the surface tension (σ, dyn/cm) of liquid iron in contact with air. From Halden and Kingery (1955). With permission from American Chemical Society

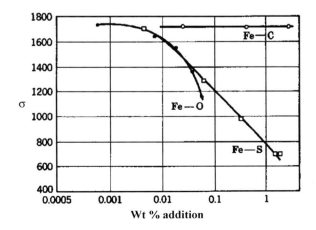

13.4   Langmuir Isotherm

A relatively simple and widely used formulation for the concentration of adsorbed species on a planar surface of a solid at a specified temperature comes from the work of Langmuir (1918) and is known as **Langmuir isotherm**. It is often derived from kinetic arguments (e.g. Lasaga 1998) by considering the rates of adsorption and desorption of a gaseous species on a planar surface, as follows.

At a given temperature, the rates of adsorption and desorption of a gaseous species, i, on a surface are given by

$$\text{Rate of adsorption} = k_{ads} P_i (1 - \theta_i)$$
$$\text{Rate of desorption} = k_{dsrp} \theta_i$$

where P_i is the partial pressure of i, the k-s represent the rate constants for adsorption (ads) and desorption (dsrp) reactions and θ_i is the fraction of total number of surface sites that are occupied by the species i. At equilibrium, the two rates are equal, and thus

$$\theta = \frac{KP_i}{1 + P_i} \qquad (13.4.1)$$

where $K = k_{ads}/k_{dsrp}$. Often K is equated to the equilibrium constant for the adsorption-desorption reaction i + s = i(s), where s and i(s) stand, respectively, for surface and adsorbed i on the surface. However, in general, this notion is erroneous. This point and other restrictions on the domain of applicability of Langmuir relation may be appreciated by considering the statistical thermodynamic derivation of the above relation that is due to Everett (1950), and discussed by Denbigh (1981).

Let $\mu_i(\theta)$ be the chemical potential of the species i in the adsorbed state (with fractional occupancy θ of the surface sites) at T, $\mu_i^o(g)$ is its chemical potential in the pure gaseous state at 1 bar and T, S_m the molar entropy of the adsorbed species at a specific configuration of the occupied and unoccupied surface sites (Fig. 13.5) and $S_m^o(g)$ is that of the pure gaseous species state at 1 bar, T. Note that for a fractional occupancy of θ of an adsorbed species, there are many possible configurations between the occupied and unoccupied sites and these lead to configurational entropy of mixing (see Sect. 2.5); for example, for the number of total and occupied surface sites (12 and 5, respectively) illustrated in Fig. 13.5, there are 198 different ways that the 5 units of the adsorbed species can be arranged on the surface sites.

Now assuming (a) the molar enthalpy and molar entropy of the absorbed species to be independent of the degree of adsorption (i.e. independent of θ) and (b) random distribution of the occupied and unoccupied sites, it can be shown from statistical thermodynamic arguments that the change of chemical potential of the species i upon its transfer from a pure gaseous state at 1 bar, T to an adsorbed state is given by (Denbigh 1981).

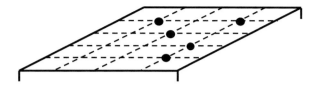

Fig. 13.5 Illustration of adsorption of planar surface sites of a solid by an adsorbate. There are 12 reactive surface sites defined by the intersection of the dashed lines and 5 units of adsorbate marked by filled circles

$$\mu_i(\theta) - \mu_i^o(g) = \Delta\overline{H}_m - T\left(S_m - S_m^o(g)\right) + RT \ln \frac{\theta}{1 - \theta} \qquad (13.4.2)$$

where $\Delta\overline{H}_m$ is the molar heat of adsorption at 1 bar, T. Now let us suppose that the pressure on the gaseous species is P_i when its equilibrium fractional occupancy in the adsorbed state is θ, and $\mu_i(P_i, T)$ is the corresponding chemical potential of i in the gaseous state. At equilibrium, $\mu_i(\theta) = \mu_i(P_i, \text{gas})$. Now, if the gas phase behaves as an ideal gas, then at equilibrium at T,

$$\mu_i(\theta) = \mu_i(P_i, \text{gas}) = \mu_i^o(g) + RT \ln P_i. \qquad (13.4.3)$$

Thus, substituting the relation $\mu_i(\theta) - \mu_i^o(g) = RT \ln P_i$ into Eq. (13.4.2), we obtain

$$P_i = \left(e^{\Delta\overline{G}_m/RT}\right)\left(\frac{\theta}{1 - \theta}\right) \qquad (13.4.4)$$

where $\Delta\overline{G}_m = \Delta\overline{H}_m - T\left(S_m - S_m^o(g)\right)$ and represents molar Gibbs energy of adsorption at 1 bar, T. Now defining K as $\exp(-\Delta\overline{G}_m/RT)$ and rearranging terms, we obtain the Langmuir relation in the form of Eq. 13.4.1.

Note that while K, as defined above, has the form of an equilibrium constant, in general it is not an equilibrium constant for the adsorption reaction (K is an equilibrium constant only when the standard state is chosen to be at 1 bar, T). Also note that the thermodynamic properties of the gaseous state that are to be used in the Langmuir relation are for an ideal gas at 1 bar, T. Nonideal interaction in the adsorbed state may be incorporated in Eq. (13.4.4) in the enthalpic term, but there is no provision for incorporating the excess entropy of mixing in the adsorbed state. In other words, the non-ideality of the adsorbed state, if any, should conform to a Regular or Sub-Regular solution model (see Sect. 9.2) for which the non-ideality is only in the enthalpy of mixing.

As an example of application of the Langmuir isotherm formulation, let us consider the problem of origin of water in the Earth that have been a highly debated issue in the field of planetary sciences, with models ranging from water delivery from exogenous sources (carbonaceous chondrites, comets, pre-solar ice) to direct accretion of water from the solar nebula by adsorption on surface of olivine grains during the accretion process. The last alternative is discussed below.

Using Density functional theory (DFT; see Appendix C.3.3), Asaduzzaman et al. (2015) found that water could have adsorbed on the olivine surfaces from the solar nebula by a mechanism of dissociative chemical adsorption, as illustrated in Fig. 13.6. In this mechanism, a H_2O molecule first gets physically adsorbed on a Mg atom on the surface of an olivine grain; it then dissociates to release a H^+ ion that gets bonded to a surface oxygen atom whereas the O from the water molecule gets bonded to an additional Mg atom on the surface (third panel). The dissociative adsorption process (i.e. the transition from the second to the third panel) is a chemical reaction process and is thus called dissociative chemical adsorption of chemisorption. Using a combination of DFT based calculation and available data on thermochemical properties, Asaduzzaman et al. (2015) estimated $\Delta \overline{G}_m \approx -101$ kJ/mol for the average Gibbs energy of chemisorption of water on olivine surfaces at 1 bar, 900 K. They chose a condition of 10^{-3} bars, 900 K for the condensation of the Earth forming minerals from the solar nebula. This condition is compatible with the calculation Saxena and Hrubiak (2014) for the formation of Earth via nebular condensation (see Sect. 10.5.1, Fig. 10.27). At this condition, $P(H_2O)$ in the solar nebula has been calculated to be $\sim 10^{-6}$ bars. Substitution of the inferred $P(H_2O)$ and $\Delta \overline{G}_m$ values in Eq. (13.4.4) yields fractional occupancy, θ, of adsorbed water on an olivine surface of 0.42.

Making now the reasonable assumption that the mass of olivine in the Earth constitutes $\sim 65\%$ of the mass of the Earth's mantle in the uncompressed state, Asaduzzaman et al. (2015) calculated the total ocean equivalent of water (OEW) in the Earth as a function of the radius of the olivine grains (1 OEW $= 1.4(10^{24})$ g). Their results are illustrated in Fig. 13.7. The dashed line in the figure shows the effect of interaction with an additional layer of water if the reactive sites were saturated by chemisorbed water that would be possible if the $\Delta \overline{G}_m \approx -93$ kJ/mol, that is ~ 8 kJ/mol higher than used in the above calculation of θ (see Asaduzzaman et al. 2015, for discussion of this mechanism).

In their calculations, Asaduzzaman et al. (2015) assumed that the number of available reactive sites equaled the number of Mg atoms on the surface. They also assumed that surfaces are free of any reactive defect sites. However, since according to Fig. 13.6, a molecule of chemisorbed water uses two Mg atoms, the number of reactive sites should be half as much, and hence the OEW of adsorbed

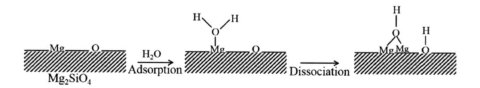

Fig. 13.6 Schematic representation of the process of physical adsorption followed by dissociative chemisorption of a water molecule on the surface of an olivine grain, as inferred from DFT (density functional theory) based calculation. From Asaduzzaman et al. (2015). With permission from Meteoritical Society

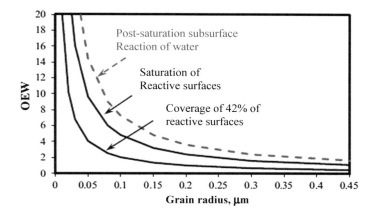

Fig. 13.7 Ocean equivalent of water (OEW) that could have been incorporated into the Earth via dissociative chemisorption of nebular water on olivine surface as a function of grain size of olivine. From Asaduzzaman et al. (2015). With permission from Meteoritical Society

water should also be half as much as what has been illustrated in Fig. 13.7. On the other hand, the presence of defects on the surface would have favored further adsorption of water. And additionally, significant amount of water might also be chemisorbed in pyroxenes. At any rate, the study of Asaduzzaman et al. (2015) suggests the intriguing possibility that a significant fraction of the Earth's water might have accreted directly from the solar nebula via chemisorption on olivine surface.

13.5   Crack Propagation

Qualitatively one may see that the stress required to propagate a crack would depend on the energy needed to create a new surface. Thus, the effect of adsorbed solutes on lowering the surface tension or surface free energy has interesting implications on crack propagation in solids. The critical stress, ξ_c, required for crack propagation in a solid is related to surface tension according to (Lawn and Wilshaw 1975)

$$\xi_c = \left(\frac{2E\sigma}{(1-v^2)\pi c}\right)^{1/2} \tag{13.5.1}$$

where E is the Young's modulus, v is Poisson's ratio and c is the sound velocity. Thus, lowering of the surface tension, σ, by chemical adsorption reduces the critical stress needed for crack propagation. This is illustrated in Fig. 13.8, in which the surface free energy values of quartz, as measured under different chemical environments, are plotted against the critical crack propagation stress in the same environment. Thus, the effect of chemical adsorption on lowering of surface free energy has potential implications on the weakening of geological materials. For example, as discussed by Dunning et al. (1984), a body of rock saturated with highly surface active brine may be substantially weaker and more prone to microfracturing than a dry rock or one saturated with only pure water. Furthermore, a locked up fault may yield to the stress as a result of lowering of surface tension or surface free energy due to chemical infiltration.

13.6   Equilibrium Shape of Crystals

As proposed independently by Gibbs (see Gibbs 1961, scientific papers) and Curie (1885), the equilibrium shape of a particle is the one that minimizes the contribution of the surface area to its free energy. Thus, the process of development of equilibrium shape seeks the minimum of the quantity $\sum G_{s(J)} A_{s(J)}$ where $G_{s(J)}$ and $A_{s(J)}$ are, respectively, the specific interfacial free energy and area of the J th surface. To simplify discussion, we focus on an one component system so that G_s is equivalent to the surface tension (Eq. 13.2.5). If σ is independent of orientation, then the equilibrium shape is a sphere since it is the geometric shape that represents the minimum area for a given volume. For the case that σ depends on orientation of the surface, the solution to the Gibbs-Curie problem is given by what is known as the **Wulff theorem**. The procedure of determining the equilibrium shape of a crystal using this theorem is as follows (see Herring 1953, for a discussion).

Construct lines radiating from the center of a crystal such that the length of a line is proportional to the surface tension of the face normal to that line, the same scaling factor or proportionality constant being maintained for all such lines. The result is a closed volume with the surface describing the variation of surface tension with orientation. Figure 13.9 represents a section through such a closed volume. This is

Fig. 13.8 The surface tension of quartz ($\times 10^{-5}$ J/cm^2) versus the crack propagation stress (MPa) in different chemical environments. From Dunning et al. (1984)

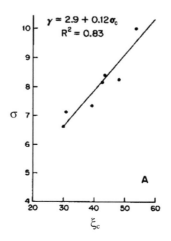

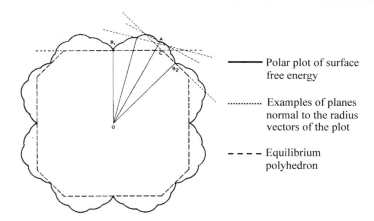

Fig. 13.9 Schematic Wulff construction to determine equilibrium shape of a crystal. Reproduced from Herring (1953). With permission from University of Chicago Press

often referred to as γ-plot, but will be a σ-plot to conform to our notation (the symbol γ has been used for activity coefficient). Now draw a plane normal to each radius vector at the point of its intersection with the enclosing surface. The planes that can be reached by the radius vectors without crossing any other plane delineate the equilibrium shape of the crystal, as illustrated for a two dimensional case in Fig. 13.9. For a normal liquid within an isotropic environment, the above procedure yields a spherical σ-plot, thus resulting in a sphere as the equilibrium shape.

Since the surface tension depends on adsorption, the equilibrium crystal shape of a solid is not unique, but depends on its environment. An interesting example is the shapes of pyrite (FeS_2) crystals that had formed under different geological environments, as shown in Fig. 13.10. Even though it can not be clearly established that these represent the equilibrium shapes under the different environments, these crystal shapes of pyrite illustrate the effect of adsorption in determining crystal morphology. Li et al. (1990) showed that the crystal habits of KCl vary with the degree of supersaturation of Pb^{2+} impurity.

Herring (1953) emphasized that in reality only small solid particles "have any hope of equilibration of shape." This is because of the fact that the energy needed to achieve any appreciable change of shape through atomic transport in a large particle is too large compared to the driving force that is given by the difference between the

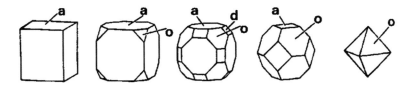

Fig. 13.10 Shapes of pyrite-type crystals in different environments. From Sunagawa (1957)

existing and equilibrium values of the surface energy. However, equilibrium shapes may be attained during geologic processes because of the very long annealing time. Heating of crystals initially ground to the shape of a sphere to sufficiently high temperature sometimes lead to the development of non-spherical equilibrium shapes.

On the basis of the requirement that the equilibrium shape of a crystal would be such that the total surface energy of the crystal is minimized, one would expect that the prominent faces of a crystal are those with relatively small surface energies. On the other hand, it is commonly found that the most prominent crystal faces are those with the highest areal density of atoms. Thus, in a given crystal the faces with the highest areal density of atoms would be expected to have the lowest surface energy. These faces would also have the largest interplanar spacing in order to preserve the bulk density of the crystal.

Curie (1885: 1859–1906) showed that for a crystal that has achieved the equilibrium shape, the ratio of the distance of the face (d_i) from the crystal center to its surface tension (σ_i) is constant. In other words,

$$\frac{d_1}{\sigma_1} = \frac{d_2}{\sigma_2} = \frac{d_3}{\sigma_3} = \dots \qquad (13.6.1)$$

where 1, 2, 3, etc. are equilibrium crystal faces. Thus, larger the surface tension of a face, the further it is from the crystal center. Because of geometrical restrictions, the area of a crystal face depends on its distance from the crystal center; further it is from the center, smaller is its area. This is illustrated in Fig. 13.11 which represents a cross section of a staurolite crystal normal to the c-axis. Staurolite is a prismatic mineral with orthorhombic symmetry so that the (001) face is farthest from the crystal center. From the relative distances and areas of the crystal faces, we conclude that in staurolite $\sigma(110) < \sigma(010) < \sigma(001)$.

It was first recognized by Becke (1913) and later by others that some metamorphic minerals tend to have better developed crystal faces than others in natural rocks. Thus the minerals in metamorphic rocks have been arranged into what is known as a crystalloblastic series according to their tendency to form euhedral

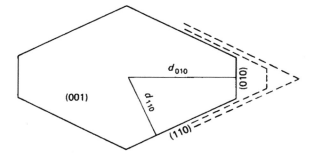

Fig. 13.11 Cross section of a common form of staurolite crystal normal to the c-axis showing the relative dimensions of the crystal faces. From the latter and the relative distances from the crystal center, we conclude that $\sigma(110) < \sigma(010) < \sigma(001)$. From Philpotts (1990)

grains (see, for example, Mueller and Saxena 1977; Philpotts 1990). This is essentially an expression of the surface energies of minerals. When minerals grow in a rock, some minerals develop prominent crystal faces while others develop as space filling crystals with anhedral shape. As discussed by Philpotts (1990), to minimize the contribution of the surface free energy to the overall free energy of the rock, a compromise is struck among the crystal habits of different minerals that lead to smaller surface areas for minerals with higher surface energies and higher surface areas for those with relatively lower surface energies.

13.7   Contact and Dihedral Angles

When the surfaces of three different crystals meet along a line, the equilibrium angles between the surfaces are governed by the interfacial tensions of the faces under hydrostatic condition (the instability of contact among a larger number of grains was demonstrated by Gibbs (1961: scientific papers). Referring to Fig. 13.12, the equilibrium relation between surface tensions and contact angles can be derived as follows. Let us consider a cross section of three grains, 1, 2 and 3, meeting at a line, and let α, β and γ be the contact angles and $\boldsymbol{\sigma_{12}}$, $\boldsymbol{\sigma_{23}}$ and $\boldsymbol{\sigma_{13}}$ be the vector representations of the interfacial tensions. Each interfacial tension exerts a force away from the point P. At equilibrium, the magnitude of the force exerted by $\boldsymbol{\sigma_{12}}$ is balanced by the net force exerted along the same straight line away from P on the right. This net force is given by the sum of normal projections of the tensions along the 1–3 and 2–3 interfaces on to the line of action of the $\boldsymbol{\sigma_{12}}$ force. Thus, if the interfacial tensions are independent of orientations, we have at equilibrium

$$\sigma_{12} = \sigma_{13} \cos(\varphi_1) + \sigma_{23} \cos(\varphi_2), \qquad (13.7.1)$$

where σ_{ij} indicates the magnitude of the vector $\boldsymbol{\sigma_{ij}}$.

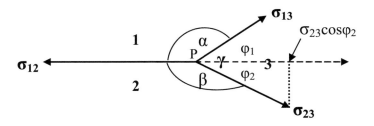

Fig. 13.12 Illustration of interfacial tension vectors acting at the junction of three grains, 1, 2 and 3. At equilibrium, the sum of the projected magnitudes of $\boldsymbol{\sigma_{13}}$ and $\boldsymbol{\sigma_{23}}$ on to the dashed line (line of action of $\boldsymbol{\sigma_{12}}$), that is $\sigma_{23}\cos\Phi_2 + \sigma_{13}\cos\Phi_1$, balances the magnitude of $\boldsymbol{\sigma_{12}}$. The angle γ is called the dihedral angle when σ_{12} represents the interfacial tension between two grains of the same phase

For the special case that all interfacial tensions are of equal magnitude, $\sigma_{12} = \sigma_{13} = \sigma_{23}$, the above expression reduces to

$$\cos(\varphi_1) = 1 - \cos(\varphi_2) \tag{13.7.2}$$

This equation has the only solution for $\varphi_1 = \varphi_2 = 60°\,(\cos(60) = 1/2)$. Thus, $\alpha = 180° - \varphi_1 = 120°$, and similarly $\beta = 120°$. Consequently, we have $\alpha = \beta = \gamma = 120°$. Three grains of the same mineral sometimes satisfy the condition of equality of interfacial tensions.

Kretz (1994) compiled the measured contact angles at the triple junctions of several minerals in granulites that show close approach to 120° in several cases of monomineralic contacts. An example is shown in Fig. 13.13. In several other cases, where one mineral is in contact with two grains of a different mineral, such as clinnopyroxene versus two grains of scapolite (Cpx vs. Scp-Scp), the angle bounded by the grains of the same mineral, which is known as the **dihedral angle**, shows departure from 120°. This is expected since the surface tensions between the like and unlike mineral pairs are different.

Using Eq. (13.7.1) and assuming that the surface tensions are independent of orientation, one can easily express the dihedral angle in terms of the surface tensions between the like and unlike minerals. Let 1 and 2 be two grains of the same phase, A, and 3 of a different phase, B (Fig. 13.12). In that case, $\alpha = \beta$ and $\varphi_1 = \varphi_2 = \gamma/2$. Thus, denoting σ_{12} as σ_{AA} and $\sigma_{13} = \sigma_{23}$ as σ_{AB}, Eq. (13.7.1) reduces to

$$\frac{\sigma_{AA}}{\sigma_{AB}} = 2\cos(\gamma/2) = 2\cos\theta \tag{13.7.3}$$

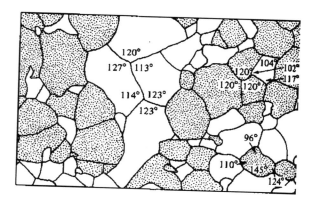

Fig. 13.13 Contact angles at the triple junctions of grains of the same mineral in a granulite. Patterned grains: Ca-pyroxene; unpatterned grains:scapolite. From Kretz (1966). With permission from Oxford University Press

where $\theta = \gamma/2$. This relation is illustrated in Fig. 13.14, which shows the variation dihedral angle, γ as a function of the ratio of surface tensions between the like and unlike pairs, σ_{AA}/σ_{AB}.

It is evident that if the interfacial tensions are independent of orientations, then a dihedral angle of 180° cannot be achieved since that would imply $\sigma_{AA} = 0$, which is physically impossible. However, 180° dihedral angles are often observed in natural rocks. An example is shown in Fig. 13.15 in which dihedral angle between hornblende and two grains of biotite, and between biotite and two grains of hornblende is 180°. The explanation of such dihedral angle lies in the orientation dependence of the interfacial tensions. Allowing for the orientation dependence of σ_{13} and σ_{23}, i.e. $\sigma_{13} = f(\varphi_1)$ and $\sigma_{23} = F(\varphi_2)$, Herring (1953) derived the following expression that must be satisfied when interfacial equilibrium is achieved.

$$\sigma_{12} - \sigma_{13}\cos\varphi_1 - \sigma_{23}\cos\varphi_2 + \frac{\partial\sigma_{13}}{\partial\varphi_1}\sin\varphi_1 + \frac{\partial\sigma_{23}}{\partial\varphi_2}\sin\varphi_2 = 0 \qquad (13.7.4a)$$

For the special case of constant interfacial tensions, this equation reduces to Eq. (13.7.1).

Analytical expressions for the temperature and pressure dependence of the dihedral angle are readily derived by differentiating Eq. (13.7.3). Thus, one obtains (Passeron and Sangiorgi 1985)

$$\sigma'_{AA} = 2\sigma'_{AB}\cos\theta - 2\sigma_{AB}\sin\theta(\theta') \qquad (13.7.4b)$$

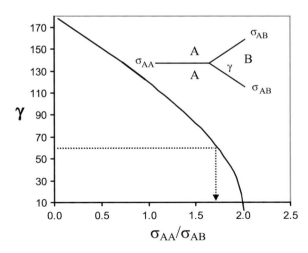

Fig. 13.14 Variation of dihedral angle, γ, as function of the ratio of surface tensions between the like and unlike pair of phases. The surface tension ratio (1.73) between solid-solid and solid-fluid interfaces that is required to develop the critical dihedral angle of 60° for fluid connectivity is shown by an arrow. There is an interconnected network of fluid channels along grain edge channels, regardless of fluid fraction, when the $\gamma \leq 60°$

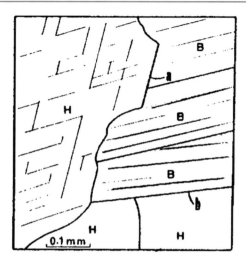

Fig. 13.15 180° contact angles between hornblende (H) and biotite (B) crystals. Interface *a* is parallel to (110) of horblende and interface *b* is parallel to (001) of biotite. From Kretz (1966). With permission from Oxford University Press

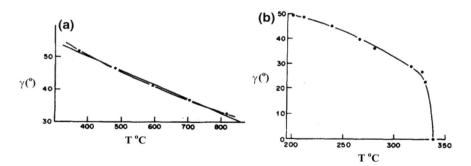

Fig. 13.16 Variation of dihedral angle in (**a**) Ni-Pb and (**b**) Zn-Sn systems as a function of temperature. The two curves in (**a**) represent two different ways (linear and nonlinear) of fitting the data. From Passeron and Sangiorgi (1985). With permission from Pergamon-Elsevier

or

$$\theta' = \frac{2\sigma'_{AB} \cos \theta - \sigma'_{AA}}{2\sigma_{AB} \sin \theta} \tag{13.7.5}$$

where the superscript ' denotes the first derivative with respect to either pressure or temperature. It is evident that the dihedral angle ($\gamma = 2\theta$) would be independent of the temperature (or pressure) change when $2\sigma'_{AB} \cos \theta = \sigma'_{AA}$, and would increase or decrease with temperature (or pressure) depending on whether $2\sigma'_{AB} \cos \theta$ is greater or less than σ'_{AA}.

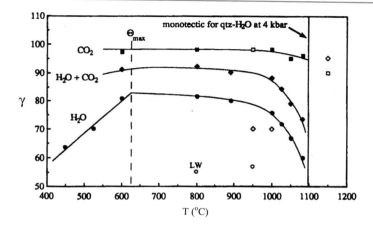

Fig. 13.17 Variation of dihedral angle of H_2O-CO_2 fluid in a matrix of quartzite as a function of temperature as determined by Holness (1993) at 4 kb (filled symbols) and complied from other sources (open symbols). From Holness (1993). With permission from Elsevier

Dihedral angle can vary both linearly and non-linearly. Figure 13.16 shows the variation of dihedral angle between melt and solid in the Ni-Pb and Zn-Sn systems as a function of temperature. In the Zn-Sn system, the dihedral angle decreases non-linearly with very rapid change above 325 °C, dropping to zero at ~ 340 °C, whereas in the Ni-Pb system, it decreases almost linearly as a function of temperature. Figure 13.17 shows the variation of dihedral angle of H_2O-CO_2 fluid in a matrix of quartzite as a function of temperature at 4 kbar pressure, as compiled by Holness (1993) from the available data. In view of the γ versus T behavior illustrated in these figures, one should be careful in extrapolating dihedral angle to temperature (or pressure) much beyond the range of experimental condition.

Passeron and Sangiorgi (1985) derived expressions for the temperature dependence of the surface tensions in terms of the change of dihedral angle, assuming that within a limited domain the dihedral angle changes linearly with respect to temperature. This assumption implies that the second derivative of the dihedral angle with respect to temperature, θ'', is zero. The procedure involves differentiation of Eq. (13.7.4) with respect to temperature and imposing the condition $\theta'' = 0$, followed by rearrangement of terms. The final expressions are

$$\sigma'_{AB} = -\sigma_{AB}\left(\frac{\theta''}{2\theta'} + \frac{\theta'}{2\tan\theta}\right) \tag{13.7.6a}$$

$$\sigma'_{AA} = -\sigma_{AB}\left(\frac{\theta''}{\theta'}\cos\theta + \frac{1+\sin^2\theta}{\sin\theta}\theta'\right) \tag{13.7.6b}$$

13.8   Dihedral Angle and Interconnected Melt or Fluid Channels

The dihedral angle at the triple junction of two crystals and a fluid phase plays a critical role in determining if a given volume fraction of a fluid would form an interconnected network or remain as isolated pockets. This, in turn, determines the extent of melt fraction that can be retained in the source region of partial melting of rocks and consequently the melt composition, the amount of melting required for the segregation of planetary cores, and the permeability of metamorphic rocks to fluid flow. The upward percolation of basaltic magma through the upper mantle requires that the lower density magma form an interconnected network. The melt fraction at which such an interconnected network would be established depends on

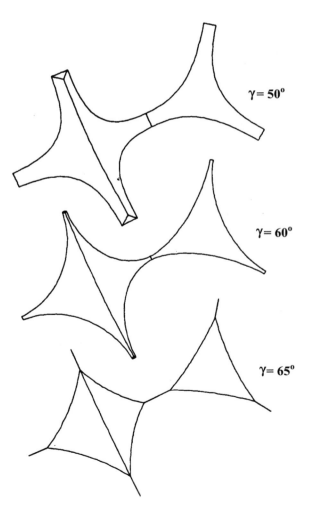

Fig. 13.18 Melt channel shapes at constant volumetric melt fraction of 0.01 (1%) and wetting angle between 50° and 65°. The change in shape between 20° and 50° is minor compared to that between 50° and 65°. From von Bergen and Waff (1986)

the dihedral angle of the magma with the mantle minerals at the mantle conditions. Furthermore, the bulk physical properties of partially molten rocks depend on the melt distribution.

The dihedral angle is also referred to as the **wetting angle** when the included phase is a melt. Smith (1964) reviewed the distribution of liquid phases in metals. He found that when the ratio of the σ_{AA} to $\sigma_{A\text{-Melt}}$ was greater than $\sqrt{3}$ (or 1.73), the liquid phase always formed a three dimensional network through the grains. On the other hand, when this ratio was less than $\sqrt{3}$, the liquid phase formed isolated pockets at the grain corners. This critical ratio of surface tensions correspond to a dihedral angle of 60° (Fig. 13.14). Later, Balau et al. (1979) and von Bergen and Waff (1986) showed that under condition of (a) hydrostatic stress, (b) constant curvature of interface between solid and melt and (c) independence of dihedral angle of orientation, a melt phase would form an interconnected channel along the grain edges, regardless of the volume fraction of the melt, when $0 < \gamma < 60°$, and would wet grain faces only when $\gamma = 0$. Figure 13.18, which is reproduced from von Bergen and Waff (1986), shows the melt channel shapes at a constant volumetric melt fraction of 0.01 and wetting angles between 50° and 65°. If the wetting angle is greater than 60°, then a critical melt fraction would be needed to establish interconnectivity. When the melt fraction for a given wetting angle falls below a critical limit, the melt gets pinched off. Figure 13.19 shows the relationship between the wetting or dihedral angle and volumetric melt fraction for the **connection** and **Pinch-off boundaries**, as calculated by von Bergen and Waff (1986). Either pinch-off geometry or connection boundary can form within the area bound by the two curves depending on the sense of curvature of the solid-melt interface. The works of Balau et al. (1979) and von Bergen and Waff (1986) have played

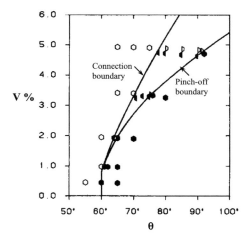

Fig. 13.19 Melt channel pinch off and connection boundaries in relation to volume percent of melt, V %, and dihedral or wetting angle, γ. Open hexagons: connected channels; filled hexagons: isolated pockets; half hexagons: one of the two possible geometries. From von Bergen and Waff (1986). With permission from Elsevier

fundamental roles in the subsequent studies of dihedral angle relations in geological and planetary systems. A few examples are discussed below. The problem of determination of true dihedral angles from the distribution of apparent dihedral angles measured on planar sections of partial melting experiments was addressed by Jurewicz and Jurewicz (1986).

13.8.1   Connectivity of Melt Phase and Thin Melt Film in Rocks

There have been many studies on the wetting angles of melts in geological systems that permit us to develop a general idea about connectivity of melt in a rock matrix. For olivine in contact with basaltic liquids ranging from tholeitic to nepheline-basaltic composition, the median wetting angle falls in the range of 20°–37° at 10–30 kb, 1230–1260 °C (Waff and Balau 1982). For a melt derived from peridotite, the median wetting angle is $\sim 45°$ for olivine-melt contact and 60°–80° for pyroxene-melt contact (Toramuru and Fuji 1986). Kohlstedt (1992) reported the average wetting angle of upper mantle rocks to be 20–50°. For granitic liquids in contact with quartz or feldspar, the median wetting angles were reported to be 60° and 50°, respectively (Jurewicz and Watson 1985). Rose and Brenan (2001) studied the wetting properties of Fe-Ni-Co-Cu-O-S melts against olivine and discussed their implications for sulfide melt mobility.

Hess (1994) presented an analysis of the thermodynamic properties of thin films. He showed that thin melt film could exist between crystal faces even when the wetting angle between the solid and bulk melt phase is greater than zero. This is because of the fact that the melt thin film has thermodynamic properties that are different from those of the bulk melt phase that forms the dihedral angle. This distinction between the thermodynamic properties of the thin film and bulk melt phases have not been made in the analysis by von Bergen and Waff (1986) on the relationship between the wetting angle and presence of a melt phase between crystal surfaces. The interested reader is referred to Hess (1994) for the details of his thermodynamic analysis and implications of the potential existence of thin film of a basaltic melt in mantle rocks.

13.8.2   Core Formation in Earth and Mars

On the basis of geophysical evidence, and phase equilibrium and density data, the cores of the Earth and Mars are generally accepted to be made of iron with some amount of dissolved sulfur and oxygen. The separation of liquid metal from a solid silicate matrix would require melt connectivity that depends on the amount of melt fraction and the dihedral or wetting angle if the stress distribution were hydrostatic.

Gaetani and Grove (1999) determined the dihedral angle between Fe-O-S melt and olivine at 1 bar total pressure with $\log f(O_2) = -7.9$ to -10.3 bars and $\log f(S_2) = -1.5$ to -2.5 bars. They found that trace amounts of oxygen dissolve in sulfide melts at $f(O_2)$ near the Fe-FeO oxygen buffer (Fe + ½ O_2 = FeO) and the wetting angle is $\sim 90°$ that does not permit melt connectivity. However, at the $f(O_2)$ condition near the quartz-fayalite-magnetite buffer (3 Fe_2SiO_4 + O_2 = 2 Fe_3O_4 + 3 SiO_2), the solubility of oxygen in the melt increases to 9 wt% and the wetting angle drops to 52°, thereby permitting melt percolation. It should be evident from Fig. 13.14 that the decrease of the wetting angle between olivine and iron melt implies an increase of the surface tension ratio $\sigma_{Ol\text{-}Ol}/\sigma_{Ol\text{-}Fe(melt)}$, and hence a decrease of $\sigma_{Ol\text{-}Fe(melt)}$ with increasing dissolution of oxygen in the liquid metal. Thus, the effect of dissolution of oxygen in Fe(melt) on the surface tension between melt and olivine is similar to that between melt and air that is illustrated in Fig. 13.4.

Since the surface tension is affected by pressure (Eq. 13.3.2), the dihedral or wetting angle between mantle silicate and Fe-O-S melt may be significantly affected by pressure. Because of the adsorption of oxygen and sulfur in the olivine-melt interface, the interfacial tension $\sigma_{Ol\text{-}Fe(metal)}$ would decrease with increasing pressure (Eq. 13.3.2), thereby leading to an increase of the dihedral angle (Fig. 13.14). Terasaki et al. (2005) determined the dihedral angle between olivine and Fe-O-S melt at pressures to 20 GPa that represent more appropriate conditions for core formation in Earth and Mars. Their results (Fig. 13.20) show a significant increase of dihedral angle from that measured by Gaetani and Grove (1999). However, the

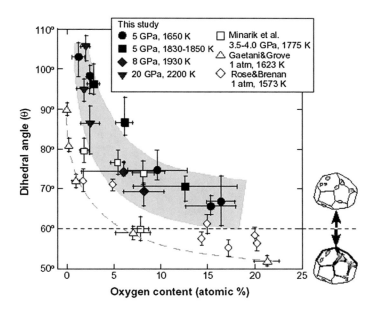

Fig. 13.20 Dihedral or wetting angle as a function of oxygen content of Fe-O-S melt in an olivine matrix at high P-T condition. Filled symbols ("This study"): experiments conducted by Terasaki et al. (2005); Open symbols: experimental data from others, as referenced in Terasaki et al. (2005). Reproduced from the latter. With permission from Elsevier

data of Terasaki et al. (2005) do not show any discernible change of dihedral angle within their P-T range of investigation of 3.5 GPa, 1377 °C to 20 GPa, 1927 °C. This observation, however, should not be taken to indicate that the dihedral angle is independent of pressure and temperature within the stated range, since the increase of pressure was accompanied by an increase of temperature and these two variables may have compensating effects. Figure 13.21 shows a bright-field transmission electron microscopic (TEM) image of Fe-S-O melt pocket in an olivine matrix. The volume percent of iron-sulfide in the starting mixture was 2%. High-resolution TEM image of an adjacent olivine grain boundary, shown by arrow, did not reveal any melt film. The dihedral angle at this grain boundary is $\sim 70°$. Thus, the absence of melt film corroborates the theoretical prediction (Fig. 13.19) of von Bergen and Waff (1986).

Following von Bergen and Waff (1986), Terasaki et al. (2005) calculated the connection and pinch-off boundaries as a function of the Fe/(Fe + Mg) molar ratio, denoted as Fe#, of olivine for S = 10 and 14 wt% that are appropriate for the core compositions of Earth and Mars, respectively (Fig. 13.22). Using the Fe# values of olivines in terrestrial and Martian mantles, which are illustrated by arrows in Fig. 13.22, we find that the connection boundary is located at 6 vol% for Martian mantle and 9.5 vol% for the terrestrial mantle. These results suggest that if the terrestrial and Martian cores formed by percolation of metallic melts and the stress regime was essentially hydrostatic, then the metallic melt fraction should have been greater than the above critical values. But once the melt was drained out and fell below the critical limits, there would have been stranded melt pockets within the silicate matrix. Geochemical evidences indicate that efficient segregation of Fe-O-S

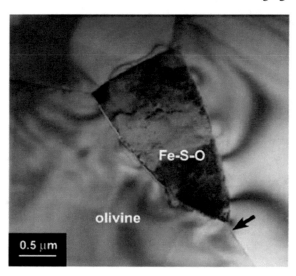

Fig. 13.21 Bright-field TEM image of a Fe-S melt pocket contained within olivine grains. High-resolution TEM image of the olivine grain boundary indicated by arrow did not reveal presence of a thin film of melt. The P-T condition of the experiment was 4.6 GPa–1960 K. From Terasaki et al. (2005). With permission from Elsevier

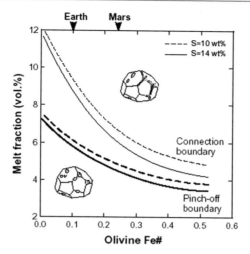

Fig. 13.22 Connection and pinch-off boundaries of iron-sulfide melt as a function of Fe/(Fe + Mg) ratio (Fe#) of olivine for sulfur contents of 10 wt% (dashed lines) and 14 wt% (solid lines) that are appropriate for the compositions of terrestrial and Martian cores, respectively. The Fe# of olivine in terrestrial and Martian mantles are indicated by arrows. From Terasaki et al. (2005). With permission from Elsevier

melt from planetary silicate mantles occurred during the core formation. Thus, some mechanism needs to be invoked to release the stranded metallic melt to the core. One possibility is to have extensive melting of the silicates leading to the formation of what has become known as the "magma ocean" (also see Sect. 11.3). A second possibility is to have non-hydrostatic stress condition in the mantle during the core formation. Bruhn et al. (2000) showed that shear deformation to large strain rates can lead to significant connectivity among metallic melt pockets that were initially stranded. Furthermore, the dihedral angle between mantle minerals and Fe-O-S melt could decrease below the critical limit of $60°$ in the presence of additional impurities (e.g. Si) and at geologically reasonable conditions that are not encompassed by the experiments of Terasaki et al. (2005).

13.9 Surface Tension and Grain Coarsening

When there is a curved interface, whether it is for a mineral grain, liquid droplet or gas bubble, there is a pressure gradient across the grain boundary and the chemical potential of a component in the surface increases with the curvature of the surface. As a result, grains with large curvature (i.e. small grains) tend to be eliminated in favor of grains with small curvature (i.e. large grains). This process of coarsening of grains through a decrease of the average surface to volume ratio is often referred to as **Ostwald ripening**.

To understand the thermodynamic basis of the above phenomenon, let us consider the growth of a spherical gas bubble in a liquid (Fig. 13.23). The system of gas bubble plus liquid has a fixed volume and temperature. The pressure inside the bubble, P_i, must be higher than the external pressure, P_{ex}, otherwise the bubble will collapse. We now want to derive an expression for this pressure difference, $P_i - P_{ex}$. Since the overall system is at a fixed T, V condition, the appropriate thermodynamic potential to consider is the Helmholtz free energy, F, which must be at minimum at equilibrium.

Let V_b and V_{ex} be the volumes of the bubble and the external medium, respectively, and A_s be the surface area of the bubble. From Eq. (13.1.3) the change in Helmholtz free energy of the bubble due to an infinitesimal increase of its volume at constant temperature and composition is given by $dF_b = -P_i dV_b + \sigma dA_s$, whereas that of the surrounding liquid is given by $dF_{ex} = -P_{ex} dV_{ex}$. The change of free energy of the external medium due to a change of surface area resulting from the expansion of the bubble is negligible. Since the overall volume of the system is kept constant, $dV_{ex} = -dV_b$ and hence $dF_{ex} = P_{ex} dV_b$. At equilibrium, we have $dF_T = dF_b + dF_{ex} = 0$, where F_T denotes the total Helmholtz free energy of the entire system. Thus, at equilibrium

$$dF_T = 0 = -dV_b(P_i - P_{ex}) + \sigma dA_s \tag{13.9.1}$$

or

$$P_i - P_{ex} = \sigma \frac{dA_s}{dV_b} \tag{13.9.2}$$

For a sphere, $A = 4\pi r^2$ and $V = 4/3(\pi r^3)$ so that $dA_s = 8\pi r dr$ and $dV_b = 4\pi r^2 dr$. Consequently, the pressure difference between the interior and exterior of a spherical bubble is given by

$$\boxed{P_i - P_{ex} = \frac{2\sigma}{r}} \tag{13.9.3}$$

This equation is also valid for spherical liquid droplets, and spherical grain of solids.

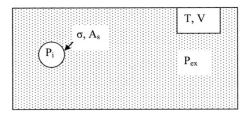

Fig. 13.23 Schematic illustration of the formation of a gas bubble with an internal pressure of P_i in a liquid medium with pressure P_{ex}. The system is maintained at a constant T, V condition. σ is the surface tension at the bubble-liquid interface and A_s is the surface area of the bubble

The last equation, which is known as the **Laplace equation**, allows us to express the chemical potential of a component within a bubble as a function of the radius of curvature, r. The chemical potential of a component within two spherical bubbles of radii r_1 and r_2, subject to a constant external pressure, are related according to

$$\mu_i(r_2) = \mu_i(r_1) + \int_{P_1}^{P_2} \left(\frac{\partial \mu_i}{\partial P}\right) dP = \mu_i(r_1) + \int_{P_1}^{P_2} v_i dP \qquad (13.9.4)$$

where P_1 and P_2 are, respectively, the pressures within the bubbles of radius r_1 and r_2, and v_i is the partial molar volume of the component i that can be safely assumed to be unaffected by the very small pressure change from P_1 to P_2, in which case the last integral in the above equation is given by $v_i(P_2 - P_1)$. Using Eq. (13.9.3), $P_2 - P_1 = 2\sigma(1/r_2 - 1/r_1)$, which, upon substitution into the last equation, yields

$$\boxed{\mu_i(r_2) = \mu_i(r_1) + 2\sigma v_i\left(\frac{1}{r_2} - \frac{1}{r_1}\right)} \qquad (13.9.5)$$

This equation is a general form of the **Gibbs-Thomson equation**. It is often presented, assuming a single component system, in terms of Gibbs free energy difference, ΔG, between a crystal of radius r and that of sufficiently large size at which its surface energy contribution is insignificant ($1/r \sim 0$) (recall that for a single component system, the chemical potential is the same as G). It is evident from the above equation that if $r_2 > r_1$, the parenthetical term is negative, and consequently $\mu_i(r_2) < \mu_i(r_1)$. In other words, the chemical potential of a component decreases with the increase of grain size. Hence, if the kinetics of chemical transfer is favorable, components would be transferred from smaller to larger grains, leading to disappearance of small grains and further growth of large grains (the rich gets richer, the poor gets poorer).

Kretz (1994) discussed the observational data on grain coarsening in metamorphic rocks, and concluded that it is a complex process in natural environments, especially if it takes place during deformation and while the temperature of the rock increases. The natural and experimental results on grain coarsening have been reviewed by Joesten (1991). Kretz (1994) concluded that coarsening driven by interfacial energy is quite conspicuous in many monomineralic rocks, such as marble and quartzite, and may also take place in rocks with two or more minerals under favorable conditions. Cashman and Ferry (1988) and Miyazaki (1991) advocated coarsening of garnet crystals in a quartz-feldspar matrix driven by interfacial energy. Kretz (1994, 2006) argued against this idea on the basis of both observational and theoretical grounds (also see Atherton, 1976, and Carlson 1999). He drew attention to the fact that the interfacial energy associated with non-coherent interfaces is a very small quantity. For example, in metals it is $\sim 1 \times 10^{-4}$ J/cm^2 (Raghavan and Cohen 1975). Kretz (2006) suggested that similar values should be expected at garnet-feldspar and garnet-quartz boundaries. In that case, using the molar volume of almandine of 115.1 cm^3

(effectively the same as its partial molar volume, Ganguly et al. 1996), we find from Eq. (13.9.5) that the chemical potential of almandine in a large crystal ($1/r \sim 0$) will be lower by only 2.3×10^{-2} J/mol from that in a crystal of radius 0.1 mm. Energy difference of this order seems to be too small to drive diffusion of almandine component from a small to a large crystal.

13.10 Effect of Particle Size on Solubility and Melting

Consider now the effect of size on solubility of particles in a solution. The solution process of a crystal of radius r can be represented by a reaction

$$i \, (\text{solid}, r) = i \, (\text{liquid}) \tag{13.10.a}$$

We now choose the standard state of solid to be pure solid at P-T of interest, but with sufficiently large radius, r^*, such that $2\sigma v_i/r^*$ term is negligible. We designate this standard state as $\mu_i^{o,s}(r*)$. Thus, from Eq. (13.9.5)

$$\mu_i^s(r) = \mu_i^s(r*) + \frac{2\sigma v_i}{r}$$

or

$$\mu_i^s(r) = \mu_i^{0,s}(r*) + RT \ln a_i^s(r*) + \frac{2\sigma v_i}{r} \tag{13.10.1}$$

For the reaction (13.10.a), we have at equilibrium, $\mu_i^s(r) = \mu_i^l$ so that

$$\mu_i^{0,s}(r*) + RT \ln a_i^s(r*) + \frac{2\sigma v_i}{r} = \mu_i^{0,l} + RT \ln a_i^l(r) \tag{13.10.2}$$

where $\mu_i^{o,l}$ represents the stated standard state of i in liquid which we choose to be pure liquid i at the P-T condition of interest, and $a_i^l(r)$ is the activity of i in liquid in equilibrium with a solid of radius r.

Writing, $\Delta_r G^o(P,T) = \mu_i^{o,l} - \mu_i^{o,s}(r*)$, and rearranging terms, we have from the last expression

$$RT \ln \frac{a_i^l(r)}{a_i^s(r*)} = -\Delta_r G^\circ(P, T) + \frac{2\sigma v_i}{r} \tag{13.10.3}$$

Denoting $K = \exp[-\Delta_r G^o(P, T)/RT]$, that is writing K for the usual equilibrium constant at T without the particle size effect,

$$\frac{a_i^l(r)}{a_i^s(r*)} = K \exp\left(\frac{2\sigma v_i}{rRT}\right) \tag{13.10.4}$$

This equation was derived earlier by Lasaga (1998). If we deal with the solubility of a pure solid, then $a_i^s(r*) = 1$. It is, thus, evident that the solubility of a solid phase would increase as its particle size gets smaller. However, there is a threshold value of r before the effect of particle size becomes significant.

From the last equation, the increase in the activity of a species in a melt due to reduction of the grain size of the pure solid from $r*$ to r is given by $\Delta a_i^l = a_i^l(r) - a_i^l(r*) = K(e^\varphi - 1)$, where $a_i^l(r*)$ is the activity of i in the liquid in equilibrium with a pure solid of radius $r*$ and φ represents the terms within the parentheses of the equation (note that for equilibrium between a liquid and a pure solid of radius $r*$, $a_i^l(r*) = K$). Thus, the fractional increase of a_i^l of a grain of radius r, which is effectively the same as its increase of solubility, is given by

$$\frac{\Delta a_i}{a_i^l(r*)} = \frac{K}{a_i^l(r*)}(e^\varphi - 1) = (e^\varphi - 1) \qquad (13.10.5)$$

The last equality is due to the fact that at $r = r*$, $a_i^l = K$ when the solid is in the pure state. As an illustrative application of this equation, let us consider the effect of grain size on the solubility of quartz, assuming that the particles are spherical in nature. Parks (1984) reported the interfacial tension of quartz in different types of environments. In contact with liquid water, the interfacial tension is 360 ($\pm$50) mJ/m^2. Figure 13.24 illustrates the results of calculation of the increase of solubility of quartz with reduction of grain size, using $\sigma = 360$ mJ/m^2 and $v = 22.7$ cm^3/mol. We see a sharp increase in the solubility of quartz as the particle radius falls below 0.25 μm.

Equation (13.10.4) describes the equilibrium solubility of a solid in a coexisting liquid as a function of its radius and temperature at a fixed pressure, assuming that the surface tension of the solid is unaffected by r and T and the liquid droplets are sufficiently large such that the effect of surface tension of liquid can be neglected. This relation can also be used to calculate the effect of grain size reduction on **melting temperature** of solids of spherical geometry under the same assumptions.

Fig. 13.24 Increase of solubility of quartz as a function of grain size at 25 °C relative to the solubility of large grains for which the effect of surface energy is negligible

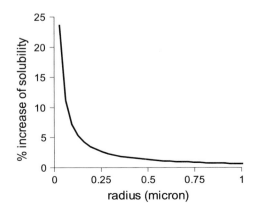

To treat the case of melting, we replace T in Eq. (13.10.3) by $T_{m,r}$, which is the melting temperature of solid of radius r, and $\Delta_r G^\circ(P, T)$ by $\Delta G^\circ_m(P, T_{m,r})$, which is the **molar** free energy of melting of pure solid of $r < r^*$ at P and $T_{m,r}$ (Fig. 13.25). Thus, from Eq. (13.10.3), we have

$$\frac{2\sigma v_i}{rRT_{m,r}} = \ln \frac{a_i^m(r)}{a_i^s(r^*)} + \frac{\Delta H^\circ_m(T_{m,r}) - T_{m,r}\Delta S^\circ_m(T_{m,r})}{RT_{m,r}} \qquad (13.10.6)$$

where $\Delta H^\circ_m(T_{m,r})$ and $\Delta S^\circ_m(T_{m,r})$ are, respectively, the molar enthalpy and molar entropy of melting.

Let us now consider the simplest case of melting of a pure solid to a liquid of its own composition so that the activity terms in the above equation are unity. Rearranging the terms, we then obtain

$$T_{m,r} = \frac{\Delta H^\circ_m(T_{m,r})}{\Delta S^\circ_m(T_{m,r})} - \frac{2\sigma V}{r\Delta S^\circ_m(T_{m,r})} \qquad (13.10.7)$$

where V is the molar volume of the pure solid. At the melting temperature $(T_{m,r*})$ of pure solid of radius r^* (or greater), $\Delta G^\circ_m(T_{m,r*}) = 0 = \Delta H^\circ_m(T_{m,r*}) - T_{m,r*}\Delta S^\circ_m(T_{m,r*})$ so that $T_{m,r*} = \Delta H^\circ_m(T_{m,r*})/\Delta S^\circ_m(T_{m,r*})$. Assuming now that the ratio of enthalpy and entropy of melting at $T_{m,r*}$ is almost the same as that at $T_{m,r}$, the above equation reduces to

$$T_{m,r} \approx T_{m,r*} - \frac{2\sigma V T_{m,r*}}{r\Delta H^\circ_m(T_{m,r*})} \qquad (13.10.8a)$$

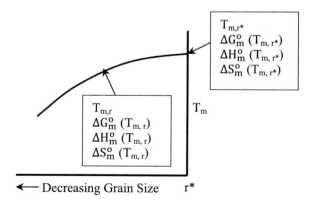

Fig. 13.25 Schematic illustration of the effect of grain size reduction on melting temperature of a solid along with the thermodynamic parameters used in the derivation of the relation between $T_{m,r}$ and r, where $T_{m,r}$ is the melting temperature at a specified grain radius (r) below the critical radius (r^*) at which the surface effect becomes negligible

or

$$\boxed{\frac{\Delta T_m}{T_{m,r*}} \approx \frac{2\sigma V}{r\Delta H_m^o\left(T_{m,r*}\right)} = \frac{2\sigma}{r\rho\Delta(H_o^m)'}}$$ (13.10.8b)

where $\Delta T_m = T_{m,r*} - T_{m,r}$, ρ is the density of the solid and $\Delta\left(H_m^o\right)'$ is the specific enthalpy (i.e. enthalpy per unit mass) of melting.

A more elaborate treatment of grain size effect on melting temperature depression is given by Buffat and Borel (1976) incorporating the effect of surface tension of the melt droplets. Their relation reduces to the one derived above if the second order terms (incorporating the temperature and pressure dependencies of the thermodynamic parameters) and the effect of surface tension of liquid droplets are neglected. Buffat and Borel (1976) also carried out experimental study of melting point depression of gold as a function of grain size down to ~25 Å and successfully modeled the data using their derived relation (Fig. 13.26). (Obviously treatment of melting point depression by grain size reduction cannot be accomplished by classical thermodynamic treatment for grain size beyond its domain of validity, i.e. macroscopic domain, but the successful modeling of the experimental results illustrated in Fig. 13.26 suggests that the limit is not exceeded at grain size probably as small as a few tens of Å or several nm. However, for large depression of melting temperature the simplifying assumptions about the constancy of $\Delta H_m^o/\Delta S_m^o$ between $T_{m,r*}$ and $T_{m,r}$ made in deriving Eq. (13.10.8) may not be valid, and the more general formulation given by Eq. (13.10.7) may provide better alternative for a single component system within a numerical scheme).

Fault movements associated with seismic slips are known to be associated with the occurrence of dark glassy material known as pseudotachylyte, so called since it resembles volcanic glass, tachylyte (Shand 1917). Lee et al. (2017) discussed geological observations in natural pseudotachylytes of quartzite that suggest melting of quartz below its accepted equilibrium melting temperature (referred

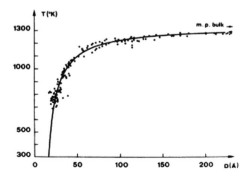

Fig. 13.26 Comparison of experimental data (symbols) on the effect of grain size reduction on the melting temperature of Gold with theoretical calculation (solid line). From Buffat and Borel (1976). With permission from American Institute of Physics

henceforth as the nominal melting temperature) determined in the laboratory under dry conditions in pure SiO_2 system. They carried out friction melting of quartzite at typical seismic slip rate (1.3 m/s) and demonstrated melting of quartz much below its nominal melting temperature. This observation was attributed to the combined effects of metastable persistence of β-quartz and depression of melting temperature due to friction induced reduction of grain size of quartz.

A first order analysis of the problem of melting temperature depression of quartz may be carried out using Eq. (13.10.8). Unfortunately, however, the required surface tension data of quartz in the presence of a melt is not available; instead there are data for σ(α-qtz) in the vacuum and in the presence of water vapor and liquid water (Parks 1984). These data could be used to calculate limiting effects, as follows, assuming that there is no significant difference between the surface tensions of α- and β-quartz.

According to Eq. (13.3.5), adsorption of excess solute on a surface ($\Gamma_2 > 0$), which is defined by zero concentration of the solvent, leads to reduction of surface tension of the solid. Consequently, calculations using σ(vacuum) for α-quartz should yield greater effect of grain size reduction on T_m relative to the situations where there are excess solutes on the surface in natural environments. Figure 13.27 illustrates the calculated T_m versus r of β-quartz using σ values in vacuum ($\sigma = 1850$ mJ/m^2) and in the presence of water vapor that causes surface adsorption of SiOH or silanol groups ($\sigma = 585$ mJ/m^2). The $T_{m,r*}$ of β-quartz at has been taken to be that at 1 bar (1470 °C), according to the data in Hudon et al. (2002), and is illustrated in the Fig. 13.27 with a diamond symbol. This temperature could be somewhat lower than the metastable melting temperature of β-quartz at the pressure (1.3 kb) of the friction melting experiments (Lee et al. 2017) since the P-T slope of melting curve is likely to be positive up to this pressure. The molar enthalpy of melting, ΔH_m^o, at the (nominal) melting temperature of β-quartz was calculated to be 12,188 J/mol using the optimized thermochemical data in Hudon et al. (2002). The temperature range (1350–1500 °C) in the friction melting experiments is shown by horizontal dashed lines in the Fig. 13.27. The inset shows the percentage reduction of T_m (with T in K) (which is independent of the value of $T_{m,r*}$) versus r for the two values of σ used to calculate the curves in the main figure and a third case in which the σ value corresponds to that in the presence of liquid water ($\sigma = 355$ mJ/m^2).

The results illustrated in Fig. 13.27 show that at the lower end of temperature range (1350 °C) in the friction melting experiments by Lee et al. (2017), melting could have taken place only if the grain size were reduced below 0.1 μm radius during the friction melting experiments. In comparison, the grain size of quartz in contact with melt in these experiments has been found to be in the range of 0.01–0.1 μm. The survival of grains smaller than 0.1 μm might have been due to local reduction of surface tension of quartz by adsorption of water.

Figure 13.27 shows that, depending on the surface tension property, there is a threshold size of quartz grains marking the transition from weak or moderate to strong effect of grain size reduction on melting. This is qualitatively analogous to the melting behavior of Au (Fig. 13.26) and is a consequence of the nature of the

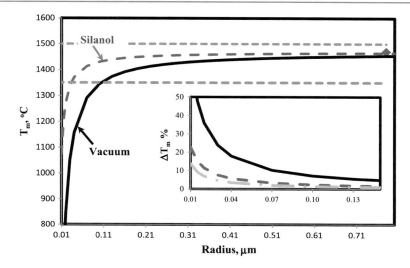

Fig. 13.27 Main frame: Calculated melting temperature, T_m, of β-quartz as function of radius using the surface energy of quartz in vacuum (solid line) and with hydroxylated surface dominated by SiOH or silanol groups (dashed line). The diamond shows the nominal metastable melting temperature of β-quartz whereas the dashed horizontal lines show the measured temperature range in the friction melting experiment. Inset: Percentage reduction of T_m (with T in Kelvin) as function of grain size for the two cases in the main frame and an additional case relating to surface adsorption in the presence of liquid water (dash-dot line)

general expression (Eq. 13.10.8) for the dependence of melting temperature on grain size. When the grain size falls below the threshold value, melt productivity increases rapidly with reduction of grain size.

From theoretical analysis, McKenzie and Brune (1972) suggested that the friction induced heating could cause melting of rocks between fault planes and the lubricating action of a molten film could allow the release of almost all the elastic strain energy in the region of shock. Spray (2005) and Di Toro et al. (2006) discussed evidence of melt lubrication and low fault strength during earthquakes. The results presented in Fig. 13.27 show that the effect of frictional heating on melt productivity becomes most efficient when the grain size falls below a threshold value. The asperities or sharp points on the rough surfaces of fault planes could be the locations for this enhanced effect of grain size reduction on melting as these would suffer the most intense grinding effects. The presence of sufficient interconnected low viscosity melt pockets could tend to increase the slip rates on fault planes and thus intensify the local stress field in the locked regions leading to crack growth and coalescence, and consequent release of elastic strain energy. The formation and freezing of melt patches that have higher viscosity relative to the country rock, however, may lead to fault strengthening.

Di Toro et al. (2006) argued that the production of pseudotachylyte from friction induced melting of rocks implies relatively dry conditions since presence of fluids would activate other weakening mechanisms preventing sufficient frictional heating. Thus, melting versus grain size relation calculated using surface tension values in vacuum, such as the curve labelled "Vacuum" in Fig. 13.27, may be used to develop an idea of the effect of grain size reduction on melting for a specific case of pseudotachylyte occurrence in nature. The threshold radial dimension for quartz for this scenario is ~ 0.1 μm or 100 nm.

Using the calculations illustrated in Fig. 13.27, it may be possible to roughly estimate the friction induced temperature increase on a fault plane in quartzite on the basis of size of quartz grains that did not experience recrystallization and are adjacent to or included in pseudotachylyte. This is an important parameter for estimating the dynamic shear resistance on a fault plane. This approach may be extended to other minerals and rock types when sufficient surface tension data become available to carry out calculations of melting temperature depression using Eq. (13.10.8).

Problem 13.1 Assuming ideal solution behavior for both solid and melt, show that Eq. (13.10.7) holds for melting (or solidus temperature) of a solid solution if the entropy of melting term is replaced by the partial molar entropy of melting, $\Delta s_m(T_{m,r})$, and also that the Eq. (13.10.8) holds if $T_{m,r*}$ is taken as the nominal melting temperature of the solid solution of the specific composition of interest.

Hint: At equilibrium, $\mu_i^m(T_{m,r*}) - \mu_i^s(T_{m,r*}) = 0 = \Delta h_m(T_{m,r*}) - T_{m,r*}\Delta s_m(T_{m,r*})$, and for ideal solutions, $\Delta h_m = H_m^o$.

13.11   Coarsening of Exsolution Lamellae

It has been observed in laboratory experimental studies that exsolution lamella in minerals coarsen when these are annealed at kinetically favorable temperatures. This phenomenon constitutes another example of "Ostwald ripening" discussed above.

Brady (1987) drew attention to the fact that continuous coarsening of perfect exsolution lamella with flat surfaces, as illustrated in Fig. 13.28, is impossible since this process does not lead to a reduction of the interfacial area to volume ratio of the lamellae. Furthermore, the chemical potential of a component on the flat surfaces of the lamella of the same composition are the same so that there is no driving force for the transfer of components from one lamellae to another. Brady (1987) proposed that lamellar coarsening takes place by the growth of those with large flat surfaces at the expense of those with wedge shaped edges (WSE), as illustrated in Fig. 13.29a. Wedge shaped edges have been observed in transmission electron microscopic images of both experimental and natural samples. An example from a meteoritic pyroxene is shown in Fig. 13.29b. The host rock is a cumulate basalt (known as cumulate eucrite) that seems to have been excavated from the asteroid Vesta (the asteroid belt lies between the orbits of Mars and Jupiter).

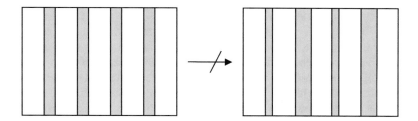

Fig. 13.28 Coarsening of perfect exsolution lamella with flat surfaces at the expense of other perfect lamella. The process is impossible since it does not lead to a reduction of the overall surface to volume ratio of the lamella

Appealing to the symmetry of the elimination process of WSE, and using a fundamental result obtained by Gibbs, Brady (1987) showed that the difference between the molar Gibbs energy of the phase α at the WSE, $G_m^\alpha(WSE)$, and that at a large flat side, $G_m^\alpha(\infty)$, is given by

$$G_m^\alpha(WSE) - G_m^\alpha(\infty) = \frac{2V_\alpha\sigma_{LFS}}{\lambda_\alpha} \qquad (13.11.1)$$

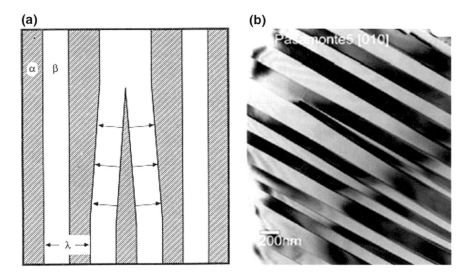

Fig. 13.29 (**a**) Illustration of the process of coarsening of exsolution lamella of a phase α with large flat surface by diffusion of material from the wedge shaped edge of a lamellae through the matrix or host phase, β (From Brady 1987). (**b**) [010] Dark-field image of (001) exsolution lamella in pyroxene from the meteorite Pasamonte that is inferred to have been excavated from the cumulate basaltic rock (or cumulate eucrite) in the Asteroid Vesta. The darker phase is augite whereas the lighter phase is pigeonite. Note the wedge shaped edge of pigeonite near the center of the figure. From Schwartz and McCallum (2005). With permission from Mineralogical Society of America

where V_α is the molar volume of the exsolved phase α, σ_{LFS} is the interfacial free energy of the large flat side of α and λ_α is the lamellar thickness of α. Thus the growth of a large face of the exsolution lamellae at the expense of a WSE, as illustrated in Fig. 13.29a, leads to a decrease of Gibbs energy of the system.

Brady (1987) assumed that both exsolution lamallae and the matrix or host phase are binary solutions, which is a good approximation of natural exsolution process, and that the matrix phase is in local equilibrium with the exsolution lamella on the two sides. Manipulating Eq. (13.11.1) and using Fick's law for the flux of a component along with geometric arguments, Brady (1987) then derived the following relation for the coarsening of exsolution lamella as a function of time.

$$\lambda^2 = \lambda_o^2 + kt \qquad (13.11.2)$$

where k is a rate constant, λ is the average wave length of the exsolution lamella (Fig. 13.29a), and λ_o is the initial value of λ, i.e. $\lambda_o = \lambda(t = 0)$.

Experimental data on lamellar coarsening kinetics show good agreement with the above relation in that the data for λ^2 versus t can be fitted well by a linear relation (Fig. 13.30). The rate constant k is a function of temperature and must be determined from the experimental λ^2 versus t relation at different temperatures. For an interesting application of the above relation to natural process, the reader is referred to Schwartz and McCallum (2005), who constrained the cooling rate of a meteorite (cumulate eucrite) excavated from the asteroid Vesta on the basis of the observed coarsening of exsolution lamella of pyroxenes (Fig. 13.29b), and the available experimental data on the lamellar coarsening kinetics.

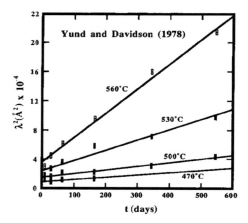

Fig. 13.30 Experimental coarsening data (symbols) of exsolved lamella in alkali feldspars (NaAlSi$_3$O$_8$-KAlSi$_3$O$_8$) as function of temperature and time, and fit of the isothermal data according to Eq. (13.10.2). The experimental data are from Yund and Davidson (1978). From Brady (1987). With permission from Mineralogical Society of America

13.12 Nucleation

13.12.1 Theory

Transformation of one phase into another requires nucleation and growth of the new phase. However, because of the surface energy associated with the formation of tiny entities or embryos of the new phase, only those embryos that exceed a critical size can grow further, leading to the formation of the new phase, while those that are below this critical size must disappear. To understand this phenomenon, consider the phase transformation $\alpha \rightarrow \beta$. The change in the Gibbs energy associated with the formation of an embryo of β from α, ΔG_e, consists of two terms,

$$\Delta G_e = V_e(\Delta G_v) + A_s \sigma,$$

where V_e and A_s are the volume and surface area of the embryo, ΔG_v is the Gibbs free energy change associated with the formation of unit volume of the embryo and σ is the interfacial tension between the embryo and the phase α. Assuming that the embryos are spheres of radius r, we then have

$$\Delta G_e = \frac{4}{3}\pi r^3(\Delta G_v) + 4\pi r^2 \sigma \tag{13.12.1}$$

The first term on the right is negative, since $\Delta G_v < 0$ (otherwise the transformation of α to β would be impossible), whereas the second right hand term is positive. The net result of combination of these two terms is the development of a maximum at a certain value of r as illustrated in Fig. 13.31 using assumed values of ΔG_v and σ. The radius corresponding to the maximum of $\Delta G_e(\beta)$ is known as the critical radius of nucleation, r_c. Embryos with $r < r_c$ disappear since further growth of these embryos raises the Gibbs free energy, whereas those with $r \geq r_c$ continue to grow and since this process lowers the free energy. Thus, the embryos with $r \geq r_c$ become the **stable** nuclei.

An expression for the critical radius, r_c, can be easily derived by imposing the condition of extremum of ΔG_e with respect to the variation of r, which yields

$$\left(\frac{\partial \Delta G_e}{\partial r}\right) = 0 = 4\pi r_c^2(\Delta G_v) + 8\pi r_c \sigma = 4\pi r_c(r_c \Delta G_v + 2\sigma)$$

so that

$$\boxed{r_c = -\frac{2\sigma}{\Delta G_v}} \tag{13.12.2}$$

(recall that $\Delta G_v < 0$ and hence $r_c > 0$).

The equilibrium number of nuclei of critical radius, N_c, is readily obtained from the Boltzmann distribution law (Sect. 14.1) that says that the fraction of particles at a certain energy level is proportional to $\exp(-E/k_B T)$, where E is the energy level of the particles above the ground state and k_B is the Boltzmann constant. Thus,

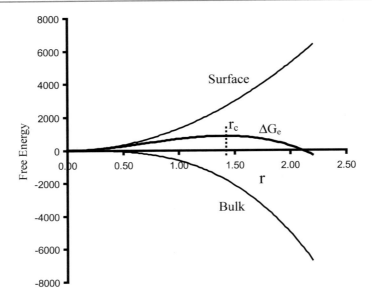

Fig. 13.31 Change of the total surface free energy ($4\pi r^2 \sigma$), total bulk free energy ($4/3\pi r^3 (\Delta G_v)$) and the net free energy (ΔG_e) of embryo as function of radius, r. The critical radius of nucleation is shown as r_c

$$N_C \propto N e^{-(\Delta G_C / k_B T)} \qquad (13.12.3)$$

where N is the number of possible nucleation sites and ΔG_c is the free energy change associated with nucleation.

It is evident from Eq. (13.12.2) that any process that leads to the reduction of the interfacial energy makes it easier for the formation of stable nucleus. Thus, adsorption of impurities, which causes reduction of σ, as discussed above, facilitates formation of a stable nucleus. Nucleation is also favored in crystal boundaries. This process can be understood in terms of the fact that the destruction of crystal boundaries (as a result of nucleation) releases some energy that provides part of the energy that is needed to overcome the nucleation energy barrier. The topic of nucleation in relation to geological processes has been treated in considerable detail by Kretz (1994) and Lasaga (1998), to which the readers are referred for further discussion of this subject.

13.12.2   Microstructures of Metals in Meteorites

Iron and stony-iron meteorites contain Fe-Ni metal alloy that shows different types of microstructures. A proper appreciation of the energetics and process of development of these microstructures is critical to the understanding of the thermal history of meteorites and the chemical interactions between the metals and silicates.

Figure 13.32 shows the equilibrium phase diagram for the Fe-Ni system (stability fields of the phases taenite and kamacite) along with some kinetic boundaries. Figure 13.33a shows an optical micrograph of a meteorite, Guarena, in which one can see two different types of metallic domains in a silicate matrix. On the left is zoned taenite plus kamacite, whereas on the right is a zoneless micron scale mixture of taenite and kamacite that is known as **plessite**. Both metallic domains developed as a result of cooling of metals of the same initial composition, ~ 10 wt% Ni. The initial T-X condition of the metals is shown by the point A in Fig. 13.32.

The development of the two types of metal microstructures can be understood in terms of the effect of crystal interface on nucleation, as discussed above (this effect is also implied by the discussions of Reisener and Goldstein 2003) and Reisener et al. 2006). In a polycrystalline aggregate of taenite, kamacite nucleates at the interfaces between taenite particles when these cool into the two phase boundary of taenite (T) + kamacite (K). For the taenite composition shown in Fig. 13.32, the upper limit of temperature at which kamacite nucleation can take place is represented by the point B. Under equilibrium condition, the Ni content of taenite evolves along the path B-C with progressive exsolution of kamacite, but the Ni content of the latter increases only slightly during cooling. However, under typical cooling rates of meteorites (degree per million years), the taenite particles do not

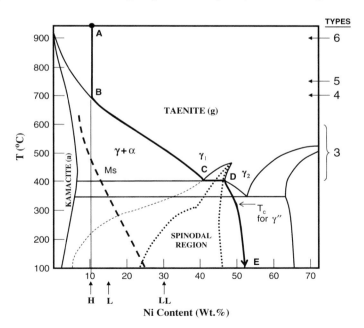

Fig. 13.32 Fe-Ni binary phase diagram showing the fields of different phases, and the equilibrium cooling path of (ABDE) of a metal grain with initial T-X value designated by the point A. H, L and LL shows the average Ni content of three types of chondritic meteorites, and numbers 3–6 on the right show the peak temperatures of different metamorphic types of meteorites. From Reisener and Goldstein (2003). With permission from Meteoritical Society

Fig. 13.33 (a) Optical photomicrograph of a meteorite, Guarena, showing zoned taenite + kamacite particle on the left and zoneless plessite on the right in a silicate matrix.
(b) Compositional profile of taneite and kamacite along the traverse shown by dashed line, as determined by an electron microprobe. From Reisener and Goldstein (2003b). With permission from Meteoritical Society

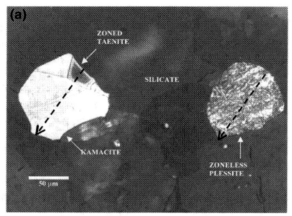

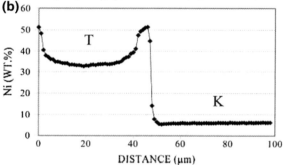

achieve complete grain scale equilibration of composition. This leads to Ni zoning profiles of the taenite grains (Fig. 13.33b) that can be used to retrieve the cooling rates of meteorites. (The method was pioneered by Wood 1964, and has been used extensively by planetary scientists. A more recent account can be found in Hopfe and Goldstein 2001). On the other hand, isolated single crystals of taenite undercool as homogeneous grains into the taenite plus kamacite (T + K) field. Upon cooling below the dashed line labeled Ms, which stands for "martensite start", the taenite grains undergo diffusionless taenite $\rightarrow$ martensite transformation, which finally decompose by the martensite $\rightarrow$ taenite + kamacite reaction with further cooling, thus leading to the formation of plessites.

13.13   Effect of Particle Size on Mineral Stability

The surface energies of phases are too small to have any perceptible effect on the stabilities of minerals unless the grain size is in the submicron domain so that the overall surface to volume ratio is sufficiently large. This may be appreciated from the effect of grain size on the solubility and melting of solids, as discussed above (Fig. 13.24). With the recent emergence of the field of nano-science that is primarily motivated by the special properties and novel industrial applications of nanometer size particles, there seems to have been a revival of interest in understanding the effect of grain size on the thermodynamic properties of minerals. In this section, I will review some of the recent developments and earlier studies that are of interest from geological and planetary science perspectives.

There has been considerable debate about the relative stabilities of hematite, Fe_2O_3, and goethite, $FeO(OH)$, in soils and sedimentary rocks. Red beds, which are sedimentary rocks stained red by hematite, have been studied in detail for use as paleoclimatic, geomagnetic and geochemical indicators. As discussed by Berner (1969), an important factor in these geological interpretations of red beds is whether or not hematite can form by the dehydration of goethite at the surface condition according to

$$\begin{array}{ccc} \text{Goethite} & \text{Hematite} & \text{liq} \\ 2FeO(OH) = & Fe_2O_3 + & H_2O \end{array} \qquad (13.13.a)$$

Since most fresh goethite crystals in sedimentary rocks are smaller than 0.1 μm, it is important to account for the surface energy of goethite in calculating its stability with respect to hematite which is typically coarse grained. As we have seen above (Figs. 13.24, 13.26 and 13.27), surface energy effects may become significant at submicron grain size. Langmuir (1971) calculated the effect of particle size on the equilibrium condition of the above reaction.

Figure 13.34 shows the effect of particle size on the Gibbs free energy change of the above reaction as a function of the particle size for three different cases: I. hematite < 1 μm, goethite > 1 μm, II. hematite and goethite equal in size and <1

Fig. 13.34 Particle size effect on Gibbs free energy change of Goethite dehydration reaction (2 Goethite = hematite + H_2O) at 1 bar, 298 K. From Langmuir (1971). With permission from American Journal of Science

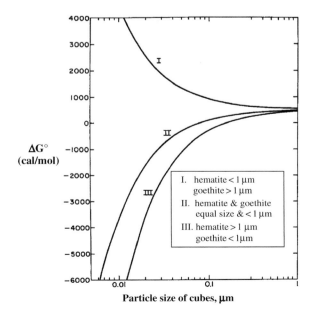

μm, and III. hematite > 1 μm and goethite < 1 μm. From these calculations, it is clear that particle size variation has no significant effect on the relative stabilities of goethite and hematite when both are ≥ 1μm. However, when the particles are a fraction of a micron, the surface energy effect would strongly influence the relative stabilities.

Figure 13.35 shows the equilibrium boundary, as calculated by Langmuir (1971) between goethite and hematite for two different particle sizes: (a) both grains are greater than 1 μm, and (b) goethite 0.1 μm, hematite > 1 μm. Most goethites in soils and sediments are fine grained, smaller than 0.1 μm. Thus, while coarse grained goethite is stable relative to coarse grained hematite plus water in the Earth's surface up to 80 °C, its stability shrinks substantially when the grains are of submicron size, and virtually disappears at 0.1 μm diameter grain size. These results reinforce the conclusion of Berner (1969) that paleoclimatic interpretation of the occurrence of hematite in soils should be made with caution, since the temperature at which hematite could form from goethite is sensitive to the grain size of the latter. In addition to particle size effect, kinetic effects are also important in controlling the natural occurrence of hematite and goethite.

Fig. 13.35 P-T stability diagram of goethite and hematite with different grains sizes. The dimensions represent the sides of cubes. $P = P_{H2O}$. From Langmuir (1971). With permission from American Journal of Science

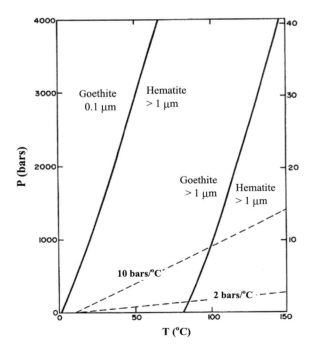

Navrotsky (2002) reviewed the available surface enthalpy values of minerals and found that the typical values are around $0.1–3$ J/m^2. Thus, as suggested by her, with surface area of $\sim 20{,}000$ m^2/mol, the enthalpy of nano-crystalline materials would be raised by 2–60 kJ/mol over that of the bulk material. The effect of grain size on the enthalpies of α-Al$_2$O$_3$ (corundum), γ-Al$_2$O$_3$ and γ-AlO(OH) (boehemite) is shown in Fig. 13.36a, and that on the enthalpies of three different polymorphs of TiO$_2$ (rutile, brookite and anatase) is shown in Fig. 13.36b. The different polymorphs show enthalpy cross-overs that could change the relative stabilities if the surface entropy contributions are negligible. Syntheses of nano-crystalline alumina usually results in the formation of γ-alumina even though corundum (α-alumina) is the stable poly-morph at the experimental P-T conditions. The enthalpy cross-over between the two polymorph, as shown in Fig. 13.36a, explains the phenomenon. Thus, surface energy studies could be of great help in guiding the syntheses of industrially important nano-crystalline materials that may not be stable as coarsely crystalline phases.

The stability of boehemite with respect to α- or γ-alumia is determined by the reaction

$$2\gamma\text{-AlO(OH)} = \text{Al}_2\text{O}_3(\alpha/\gamma) + \text{H}_2\text{O} \qquad (13.13.b)$$

Decreasing particle size raises the overall enthalpy of the Al$_2$O$_3$ phases faster than that of boehmite. Thus, if the particle sizes of both the hydrous and anhydrous phases decrease at the same rate, then the decrease of particle size would favor the stability of the hydrous phase. However, as we have seen in the case of goethite stability, the particle sizes of the hydrous and anhydrous phases are not, in general, the same.

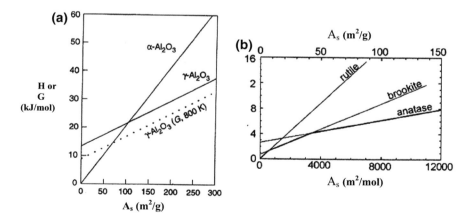

Fig. 13.36 (a) Enthalpy (H) of Al$_2$O$_3$ polymorphs relative to coarse grained corundum (α-Al$_2$O$_3$) as a function surface area. The dotted line represents free energy (G) of γ-Al$_2$O$_3$ relative to corundum at 800 K. (b) Enthalpy of TiO$_2$ polymorphs with respect to bulk rutile as a function of surface area. From McHale et al. (1997) and Navrotsky (2002), respectively. With permission from Science

In **carbonaceous chondrites** the hydrous phyllosilicates (e.g. serpentine) are often very fine grained, of the order of 100 Å (0.01 μm). One would expect that at this small particle size, the surface energy would have a significant influence on the condensation temperature of these minerals, if these represent primary condensates from the solar nebula, or on the temperature of their formation by secondary alteration process in the parent bodies. Specifically, the condensation or alteration temperature would decrease relative to that of coarse grained particles if the precursor anhydrous minerals which reacted with water to from the phyllosilicates were coarse grained. The effect may be opposite if the precursor minerals were also fine grained.

References

Asaduzzaman A, Muralidharan K, Ganguly J (2015) Incorporation of water into olivine during nebular condensation: insights from density functional theory and thermodynamics, and implications for phyllosilicate formation and terrestrial water inventory. Meteorit Planet Sci 50:578–589

Atherton MP (1976) Crystal growth models in metamorphic tectonites. Phil Trans Royal Soc London A 283:255–270

Balau JR, Waff HS, Tyburczy JA (1979) Mechanical and thermodynamic constraints on fluid distribution in partial melts. J Geophys Res 84:6102–6114

Becke F (1913) Uber Mineralbestand und Structur der Krystallinischen Schiefer. Ksehr Akad Wiss Wien 78:1–53

Berner RA (1969) Goethite stability and the origin of red beds. Geochim Cosmochim Acta 33:267–273

Brady J (1987) Coarsening of fine-scale exsolution lamellae. Am Min 72:697–706

Bruhn D, Groebner N, Kohlstedt DL (2000) An interconnected network of core-forming melts produced by shear deformation. Nature 403:883–886

Buffat P, Borel JP (1976) Size effect on melting temperature of gold particles. Phys Rev A13:2287–2298

Cahn JW (1979) Thermodynamics of solid and fluid surfaces. In: Johnson WC, Blakely JM (eds) Interfacial segregation. American Inst Metals, Ohio, pp 3–24

Carlson WD (1999) The case against Ostwald ripening of porphyroblasts. Canad Miner 37:403–413

Cashman KV, Ferry JM (1988) Crystal size distribution (CSD) in rocks and the kinetics and dynamics of crystallization. 3. Metamorphic crystallization. Contrib Miner Petrol 99:401–415

Curie P (1885) Sur la formation des crustaux et sur les constantes capillaires de leurs différentes faces. Soc Minéral France Bull 8:145–150

Denbigh K (1981) The principles of chemical equilibrium, Dover

Di Toro G, Hirose T, Nielson S, Pennachioni G, Shimamoto T (2006) Natural and experimental evidence of melt lubrication of faults during earthquakes. Science 311:647–649

Dunning JD, Petrovski D, Schuyler J, Owens A (1984) The effects of aqueous chemical environments on crack propagation in quartz. J Geophys Res 89:4115–4123

Gaetani GA, Grove TL (1999) Wetting of mantle olivine by sulfide melt: implications for Re/Os ratios in mantle peridotite and late-stage core formation. Earth Planet Sci Lett 169:147–163

Ganguly J, Cheng W, Tirone M (1996) Thermodynamics of aluminosilicate garnet solid solution: new experimental data, an optimized model, and thermometric applications. Contrib Miner Petrol 126:137–151

Halden FH, Kingery WD (1955) Surface tension at elevated temperatures. II. Effect of carbon, nitrogen, oxygen, and sulfur on liquid-iron surface tension and interfacial energy with alumina. J Phys Chem 59:557–559

Herring C (1953) The use of classical macroscopic concepts in surface-energy problems. In: Gomer G, Smith CS (eds) Structure and properties of solid surfaces. The University of Chicago Press, pp 5–72

Hess P (1994) Thermodynamics of thin films. J Geophys Res 99:7219–7229

Holness MB (1993) Temperature and pressure dependence of quartz-aqueous fluid dihedral angles: the control of adsorbed H_2O on the permeability of quartzites. Earth Planet Sci Lett 117:363–377

Hopfe WD, Goldstein JI (2001) The metallographic cooling rate method revised: application to iron meteorites and mesosiderites. Meteorit Planet Sci 36:135–154

Hudon P, Jung I-H, Baker DR (2002) Melting of β-quartz up to 2.0 GPa and thermodynamic optimization of silica liquidus up to 6.0 GPa. Phys Earth Planet Inter 130:159–174

Joesten RL (1991) Kinetics of coarsening and diffusion-controlled mineral growth. Rev Min 26:507–582

Jurewicz SR, Jurewicz JG (1986) Distribution of apparent angles on random sections with emphasis on dihedral angle measurements. J Geophys Res 91:9277–9282

Jurewicz SR, Watson EB (1985) Distribution of partial melt in a granitic system. Geochim Cosmochim Acta 49:1109–11121

Kohlstedt DL (1992) Structure, rheology and permeability of partially molten rocks at low melt fractions. In: Morgan JP (ed) Mantle flow and melt generation. Geophys Monograph 71, Am Geophys Union, pp 103–121

Kretz R (1966) Interpretation of shapes of mineral grains in metamorphic rocks. J Petrol 7:68–94

Kretz R (1994) Metamorphic crystallization. Wiley, p 507

Kretz R (2006) Shape, size, spatial distribution and composition of garnet crystals in highly deformed gneiss of the Otter Lake area, Québec, and a model for garnet crystallization. J Met Geol 24:431–449

Langmuir D (1971) Particle size effect on the reaction goethite = hematite + water. Am J Sci 271:147–156

Lasaga AC (1998) Kinetic theory in earth sciences. Princeton University Press, New Jersey, p 811

Lawn B, Wilshaw T (1975) Fracture of brittle solids. Canbridge, New York

Lee SK, Han R, Kim EJ, Jeong GJ, Khim H, Hirose T (2017) Quasi-equilibrium melting of quartzite upon extreme friction. Nat Geosci 10:436–441

Li L, Tsukamoto K, Sunagawa I (1990) Impurity adsorption and habit changes in aqueous solution grown KCl crystals. J Cryst Growth 99:150–155

McHale JM, Auroux A, Perrotta AJ, Navrotsky A (1997) Surface energies and thermodynamic phase stabilities in nanocrystalline aluminas. Science 277:788–791

McKenzie D, Brune JN (1972) Melting on fault planes during large earthquakes. Geophys J Roy Astron Soc 29:65–78

Miyazaki K (1991) Ostwald ripening of garnet in high P/T metamorphic rocks. Contr Miner Petrol 108:118–128

Mueller RF, Saxena SK (1977) Chemical petrology. Springer, Berlin, p 394

Navrotsky A (2002) Thermochemistry, energetic modelling, and systematics. In: Gramaciolli CM (ed) Energy modelling in minerals, vol 14. European mineralogical union. Eötvös University Press, Budapest, Hungary, pp 5–26

Parks GA (1984) Surface and interfacial free energies of quartz. J Geophys Res 89:3997–4008

Passeron A, Sangiorgi R (1985) Solid-liquid interfacial tensions by the dihedral angle method. A mathematical approach. Acta Metall 33:771–776

Philpotts AR (1990) Principles of igneous and metamorphic petrology. Prentice Hall, New Jersey, p 498

Raghaven V, Cohen M (1975) Solid state phase transformations. In: Hannay NB (ed) Treatise on solid state chemistry, vol 5. Plenum Press, New York

Reisener RJ, Goldstein JI (2003) Ordinary chondrite metallography: Part 1. Fe-Ni taenite cooling experiments. Meteorit Planet Sci 38:1669–1678

Reisener RJ, Goldstein JI, Pataev MI (2006) Olivine zoning and retrograde olivine - orthopyroxene-metal equilibration in H5 and H6 chondrites. Meteorit Planet Sci 41:1839–1852

Rose LA, Brenan JM (2001) Wetting properties of Fe-Ni-Co-O-S melts against olivine: implications for sulfide melt mobility. Econ Geol Bull Soc Econ Geol 145–157

Rosenberg R (2005) Why ice is slippery? Phys Today 58: 50–55

Saxena SK, Hrubiak R (2014) Mapping the nebular condensates and chemical composition of the terrestrial planets. Earth Planet Sci Lett 393–119

Schwartz JM, McCallum IS (2005) Comparative study of equilibrated and unequilibrated eucrites: subsolidus thermal histories of Haraiya and Pasamonte. Am Miner 90:1871–1886

Shand SJ (1917) The pseudotachylite of Parijs (Orange Free state), and its relation to "trapp-shotten-gneiss" and "flinty-crush-rock". Geol Soc London Q J 72:198–221

Smith CS (1964) Some elementary principles of polycrystalline microstructures. Met Rev 9:1–48

Spray JG (2005) Evidence for melt lubrication during large earthquakes. Geophys Res Lett 32: L07301–L07305

Sunagawa I (1957) Variation in crystal habit of pyrite. Jap Geol Surv Rep 175:41

Terasaki H, Frost DJ, Rubie DC, Langenhorst F (2005) The effect of oxygen and sulfur on the dihedral angle between Fe-O-S melt and silicate minerals at high pressure: implications for Martian core formation. Earth Planet Sci Lett 232:379–392

Toramuru A, Fuji N (1986) Connectivity of melt phase in a partially molten peridotite. J Geophys Res 91:9239–9252

von Bergen N, Waff HS (1986) Permeabilities, interfacial areas and curvatures of partially molten systems: results of numerical computations and equilibrium microstructures. J Geophys Res 91:9261–9276

Waff HS, Bulau JR (1982) Experimental determination of near equilibrium textures in partially molten silicates at high pressures. In: Akimoto S, Manghnani MH (eds) High pressure research in geophysics. Center for Academic Publications, Tokyo, pp 229–236

Wood JA (1964) The cooling rates and parent planets of several meteorites. Icarus 3:429–459

Yund RA, Davidson P (1978) Kinetics of lamellar coarsening in cryptoperthites. Am Miner 63:470–477

Statistical Thermodynamics Primer

14

As discussed in the Introductory Sect. 1.1, classical thermodynamics deals with equilibrium properties of a system at the macroscopic scale, but without any concern or appeal to the microscopic states that are ultimately responsible for the macroscopic properties; it cannot throw any fundamental insight about why something has the thermodynamic property that it has. A major goal of Statistical Thermodynamics is to fill this deficiency in our knowledge by developing formal relationships between the macroscopic properties of a system and its microscopic states and thus to enable, at least in principle, calculation of thermodynamic properties of a system from energetic considerations at the molecular or atomic scale. However, this has been a dream for the complex systems of geological interest, but is now becoming a "dream come true" with the advent of modern quantum chemical software packages, especially those based on density functional theory in quantum chemistry (see Appendix C) and computer power. Like Thermodynamics, Statistical Thermodynamics also has two branches dealing with equilibrium and non-equilibrium problems. The objective of this chapter is to provide a brief overview of Equilibrium Statistical Thermodynamics and show some applications to systems of geochemical interest.

14.1 Boltzmann Distribution and Partition Function

In Sect. 1.6.2 we discussed the distribution of phonons over the quantized vibrational energy levels of a crystal. This discussion can be generalized to distribution of *non-interacting* particles over different types of quantized energy states, viz., vibrational, rotational, translational and electronic. When subject to the constraints that the total number of particles and the total energy are conserved, the most probable fraction of particles in an allowed energy state, $\in_i$, is given by the **Boltzmann distribution law**, which is as follows.

© Springer Nature Switzerland AG 2020
J. Ganguly, *Thermodynamics in Earth and Planetary Sciences*,
Springer Textbooks in Earth Sciences, Geography and Environment,
https://doi.org/10.1007/978-3-030-20879-0_14

$$\frac{n_i}{N} = \frac{\alpha_i e^{-\epsilon_i/k_B T}}{\sum_i \alpha_i e^{-\epsilon_i/k_B T}} \qquad (14.1.1)$$

where n_i is the number of particles in the energy state ϵ_i, N is the total number of particles and k_B is the Boltzmann constant; the factor α_i stands for the degeneracy of state i or the number of states with energy ϵ_i. If there is only one state with energy ϵ_i, then $\alpha_i = 1$, for two states with the same energy, $\alpha_i = 2$, and so on. The summation term in the denominator is called a **partition function** and has a profound role in the development of relations between macroscopic (i.e. thermodynamic) and microscopic properties of a substance. The term $e^{-\epsilon_i/k_B T}$ on the right hand side is called the **Boltzmann factor**. Thus, the probability that a system is in some particular state ϵ_i is proportional to the Boltzmann factor of the state.

Consider now a *closed* macroscopic system with a fixed volume, V, and temperature, T. In order to calculate the internal energy of such a system, we consider a collection of large number of systems that are identical to the one of specific interest but are separated from one another by diathermal walls (i.e. walls that permit heat transfer but not mass transfer). We then immerse the entire ensemble within a heat bath that is maintained at a temperature T and is isolated from the surrounding (Fig. 14.1). This is known in statistical thermodynamics as a **canonical ensemble**. (There are also other types of ensembles, such as a microcanonical ensemble, in which each unit has a fixed energy instead of a fixed temperature; however, the canonical ensemble is the one of interest to us in this chapter and also is the one most commonly used in statistical thermodynamics).

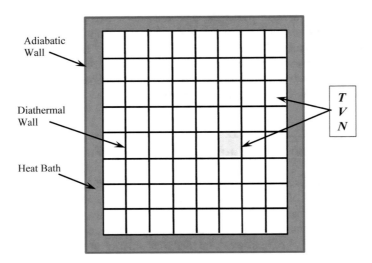

Fig. 14.1 Schematic illustration of a canonical ensemble. Shaded block is the unit of interest with fixed V and N at temperature T. All other blocks are replica of the shaded block. A uniform temperature is maintained throughout the ensemble by immersing it in a heat bath

When equilibrium is attained, the measured temperature must be uniform throughout the ensemble and be the same as that of the heat bath. The energy of each subsystem will, however, fluctuate over the allowed energy states that we now label as E_1, E_2, E_3 etc. If we are to measure the energy or any thermodynamic property of a single subsystem in the ensemble, it has to be carried out over a finite interval of time during which it would have passed through many energy states. The measured value of energy of the system will be its time averaged value. However, time averaging is an impossible task for a system consisting of the extremely large number of particles ($\sim 10^{23}$) contained in a molar quantity of material. This problem was circumvented by Gibbs who introduced the concept of ensemble and hypothesized that the time averaged value of a thermodynamic property of a single system is the same as the instantaneous average of the same property of the subsystems in the canonical ensemble (this is known as **ergodic hypothesis**). Thus,

$$U = \bar{E} = \frac{\sum N_i E_i}{N_c} \qquad (14.1.2)$$

where N_i is the number of subsystems at a state with energy E_i and N_c is the total number of subsystems in the canonical ensemble. Combining the Boltzmann distribution (i.e. Eq. 14.1.1 with n_i, $\in_i$ and N being replaced by N_i, E_i and N_c, respectively) with the last expression and setting $\alpha_i = 1$ in order to make the formulations less cumbersome, we obtain

$$U = \frac{\sum E_i e^{-E_i/k_B T}}{\sum e^{-E_i/k_B T}} = \frac{\sum E_i e^{-E_i/k_B T}}{Q} \qquad (14.1.3)$$

where $Q = \sum e^{-E_i/k_B T}$ (or $= \sum \alpha_i e^{-E_i/k_B T}$) is the **partition function** of the canonical ensemble.

14.2 Thermodynamic Properties

From the above definition of canonical partition function, Q, it follows that

$$\left(\frac{\partial Q}{\partial (1/T)} \right)_V = -\frac{1}{k_B} \left(\sum E_i e^{-E_i/k_B T} \right)$$

so that

$$\left(\frac{\partial Q}{\partial T} \right)_V = \frac{1}{k_B T^2} \left(\sum E_i e^{-E_i/k_B T} \right) \qquad (14.2.1)$$

Expressing the numerator of the Eq. (14.1.3) in terms of this relation we now obtain an expression of internal energy as

$$U = \frac{k_B T^2}{Q}\left(\frac{\partial Q}{\partial T}\right)_V = k_B T^2 \left(\frac{\partial \ln Q}{\partial T}\right)_V \qquad (14.2.2)$$

Using his expression as a starting point, we develop below expressions of several other thermodynamic properties in terms of Q using classical thermodynamic relations among those properties.

From the thermodynamic relation between **molar heat capacity at constant volume**, C_V, and U (Eq. 3.7.4b), we obtain

$$C_V = \left(\frac{\partial U}{\partial T}\right)_V = k_B\left[2\left(\frac{\partial \ln Q}{\partial \ln T}\right)_V + T^2\left(\frac{\partial^2 \ln Q}{\partial T^2}\right)_V\right] \qquad (14.2.3)$$

This expression can be shown to be equivalent to

$$C_V = \frac{k_B}{T^2}\left(\frac{\partial^2 \ln Q}{\partial (1/T)^2}\right)_V \qquad (14.2.4)$$

It is left to the reader to verify the equivalence (Hint: write the parenthetical quantity in the last equation as $\partial/\partial(1/T)(\ldots)$ and use the relation $d(1/T) = -dT/T^2$ to derive Eq. 14.2.3).

The **molar entropy** can now be expressed in terms of Q according to the relationship between S and C_V (Eq. 3.7.6), as follows.

$$S(T) = \int \frac{C_V}{T}dT = 2k_B\int\left(\frac{\partial \ln Q}{\partial T}\right)_V dT + k_B\int T\frac{\partial}{\partial T}\left(\frac{\partial \ln Q}{\partial T}\right)_V dT \qquad (14.2.5)$$

Using integration by parts[1], the second integral on the right can be reduced to:

$$\int T\frac{\partial}{\partial T}\left(\frac{\partial \ln Q}{\partial T}\right)_V dT = T\left(\frac{\partial \ln Q}{\partial T}\right)_V - \int\left(\frac{\partial \ln Q}{\partial T}\right)_V dT \qquad (14.2.6)$$

Thus,

$$\begin{aligned} S(T) &= k_B\int\left(\frac{\partial \ln Q}{\partial T}\right)_V dT + K_B T\left(\frac{\partial \ln Q}{\partial T}\right)_V \\ &= k_B\ln Q + k_B T\left(\frac{\partial \ln Q}{\partial T}\right)_V + I_S \end{aligned} \qquad (14.2.7)$$

where I_S is an integration constant. Combining this expression with that for the internal energy, U, (Eq. 14.2.2), we obtain

$$S(T) = k_B\ln Q + \frac{U}{T} + I_S \qquad (14.2.8)$$

---

[1]Integration by parts: $\int u\,dw = uw - \int w\,du$ or $\int u(dw/du)du = uw - \int w\,du$; set $w = \partial \ln Q/dT$ and $u = T$.

The integration constant I_S drops out in thermodynamic calculations of the entropy change of a reaction, $\Delta_r S$. However, it can also be evaluated by invoking the third law of thermodynamics. It can be shown (McQuarrie 2003, p. 47) that as $T \rightarrow 0$, the first two terms also tend to zero if there is no degeneracy of states so that S $(T = 0) = I_S$. But according to the third law, $S(T = 0) = 0$. Thus, $I_S = 0$.

The expression for the **Helmholtz free energy**, F, follows from combining the thermodynamic relation $F = U - TS$ (Box 3.1.1) with the expressions U and S derived above (Eqs. 14.2.2 and 14.2.8), and some straightforward mathematical manipulations:

$$F = -k_B T \ln Q \qquad (14.2.9)$$

Pressure is related to F according to $P = -(\partial F / \partial V)_T$ (Box 3.1.2) and thus, from the above relation,

$$P = k_B T \left(\frac{\partial \ln Q}{\partial V} \right)_T \qquad (14.2.10)$$

The expressions for enthalpy, H, and Gibbs free energy, G, now follow from using the thermodynamic relations $H = U + PV$ and $G = F + PV$ (Box 3.1.1). The expressions of different thermodynamic functions in a closed system in terms of the canonical partition function are summarized in the Box (14.2.1). These relations form the basis of calculations of the thermodynamic properties in the computational quantum chemistry software packages (see Appendix C). Note that although we have derived the relations assuming that there are no degeneracies of states ($\alpha_i = 1$), it can be easily verified that they are not affected if we allow for degeneracies.

$$U = k_B T^2 \left(\frac{\partial \ln Q}{\partial T} \right)_V$$

$$C_v = k_B \left[2 \left(\frac{\partial \ln Q}{\partial \ln T} \right)_v + T^2 \left(\frac{\partial^2 \ln Q}{\partial T^2} \right)_v \right] = \frac{k_B}{T^2} \left(\frac{\partial^2 \ln Q}{\partial (1/T)^2} \right)_v$$

$$S(T) = \frac{U}{T} + k_B \ln Q$$

$$F = -k_B T \ln Q$$

$$H = k_B T \left[\left(\frac{\partial \ln Q}{\partial \ln T} \right)_V + \left(\frac{\partial \ln Q}{\partial \ln V} \right)_T \right]$$

$$G = k_B T \left[\left(\frac{\partial \ln Q}{\partial \ln V} \right)_T - \ln Q \right]$$

Box (14.2.1)

14.3   Expressions of Partition Functions

Let us first consider a system consisting of two distinguishable but *non-interacting* particles of the same type, *a* and *b*, such as two particles of an ideal gas, distributed between two energy levels ε_1 and ε_2. There are four distinguishable arrangements of the particles in the system: (1) both particles at the energy level ε_1; (2) both at the energy level ε_2; (3) particle *a* at ε_1 and *b* at ε_2; (4) reverse of (3). The associated two particle energies of the system are: $E_1 = 2\varepsilon_1$; $E_2 = 2\varepsilon_2$; $E_3 = \varepsilon_1 + \varepsilon_2$; $E_4 = \varepsilon_1 + \varepsilon_2$. Thus, denoting $-1/k_BT$ as β, the canonical partition function of the system is given by

$$Q = \sum e^{\beta E_i} = e^{2\beta\varepsilon_1} + e^{2\beta\varepsilon_2} + 2e^{\beta(\varepsilon_1 + \varepsilon_2)}$$
$$= \left(e^{\beta\varepsilon_1}\right)^2 + \left(e^{\beta\varepsilon_2}\right)^2 + 2(e^{\beta\varepsilon_1}e^{\beta\varepsilon_2}) = \left(e^{\beta\varepsilon_1} + e^{\beta\varepsilon_2}\right)^2 \qquad (14.3.1)$$

This expression may be written in the form $Q = \left(\sum e^{\beta\varepsilon_i}\right)^2$. Since the exponent 2 equals the number of distinguishable particles, we should anticipate that, in general, $Q = \left(\sum e^{\beta\varepsilon_i}\right)^{N'}$ where N' is number of distinguishable particles. This anticipated relation can be shown to be rigorously correct. The summation term denotes the partition function arising from the occupancy of a single particle in different energy levels, and is thus usually referred to as a **single particle partition function** (or **molecular partition function** if the particle is a molecule) that we would denote by the symbol q. Thus,

$$Q = (q)^{N'} \qquad (14.3.2a)$$

If, however, the particles are *indistinguishable*, then interchanging the position of the particles between different energy levels does not create a new state. For example, if the two particles (*a* and *b*) in the above analysis are indistinguishable, then the states (3) and (4) do not represent two different states, but just one state. If all N' particles are indistinguishable and there is one particle per energy level, then there will be $N'!$ permutations among the particles. Hence, for a system that has many more available energy levels than the number of particles so that "crowding" of the energy levels is avoided

$$Q = \frac{(q)^{N'}}{N'!} \qquad (14.3.2b)$$

where the factor $1/N'!$ corrects for the error introduced by over-counting the states of N' indistinguishable particles, resulting from their treatment as distinguishable particles. The above equation is typically applicable to an ideal gas at ordinary temperatures and densities. Equation (14.3.2a), on the other hand, would apply to crystals since identical particles (e.g. Mg atoms in forsterite, Mg_2SiO_4) are localized at specific lattice sites and hence are distinguishable.

The central problem in the derivation of the expressions of different types of partition functions of a molecule is finding analytical expressions for the allowed energy levels. This is accomplished by solving the famous Schrödinger wave equation in quantum mechanics, where such solutions are possible, and making modifications to these solutions, if the nature of the problem demands it, using other methods such as quantum mechanical perturbation theory.

As an example, for one dimensional motion within a bounded domain of length a, the solution of the wave equation yields

$$\varepsilon_n = n^2 h^2 / 8ma^2 \qquad (14.3.3)$$

for the allowed energy levels, where $n = 1, 2, 3 \ldots$ and m is the mass of the particle. Thus, the molecular or single particle partition function for one dimensional translational motion becomes

$$q_{tr:a} = \sum \exp\left(-\frac{n^2 h^2}{k_B T (8ma^2)}\right) \qquad (14.3.4)$$

However, the translational energy levels are so closely spaced that the summation of the exponential term may be replaced by integration of the term so that $q_{tr:a} = \int \exp(\ldots) dn$. Working through this integration, one finally obtains the following expression for one dimensional **translational partition function** of *non-interacting* particles

$$q_{tr:a} = \left(\frac{(2\pi M k_B T)^{1/2}}{h}\right) a \qquad (14.3.5)$$

where M is the atomic or molecular mass. For three degrees of translational freedom,

$$q_{tr} = z'_{tr:a} z'_{tr:b} z'_{tr:c} = \left(\frac{(2\pi M k_B T)^{1/2}}{h}\right)^3 v \qquad (14.3.6)$$

where v is the volume (abc). The inverse of the collection of terms within the parentheses in Eq. (14.3.5) is known as the **thermal de Broglie wavelength**, Λ (i.e. $\Lambda = h/(2\pi M k_B T)^{1/2}$). The condition of non-interactive behavior of the particles is satisfied if Λ is much less than the average spacing between the particles. And this is the case for all molecular species except for H_2 and He for which the molecular masses are so small that inter-particle interaction needs to be accounted for at temperature near absolute zero.

Except for hydrogen, the rotational energy levels are also effectively continuous (Bigeleisen and Mayer 1947). If the rotational motion is decoupled from the vibrational motions, as in a rigid rotor, then the **rotational partition function of a linear molecule** is given by

$$q_{r:l} = \frac{8\pi^2 I\, k_B T}{\sigma' h^2} \qquad (14.3.7a)$$

where σ' is the symmetry number and I is the moment of inertia. The symmetry number indicates the number of discrete repetitions of the same molecular configuration upon $360°$ rotation about symmetry axis axes. Thus $\sigma = 1$ for a heteronuclear diatomic molecule, such as HD, $\sigma' = 2$ for a homonuclear diatomic molecule, such as HH (one two-fold axis of symmetry) and $\sigma' = 12$ for CH_4 (tetrahedral configuration with a central C atom and four apical H atoms so that there are four three-fold axes of symmetry coincident with the four CH bonds) etc. The above expression for partition function applies to both diatomic and linear polyatomic molecules such as CO_2 and C_2H_2. For nonlinear molecules such as H_2O, the rotational partition function is

$$q_{r:nl} = \frac{8\pi^2 (8\pi^3 I_A I_B I_C)^{1/2} (k_B T)^{3/2}}{\sigma' h^3} \qquad (14.3.7b)$$

where I_A, I_B and I_C are moments of inertia about three special axes in the Cartesian coordinates that are known as the principal axes. The treatment of moments of inertia about any arbitrary set of Cartesian axes with origin at the center of mass of the rigid body requires a formulation with 3×3 matrix with diagonal elements I_{xx}, I_{yy} and I_{zz} plus off-diagonal elements I_{xy}, I_{xz} and so on. For the principal axes, the off-diagonal elements are zero. The moments of inertia about the principal axes are known as the **principal moments of inertia**. An axis of symmetry of a molecule constitutes a principal axis. (Petrologists and Geochemists who deal with diffusion kinetics would note the similarity between the treatments of moments of inertia and diffusion kinetics in anisotropic crystals in terms of principal diffusion coefficients and axes). The above expressions for translational and rotational partition functions, which were derived by assuming the translational and rotational energy levels to be effectively continuous, are the same as those derived in classical mechanics.

For hydrogen, the rotational partition function may be expressed by introducing correction terms to the classical expression (Eq. 14.3.7a) as follows (Bigeleisen and Mayer 1947)

$$q_{r:l} = \frac{8\pi^2 I\, k_B T}{\sigma' h^2} \left(1 + \frac{\varphi}{3} + \frac{\varphi^2}{15} + \frac{4\varphi^3}{315} + \cdots \right) \qquad (14.3.8)$$

where φ is the inverse of the leading term. Thus, with increasing temperature, the rotational partition function of hydrogen approaches the classical value; even at room temperature, the correction term is found to be small (Bigeleisen and Mayer 1947).

For the **vibrational partition function**, we consider the simple case of harmonic oscillator model (see Sect. 1.6) and decoupling from the rotational motion. Since anharmonicity is the reason behind positive thermal expansion (see Sect. 3.7.1 for negative thermal expansion), the harmonic oscillator model breaks down above certain temperature that depends on the system. Still, this model yields useful results. And, as we would see later, sometimes there are self-cancellations of anharmonic effects in the statistical thermodynamic calculation of the equilibrium constants of isotopic exchange reactions. Also in a comprehensive review of the field of oxygen isotope thermometry that is based on isotopic exchange equilibrium between minerals, Javoy (1977) concluded that harmonic approximation usually gives satisfactory results when compared with experimental data.

For vibrational energy levels, solution of the Schrödinger equation yields the relation

$$\varepsilon_n = (n + 1/2)h\nu \tag{14.3.9}$$

(with n being 0, 1, 2, 3 …), for the allowed energy levels of a harmonic oscillator. Hence, according to Eq. (14.1.1), the partition function for each normal mode of vibration j is given by

$$q_{\nu_j} = \sum_n e^{-(n + 1/2)h\nu_j/k_BT} = e^{-h\nu_j/2k_BT} \sum_n e^{-nh\nu_j/k_BT} \tag{14.3.10}$$

The summation term after the second equality constitutes a geometric series[2]. Thus, the partition function reduces to

$$q_{\nu_j} = \frac{e^{-h\nu_j/2k_BT}}{1 - e^{-h\nu_j/k_BT}} \tag{14.3.11}$$

The term in the numerator arises solely from the zero point energy (ZPE). If ZPE contribution is neglected (i.e. $\varepsilon_n = nh\nu$), as is done in some problems, then the term in the numerator becomes unity. (It should be noted that the vibrational frequencies are typically listed in terms of wave numbers, ω, which is the inverse of wave length, λ, and often with units of cm^{-1}. Thus, these numbers need to be multiplied by the velocity of light (c = 299,792,458 m/s) for use in the calculation of hν). The effect of anharmonicity on the vibrational energy levels may be accounted for by the following modification of the harmonic approximation, as obtained from the quantum mechanical perturbation theory.

$$\varepsilon_n = (n + 1/2)h\nu - x_e(n + 1/2)^2 h\nu + y_e(n + 1/2)^3 h\nu - \cdots . \tag{14.3.12}$$

where x_e, y_e, … are constants. One can also incorporate into this expression the effect of coupling between the rotational and vibrational motions (e.g. McQuarrie 2003).

---

[2]In a simple geometric summation series, each term can be obtained by multiplying the preceding term by a constant factor: $\sum_{j=1}^{\infty} a_1 r^{j-1}$, where r is a constant term; the summation of such a series is given by $S_\infty = \frac{a_1}{1-r}$. In the summation series in Eq. (14.3.10), $a_1 = 1$ and r = $\exp(-h\nu_j/k_BT)$.

A diatomic molecule has only one vibrational mode. A polyatomic molecule, on the other hand, has a number of normal modes of vibration. In general, if n is the number of atoms in a molecule, then a linear molecule, such as CO_2, has $3n - 5$ and a nonlinear molecule, such as H_2O, has $3n - 6$ normal vibrational modes. (Since it takes three coordinates to locate each atom, there are 3n coordinates for a polyatomic molecule with n atoms. Of these, three are used to locate the center of mass of the molecule; additionally two are needed to specify the angular orientation of a linear molecule and three for a nonlinear molecule. Consequently, there are $3n - 5$ and $3n - 6$ coordinates available as internal coordinates for linear and non-linear molecules, respectively, to specify the positions of the n atoms). Thus, CO_2 has four normal modes $(3 \times 3 - 5)$ of vibration, viz. symmetric stretch ($\omega_1 = 1340$ cm^{-1}), asymmetric stretch ($\omega_2 = 2349$ cm^{-1}) and two degenerate bending modes ($\omega_3 = 667$ cm^{-1}), which are illustrated in Fig. 14.2, whereas H_2O has three normal modes. The vibrational partition function for a polyatomic molecule is a product of the partition functions for each normal mode:

$$q_v = q_{v_1} q_{v_2} \cdots = \prod q_{v_j} \qquad (14.3.13)$$

where v_j is the jth normal vibrational mode. If there are k-modes of the same frequency v* (k-fold degeneracy), then the corresponding q_{v*} must be raised to the power k in the above product function or, in other words, each degenerate mode should be counted as a separate entity.

The Box (14.2.2) shows a summary of the expressions of single particle partition functions for different types of motion of a molecule behaving as a rigid rotor and in which the translational and rotational energy levels have been treated as continuous. In addition to these, there are also **electronic** and **nuclear partition functions**. It, however, takes exceedingly high temperatures that far exceed anything attainable in terrestrial processes to achieve nuclear excitation since the nuclear energy levels are separated by energies of the order of 10^6 eV. The separation between the electronic energy levels is usually of the order of eV, and thus it takes a few thousand degrees

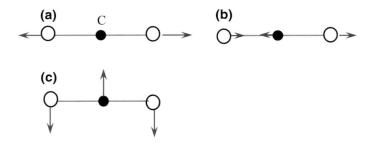

Fig. 14.2 Vibrational modes of CO, (**a**) symmetric stretch ($\omega_1 = 1340$ cm^{-1}), (**b**) asymmetric stretch ($\omega_1 = 2349$ cm^{-1}), and (**c**) two degenerate (one in the plane of the paper and the other normal to it) bending modes ($\omega_3 = 667$ cm^{-1})

in most cases for any significant contribution to the partition function from excitation of electrons. To see this, let us consider the situation with $\varepsilon = 1$ eV and $T = 3000$ K. These values yield $e^{-\varepsilon/kT} = 1.9(10^{-7})$ that is vanishingly small. This means that if the first electronic energy level is 1 eV above the ground state, it makes no significant contribution to the electronic partition function. The energy ε_o of the ground state may be arbitrarily set to zero so that

$$q_e = \sum \alpha_j e^{-\varepsilon_i/k_B T} = \alpha_o + \sum \alpha_i e_1^{-\Delta\varepsilon_i/k_B T} + \ldots \approx \alpha_0 \qquad (14.3.14)$$

where $\Delta\varepsilon_i$ is the energy of an excited state with respect to the ground state. For most molecules, there is no degeneracy of the ground state so that the electronic partition function may be usually taken to be unity. Even when the first excited electronic state is close to the ground state, it is generally sufficient to include only the first two terms in the expression of q_e.

The contributions of different types of motions to thermodynamic properties of a substance can now be calculated from the relations summarized in the Box (14.2.1). As an example, let us consider the translational entropy of a monatomic gas for which the translational motion is effectively the only source of entropy. We set $q_{elec} = 0$, and proceed in two steps: (a) first we calculate the internal energy U and (b) then use the relationship between S and U in the Box (14.2.1) to calculate S.

Translational

$$q_{tr:a} = \left(\frac{(2\pi M k_B T)^{1/2}}{h}\right) a \qquad 1-d$$

$$q_{tr} = \left(\frac{(2\pi M k_B T)^{1/2}}{h}\right)^3 V \qquad 3-d$$

Rotational

$$q_{r:l} = \frac{8\pi^2 I\, k_B T}{\sigma' h^2} \qquad \begin{array}{l}\text{linear molecule} \\ (\sigma'\text{: symmetry no.})\end{array}$$

$$q_{r:nl} = \frac{8\pi^2 \left(8\pi^3 I_x I_y I_z\right)^{1/2} (k_B T)^{3/2}}{\sigma' h^3} \qquad \text{non-linear molecule}$$

Box (14.2.2)

Vibrational

$$q_{v_j} = \frac{e^{-h v_j/2 k_B T}}{1 - e^{-h v_j/k_B T}} \qquad \text{for a normal mode } j$$

$$q_v = \prod q_{v_j} \qquad \begin{array}{l}\text{overall: product over} \\ \text{all normal modes}\end{array}$$

From the expression of translational partition function,

$$\frac{\partial \ln q_{tr}}{\partial T} = \frac{3}{2}\left(\frac{\partial \ln T}{\partial T}\right) = \frac{3}{2}\left(\frac{1}{T}\right) \tag{14.3.15}$$

Taking the number of particles in the system to be L, where L is the Avogadro's number, $Q_{tr} = (q_{tr})^L/L!$ (Eq. 14.3.2b), and hence

$$\frac{\partial \ln Q_{tr}}{\partial T} = L\frac{\partial \ln(q_{tr})}{\partial T}$$

Thus, using Eq. (14.3.15) and the expression for U in the Box (14.2.1), we have for the molar internal energy of a **monatomic ideal gas**

$$U_{tr} = (Lk_B)T^2\left(\frac{\partial \ln q_{tr}}{\partial T}\right) = RT^2\left(\frac{3}{2T}\right) = \frac{3}{2}RT \tag{14.3.16}$$

This equation can also be derived from consideration of kinetic energy of an ideal gas. Using now the expression for S in terms of U in the Box (14.2.1) and Sterling's approximation for the factorial of large numbers,

$$S_{tr} = \frac{U_{tr}}{T} + k_B \ln Q_{tr} = \frac{3}{2}R + k_B \ln\left(\frac{q_{tr}^L}{L!}\right)$$

$$= \frac{3}{2}R + R\ln q_{tr} - k_B(L\ln L - L)$$

$$= \frac{5}{2}R + R\ln\frac{q_{tr}}{L}$$

Now substituting the expression of q_{tr} from the Box (14.2.2), we obtain, after rearranging the terms,

$$S_{tr} = \frac{5}{2}R + R\ln\left[\frac{(2\pi M k_B T)^{3/2}V}{Lh^3}\right] \tag{14.3.17a}$$

or, writing the first right hand term as $R\ln(e^{5/2})$,

$$S_{tr} = R\ln\left[\frac{(2\pi M k_B T)^{3/2}e^{5/2}V}{Lh^3}\right] \tag{14.3.17b}$$

This is known as the **Sackur-Tetrode equation** that has been widely used to calculate molar entropy of monatomic ideal gases. If electronic contribution is to be included, then one needs to add a S(elec) term on the right (e.g. McQuarrie 2003). Figure 14.3 shows a comparison of the calculated molar entropy values of 10 monatomic gases (Hg, Ne, Ar, Kr, Xe. C, Na, Al, Ag and Hg) with the experimentally determined values at 1 bar, 273 K. The excellent agreement is obvious (the average disagreement being within 0.04%).

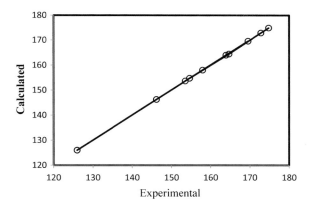

Fig. 14.3 Comparison between the calculated and experimental values of entropy (J/mol K) of 10 monatomic gases (Hg, Ne, Ar, Kr, Xe, C, Na, Al, Ag and Hg) at 1 bar, 298 K. The calculations are based on Sackur-Tetrode equation, Eq. (14.3.8). Data from McQuarrie (2003)

Problem 14.1 Derive an expression for the vibrational contribution to the molar internal energy of a diatomic gas, assuming that there is a single vibrational mode, as in a linear molecule.

Problem 14.2 Using the Sackur-Tetrode relation calculate the molar entropy of He at 1 bar, 298 K. The experimental value is 126.0 J/mol K.

*Problem 14.3 Show that the chemical potential of a volatile species at constant T-V condition is given by the expression

$$\mu_i = -k_B T \left(\frac{\ln q_i}{n_i} \right) \tag{14.3.18}$$

where q_i and n_i are the single particle partition function and number of moles of the species i.

14.4 Heat Capacity of Solids

An introductory and elementary discussion of the heat capacity of solids has been given Sect. 1.6 involving the concept of phonons and density of states. Additionally, the experimental observations and the important theoretical and empirical developments on the temperature dependence of heat capacity of solids have been summarized in Sect. 4.2. In this section, we would confine ourselves primarily to the derivation of Einstein model of heat capacity of solids that not only marks the beginning, but arguably constitutes the most important theoretical contribution in this field, and make some explanatory comments about the Debye model that was also presented in the Sect. 4.2. These discussions should be easy to understand (hopefully) after learning the formal concepts of statistical thermodynamics discussed above. An overview of other important advancements in this field has been given in Sect. 4.2.

To recapitulate, Einstein (1907) introduced the ground breaking idea of quantization of elastic waves in a crystal, and named the quantized waves as **phonons** (as he also did for the electromagnetic waves, introducing the concept of **photons**). Einstein proposed that atoms in a crystal vibrate around their equilibrium positions and made the simplifying assumptions that (a) they behave as independent harmonic oscillators and (b) that all oscillators vibrate with the same frequency. This frequency is usually labeled as v_E. As discussed above, in a three dimensional crystal of n atoms, there are $3n - 6$ vibrational modes (here 6 comes from three translational and 3 rotational degrees of freedom of the whole crystal), but since n is very large for a macroscopic crystal, we can set $3n - 6 \approx 3n$ for thermodynamic treatment. Thus, using 3n as the number of distinguishable particles (phonons) in Eq. (14.3.2a), we have $\ln Q_v = 3n(\ln q_v)$ where q_v is given by Eq. (14.3.11). Using now the relation between C_v and the canonical partition function Q (which is Q_v for a crystal), as given by Eq. (14.2.3), and some mathematical manipulations leads to the famous Einstein relation of heat capacity of a crystal

$$C_v = 3R \frac{x^2 e^x}{(e^x - 1)^2} \tag{14.4.1}$$

Here $x = \theta_E / T$, where θ_E is the so called Einstein temperature introduced in the Sect. 1.6 ($\theta_E = h v_E / 2\pi k_B$, with v_E being the fixed vibrational frequency of the atomic oscillators). The term R comes from setting the number of atoms n equal to the Avogadro's number L so that $n k_B = R$.

The Einstein relation predicts a high temperature limiting value of molar heat capacity C_v of a monatomic solid equal to 3R that is in agreement with the classical thermodynamic Dulong-Petit limit (Sect. 4.2). However, the low temperature limiting behavior of the Einstein relation does not agree well with experimental data. Debye (1912) refined Einstein's model by allowing for variation of vibrational frequencies, but conforming to a specific form in which the density of vibrational states versus frequency relation, g(v) versus v, has a smoothly increasing form up to a cut off limit of v_D (known as the Debye frequency), as illustrated schematically in Fig. 14.4. Thus, in Debye's model

$$\int_0^{v_D} g(v) dv = 3n \quad \text{(i.e. total number of normal vibrational modes)} \tag{14.4.2}$$

Debye further assumed that all lattice vibrational waves move with the speed of sound in the solid. Hence, the Debye relation describes the heat capacity of a solid due to acoustic modes (i.e. the vibrational modes excited by acoustic waves).

Within the framework of the above assumptions, Debye showed that the heat capacity C_v of a monoatomic solid, can be expressed as a function of the ratio T/θ_D, that is $C_v = f_D(T/\Theta_D)$. Writing T/Θ_D as χ, the Debye function, $f_D(\chi)$ is given by (Lewis and Randall 1961)

$$f_D(\chi) = 9R\chi^3 \int_0^{\chi} \frac{u^4 e^u du}{(e^u - 1)^2} \tag{14.4.3}$$

where u is a dummy integration variable.

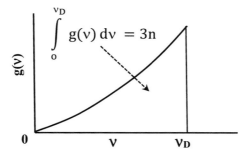

Fig. 14.4 Schematic illustration of density of vibrational states, as assumed in the Debye model of heat capacity of solids

   The Debye model correctly yields the high temperature Dulong-Petit relation and the low temperature "T to the power third" limiting relation (see Sect. 4.2) of C_v. A later formulation due to Kieffer (1979) that is usually referred to as **Kieffer model** has the characteristics of both Einstein and Debye models; it has been widely used in mineralogical literature and discussed briefly in the Sect. 4.2. With the advent of powerful software packages in computational quantum chemistry, the emphasis now seems to have shifted to the calculation of heat capacity of solids from first principles calculation of density of states, not only of vibrational but also of other types (Appendix C.3.3).

14.5   Chemical Equilibria and Stable Isotope Fractionation

Statistical thermodynamic calculations of reactions of geochemical interest go back to the pioneering study of Harold Urey (1893–1981; Nobel Prize: 1934) in 1947 on "The Thermodynamic Properties of Isotopic Substances" followed quickly by another foundational paper by Bigeleisen and Mayer (1947). Urey (1947) showed that the difference between the thermodynamic properties of molecules differing only by the isotopic nature of an element with low mass numbers, such as $C^{16}O_2$ and $C^{18}O_2$, become sufficiently different at low temperatures to cause temperature dependent isotopic fractionation that should be measurable by a sensitive mass spectrometer. Urey (1947) remarked that "These calculations suggest investigations of particular interest to geology." Specifically, he discussed the possibility of using oxygen isotopic ratio of calcite relative to the ocean water to determine the temperature at which they formed; this marked the beginning of the field of **paleothermometry** and **stable isotope geochemistry**. (Quite interestingly, four years later Hans Ramberg and George DeVore, who were in the Geology department of the University of Chicago where Harold Urey was a professor in the Chemistry department, came up with the idea of high temperature geothermometry (Ramberg and DeVore 1951) on the basis of element fractionation between coexisting minerals – see Sect. 11.1). Urey's work was complemented by the development of mass spectrometer by the physicist Alfred Nier (1911–1994) at the University of Minnesota.

The first paleothermometric measurements were made by Urey and co-workers (1951), who determined the $^{18}O/^{16}O$ ratio of $CaCO_3$ of a Jurassic **belemnite** from the Isle of Skye, Scotland, at various distances from the axis of the sample. Using their data and showing that there could not have been any significant diffusive resetting of the isotopic ratio of the extracted calcite samples over a period of 100 million years, Urey et al. (1951) deduced (Fig. 14.5) the difference between the summer and winter temperatures (21 °C vs. 15 °C; mean 17.6 °C) during the Jurassic period and progressive drop of winter temperature within the life time of the belemnite (three summers and four winters). Study of additional samples of belemnite from the Upper

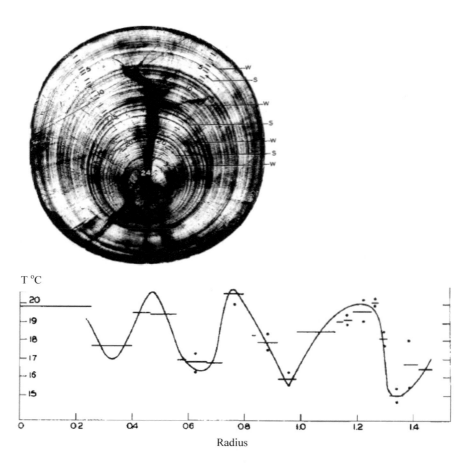

Fig. 14.5 A Jurassic belemnite (2.5 cm in diameter) from the Isle of Skye, Scotland along with paleotemperatures recorded by the oxygen isotopic composition of calcite crystals as function of radial distance. From Urey et al. (1951). With permission from the Geological Society of America

Cretaceous of United States, England and Denmark revealed a surprisingly uniform temperature of about 15–16 °C over this latitudinal belt.

The primary focus of this section is to show the role of statistical thermodynamics in the calculation of stable isotope fractionation using selected problems and cases. The aim is not an exhaustive survey of the field for which there are excellent review articles, notably by Richet et al. (1977), Chacko et al. (2001) and Blanchard et al. (2017), spanning a period of over 40 years that, in itself, is a reflection of the sustained interest in the field. The statistical thermodynamic treatment of stable isotope fractionation is, however, a special case of the general treatment of chemical reaction in terms of partition functions. Thus, we first discuss the general case before proceeding to isotopic fractionation.

14.5.1 General Treatment of Chemical Reaction

Let us first write a generic chemical reaction, as we have done earlier (Sect. 10.4), as

$$\sum_i \nu_i C_i = 0 \tag{14.5.a}$$

where, as before, the stoichiometric coefficient $\nu_i > 0$ for products and < 0 for reactants. Choosing pure component at P-T as the standard states of the respective components, the equilibrium condition of any reaction at a specified P-T condition must satisfy the relation $\Delta_r G(P', T) = 0 = \Delta_r G^\circ(P', T) + RT\ln \Pi(a_i)^{\nu_i}$ where a_i is the activity of the component i (unity in the pure state at any P-T). Now, since G(P', T) = G(1 bar, T) + $\int$ vdP over the interval 1 bar to P', and G = F + PV, we can write that at equilibrium at a given P'-T condition,

$$RT\ln\Pi(a_i)^{\nu_i} = RT\ln K_{5a} = -\left[\Delta_r F^\circ(1, T) + P_1\Delta_r V^\circ(1, T) + \int_1^{P'} \Delta_r V^\circ dP\right] \tag{14.5.1}$$

where P_1 stands for 1 bar pressure and K_{5a} stands for the equilibrium constant of the reaction (14.5.a). (The pressure term is retained in the second right hand term even though it is unity, since if it is written as $\Delta_r V^\circ$ the unit of the second right hand term, bar(cm^3), may get overlooked). The collection of terms within the square brackets equals $\Delta_r G^\circ(P', T)$ whereas the first two terms equal $\Delta_r G^\circ(1, T)$, with 1 bar being written as 1 for brevity.

The $\Delta_r F^\circ(1, T)$ term is amenable to evaluation from statistical thermodynamics through its relation to partition function, Q (Box 14.2.1). However, in order to evaluate Q from molecular partition functions (see Eq. 14.3.2 for the relationship between Q and q), we need to find a way to relate the overall molecular partition function of a species, q_i, to its translational and internal partition functions. To do this, we first note that the translational energy levels of a species are not affected by its internal properties or internal degrees of freedom so that $\varepsilon_i = \varepsilon_{i:tr} + \varepsilon_{i:int}$. With respect to the latter, it is commonly assumed that the rotational and vibrational states are decoupled (rigid rotor model), as has been done in the derivation of expressions of the vibrational and rotational partition functions. Thus, we write for the energy of a species i, $\varepsilon_i = \varepsilon_{i:tr} + \varepsilon_{i:r} + \varepsilon_{i:v} + \varepsilon_{i:e}$, in which case the overall molecular partition function becomes a product of the translational and internal partition functions, viz.

$$q_i = \sum e^{-\varepsilon_i/kT} = q_{i:tr}q_{i:r}q_{i:v}q_{i:e} \tag{14.5.2}$$

(the nuclear partition function has been ignored for reasons discussed above, but if it is to be included, it can be written as a product term on the right). It now follows from Eq. (14.3.2a, b) that for a mole of a gaseous species,

$$Q_i = \frac{(q_{tr}q_rq_vq_e)^L}{L!}, \tag{14.5.3a}$$

whereas for that of a crystalline material with distinguishable particles,

$$Q_j = (q_{tr}q_rq_vq_e)^L. \tag{14.5.3b}$$

For reasons discussed above, we will set $q_e = 1$ in the following developments.

In order to calculate the free energy change of a reaction from partition functions, the latter need to be referred to a common zero energy level for both reactants and products (one may conceptualize this by recalling that the enthalpy change of a reaction, Δ_rH, in thermochemical calculations are calculated by using enthalpies of formation from a common base, either elements ($\Delta_{f,e}H$) or oxides ($\Delta_{f,o}H$), as discussed in Sect. 4.5). Thus, according to Eq. (14.2.9), the molar Helmholtz free energy change of a reaction in an ideal gas phase (for which Q is expressed in terms of q according to Eq. 14.3.2b) is given by

$$\Delta_rF^\circ = -RT\ln \prod \left(\frac{q_i}{L!}\right)^{v_i} + \Delta_rU_o \tag{14.5.4a}$$

where Δ_rU_o, which arises from reconciling the quantum energy levels to a common zero, is the net change of internal energy at absolute zero (see Moore 1972, 8.13, for further discussion). For a solid state reaction, for which Q_j and q_j are related according to Eq. (14.3.2a),

$$\Delta_rF^\circ = -RT\ln \prod (q_i)^{v_i} + \Delta_rU_o \tag{14.5.4b}$$

14.5.2 Stable Isotope Fractionation: Theoretical Foundation

We now consider the case of equilibrium fractionation of two isotopes, j and j*, of an element (e.g. ^{16}O and ^{18}O) between two phases A and B, with j* being designated as the heavier isotope. Like the case of element fractionation that has been discussed in the Sect. 11.1.1, the isotopic fractionation is also treated in terms of an isotope exchange reaction, such as, for example,

$$C^{18}O + 1/2\ C^{16}O_2 \leftrightarrow C^{16}O + 1/2\ C^{18}O_2 \tag{14.5.b}$$

In a general form, we write

$$\nu_a(i\text{-}A) + \nu_b(i^*\text{-}B) \leftrightarrow \nu_a(i^*\text{-}A) + \nu_b(i\text{-}B) \tag{14.5.c}$$

As discussed in the Sect. 11.1.1, to deal with an exchange equilibrium, it is convenient to balance the reaction in such a way that there is one mole of exchangeable species involved in the reaction, as shown in the above example.

According to Eq. (14.5.1), the equilibrium constant for the reaction (14.5.c) at 1 bar, T condition may be written as,

$$K_{5c}(1, T) = e^{-\Delta_r F^\circ(1,T)/RT} e^{-P_1 \Delta_r V^\circ(1,T)/RT} \tag{14.5.5}$$

where P_1 stands for 1 bar pressure. For isotopic exchange reaction, the second exponential term is sufficiently close to unity at 1 bar pressure and temperature of the order of $10^2\ ^\circ C$ that we can write $K_{5c}(1, T) = \exp[-\Delta_r F^\circ(1, T)/RT]$ for practical purposes of geological interest. A further simplification stems from the fact that $\Delta_r U_o$ is effectively zero for an isotopic exchange reaction. Combining Eqs. (14.5.4) and (14.5.5) it then follows that

$$K_c(1, T) = \prod (q_i)^{\nu_i} \tag{14.5.6}$$

Note that it does not matter whether we use Eq. (14.5.4a) or (14.5.4b) since the L! terms cancel out in an isotopic exchange reaction.

To deal with an exchange equilibrium, it is convenient to deal with partition function ratio of the isotopic components of a molecule and relate the equilibrium constant to these ratios. For example, using the last expression, we write the equilibrium constant for the reaction (14.5.b) as

$$k_{5b} = \frac{(q^*/q)^{1/2}_{CO_2}}{(q^*/q)_{CO}} \tag{14.5.7}$$

where q* and q are the molecular partition functions of the heavy and light isotopic components, respectively, of the specified molecule. In the treatment of isotopic exchange equilibrium, the molecular partition function of a compound may be expressed, to a very good approximation, as a product of its translational, rotational and vibrational partition functions. Substituting the expressions for these partition functions from the Box (14.2.2), the molecular partition function ratios are given by the following relations

$$\text{Linear molecule}: \frac{q^*}{q} = \left[\left(\frac{M^*}{M}\right)^{3/2}\right]\left[\left(\frac{I^*}{I}\right)\left(\frac{\sigma'}{\sigma'^{(*)}}\right)\right]\left[\Pi\left(\frac{e^{-u_j^*/2}}{e^{-u_j/2}}\right)\left(\frac{1-e^{-u_j}}{1-e^{-u_j^*}}\right)\right]$$

$$(14.5.8)$$

Nonlinear molecule:

$$\frac{q^*}{q} = \left[\left(\frac{M^*}{M}\right)^{3/2}\right]\left[\left(\frac{I_A^* I_B^* I_C^*}{I_A I_B I_C}\right)^{1/2}\left(\frac{\sigma'}{\sigma'^{(*)}}\right)\right]\left[\Pi\left(\frac{e^{-u_j^*/2}}{e^{-u_j/2}}\right)\left(\frac{1-e^{-u_j}}{1-e^{-u_j^*}}\right)\right] \quad (14.5.9)$$

where u_j stand for hv_j/k_BT for the vibrational frequency of the j th mode. The terms within the first, second and third square brackets represent the ratios of the translational, rotational and vibrational partition functions, respectively, of the heavy and light isotopic molecules.

It should be evident from the last two equations that the temperature dependence of equilibrium constant is solely due to that of the vibrational term, if the effect of electronic excitation is negligible. The first parenthetical term within the last square brackets, which we shall denote by φ_{zpe}, denote the zero point energy (ZPE) effect and constitutes the most dominant vibrational contribution at low temperature while the contribution from the second parenthetical term, φ_2, increases with increasing temperature. To illustrate this, let us consider the case of CO whose vibrational frequencies are 2141 and 2089 cm^{-1} in the end-member isotopic molecules C^{16}O and C^{18}O, respectively. These data yield values of 1.14592 and 1.00000 for φ_{zpe} and φ_2, respectively, at 273 K; at 773 K, we get $\varphi_{zpe}=1.04930$ and $\varphi_2=1.0019$.

The treatment of isotopic exchange equilibrium in the harmonic approximation is greatly simplified, eliminating the need for calculation of the moments of inertia, by making use of a theorem due to Teller and Redlich (Urey 1947), often referred to as the **Teller-Redlich product rule**. According to this theorem, the products of the mass and moment of inertia terms in the last two expressions is given by (McQuarrie 2003)

$$[\text{mass terms}][\text{moment of inertia terms}] = \Pi\left(\frac{m_i^*}{m_i}\right)^{3/2}\Pi\frac{v_i^*}{v_j} \quad (14.5.10)$$

where m_i is the mass of the ith atom in a molecule and the first product is taken over $i = 1$ to n (the number of atoms in a molecule), and the second product over $j = 1$ to $3n - 5$ for a linear and 1 to $3n - 6$ for a nonlinear molecule. Combining the last relation with the expressions for partition function ratios of linear and nonlinear molecules (Eqs. 14.5.8 and 14.5.9), we get

$$\frac{q_i^*}{q_i} = \left(\frac{m_i^*}{m_i}\right)^{3/2} \left\{ \left(\frac{\sigma}{\sigma^*}\right) \Pi \left[\frac{v_i^*}{v_i} \left(\frac{e^{-u_j^*/2}}{e^{-u_j/2}}\right) \left(\frac{1-e^{-u_j}}{1-e^{-u_j^*}}\right) \right] \right\} \tag{14.5.11}$$

$$\xleftarrow{\hspace{2cm}} \text{RPFR:} f \xrightarrow{\hspace{2cm}}$$

where the product convention for linear and non-linear molecules is the same as above. If the number of isotopes being exchanged exceeds one, then the exponent for the mass term becomes $3\eta/2$ where η is the number of isotopes.

It can be easily verified that in the expression of equilibrium constant of an isotopic exchange equilibrium, the mass terms cancel out so that the terms within the curly bracket in Eq. (14.5.11) are the only useful terms for the statistical thermodynamic treatment of isotopic fractionation. This collection of terms is often called the **reduced partition function ratio (RPFR)** of a molecule and referred to by the symbol f, following Bigeleisen and Mayer (1947), who first introduced this function, although they did not name it this way. The advantage of using f stems from the fact that it is a molecule specific property and thus can be used in the calculation of equilibrium constant of any isotopic exchange equilibrium in which it is involved. For example, if one calculates $f(CO)$ or the product term (Π ...) for CO in the above equation in the course of calculation of K_{5b}, one can use it later for the calculation of the equilibrium constant for, say, $H_2^{18}O + C^{16}O = H_2^{16}O + C^{18}O$. When all lighter isotopes of an element in a molecule are replaced by heavier isotopes in the exchange equilibrium, the symmetry numbers would also cancel out in the expression of the equilibrium constant, such as for the case of equilibrium (14.5.b), for which the equilibrium constant (Eq. 14.5.7) can now be written as

$$K_{5b} = \frac{(q^*/q)_{CO_2}^{1/2}}{(q^*/q)_{CO}} = \frac{(f_{CO_2})^{1/2}}{f_{CO}} = \frac{\left\{ \Pi \left[\frac{v_j^*}{v_j} \left(\frac{e^{-u_j^*/2}}{e^{-u_j/2}}\right) \left(\frac{1-e^{-u_j}}{1-e^{-u_j^*}}\right) \right]_{CO_2} \right\}^{1/2}}{\left[\frac{v_i^*}{v_i} \left(\frac{e^{-u_j^*/2}}{e^{-u_j/2}}\right) \left(\frac{1-e^{-u_j}}{1-e^{-u_j^*}}\right) \right]_{CO}} \tag{14.5.12}$$

A remarkable aspect of expression of equilibrium constant of an isotope exchange equilibrium in the framework of Teller-Redlich theorem is the fact that it can be calculated only from knowledge of the vibrational frequencies of the species plus/minus their symmetry numbers. There has been many applications of RPFR approach to calculate equilibrium constants for stable isotope fractionations in systems of geochemical interests, and in some of the modern applications (e.g. Schauble et al. 2006; Rustad et al. 2008, 2010) the required vibrational frequencies were calculated from quantum chemical methods within the framework of density functional theory (see Appendix C).

As an illustrative example, let us calculate the equilibrium constant for the hydrogen isotopic reaction

$$H_2O + D_2 = D_2O + H_2 \qquad (14.5.d)$$

as a function of temperature at 1 bar pressure using the vibrational frequencies of the species and compare the results with that calculated from the available Gibbs free energy of formation values. We use two sets of vibrational frequency data, one derived from spectroscopic studies (Huber et al. 1979) and the other calculated from the Density Functional Theory (DFT) based computational quantum chemistry program package, Gaussian (see Appendix C). The frequency data are listed below in terms of wave numbers, ω (cm^{-1}).

	Spectroscopic data	Gaussian data
H_2	4401	4421
D_2	3156	3128
H_2O	1595, 3657, 3756	1603, 3817, 3922
D_2O	1178, 2671, 2788	1174, 2751, 2874

Writing $K_{5d} = f_{water}/f_{hyd}$ and using the above vibrational frequencies, K_{5d} can now be calculated as a function of temperature. Using the Gaussian frequency values, this exercise yields $f_{water} = 37.846$ and $f_{hyd} = 7.244$ at 1 bar, 400 K, so that $K_{5d}(1; 400\,K) = f_{water}/f_{hyd} = 5.22$. At the same condition, thermochemical calculation yields, using the data listed in JANAF (Joint Army, Navy and Air Force: Malcolm 1998) Tables ($\Delta_f G°(H_2O) = -223.901 kJ/mol$; $\Delta_f G°(D_2O) = -229.379 kJ/mol$, so that $\Delta_r G_{5d}° = -5.478 kJ/mol$), $K_{5d} = \exp(-\Delta_r G°/RT) = 5.19$, which is almost exactly the same as the result from RPFR based calculation. Figure 14.6 shows a comparison of the values of K_{5d} in the range of 400–2000 K calculated from thermochemical and vibrational frequency data. There is practically no difference between the results from the Gaussian frequency based and thermochemical calculations. The spectroscopic frequencies yield slightly smaller values of K_{5d} at T < 600 K (at 400 K, K_{5d}(Spect) = 4.27). An important conclusion to be derived from the excellent agreement between the calculations based on harmonic RPFR formulation and thermochemical data over such a large temperature range (400–2000 K) is that the *anharmonicity effects* have cancelled out in the isotopic exchange equilibrium. Also, the non-classical behavior of the rotational partition function of H_2 and D_2 has no effect at least down to 400 K, further corroborating the conclusion of Bigeleisen and Mayer (1947), as discussed above.

With respect to **graphical representation** of stable isotope fractionation data between two phases, it is a common practice to use a lnK versus $1/T^2$ plot for an isotopic exchange equilibrium, which is in contrast to that of using lnK versus 1/T plot for any other type of reaction (see, for example, Fig. 11.1). It has been shown in Sect. 10.4.2 that lnK versus 1/T relation should be linear over a restricted temperature interval within which $\Delta_r H°$ and $\Delta_r S°$ do not change significantly as a function of temperature, which is usually the case for common chemical reactions. However, Clayton and Epstein (1961) first observed that experimental data on

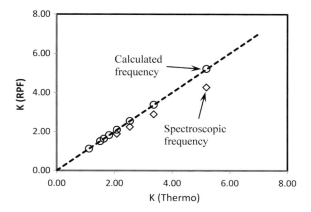

Fig. 14.6 Comparison of equilibrium constant, K, of the isotopic exchange reaction $H_2O + D_2 = D_2O + H_2$ calculated from (harmonic) reduced partition function ratio (RPFR) formulation with the results from thermochemical calculations at 1 bar, 400–2000 K. Circles: RPFR calculations using "Gaussian" vibrational frequencies; diamonds: those calculated from spectroscopic frequencies. The dashed line represents perfect agreement between the RPFR and thermochemical calculations

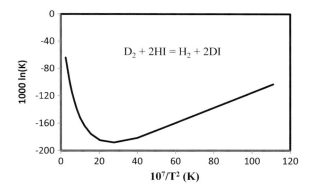

Fig. 14.7 Equilibrium constant of the reaction $D_2 + 2HI = H_2 + 2DI$ as a function of temperature. Data from Bigeleisen and Mayer (1947)

oxygen isotope fractionation in the calcite–water system and also between coexisting minerals conform to linear relations of the form of $\ln\alpha = A + B/T^2$ where $A \geq 0$, with K being equated to the isotopic fractionation factor, $\alpha = (i^*/i)_\beta/(i^*/i)_\phi$, between the phases β and ϕ. Since then, temperature dependence of α has been most frequently illustrated using $\ln\alpha$ versus $1/T^2$ plot even when the relationship is not linear. Bigeleisen and Mayer (1947) illustrated the temperature dependence of D–H fractionation between hydrogen and hydrogen iodide using a K versus $1/T^2$ plot, and this seems to be the first use of $1/T^2$ as the x-axis to illustrate temperature dependence of equilibrium constant of isotopic exchange reaction. The data calculated by them for the equilibrium $D_2 + HI = H_2 + DI$ are illustrated in Fig. 14.7 to show the variation of $\ln K$ versus $1/T^2$. An important point to notice is the presence of a minimum. This must be a general feature of any isotopic exchange equilibrium for which K is less than unity ($\ln K < 0$) at some temperature and decreases initially with increasing temperature since at some high temperature, the isotopic fractionation must disappear ($K = 1$, $\ln K = 0$). Calculated $\ln K$ versus $1/T^2$ relation for hydrogen isotopic fractionation between a number of hydrous minerals and water show the presence of minimum in the temperature range 135–500 °C (Méheut et al. 2010; Asaduzzaman and Ganguly 2018). For nonlinear behavior of $\ln K$ versus $1/T^2$, the form $\ln K = A + B/T + C/T^2$ seems to fit the data well.

*Problem 14.4 Using reduced partition function formulation, calculate the D–H fractionation between methane and hydrogen bromide as a function of temperature in the range of 300–800 K, using the following values of the vibrational frequencies (McQuarrie 2003).

CH_4: 2917, 1534(2), 3019(3), 1306(3) cm^{-1}
CH_3D: 2200, 2945, 1310, 1471(2), 3021(2), 1155(2) cm^{-1}
HBr: 2650; DBr: 1880 cm^{-1}

Plot the $\ln K$ value versus $1/T$ and $1/T^2$ and comment on the expectation of conformity to linear relationship.

14.5.3 Stable Isotope Fractionation: Some Geochemical Applications

14.5.3.1 "Clumped Isotope" Thermometry

Application of the oxygen isotope paleo-thermometer based on $^{18}O/^{16}O$ fractionation between ocean water and carbonate minerals, which began with the seminal work of Urey (1947) and Urey et al. (1951), requires data not only for the isotopic ratio of the carbonate minerals, but also of the ocean water from which the minerals precipitated in the geologic past. The ocean chemistry today is not necessarily the same as it was in an ancient geologic period. For example, the $^{18}O/^{16}O$ ratio of ocean water during the glacial periods were higher than in the modern ocean owing to isotopic fractionation between ocean water and ice, with the latter having a preference for ^{16}O. This problem led to the development of a new type of paleothermometer (Ghosh et al. 2006; Schauble et al. 2006) that is based on redistribution of isotopes as a function of temperature by homogeneous equilibria within a carbonate mineral, such as

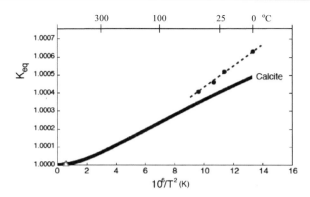

Fig. 14.8   Equilibrium constant of the homogeneous clumped isotope exchange reaction (14.5.e) in calcite, as calculated from DFT based quantum chemical calculation of vibrational frequencies of the isotopologues and reduced partition function formulation. The filled circles show the experimental data of Ghosh et al. (2006). From Schauble et al. (2006). With permission from the Geochemical Society

$$M\left(^{12}C^{18}O^{16}O_2\right)^{2-} + M\left(^{13}C^{16}O_3\right)^{2-} \leftrightarrow M\left(^{13}C^{18}O^{16}O^{16}O\right)^{2-} + M\left(^{12}C^{16}O_3\right)^{2-}$$

$$(14.5.e)$$

where M^{2+} is a metal cation in the carbonate. This type of homogeneous equilibrium is independent of the chemistry of the water from which the carbonate minerals precipitated. One can write many reactions of this type involving isotopic variants or **isotopolgues** of the carbonate ion, but the primary focus has been on the above equilibrium since it involved the most abundant (67 ppm) doubly substituted carbonate ion, $\left(^{13}C^{18}O\right)^{16}O_2^{2-}$, which is advantageous for its mass spectroscopic determination (Ghosh et al. 2006).

Schauble et al. (2006) calculated the vibrational frequencies of the isotopically substituted carbonate molecules using quantum chemical methods within the framework of density functional theory (see Sect. C.3.3 in Appendix C), and from those the reduced partition functions of the molecules and the equilibrium constant of the reaction (14.5.e), K_{5e}, as a function of temperature, according to the methodology discussed in the last Sect. (14.5.2). Their result for $CaCO_3$ is illustrated in Fig. 14.8 that also shows a comparison with the experimental data of Ghosh et al. (2006).

14.5.3.2   Mineral-Hydrogen/Water Stable Isotope Fractionation

There have been extensive experimental and theoretical studies of stable isotope fractionations between mineral-water systems (see Valley and Cole 2001). One of the reasons for the sustained interest in this field in the geochemical community lies in the determination of D/H ratio, and hence the origin of the fluid phase in ancient geological environments from measurement of that in minerals with which the fluid had interacted. The mineral-water $^{18}O/^{16}O$ ratios have also been utilized to generate calibration of mineral-mineral isotopic fractionations as function of temperature and their applications as geothermometers (($^{18}O/^{16}O)_{mineral-A}/(^{18}O/^{16}O)_{mineral-B}$ equals the ratio of the mineral-A/water and mineral-B/water isotopic fractionation data); a comprehensive review of this topic may be found in Valley (2001).

The experimental studies on mineral-water isotopic fractionation, however, suffer from two important problems, namely, failure to attain equilibrium, especially at low temperatures, and solution-reprecipitation of minerals within the experimental charges. The latter process compromises the results of mass spectrometric analyses of isotopic ratios of mineral separates and water, intended to determine those developed by equilibrium fractionation. To circumvent the problem of dissolution of mineral in hydrothermal experiments, a method was developed in which D–H fractionation could be measured between a mineral and hydrogen gas (Vennemann and O'Neil 1996), which may then be combined with D–H fractionation data between hydrogen and water (Fig. 14.4) to derive mineral-water D–H fractionation as a function of temperature.

Asaduzzaman and Ganguly (2018) calculated the equilibrium constant for D–H fractionation between epidote and hydrogen/water on the basis of molecular partition functions, q_i, of the end member species in the following reactions (Eq. 14.5.6).

$$x'1(OH) + 1/2\ D_2O = x'1(OD) + 1/2\ H_2O \qquad (14.5.f)$$

and

$$x'1(OH) + 1/2\ D_2 = x'1(OD) + 1/2\ H_2 \qquad (14.5.g)$$

The molecular partition functions of (OH) and (OD) in epidote equal their respective vibrational partition functions (since there are no rotational and translational motions within a crystal lattice) whereas that for a gaseous species equal the product of q_{tr}, q_v, q_r (Eq. 14.5.2). Asaduzzaman and Ganguly (2018) calculated the partition functions of (OH) and (OD) in epidote and the gaseous species, using a density functional theory based quantum chemical approach, and used these data to calculate the D–H fractionation between epidote and hydrogen/water systems. Their results for the epidote-hydrogen system are found to be in excellent agreement with the experimental data of Vennemann and O'Neil (1996) (Fig. 14.9). (The K_{5g} vs. T for the epidote-hydrogen system has been recalculated using the RPFR formulation (see solution to Problem 14.4) and the results are in almost exact agreement with those of Asaduzzaman and Ganguly 2018).

In Fig. 14.9 the "experimental" curve represents the bulk D–H fractionation between epidote and hydrogen, $\alpha(x'l\text{-hyd})$, determined on the basis of analysis of (D/H) ratio in the gas and mineral grains in a mass spectrometer, and defined as follows.

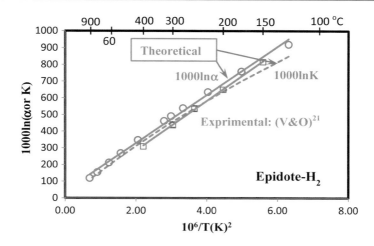

Fig. 14.9 Comparison of the calculated D–H fractionation data between epidote and hydrogen (lnK: dashed line and ln α: solid line with circles representing calculated points) versus $1/T^2$ with the experimental data of Vennemann and O'Neil (1996). From Asaduzzaman and Ganguly (2018). With permission from American Chemical Society

$$\alpha(x'l - hyd) = \frac{(D/H)_{x'l}}{(D/H)_{hyd}} \qquad (14.5.13)$$

The equilibrium constant for the isotopic exchange reaction (14.5.g) is, on the other hand, defined as

$$K_{5g} = \frac{(D/H)_{x'l}}{(D_2/H_2)_{hyd}^{1/2}} \qquad (14.5.14)$$

where the parenthetical terms indicate activities (a) of the specified components. This is equivalent to α(ep-hyd) if the D_2 and H_2 are the only deuterium and hydrogen bearing species in the gas phase because, in that case, $a(D_2) = (X_D)^2$ and $a(H_2) = (X_H)^2$ for ideal mixing (indeed a very good assumption for mixing of isotopes), with X denoting the mole fraction of the specified species. However, the vapor phase may have non-negligible content of DH, especially at low temperature, causing a significant difference between α and K_{5g}. This type of problem also needs to be addressed in the case of other systems in which the vapor phase may contain significant amount of isotopic species beyond those written in the exchange equilibrium. Thus, for example, a H–D–O vapor phase may have HDO, H_2O, D_2O in which case the equilibrium constant of the exchange reaction $x'1(OH) + 1/2D_2O = x'1(OD) + 1/2H_2O$ may differ from the experimentally determined fractionation factor, α, calculated on the basis of mass spectroscopic data of bulk isotopic ratios of D to H in the crystal (x'l) and the vapor phase.

The procedure for relating K (calculated from statistical thermodynamics) to α is discussed below with the D–H fractionation between a crystal and hydrogen gas as an example.

The equilibrium proportions of D_2, H_2 and HD in the gas phase are governed by the homogeneous reaction

$$1/2H_2 + 1/2D_2 = HD \tag{14.5.h}$$

so that $(HD) = K_{5h}[(D_2)(H_2)]^{1/2}$, where the parentheses imply the number of moles of a species. Thus, for the molar abundances of D and H in the gas phase, we can write

$$\begin{aligned}(D) \approx (HD) + 2(D_2) &= K_{5h}[(D_2)(H_2)]^{1/2} + 2D_2 \\ &= 2(D_2)^{1/2}\left[K_{5h}/2(H_2)^{1/2} + (D_2)^{1/2}\right]\end{aligned} \tag{14.5.15}$$

and similarly

$$(H) \approx (HD) + 2(H_2) = 2(H_2)^{1/2}\left[K_{5h}/2(D_2)^{1/2} + (H_2)^{1/2}\right] \tag{14.5.16}$$

(the factor 2 comes from the fact that one mole of D_2 or H_2 yields 2 mol of D or H, respectively). Now dividing Eq. (14.5.15) by Eq. (14.5.16),

$$\left(\frac{D}{H}\right)_{hyd} = \left(\frac{D_2}{H_2}\right)_{hyd}^{1/2} \varphi \tag{14.5.17}$$

where

$$\varphi = \frac{(D_2)^{1/2} + (K_{5h}/2)(H_2)^{1/2}}{(H_2)^{1/2} + (K_{5h}/2)(D_2)^{1/2}} \tag{14.5.18}$$

Thus,

$$\alpha(x'1 - hyd) \equiv \frac{(D/H)_{x'1}}{(D/H)_{hyd}} = \left[\frac{(D/H)_{x'1}}{\left(\frac{D_2}{H_2}\right)_{hyd}^{1/2}}\right] \frac{1}{\varphi} \tag{14.5.19}$$

or

$$\alpha(x'1 - hyd) = \frac{K_{5g}}{\varphi} \tag{14.5.20}$$

The equilibrium constant K_{5h} varies between 1.806 and 1.985 in the temperature range 300–1200 K. Therefore, since $(H_2) \gg D_2$, we may, to a very good approximation, write from Eq. (14.5.18)

$$\varphi \approx \frac{(K_{5h}/2)(H_2)^{1/2}}{(H_2)^{1/2}} = \frac{K_{5h}}{2} \qquad (14.5.21)$$

Using the last two equations, Asaduzzaman and Ganguly (2018) calculated the bulk hydrogen isotopic fractionation between epidote and hydrogen gas, $\alpha(x'l\text{-hyd})$, from the equilibrium constant of the exchange reaction, K_{5g}, that they calculated from statistical thermodynamic treatment of the isotopic exchange reaction (Fig. 14.9). Following similar procedure, it can be shown that hydrogen isotopic exchange between epidote and water is governed by analogous relations if H_2O, D_2O and HD are the only important species in the vapor phase, as would be the case if $H_2/O_2 \sim 2$, with K_{5h} in the last equation being replaced by the equilibrium constant of the gas phase reaction $1/2H_2O + 1/2D_2O = HDO$.

Problem 14.6 Consider the oxygen isotopic exchange reaction between CO and CO_2:

$C^{18}O + 1/2C^{16}O_2 = C^{16}O + 1/2C^{18}O_2$ (i), for which the experimentally determined fractionation factor is defined as $\alpha = \frac{(18_O/16_O)_{CO_2}}{(18_O/16_O)_{CO}}$. The isotopic species in the gas phase are also contained in the mixed isotopic molecule $C^{16}O^{18}O$ and hence one can write another homogeneous reaction $C^{16}O_2 + C^{18}O_2 = 2C^{16}O^{18}O$ (j). Show that $\alpha = K_i$ if $K_j = 4$.

Hint: Develop equations similar to (14.5.15) through (14.5.19).

References

Asaduzzaman A, Ganguly J (2018) Hydrogen isotope fractionation between epidote-hydrogen and epidote-water systems: theoretical study and implications. ACS Earth Space Chem 2:1029–1034

Bigeleisen J, Mayer MG (1947) Calculation of equilibrium costants for isotopic exchange reactions. J Chem Phys 15:261–267

Blanchard M, Balan E, Schauble EA (2017) Equilibrium fractionation: a molecular modeling perspective. Rev Mineral Geochem 82:27–63

Chacko T, Cole DR, Horita J (2001) Equilibrium oxygen, hydrogen and carbon isotope fractionation factors applicable to geologic systems. In: Valley JW, Cole DR (eds) Stable isotope geochemistry. Rev Mineral Geochem 43, Mineral Soc A, 1–82

Clayton RN, Epstein S (1961) The use of oxygen isotopes in high temperature geological thermometry. J Geol 69:447–452

Debye P (1912) Zur Theorie der spezifischen wärmen. Ann der Physik 39:789–839

Einstein A (1907) Die Plancksche theorie der strahlung und de theorie der spezifischen Warmen. Annal der Physik 22:180–190

Ghosh P, Adkins J, Affek H, Balta B, Guo W, Schauble EA, Shrag D, Eiler JM (2006) ^{13}C-^{18}O bonds in carbonate minerals: a new kind of paleothermometer. Geochim Cosmochim Acta 70:1439–1456

Huber KP, Herzberg G (1979) Molecular spectra and molecular structure. Van Nostrand, New York

Javoy M (1977) Stable isotopes and geothermometry. J Geol Soc Lond 133

Kieffer SW (1979) Thermodynamics and lattice vibrations of minerals: 1. Mineral heat capacities and their relationships to simple lattice vibrational models. Rev Geophys Space Phys 17:1–19

Lewis GN, Randall M (1961) Thermodynamics. McGraw-Hill, New York, p 723

Malcolm WJC (1998) NIST-JANAF thermochemical tables, 4th edn. National Institute of Standard and Technology (NIST), Gaithersburg, MD. https://janaf.nist.gov/

McQuarrie DA (2003) Statistical mechanics. Viva Books, New Delhi, p 639

Méheut M, Lazzeri M, Balan E, Mauri F (2010) First-principles calculation of H/D fractionation between hydrous minerals and water. Geochim Cosmochim Acta 74:3874–3882

Moore WJ (1972) Physical chemistry, 4th edn. Prentice-Hall, p 977

Ramberg H, DeVore G (1951) The distribution of Fe^{++} and Mg^{++} in coexisting olivines and pyroxenes. J. Geol. 59:193–210

Richet P, Bottinga Y, Javoy M (1977) A review of hydrogen, carbon, nitrogen, oxygen sulphur, and chlorine stable isotope fractionation among gaseous molecules. Ann Rev Earth Planet Sci 5:65–110

Rustad JR, Nelmes AL, Jackson VE, Dixon DA (2008) Quantum-chemical calculations of carbon-isotope fractionation in $CO_2(g)$, aqueous carbonate species, and carbonate minerals. J Phys Chem A 112:542–555

Rustad JR, Casey WH, Yin Q-Z, Bylaska EJ, Felmy AR, Bogotka SA, Jackson VE, Dixon DA (2010) Isotopic fractionation of $Mg^{2+}(aq)$, $Ca^{2+}(aq)$, and $Fe^{2+}(aq)$ with carbonate minerals. Geochim Cosmochim Acta 74:6301–6323

Schauble EA, Ghosh P, Eiler JM (2006) Preferential formation of ^{13}C-^{18}O bonds in carbonate minerals, estimated first-principles lattice dynamics. Geochim Cosmochim Acta 70:2510–2529

Urey HC (1947) The thermodynamic properties of isotopic substances. J Chem Soc, London, pp 562–581

Urey HC, Lowenstam HA, Epstein S, McKinney CR (1951) Measurement of paleo-temperatures and temperatures in the Upper Cretaceous of England, Denmark, and the Southwestern United States. Bull Geol Soc America 62:399–416

Valley J (2001) Stable isotope thermometry at high temperatures. In: Valley JW, Cole DR (eds) Stable isotope geochemistry. Rev Mineral Geochem 43, Mineral Soc Am 365–414

Valley JW, Cole DR (2001) Editors: stable isotope geochemistry. Rev Mineral Geochem 43, Mineral Soc Am 662

Vennemann TM, O'Neil JR (1996) Hydrogen isotope exchange reactions between hydrous minerals and molecular hydrogen: I. A new approach for the determination of hydrogen isotope fractionation at moderate temperatures. Geochim Cosmochim Acta 60:2437–2451

Appendix A
Rate of Entropy Production and Kinetic Implications

According to the second law of thermodynamics, the entropy of an isolated system can never decrease on a macroscopic scale; it must remain constant when equilibrium is achieved or increase due to spontaneous processes within the system. The ramifications of entropy production constitute the subject of irreversible thermodynamics. A number of interesting kinetic relations, which are beyond the domain of classical thermodynamics, can be derived by considering the rate of entropy production in an isolated system. The rate and mechanism of evolution of a system is often of major or of even greater interest in many problems than its equilibrium state, especially in the study of natural processes. The purpose of this Appendix is to develop some of the kinetic relations relating to chemical reaction, diffusion and heat conduction from consideration of the entropy production due to spontaneous processes in an isolated system.

A.1. Rate of Entropy Production: Conjugate Flux and Force

In order to formally treat an irreversible process within the framework of thermodynamics, it is important to identify the force that is conjugate to the observed flux. To this end, let us start with the question: what is the force that drives heat flux by conduction (or heat diffusion) along a temperature gradient? The intuitive answer is: temperature gradient. But the answer is not entirely correct. To find out the exact force that drives diffusive heat flux, let us consider an isolated composite system that is divided into two subsystems, I and II, by a rigid diathermal wall, the subsystems being maintained at different but uniform temperatures of T^I and T^{II}, respectively. It was shown in Sect. 2.8 that the entropy change of the composite system is given by (Eq. 2.8.2)

$$dS = \delta q^I \left(\frac{1}{T^I} - \frac{1}{T^{II}} \right)$$

© Springer Nature Switzerland AG 2020
J. Ganguly, *Thermodynamics in Earth and Planetary Sciences*,
Springer Textbooks in Earth Sciences, Geography and Environment,
https://doi.org/10.1007/978-3-030-20879-0

where δq^I is the heat absorbed by the subsystem I (It was also shown in Sect. 2.8 that because of the second law, $\delta q^I > 0$ if $T^{II} > T^I$, and vice versa). Thus, the rate of entropy production is given by

$$\frac{dS}{dt} = \frac{\delta q^I}{dt}\left(\frac{1}{T^I} - \frac{1}{T^{II}}\right) \qquad (A.1.1)$$

If Δx is the length of the composite system and A is the cross sectional area of the wall separating the two subsystems (Fig. 2.9), then $A(\Delta x)$ is the total volume of the system. Thus the rate of entropy production per unit volume, commonly denoted by σ, is given by

$$\sigma = \frac{1}{A(\Delta x)}\left(\frac{dS}{dt}\right) \qquad (A.1.2)$$

so that

$$\sigma = \left(\frac{\delta q^I}{Adt}\right)\left(\frac{1/T^I - 1/T^{II}}{\Delta x}\right) \qquad (A.1.3)$$

Considering infinitesimal change, the quantity within the second parentheses can be written as $d(1/T)/dX$. The term within the first parentheses represents the heat flux, J_Q, (i.e. rate of heat flow per unit area) across the wall separating the subsystems I and II. Thus,

$$\boxed{\sigma = J_Q\left(\frac{d(1/T)}{dx}\right)} \qquad (A.1.4)$$

The parenthetical derivative term in the above equation represents the **force driving the heat flux**, J_Q. Note that the force is the spatial gradient of the **inverse** temperature instead of that of temperature itself. Noting that $d(1/T) = -dT/T^2$, the last equation can be written as

$$\sigma = -\frac{J_Q}{T^2}\frac{dT}{dx} \qquad (A.1.5)$$

Equation (A.1.4) shows a general property of the rate of entropy production in a system due to an irreversible process or several irreversible processes. For each process, k, the rate of entropy production per unit volume, σ_k, is given by the product of a flux term, J_k, associated with the process k (or a rate per unit volume for chemical reaction, as shown below), and the force, χ_k, driving the flux. That is

$$\sigma = (\text{Either flux or rate per unit volume})(\text{conjugate force})$$
$$= J_k \chi_k \tag{A.1.6}$$

Thus, consideration of the entropy production during a process leads to the proper identification of the **conjugate** force and flux terms for the process. The rate of total entropy production per unit volume in a closed system due to several irreversible processes is given by

$$\sigma = \sum_k \sigma_k = \sum_k J_k \chi_k \tag{A.1.7}$$

Let us now try to identify the appropriate forces that drive **diffusion** and **chemical reaction** by considering the rate of entropy production in a system due to these processes. For the first problem, let us again consider an isolated composite system, as above, but make the wall separating the two subsystems porous to the diffusion of the component i, and denote dn_i^I and dn_i^{II} as the change in the number of moles of i in the subsystems I and II, respectively. Since the total number of moles of a component (n_i) in the overall composite system is conserved, we obviously have $dn_i^I = -dn_i^{II}$. For simplicity, we assume that the two subsystems are at the same temperature.

The change of total entropy, S, of the composite isolated system due to a change in the number of moles of the component i in the subsystem I is given by

$$\left(\frac{dS}{dn_i^I}\right)_{n_i} = \frac{dS^I + dS^{II}}{dn_i^I} = \frac{dS^I}{dn_i^I} - \frac{dS^{II}}{dn_i^{II}} \tag{A.1.8}$$

where S^I and S^{II} are the entropies of the subsystems I and II, respectively. Now, since the volumes of the subsystems are constant, we have, from Eq. (8.1.4) (i.e. $dU = TdS - PdV + \Sigma\mu_i dn_i$)

$$\frac{dS^I}{dn_i^I} = \frac{dU^I}{Tdn_i^I} - \frac{\mu_i^I}{T} \tag{A.1.9a}$$

and

$$\frac{dS^{II}}{dn_i^{II}} = \frac{dU^{II}}{Tdn_i^{II}} - \frac{\mu_i^{II}}{T}$$
$$= -\frac{dU^{II}}{Tdn_i^I} - \frac{\mu_i^{II}}{T} \tag{A.1.9b}$$

Substituting the last two relations in Eq. (A.1.8), and noting that $d(U^I + U^{II}) = 0$, we obtain

$$\frac{dS}{dn_i^I} = \left(\frac{\mu_i^{II} - \mu_i^I}{T}\right) \tag{A.1.10}$$

Using chain rule, Eq. (A.1.2) can be written as

$$\sigma = \frac{1}{A\Delta x}\left(\frac{dS}{dn_i^I}\right)\left(\frac{dn_i^I}{dt}\right) \tag{A.1.11}$$

Substituting Eq. (A.1.10) into the last expression and rearranging terms, we obtain

$$\sigma = \frac{\mu_i^{II} - \mu_i^I}{T\Delta x}\left(\frac{dn_i^I}{Adt}\right) \tag{A.1.12}$$

Thus, we can write the following expression for the **rate of entropy production due to chemical diffusion**:

$$\boxed{\sigma = -\frac{\partial(\mu_i/T)}{\partial x}J_{d,i}} \tag{A.1.13}$$

where $J_{d,i}$, which stands for the parenthetical term in Eq. (A.1.12), is the diffusive flux of the component i (rate of diffusive transfer of the component i per unit surface area). From the above equation, we see that, in the absence of an external force, the **driving force for diffusion of i is the negative gradient of μ_i/T**.[1] As we shall see later, recognition of the appropriate driving force leads to the formal understanding of how diffusion within a solution is affected by its thermodynamic mixing property.

Let us now consider a system undergoing irreversible chemical reaction under constant P-T condition within an isolated system. From Eq. (10.2.5)

$$\frac{dS_{int}}{d\xi} = \frac{A}{T}$$

where S_{int} is the internal entropy production due to chemical reaction, and A is the affinity of the reaction, as defined by the Eq. (10.2.6), and equals $-\Delta_r G$. Thus,

---

[1]If there is an external force acting on the diffusing species i, then the driving force becomes $-$grad $(\mu_i/T) + F_i$, where F_i is the force per unit mass acting on i. For example, if there is an electrical field E_d, then there is a driving force zE_d, where z is the charge of the diffusing species.

$$\frac{dS}{dt} = \left(\frac{dS}{d\xi}\right)\left(\frac{d\xi}{dt}\right) = \frac{A}{T}\left(\frac{d\xi}{dt}\right) \tag{A.1.14}$$

Consequently, **the rate of entropy production per unit volume due to chemical reaction** is given by

$$\boxed{\sigma = \frac{1}{V}\frac{dS}{dt} = \frac{A}{T}\left(\frac{d\xi}{Vdt}\right)} \tag{A.1.15}$$

where V is the volume of the system. The parenthetical quantity is the reaction rate per unit volume, which we would denote by R_i for the i th reaction. Thus, the term A/T (or $-\Delta_r G/T$) represents the **driving force for a chemical reaction**.

The fluxes and conjugate forces for heat conduction (or heat diffusion), chemical diffusion and chemical reaction are summarized in Table A.1.

Table A.1 Summary of fluxes and conjugate forces for some irreversible processes

Process	Flux	Conjugate force
Heat conduction (or heat diffusion)	J_Q (J/s-cm^2)	grad (1/T)
Chemical diffusion	$J_{d,i}$ (mol/s-cm^2)	$-[\text{grad }(\mu_i/T) - F_i]^a$
Chemical reaction	Rate of reaction per unit volume	A/T or $-\Delta_r G/T$

^{a}F: External force per unit mass acting on i

Using Eq. (A.1.7) and the conjugate flux and force terms tabulated above, the total rate of entropy production in a system within which both heat and chemical diffusion and irreversible reactions are taking place, is given by

$$\boxed{\sigma_T = -\frac{J_Q}{T^2}\frac{dT}{dx} - \sum_i J_{d,i}\left(\frac{d(\mu_i/T)}{dx} - F_i\right) + \sum_i R_i\frac{A_i}{T}} \tag{A.1.16}$$

A.2. Relationship Between Flux and Force

A flux J_k depends on the conjugate force, χ_k. If there are other forces in the system, then J_k depends also on those other forces. In general, J_k could become a complicated function of the forces. In the simplest approximation, one can write that

$$J_1 = L_{11}\chi_1 + L_{12}\chi_2 + L_{13}\chi_3 + \ldots$$

or

$$\boxed{J_1 = \sum_{j=1}^{n} L_{1j}\chi_j} \tag{A.2.1}$$

where it is assumed that there are n **independent** forces in the system. The L terms represent phenomenological coefficients. Because of the linear nature of the above equation, it is called an expression of flux in the domain of **linear irreversible thermodynamics**. As remarked by Lasaga (1998), the linear approximation works quite well for heat and chemical diffusion processes, but may not work well for all problems of fluid flow of geological interest and fails for chemical reactions far from equilibrium for which one needs to introduce higher order terms.

In a system subjected to **both** heat conduction and chemical diffusion of a species i, we have according to the last expression

$$\begin{aligned} J_Q \equiv J_1 = L_{11}\chi_Q + L_{12}\chi_{d,i} \\ J_{d,i} \equiv J_2 = L_{21}\chi_Q + L_{22}\chi_{d,i} \end{aligned} \tag{A.2.2}$$

Physically, these equations mean that chemical diffusion affects heat conduction and vice versa. These processes are known, respectively, as **Dufour effect** and **Soret effect**. Lesher and Walker (1991) have explored the importance of Soret effect (also known as thermal diffusion) in petrological problems.

A.3.   Heat and Chemical Diffusion Processes

The classic Fourier's law of heat conduction (or heat diffusion) and Fick's law of chemical diffusion are empirical laws, but these laws of heat and chemical diffusion can be derived rigorously from irreversible thermodynamics. Comparison between the empirical and theoretical laws provides important insights about the nature of thermal conductivity and thermodynamic effect on chemical diffusion coefficient in the Fick's law, as discussed below.

A.3.1.   Comparison with Empirical laws

Let us consider a planar section that has a fixed position in an isotropic medium with respect to a Cartesian coordinate system section. Assuming that T decreases as x increases, the heat flux resulting from conductive heat transfer across the plane is conventionally expressed by the **Fourier's law** as

$$J_q = -K\frac{dT}{dx} \tag{A.3.1}$$

where K is the thermal conductivity that has a dimension of energy/(t-L-K) and dT/dX is the temperature gradient across the plane (Fig. A.1) (The thermal conductivity of common upper mantle and crustal rocks is $\sim$ 3 to 4 W/(m-K); note that J/s $\equiv$ W).

Fig. A.1 Illustration of heat flux, according to Fourier's law, across a plane normal to the X-axis at the point X_1 due to one dimensional temperature gradient in an isotropic medium, the constant K being the thermal conductivity of the medium

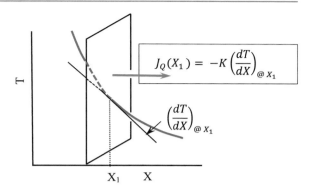

$$J_Q(X_1) = -K\left(\frac{dT}{dX}\right)_{@\,X_1}$$

The relationship between heat flux and temperature gradient is easily derived from Eq. (A.2.1) by assuming that the only driving force in the system is that which drives heat conduction, namely $d(1/T)/dx$ (Table A.1). Thus,

$$J_Q = L_Q \frac{d(1/T)}{dx}$$

so that

$$J_Q = -\frac{L_Q}{T^2}\left(\frac{dT}{dx}\right) \tag{A.3.2}$$

which has the same form as the Fourier law, Eq. (A.3.1).

The flux of a diffusing species is generally described by **Fick's law**, which was formulated by analogy with the Fourier's law of heat conduction. Diffusion processes are affected by thermodynamic mixing properties of the system. However, this effect is not apparent in the empirical formulation of diffusion flux given by the Fick's law, but becomes transparent when one develops the expression of diffusion flux from Eq. (A.2.1), as shown below.

Let us again consider a planar section that has a fixed position in an isotropic medium with respect to a Cartesian coordinate system, and assume, for simplicity, that there is no external force (such as electrical and gravitational forces) acting on the diffusing species i. According to Fick's law, the diffusive flux of a component i, $J_{d,i}$, through this planar section is proportional to its **local** concentration gradient, and is given by an equation of the same form as the Fourier law of heat conduction, Eq. (A.3.1), viz.

$$J_{d,i} = -D_i \frac{dC_i}{dx} \tag{A.3.3}$$

where C_i is the concentration of i, in moles per unit volume, decreasing in the direction of increasing x and D_i is the diffusion coefficient of i (with dimension of L^2/t). Since $D_i > 0$, the negative sign in the above expression is introduced to make the flux positive in the direction of decreasing C_i. The flux of a component may also be affected by the concentration gradient of other diffusing components, but we ignore this cross-coupling effect at this stage.

Let us now derive the expression of flux from Eq. (A.2.1) that follows from consideration of entropy production in the system due to irreversible process. With the driving force given by $-\text{grad}\,(\mu_i/T)$ (Table A.1; Eq. A.1.13), we have

$$J_{d,i} = -L_i \frac{d\mu_i/T}{dx}$$

or

$$J_{d,i} = -\frac{L_i}{T}\left(\frac{d\mu_i}{dx}\right) = -\frac{L_i}{T}\left(\frac{d\mu_i}{dC_i}\right)\left(\frac{dC_i}{dx}\right) \qquad (A.3.4)$$

Since at constant P-T condition, $d\mu_i = RTd\ln a_i = RTd\ln(C_i\gamma_i)$, the above expression transforms to

$$J_{d,i} = -\underbrace{\frac{RL_i}{C_i}\left(1 + \frac{d\ln\gamma_i}{d\ln C_i}\right)}_{D_i}\left(\frac{dC_i}{dx}\right) \qquad (A.3.5)$$

A.3.2. Non-ideal Mixing Effect on Chemical Diffusion

The first parenthetical term on the right in Eq. (A.3.5) shows the effect of thermodynamic mixing property on diffusion flux, and is often referred to as the **thermodynamic factor**. Comparing this equation with the Fickian expression of diffusion flux, Eq. (A.3.3), we find that at constant P-T condition

$$D_i = D_i^+\left(1 + \frac{d\ln\gamma_i}{d\ln C_i}\right) \qquad (A.3.6)$$

where

$$D_i^+ = \frac{RL_i}{C_i} \qquad (A.3.7)$$

The quantities D_i^+ and D_i are known, respectively, as the self- and chemical-diffusion coefficient of the component i.

From Eqs. (A.3.6) and (9.2.10), we obtain the TF for a sub-regular binary solid solution as

$$TF = 1 + \frac{X_iX_j}{RT}\left[W_{ij}\left(2X_i - 4X_j\right) + W_{ji}\left(2X_j - 4X_i\right)\right] \qquad (A.3.8)$$

For a regular solution, which represents the simplest form of deviation from ideality of mixing, $W_{ij} = W_{ji} = W$, so that

$$TF = 1 - \frac{2X_iX_jW}{RT} \qquad (A.3.9)$$

As an example of the effect of non-ideal mixing on the inter-diffusion coefficient of a binary solid solution, let us consider the clinopyroxene solid solution, $^{M2}Ca\ ^{M1}MgSi_2O_6$ (Diopside: Di)–$^{M2}Mg\ ^{M1}MgSi_2O_6$ (Enstatite: En) in which Ca and Mg mix in the larger of the two octahedral sites, M2. In a binary solid solution, the expression for the flux of a component has the same form as Eq. (A.3.3) but the diffusion coefficient D_i is replaced by what is known as an inter-diffusion coefficient, $D(i - j)$ (the term **inter-diffusion** implies simultaneous diffusion of two species in opposite directions). For a system with constant molar volume, it can be shown that

$$D(i - j) = D^+(i - j)\frac{d\ln a_i}{d\ln X_i} = D^+(i - j)\underbrace{\left(1 + \frac{d\ln\gamma_i}{d\ln X_i}\right)}_{TF} \qquad (A.3.10)$$

where $D^+(i - j)$ is the ideal part of the inter-diffusion coefficient, the explicit form of which depends on whether one is dealing with inter-diffusion of neutral species, as in metallic systems, or of charged species, and in the latter case, if charges are the same or different. Note that the TF in Eq. (A.3.6) is written in terms of concentration, C_i, whereas in the above expression it is written in terms of X_i. This is because of the fact that at constant volume, $d\ln X_i = d\ln C_i$. It is left to the reader to prove that. For the general case, $d\ln X_i$ in Eq. (A.3.10) should be replaced by $d\ln C_i$.

For the inter-diffusion of equally charged species, as in Ca-Mg inter-diffusion in clinopyroxene, $D^+(i - j)$ has the following form (Barrer et al. 1963; Manning 1968).

$$D^+(i - j) = \frac{D_i^+ D_j^+}{X_iD_i^+ + X_jD_{ji}^+} \qquad (A.3.11)$$

(see Ganguly 2002 for further discussion). Note, incidentally, that as $X_i \rightarrow 0$, $D^+(i-j) \rightarrow D_i^+$. The same is also true for a binary metallic system involving inter-diffusion of neutral species, for which (Darken 1948)

$$D^+(i-j) = X_j D_i^+ + X_i D_j^+ \tag{A.3.12}$$

According to Lindsley et al. (1981), the Di–Enst system behaves as a Sub-Regular solution (Sect. 9.2.2) with $W_{CaMg} = 31{,}217 + 0.0061(P)$ and $W_{MgCa} = 25{,}485 + 0.0812(P)$ J/mol. The self-diffusion coefficients for Ca and Mg were determined by Zhang et al. (2010) and presented in the form of Arrhenius relation, $D_i^+ = D_i^{+,o} \exp\left(-\dfrac{Q_i}{RT}\right)$, where Q_i is the activation energy of diffusion. The diffusion parameters parallel to the c-axial direction are $D_{Ca}^{+,o} = 0.24(10^{-9})\,\text{m}^2/\text{s}$, $Q_{Ca} = 264.96$ kJ/mol, $D_{Mg}^{+,o} = 1.44(10^{-13})\,\text{m}^2/\text{s}$, $Q_{Mg} = 176.15$ kJ/mol.

Using the above sub-regular parameters, Zhang et al. (2010) calculated the (non-coherent) TF, according to Eq. (A.3.8), as a function of composition, temperature and pressure. Their results are illustrated in Fig. A.2. At each temperature, the TF versus X(Di), curve has been terminated at the onset of miscibility gap (Fig. 8.19). Figure A.3 shows an Arrhenius plot of D(Ca-Mg) that has been calculated using the above diffusion kinetic and thermodynamic data and compared with those for the self-diffusion data of the individual species and D(Ca-Mg) for the ideal mixing model (i.e. setting the TF = 1). The inter-diffusion coefficient has been calculated for two different compositions, X(Ca:Di) = 0.8 and 0.7 and the lnD(Ca-Mg) trends have been terminated when these encounter the solvus conditions at the low temperatures.

There are two important points to notice in Fig. A.3: (a) in a non-ideal solution, an inter-diffusion coefficient does not necessarily lie within the limits defined by the corresponding self-diffusion coefficients and (b) log D(i − j) versus 1/T relation may become quite strongly non-linear so that the usual practice of linear extrapolation of experimental data in an Arrhenius plot may lead to significant error, aside from the problem associated with a change of diffusion mechanism within the extrapolated thermal regime.

Fig. A.2 Illustration of the dependence of non-coherent thermodynamic factor affecting diffusion coefficient in the Diop-Enst solid solution as a function of composition, temperature and pressure; the calculations terminate at the binary solvus illustrated in Fig. 8.19 (from Zhang et al. 2010)

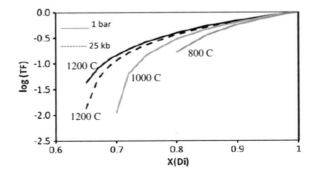

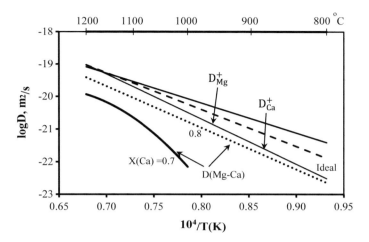

Fig. A.3 Arrhenius plot of the Ca-Mg self (D^+) and inter-diffusion ($D(Ca-Mg)$) coefficients in clinopyroxene as a function of temperature and composition at 1 bar pressure. The dotted and curved lines show the inter-diffusion coefficients for $X(Ca) = 0.8$ and 0.7, taking into account the effect of non-ideal mixing of Ca and Mg in clinopyroxene, whereas the dashed line show $D(Ca-Mg)$ ignoring the thermodynamic effect for the two compositions that happen to be indistinguishable in the plot

Problem A.1 In Eq. (A.3.10), the thermodynamic factor, TF, has been written in terms of the derivative of the activity coefficient of the component i, γ_i. Show that the TF does not change if the derivative is taken for γ_j (Hint: use Gibbs-Duhem relation).

The extension of the above developments to binary and multicomponent systems may be found in a number of publications, such as Chakraborty and Ganguly (1991) and Ganguly (2002).

A.4. Onsager Reciprocity Relation and Thermodynamic Applications

Within the domain of validity of linear irreversible thermodynamics, expressions for different types of fluxes in a system subject to several irreversible processes can be written, in a generalized form, as

$$
\begin{aligned}
J_1 &= L_{11}\chi_1 + L_{12}\chi_2 + L_{13}\chi_3 + \ldots \\
J_2 &= L_{21}\chi_1 + L_{22}\chi_2 + L_{23}\chi_3 + \ldots \\
J_3 &= L_{31}\chi_1 + L_{32}\chi_2 + L_{33}\chi_3 + \ldots
\end{aligned}
\qquad (A.4.1)
$$

and so on. In a matrix notation, we can then write

$$\mathbf{J} = [\mathbf{L}][\chi] \tag{A.4.2}$$

where $\mathbf{J}$ and χ represent column vectors of fluxes and forces, respectively, and $[\mathbf{L}]$ is a matrix of the L coefficients. It was shown by Onsager (1945) (1903–1976) that if the above equations are written in terms of **independent** fluxes and forces, then the $[\mathbf{L}]$ matrix becomes symmetric, i.e. $L_{ij} = L_{ji}$. This is known as the **Onsager reciprocity relation** (ORR), for which Onsager was awarded the Nobel Prize in Chemistry in 1968. Physically, the reciprocity relation means that the effect of a force χ_1 in driving the flux J_2 is the same as the effect of the force χ_2 in driving the flux J_1. The reciprocity relation greatly simplifies the treatment of cross-coupling of different forces in a system, and also provides upper bounds on the sizes of the L coefficients. Lasaga (1998), Chakraborty and Ganguly (1994) and Chakraborty (1995) have discussed applications of the ORR to geological problems.

Onsager's reciprocity relation provides a means to test the mutual compatibility of the diffusion and thermodynamic mixing properties of species in a system in which several components are diffusing simultaneously. We provide here a brief exposition of this compatibility criterion. In a system with n diffusing components, there are $n - 1$ independent components, since the flux of the nth component, which may be chosen arbitrarily, is determined by mass balance. In multicomponent diffusion, the flux of each component is affected not only by its own concentration or chemical potential gradient, but also by those of others. Within the domain of validity of linear irreversible thermodynamics, the fluxes of the independent components in one-dimensional flow are given, taking into consideration the cross-coupling of the diffusing species, by the **Ficks-Onsager relation** (Onsager 1945) as follows

$$J_1 = -D_{11}\frac{\partial C_1}{\partial X} - D_{12}\frac{\partial C_2}{\partial X} - \cdots\cdots\cdots\cdots\cdots - D_{1(n-1)}\frac{\partial C_{n-1}}{\partial X}$$

$$J_2 = -D_{21}\frac{\partial C_1}{\partial X} - D_{22}\frac{\partial C_2}{\partial X} - \cdots\cdots\cdots\cdots\cdots - D_{2(n-1)}\frac{\partial C_{n-1}}{\partial X} \tag{A.4.3}$$

$$\cdots\cdots\cdots\cdots\cdots\cdots\cdots\cdots\cdots\cdots\cdots\cdots\cdots\cdots\cdots$$

$$J_{n-1} = -D_{(n-1)1}\frac{\partial C_1}{\partial X} - D_{(n-1)2}\frac{\partial C_2}{\partial X} - \cdots\cdots\cdots - D_{(n-1)(n-1)}\frac{\partial C_{n-1}}{\partial X}$$

Using the principle of matrix multiplication, this can be written as

$$\begin{bmatrix} J_1 \\ J_2 \\ \vdots \\ J_{n-1} \end{bmatrix} = - \begin{bmatrix} D_{11} & D_{12} & \cdots & D_{1(n-1)} \\ D_{21} & D_{22} & \cdots & D_{2(n-1)} \\ \vdots & & \vdots & \vdots & \vdots \\ D_{(n-1)1} & \cdots & \cdots & D_{(n-1)(n-1)} \end{bmatrix} \cdot \begin{bmatrix} \partial C_1/\partial X \\ \partial C_2/\partial X \\ \vdots \\ \partial C_{n-1}/\partial X \end{bmatrix} \tag{A.4.4}$$

or

$$\mathbf{J} = -\mathbf{D}(\partial \mathbf{C}/\partial \mathbf{X}) \qquad \text{(A.4.5a)}$$

where $\mathbf{J}$ and $\mathbf{C}$ are $(n-1)$ column vectors and $\mathbf{D}$ is an $(n-1) \times (n-1)$ matrix of diffusion coefficients that is commonly referred to as the $\mathbf{D}$-matrix. Similarly, using $\mathbf{L}$ instead of $\mathbf{D}$ coefficients, one may write

$$\mathbf{J} = -\mathbf{L}(\partial(\boldsymbol{\mu}/T)/\partial \mathbf{X} \qquad \text{(A.4.5b)}$$

where $\boldsymbol{\mu}$ is an $(n-1)$ column vector of chemical potentials and $\mathbf{L}$ is an $(n-1) \times (n-1)$ matrix of the L_{ij} coefficients.

Comparing the last two expressions, one obtains a relationship between the $\mathbf{D}$ and $\mathbf{L}$ matrices as follows (Onsager 1945)

$$\mathbf{D} = \mathbf{LG} \qquad \text{(A.4.6)}$$

or

$$\mathbf{DG}^{-1} = \mathbf{L} \qquad \text{(A.4.6}')$$

where an element of the $\mathbf{G}$-matrix, commonly referred to as a thermodynamic matrix, is given by

$$G_{ij} = \frac{\partial}{\partial X_j} (\mu_i - \mu_n), \qquad \text{(A.4.7)}$$

μ_n being the chemical potential of the chosen dependent component, and $\mathbf{G}^{-1}$ the inverse of the $\mathbf{G}$ matrix. Evaluation of the G_{ij} elements requires data on the thermodynamic mixing properties of the components in the system.

According to Eq. (A.4.6$'$), the product of the $\mathbf{D}$ and $\mathbf{G}^{-1}$ matrices must be symmetric since, according to the ORR, $\mathbf{L}$ matrix is symmetric (it is also positive definite). Thus, Eq. (A.4.6$'$) provides a test of the mutual compatibility of the data on diffusion coefficients and thermodynamic mixing properties of components in a system. This criterion was used by Spera and Trial (1993) and Chakraborty (1994) to test the mutual compatibility of the data on the diffusion kinetic and thermodynamic mixing properties of components in the $CaO-Al_2O_3-SiO_2$ and $K_2O-Al_2O_3-SiO_2$ melts, respectively. Using Eq. (A.4.6$'$), one can also extract an unknown thermodynamic mixing parameter or a diffusion coefficient if the other parameters are known (Chakraborty and Ganguly 1994). In this approach, one varies the value of the unknown parameter until the product of $\mathbf{D}$ and $\mathbf{G}^{-1}$ matrices become symmetric and positive definite.

Appendix B
Review of Some Mathematical Relations

> *One reason why the study of thermodynamics is so valuable to students of chemistry and chemical engineering is that it is a theory which can be developed in its entirety, without gaps in the argument, on the basis of only a moderate knowledge of mathematics*
>
> Kenneth Denbigh

The purpose of this Appendix is not to present a comprehensive review of the mathematical methods used in thermodynamics, but to primarily review some of the concepts of calculus that are used frequently in the development of classical thermodynamics. In addition, I have tried to collect together some mathematical techniques that are often forgotten by the not-so-mathematically oriented reader so that they can read the book without having to take the trouble of consulting mathematics books.

B.1. Total and Partial Differentials

Consider a function Z that can be expressed as a function of the real variables x and y, $Z = f(x,y)$. The total change of Z, dZ, corresponding to the changes dx and dy of x and y, respectively, is given by

$$dZ = \left(\frac{\partial Z}{\partial x}\right)_y dx + \left(\frac{\partial Z}{\partial y}\right)_x dy$$

$$= Z_x dx + Z_y dy$$

(B.1.1)

Here the parenthetical quantities on the right hand side are called **partial derivatives** or **partial differentials**. Commonly the symbol ∂ is used to indicate a partial derivative. The first partial derivative on the right, which is also denoted by the short-hand notation Z_x, indicates the rate of change of Z with respect to the variable x, when y is held constant. Analogous statement also applies to the second partial derivative. The differential dZ is called the **total derivative** or **total differential** of Z.

© Springer Nature Switzerland AG 2020
J. Ganguly, *Thermodynamics in Earth and Planetary Sciences*,
Springer Textbooks in Earth Sciences, Geography and Environment,
https://doi.org/10.1007/978-3-030-20879-0

In general, if $Z = f(x_1, x_2, x_3 \dots x_n)$, then one can express the total derivative of Z as

$$dZ = \sum_i^n \left(\frac{\partial Z}{\partial x_i}\right)_{j \neq i} dx_i \qquad (B.1.2)$$

in which the subscript $j \neq i$ indicates that all variables, except i, are held constant.

The above equations are valid regardless of whether x and y are independent variables or not. It follows that if x and y depend on a single independent variable, say m, then from Eq. (B.1.1), we have

$$\frac{dZ}{dm} = \left(\frac{\partial Z}{\partial x}\right)_y \frac{dx}{dm} + \left(\frac{\partial Z}{\partial y}\right)_x \frac{dy}{dm} \qquad (B.1.3)$$

For the special case m = y, the above relation reduces to

$$\frac{dZ}{dy} = \left(\frac{\partial Z}{\partial x}\right)_y \frac{dx}{dy} + \left(\frac{\partial Z}{\partial y}\right)_x \qquad (B.1.4)$$

Although it should be obvious, it is reiterated that the two differentials dZ/dy and $(\partial Z/\partial y)_x$ must not be confused. The first one indicates the **total** rate of change of Z with respect to y, taking into account the rate of change of x with respect to the latter, whereas the partial derivative indicates the rate of change of Z with respect to y when x is held constant. The extension of the last two relations for a function that involves more than two variables should be obvious.

B.2. State Function, Exact and Inexact Differentials, and Line Integrals

When Z can be expressed as function of the variables $x_1, x_2, x_3 \dots x_n$, as in the above examples, then integration of dZ between two states a and b is given by

$$\int_a^b dZ = Z(b) - Z(a) \qquad (B.2.1)$$

In other words, the result of integration depends only on the values of Z in the final and initial states, and not on the path by which the change of state is brought about. A function $Z = f(x_1, x_2, x_3 \dots x_n)$ is, thus, often referred to as a **state function**. As an example, consider the change in the volume of a gas between two arbitrary states marked as A and D in Fig. B.1.We know from experience that the change in the

Fig. B.1 Schematic illustration of the change of state of a gas from A to D along two different paths in the P-T space, A → B → D and A → C → D. The figure is also repeated as Fig. 1.4 in Chap. 1

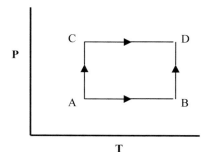

volume of the gas depends only on the P, T conditions of the initial and final states, and not on the path in the P-T space along which the change of state is achieved. Thus, V is a state function.

Let us test the validity of the above statement for an ideal gas, for which the equation state is given by $V_m = RT/P$, where V_m is the molar volume. Using Eq. (B.1.1),

$$dV_m = \frac{R}{P} dT - \frac{RT}{P^2} dP \qquad (B.2.2)$$

Now for the change of state of the gas from A to D along the path ABD, we have

$$\int_A^D dV_m = \int_A^B dV_m + \int_B^D dV_m = \frac{R}{P_1}(T_2 - T_1) + RT_2\left(\frac{1}{P_2} - \frac{1}{P_1}\right) \qquad (B.2.3)$$

Substitution of the relation $V_m = RT/P$ then yields

$$\int_A^D dV_m = V_m(P_2, T_2) - V_m(P_1, T_1) \qquad (B.2.4)$$

Following the above procedure, exactly the same expression is obtained for the integral of dV along the path ACD.

The differential dZ where $Z = f(x_1, x_2, x_3 \ldots x_n)$ is also called an **exact or perfect** differential, especially in thermodynamic discussions. Obviously, total differential and exact differential are synonymous. Now, let us consider a differential quantity

$$d\phi = M_1 dx_1 + M_2 dx_2 + M_3 dx_3 \qquad (B.2.5)$$

This expression does not necessarily imply that ϕ is a function of the variables x_1, x_2 and x_3. For example, if $d\phi = (8y)dx + (4x)dy$, it is not possible to express ϕ as a function of x and y. The differential $d\phi$ is then not a differential of a function ϕ, but simply a differential quantity that stands for $\Sigma M_i dx_i$. In that case, $d\phi$ is called an **inexact or imperfect** differential. To distinguish it from an exact or total differential, which is commonly denoted by the symbol d, and inexact differential is indicated by a symbol that looks like d, but is not d, e.g. đ or δ. In this book, we have chosen to use the latter symbol for an inexact differential.

Integration along a specific path, as we have done above for the integration of dV, is known as **line integration**. If the differential that is being integrated is exact, then line integration is redundant. All that we need to know are the limits of integration. However, the result of integration of an inexact differential not only depends on the terminal states but also on the path connecting those states. Consider, for example, the work done by an ideal gas as its volume changes from the state A to state D along the paths ABD and ACD in Fig. B.1. Integration of $\delta w^+ = PdV$ along the two paths, which is posed as a problem in Chap. 1 (Problem 1.1), yields different results. Line integral along a closed loop, that is a path for which the initial and final states are the same, is indicated by the symbol $\oint$. Evidently, if dZ is exact, then $\oint dZ = 0$, whereas, if it is inexact, then $\oint \delta Z \neq 0$.

B.3.   Reciprocity Relation

From Eq. (B.1.1), we derive an important property of an exact differential. Differentiating Z_x with respect to y and Z_y with respect to x, we get, respectively,

$$\frac{\partial Z_x}{\partial y} \equiv \frac{\partial}{\partial y}\left(\frac{\partial Z}{\partial x}\right) = \frac{\partial^2 Z}{\partial y \partial x} \qquad (B.3.1)$$

and

$$\frac{\partial Z_y}{\partial x} \equiv \frac{\partial}{\partial x}\left(\frac{\partial Z}{\partial y}\right) = \frac{\partial^2 Z}{\partial x \partial y} \qquad (B.3.2)$$

However, since the order of differentiation is immaterial,

$$\frac{\partial Z_x}{\partial y} = \frac{\partial Z_y}{\partial x} \qquad (B.3.3)$$

This is known as the **reciprocity relation**. If a differential satisfies this relation, then it must be an exact differential. An inexact differential will not satisfy the reciprocity relation. Consider, for example, the relation $dZ = y^2 dx + (2xy)dy$. Here dZ is exact since the terms on the right satisfy the reciprocity relation:

$(\partial(y^2)/\partial y) = 2y = \partial(2xy)/\partial x$. However, $\delta\phi = y^2 dx - (2xy)dy$ is inexact since it does not satisfy the reciprocity relation. In thermodynamics, we do not test if a differential is exact or not by using the reciprocity relation. We already know when a differential is exact, and thus use the reciprocity relation to obtain useful relation among the variables. A set of important relations derived in this manner, which are known as the **Maxwell relations**, are summarized in Sect. 3.4, Box (3.4.1).

If $Z = f(x_1, x_2, x_3 \ldots x_n)$ so that

$$dZ = N_1 dx_1 + N_2 dx_2 + N_3 dx_3 + \ldots N_n dx_n = \sum_i N_i dx_i, \qquad (B.3.4)$$

where $N_i = \partial Z/\partial x_i$, then the reciprocity condition of exact differential is given by the generalization of Eq. (B.3.3) as follows.

$$\left(\frac{\partial N_i}{\partial x_j}\right)_{x_i \neq x_j} = \left(\frac{\partial N_j}{\partial x_i}\right)_{x_j \neq x_i}, \qquad (B.3.5)$$

where the subscripts indicate that all variables, other than the one with respect to which the differentiation is carried out are held constant.

B.4. Implicit Function

Let us consider a simple thermodynamic system consisting of a chemically homogeneous fluid or solid, which is subjected to a temperature, T, and a hydrostatic pressure, P. For a given amount of substance in the system, P, T and V are not independent quantities, but are related by an equation of state (EoS), which can be expressed by a relation of the type

$$f(P, V, T) = 0 \qquad (B.4.1)$$

For example, consider an ideal gas, which obeys the equation of state

$$PV = nRT \qquad (B.4.2)$$

where n is the number of moles and R is the gas constant. This can be written as $f(P,V,T) = 0$ where $f(P,V,T) = PV - nRT$.

A function that can be written in the form of Eq. (B.4.1) is known as an **implicit function**. By taking the total derivative of an implicit function $f(x,y,z)$, and some manipulations, as shown below, we obtain two useful relations

$$\left(\frac{\partial x}{\partial y}\right)_z \left(\frac{\partial y}{\partial z}\right)_x \left(\frac{\partial z}{\partial x}\right)_y = -1 \qquad (B.4.3)$$

or

$$\left(\frac{\partial x}{\partial y}\right)_z = -\frac{(\partial x/\partial z)_y}{(\partial y/\partial z)_x} \qquad (B.4.4)$$

In the first equation, note the order of appearance of the variables in the set (x,y,z) as $(x,y)z$, $(y,z)x$, $(z,x)y$ that is easy to remember, with the variable outside the parentheses being held constant. For an example of thermodynamic application of these relations, see Sect. 3.7.

To derive the above relations, we first take the total derivative of the relation $f(x,y,z) = 0$, which yields

$$df = f_x dx + f_y dy + f_z dz = 0 \qquad (B.4.5)$$

where, as before, f_x is the partial derivative of f with respect to x, and so on. Now, differentiating with respect to y at constant z, we get

$$\left(\frac{\partial f}{\partial y}\right)_z = \left(\frac{\partial f}{\partial x}\right)_{y,z} \left(\frac{\partial x}{\partial y}\right)_z + \left(\frac{\partial f}{\partial y}\right)_{x,z} = 0$$

or

$$\left(\frac{\partial x}{\partial y}\right)_z = -\left(\frac{\partial f}{\partial y}\right)_{x,z} \left(\frac{\partial x}{\partial f}\right)_{y,z} \qquad (B.4.6)$$

Similarly, by successively differentiating Eq. (B.4.5) with respect to z at constant x, and x at constant y, we get

$$\left(\frac{\partial y}{\partial z}\right)_x = -\left(\frac{\partial f}{\partial z}\right)_{x,y} \left(\frac{\partial y}{\partial f}\right)_{x,z} \qquad (B.4.7)$$

and

$$\left(\frac{\partial z}{\partial x}\right)_y = -\left(\frac{\partial f}{\partial x}\right)_{y,z} \left(\frac{\partial z}{\partial f}\right)_{x,y} \qquad (B.4.8)$$

Multiplication of the left hand terms in the last three equations then yields

$$\left(\frac{\partial x}{\partial y}\right)_z \left(\frac{\partial y}{\partial z}\right)_x \left(\frac{\partial z}{\partial x}\right)_y = -1$$

B.5. Integrating Factor

In some cases, an inexact differential may be multiplied by what is known as an integrating factor to make it exact. Consider, for example, the inexact differential

$$\delta Z = (8y)dx + (4x)dy \tag{B.5.1}$$

that does not satisfy the reciprocity relation. However, multiplying both sides by x, we get

$$x\delta Z = (8xy)dx + (4x^2)dy, \tag{B.5.2}$$

It can be easily verified that the expression of $x\delta Z$ satisfies the reciprocity relation. Thus, the inexact differential has been transformed to an exact differential by multiplication by x. A term that, upon multiplication, **transforms** an inexact differential to an exact differential is known as an **integrating factor**. Note, however, that an integrating factor does not exist for any arbitrary inexact differential. As discussed in Sect. 2.3, the conversion of an inexact differential, δq_{rev}, that is the heat absorbed by a system in a reversible process, to an exact differential, dS, by an integrating factor (1/T) constitutes the fundamental basis of the second law of thermodynamics. Although the second law was derived from logical analysis of experimental observations, specifically the failure to completely convert energy to work without any wastage or dissipation, as discussed in Sect. 2.3, Carathéodery in 1910 demonstrated the existence of an integrating factor that converts δq_{rev} to an exact differential, and thus derived the second law without any recourse to any experimental data.

B.6. Taylor Series

In the earlier chapters, we have used different types of power series expressions, viz. Binomial series, Geometric series and Taylor series. Here we discuss the last one in greater detail because of its wide range of applications.

Consider a function f(x) that is graphically illustrated in the Fig. B.2. Suppose that we do not know the explicit relation between the function and the variable x that is required to calculate f(x) at any value of x, but we know f(x) at some point x = a, and its first and higher order derivatives at this point. Now, if we want to calculate f(x) at another value of x that is very close to x = a, we can make the following linear approximation, as illustrated in Fig. B.1.

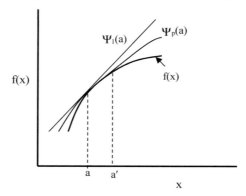

Fig. B.2 Schematic illustration of the approximation of the value of a function f(x) by Taylor series using the first and higher order derivatives of the function at x = a. When the value of x is close to a, the function may be approximated by $\Psi_1(a)$, which is given by the linear equation (B.6.1) in the text containing only the first derivative of f(x) at x = a. However, when x is sufficiently removed from the point a for the linear approximation to be valid, such as when x = a′, higher order derivatives of f(x) at x = a are needed to approximate the function, as given by Eq. (B.6.3). $\Psi_p(a)$ is the approximation function containing up to p th derivatives of f(x) at x = a

$$f(x) \approx \Psi(x) = f(a) + f'(a)(x - a) \tag{B.6.1}$$

where f′(a) is the first derivative of f(x) at x = a, and $\Psi(x)$ is the approximation function. However, if the point of interest is sufficiently removed from x = a for the linear approximation to be valid, then the value of the function can be approximated by including higher derivatives of f(x) at the point x = a. This method of approximation is given by Taylor series, as follows

$$f(x) \approx \psi(x) = f(a) + f'(a)(x - a) + \frac{f''(a)}{2}(x - a)^2$$
$$+ \frac{f'''(a)}{3!}(x - a)^3 + \cdots + \frac{f^{(n)}(a)}{n!}(x - a)^n \tag{B.6.2}$$

where the superscript of f indicates the order of the derivative of f(x) at x = a (′: first derivative; ″: second derivative etc. [(n)] nth derivative). The nature of Taylor series approximation of the function f(x) is schematically illustrated in Fig. B.2, in which $\Psi_1(a)$ denotes the approximation function containing only the first derivative of f(x) at x = a, and $\Psi_p(a)$ denotes that containing up to the pth derivatives of f(x) at x = a.

In a compact form, the Taylor series can be expressed as

$$f(x) \approx \psi(x) = f(a) + \sum_{m=1}^{n} \frac{f^{(m)}(a)}{m!}(x - a)^m \tag{B.6.3}$$

where $f^{(m)}(a)$ is the m th derivative of the function at $f(x)$ at $x = a$.

As an example of the application of Taylor series, consider the function e^x. We can calculate the value of this function at an arbitrary value of x by knowing that $e^0 = 1$ and finding the successively higher derivatives of e^x at $x = 0$. This procedure yields

$$e^x = 1 + x + \frac{x^2}{2} + \frac{x^3}{3!} + \frac{x^4}{4!} + \ldots$$

B.7. Sterling's Approximation

The Sterling's approximation is a method used to calculate factorials of large numbers. It is based on the integral expression of the factorial of a number N, which is as follows.

$$N! = \int_0^\infty x^N e^{-x} dx \qquad (B.7.1)$$

To a very good approximation, the integral on the right can be written as $\exp[N \ln N - N](2\pi N)^{1/2}$ so that

$$\ln N! \approx [N \ln N - N] + 1/2 \ln(2\pi N) \qquad (B.7.2)$$

This expression yields almost the exact value for lnN! when $N \geq 5$, and the importance of the last term decreases with increasing N. At $N = 100$, the exact value of lnN! = 363.7 whereas the terms within the square brackets equal 360.5 and the last term equals 3.2. Thus, in thermodynamic studies, in which N is very large, it is customary to use the relation

$$LnN! = N \ln N - N \qquad (B.7.3)$$

Problem B.1 Consider the differential $dZ = (xy^2 - x^2)dx + (x^2y + y^2)dy$. From its very representation, you would conclude that dZ is an exact differential. However, satisfy yourself that it is indeed so. Then find the function $Z = f(x, y)$ and evaluate its integral over a semicircular path of radius $a/2$ from $(0,0)$ to $(0,a)$.

Appendix C
Estimation of Thermodynamic Properties of Solids

There is more joy in heaven in a good approximation than in an exact solution

Julian Schwinger

Despite remarkable progress in the Earth science community in the measurement of thermodynamic properties of end-member minerals and mixing properties of solid solutions, and development of self-consistent data bases, important gaps still persist that sometimes force the practitioners to estimate the missing properties through some empirical scheme or fundamental theoretical method. Several methods have been developed for the estimation of thermochemical properties of end-member minerals and solid solutions that may be used judiciously when the required data are not available in the thermodynamic data bases. In this section, I summarize some of these estimation methods that have been found to be relatively successful.

C.1. Estimation of C_p and S of End-Members from Constituent Oxides

C.1.1. Linear Combination of Components

At the outset we note that properties such as entropy and heat capacity of end-member phases that depend primarily on vibrational properties could be usually approximated much better than enthalpy that depends on the bonding and potential energies. To develop the estimation schemes, let us consider the estimation of C_p and S of forsterite, Mg_2SiO_4. We may begin by using a linear combination of the corresponding properties of MgO and SiO_2 so that

$$C_p(For) \approx 2C_p(MgO) + C_p(SiO_2), \qquad (C.1.1)$$

and similarly for S(For). However, if data for an isostructural compound are available, then it is usually better to utilize those data in the estimation scheme since

© Springer Nature Switzerland AG 2020
J. Ganguly, *Thermodynamics in Earth and Planetary Sciences*,
Springer Textbooks in Earth Sciences, Geography and Environment,
https://doi.org/10.1007/978-3-030-20879-0

this procedure provides a better approximation of the vibrational properties of the compound for which one wishes to estimate C_p and S. Thus, for example, if the data on Mg_2SiO_4 (Fo) are available, and we are to estimate the C_p and S values for Fayalite (Fa), Fe_2SiO_4, then it is better to use the following scheme instead of a linear combination of the corresponding values for FeO and SiO_2. Writing a balanced reaction between Fa and For as $Fe_2SiO_4 + 2\,MgO = Mg_2SiO_4 + 2\,FeO$, or

$$Fe_2SiO_4 = Mg_2SiO_4 + 2\,FeO - 2\,MgO, \qquad (C.1.a)$$

we approximate the $C_p(Fay)$ as

$$C_p(Fa) \approx C_p(For) + 2C_p(FeO) - 2\,C_p(MgO) \qquad (C.1.2)$$

and similarly for S(Fa).

$$S(Fa) \approx S(For) + 2\,S(FeO) - 2\,S(MgO) \qquad (C.1.3)$$

The entropy estimates can be improved by making corrections for the volume and electronic effects, as discussed below.

C.1.2. Volume and Coordination Effects on Entropy

The effect of volume change of a compound on its entropy is given by the Maxwell relation (Eq. 3.4.3) $(\partial S/\partial V)_T = (\partial P/\partial T)_V$. The last derivative equals the ratio α/β_T (Eq. 3.7.9). Thus, to estimate entropy for a compound that does not require any other type of correction, we proceed in two steps: first by creating a compound (j) that has the same volume as the sum of its components used in the linear combination scheme (e.g. Eq. C.1.a), and then relaxing the volume of the compound to its equilibrium value at the given P-T condition according to $\Delta S = (\alpha/\beta_T)$ ΔV. Here $\Delta V = V_j - \Sigma n_i V_i$, where V_j is the equilibrium volume of the compound, V_i is the volume of a component in the linear combination scheme and n_i is its number of moles. Thus, incorporating the volume correction term

$$S_j \approx \sum_i n_i S_i + \frac{\alpha}{\beta_T}\left(V_j - \sum_i n_i V_i\right) \qquad (C.1.4a)$$

This method of correction for the volume effect was introduced by Fyfe et al. (1958), and has since been used widely. Holland (1989) derived a value of α/β_T of ~ 1.0 J/(K-cm^{-3}) for silicate and oxide minerals from theoretical arguments and found this value to be in excellent agreement with that derived from statistical regression of S-V values of 60 solids. Setting α/β_T of ~ 1.0 J/(K-cm^{-3}), the above expression may be rewritten as

$$S_j \approx V_j + \sum_i n_i(S_i - V_i) \qquad (C.1.4b)$$

Holland (1989) derived the values of $(S_i - V_i)$ of different oxide units, such as $^{[4]}SiO_2$, $^{[4]}Al_2O_3$, $^{[6]}Al_2O_3$ etc. from statistical treatment of the experimentally determined values of S and V of oxides and silicates, and presented the data, along with V_j, in a tabular form.

For silicates, the data of Holland (1989) may be used to estimate, according to the above equation, the unknown entropy of a phase in which Si is in a tetrahedral site. However, at high pressure corresponding to the Earth's lower mantle (Fig. 7.7), Si sometimes occupies an octahedral site such as in the minerals stishovite (Stv), perovskite (Prv), post-perovskite (PPrv) phase and anhydrous-B (AnhB). To estimate the entropy of such phases, one needs to account for the effect of increase in the coordination number of Si from 4 to 6. This may be done by adding to the right hand side of Eq. (C.1.4) the entropy change due to the change of coordination number of Si from 4 to 6. An approximate value of this entropy change can be derived as follows from the entropy difference between α-quartz and Stv, in which Si is present in 4- and 6-coordinated positions, respectively:

$$\Delta S(^{[4]}Si \rightarrow ^{[6]} Si) = \Delta S(total) + \alpha/\beta_T(\Delta V)_{Qtz}$$

where $\Delta S(total) = S(Stv) - S(\alpha\text{-Qtz})$, and the last term on the right accounts for the effect of compression of α-quartz to the volume of Stv $(\Delta V_{Qtz} = V_{Stv} - V_{Qtz})$, according to Eq. (C.1.4a). Setting $\alpha/\beta_T = 1$ J/(K-cm^3), as above, and using the values in the optimized databases of Fabrichnaya et al. (2004) and Holland and Powell (2011), we obtain $\Delta S(^{[4]}Si \rightarrow ^{[6]}Si) = -8.6(\pm 0.5)$ J/mol-K of Si.

Another element that can be present in both 4 and 6 coordinated positions in oxides and silicates is Al, and as for Si, the data in Holland (1989) shows also a decrease of entropy due to the increase of coordination number of Al (also see Table C.1). However, the *opposite* effect is found for the increase of coordination number of Mg from 6 (common coordination) to 8, as in pyrope-garnet $(Mg_3Al_2Si_3O_{12})$, for which S equals 266–269 J/mol-K (Fabrichnaya et al. 2004; Holland and Powell 2011), as compared to the estimated value of 235 J/mol-K using S-V data of $^{[6]}MgO$, $^{[6]}Al_2O_3$, $^{[4]}SiO_2$ in Holland (1989) and Eq. C.1.4b. The high entropy value of Prp was explained by Newton and co-workers (Newton et al. 1977b; Haselton and Newton 1980), who first noticed this problem by comparing their calorimetric entropy data to the oxide sum approximation, to be a consequence of the presence of a small cation in a large site. Mg^{2+} is the smallest octahedral cation in garnet, and has an 8-coordinated ionic radius of 0.89 Å that is ~20% smaller than that of a cation contacting 8 oxygen ions (this is one of the five Pauling's rules (Pauling 1960) for structural stability that were formulated assuming hard sphere ionic radius model). According to the data in Holland (1989), the entropy of MgO increases by 7.2 J/mol-K [(S-V) by 10.31 J/mol-K] for a change in the oxygen coordination number of Mg from 6 to 8.

C.1.3. Electronic Ordering Effect on Entropy

Wood (1981) drew attention to the fact that the removal of internal degeneracies of t_{2g} and e_g orbitals of a transition metal ion by the distortion of the coordinating polyhedron may have significant effect on the entropy of the compound hosting the cation. In the entropy estimation scheme involving FeO, such as given by Eq. (C.1.3), there may be a change in the electronic configuration of Fe^{2+} if there is a change in its coordination environment, as it is transferred from the lattice site in FeO to that in another solid. In FeO, Fe^{2+} occupies a regular octahedral site in which the d-orbitals are split into two energy levels e_g and t_{2g}, consisting respectively of two and three degenerate d-orbitals (Sect. 1.7.2 and Fig. 1.10). Since there are six d-electrons in Fe^{2+}, one d-electron must reside in one of the three t_{2g} orbitals in a spin-paired state. But since these three orbitals are degenerate, there is equal probability of the existence of the sixth d-electron in any of these orbitals (that is it is disordered among three different configurational states), leading to an electronic configurational entropy given by the Boltzmann relation (Eq. 2.5.1)

$$S = R \ln \Omega_{conf} = R \ln(3) = 9.134 \ J/mol-K \qquad (C.1.5)$$

where R(gas constant) = LK_B. Now suppose that Fe^{2+} is transferred into a distorted octahedral site in a crystalline lattice. The distortion of the site may completely or partially remove the degeneracy of the t_{2g} orbitals (Fig. 1.11). When there is only one orbital in the lowest energy level, the sixth d-electron exists in a spin-paired state in that orbital in the ground state, in which case $\Omega_{conf} = 1$. Consequently, there is a reduction of configurational entropy of 9.134 J/mol-K in transferring Fe^{2+} from FeO into a distorted octahedral crystalline lattice that has only one orbital in the lowest energy state.

For **stoichiometric** FeO, Wood (1981) recommends entropy value of 56.30 J/mol-K at 1 bar, 298 K. This value, along with the change of electronic entropy due to crystal field effect, should be used in the entropy estimation of a ferrous compound at 298 K using linear combination scheme. For example,

$$S(Fa : Fe_2SiO_4) = S(Fo : Mg_2SiO_4) + 2[(56.30) - S(MgO)] - 2(9.134) \ J/mol-K$$

C.2. Polyhedral Approximation: Enthalpy, Entropy and Volume

Mueller (1962) introduced the idea that a particular polyhedron, such as $^{[6]}MgO$ in which a Mg ion is coordinated to 6 oxygen ions located at the apices of an octahedron, should have distinctive thermodynamic properties that are not very significantly affected by the rest of the structural environments in silicate minerals.

Several workers pursued this idea to estimate the bulk modulus (Hazen 1985), enthalpy, entropy and Gibbs free energy (Chermak and Rimstidt 1989, 1990; Helgeson et al. 1978; Van Hinsberg et al. 2005) of silicate minerals. Cherniak and Rimstidt (op. cit) expressed silicate minerals as linear combinations of constituent polyhedra, and utilized the known properties of the silicates to retrieve the polyhedral properties using statistical regression. To clarify the basic concept, let us consider two minerals, forsterite (For: Mg_2SiO_4) and enstatite (En: $MgSiO_3$). The properties of these two minerals can be expressed as linear combination of those of $^{[6]}MgO$ and $^{[4]}SiO_2$ according to $Mg_2SiO_4 = 2(^{[6]}MgO) + {}^{[4]}SiO_2$, and $MgSiO_3 = {}^{[6]}MgO + {}^{[4]}SiO_2$. Consequently, if the polyhedra have properties that do not depend significantly on their structural arrangements, then a particular type of polyhedral property, such as S, may be retrieved from simultaneous solution of the two linear equations if the properties of For and En are known.

Van Hinsberg et al. (2005) determined the properties of 30 polyhedral units by statistical regression of the thermodynamic properties of 60 minerals each of which can be expressed as linear combination of the chosen polyhedral units. The retrieved polyhedral properties are reproduced in Table C.1. The agreement between the input and estimated values for enthalpy and entropy are shown in

Table C.1 Thermochemical properties of polyhedral units derived from statistical regression of the thermochemical properties of minerals that are expressed as linear combinations of polyhedral units

	Δh_i (J mol^{-1})	SD	s_i (J mol^{-1} K^{-1})	SD	v_i (J bar^{-1})	SD	$(s-v)_i$	SD
Si-tet	−921,484	5256	39.8	1.07	2.45	0.05	15.9	1.03
Al-tet	−816,087	5581	40.3	1.16	2.17	0.06	19.8	1.12
Al-oct	−852,961	6012	22.2	1.38	0.75	0.07	13.5	1.32
Al-OHO	−1,049,365	9787	38.9	2.03	1.45	0.10	22.7	2.07
Al-OH	−1,170,579	17,017	57.3	6.90	2.83	0.23	27.4	6.19
Mg-tet	−633,580	29,576	53.6	5.49	2.44	0.29	32.6	5.04
Mg-oct	−625,422	5274	28.3	1.37	0.91	0.06	19.2	1.30
Mg-OHO	−764,482	7810	35.8	1.90	1.43	0.08	21.1	1.80
Mg-OH	−898,776	11,242	48.0	2.01	2.19	0.11	27.2	1.96
Fe-oct	−269,316	11,259	43.0	1.93	1.03	0.11	32.6	1.79
Fe-OHO	−385,309	12,794	50.7	3.49	1.48	0.12	35.0	3.00
Mn-oct	−403,304	17,588	46.1	2.38	1.13	0.10	33.8	2.00
Ti-oct	−955,507	35,677	55.4	6.03	1.99	0.19	31.4	5.15
Fe3-oct	−404,103	18,177	30.7	4.32	0.99	0.17	19.7	3.66
K-multi	−354,612	19,274	56.0	4.80	1.31	0.18	43.6	4.15
Na-multi	−331,980	18,944	38.3	3.17	0.85	0.15	27.9	2.97
Ca-oct	−703,920	12,247	42.0	1.73	1.33	0.09	27.9	1.68
Ca-multi	−705,941	9516	38.8	1.63	1.36	0.09	26.1	1.55
H_2O free	−306,991	12,836	44.1	1.97	1.46	0.09	28.3	1.93

Δh_i, s_i and v_i are, respectively, the enthalpy of formation from elements, the third law entropy and volume of the polyhedral unit i. SD is the standard deviation. From Van Hinsberg et al. (2005)

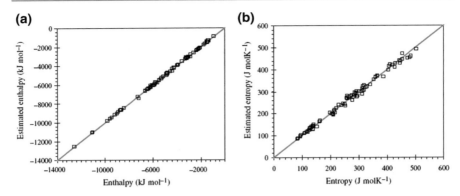

Fig. C.1 Comparison between the estimated and input values for enthalpy and entropy of minerals. The estimated values are obtained from linear combination of polyhedral properties in Table C.1 whereas the input values are those used to retrieve the polyhedral properties. From Van Hinsberg et al. (2005)

Fig. C.1. The entropy estimates represent simple linear combination of the polyhedral entropies without correction for the volume effect. This is because linear combination of the volumes of the polyhedral units yields the volume of the compound for which the entropy needs to be estimated. If this criterion is not satisfied, then one needs to correct for the volume effect according to Eq. (C.1.4).

When there is a change of oxygen coordination of Si from four to six due to increase of pressure, the estimated entropy of a mineral using the entropy of $^{[4]}SiO_2$ (Table C.1) needs to be corrected for the effect of coordination change. This may be accomplished by using the entropy difference between stishovite, in which Si is present in octahedral site, and the entropy of $^{[4]}SiO_2$ or α-quartz adjusted, according to Eq. (C.1.4), for the decrease of volume to the level of stishovite. This procedure suggests a reduction of the estimated entropy of the compound by ~ 8 J/mol-K of Si to account for the effect of change of coordination number of Si from 4 to 6. Alternatively, one may use S(stishovite) as $S(^{VI}Si)$ instead of $S(^{IV}Si)$ from Table C.1, and use Eq. (C.1.4) to correct for the effect of volume change between the polyhedral sum and the compound with 6-coordinated Si.

The effective thermochemical properties of polyhedral units depend on how these are connected with one another. In order to minimize the destabilizing effect of cation-cation repulsions, the polyhedral units share corners as much as possible. However, the requirement of volume reduction at high pressure often leads to considerable face- and edge-sharing of polyhedra. The polyhedral properties in the Table C.1 have been derived from thermochemical properties of minerals that are stable at the P-T conditions of the Earth's crust. Therefore, linear combination of these properties may not lead to useful estimate of thermochemical properties of high pressure minerals for which the polyhedra are linked in significantly different way than in the minerals from which the effective polyhedral properties have been derived.

Problem C.1 Calculate the correction that is to be made due to the change in the crystal field environment of Fe^{2+} if the entropy of spinel, $FeAl_2O_4$, in which Fe^{2+} occupies tetrahedral site, is to be estimated from summation of the entropies of component oxides (see Sect. 1.7.2 for crystal field effect).

Problem C.2 Using the data of polyhedral entropies in Table C.1, calculate the entropies of the high pressure Mg_2SiO_4 polymorphs wadsleyite and ringwoodite at 1 bar, 298 K, as well as you can. These will be the entropies for the metastable persistence of the high pressure forms that can be used to estimate the entropies at other conditions (The data in Table C.1 yield good estimate of entropy of olivine (Mg_2SiO_4) at 1 bar, 298 K). For comparison, the calorimetric third law entropy values are as follows. Wadsleyite: 86.4($\pm$0.4) J/mol-K; Ringwoodite: 82.7 ($\pm$0.5) J/mol-K. Your entropy estimates should be in good agreement with the calorimetric values.

Problem C.3 One of the most important mineralogical transitions in the Earth's mantle is that of γ-Mg_2SiO_4 (or ringwoodite: Rng) to $MgSiO_3$-perovskite (Prv) and MgO (pericalse: Per) marking the boundary between the mantle transition zone and lower mantle (Box 7.1). Experimental data show the P-T slope to be slightly negative at high P-T conditions. This is a rather unusual behavior and whether the slope is positive or negative has important consequence for plate tectonics. Discuss what should be the sign of $\Delta_r S$ for the equilibrium Rng = Prv + Per if the P-T slope is indeed negative, and evaluate if it is likely to be negative on the basis of estimated entropy change of this reaction according to the procedure and data in Sect. C.1.2 (The slope of the reaction is known to decrease with increasing pressure).

Hint: First determine the sign of $\Delta_r V$ on the basis of the fact that the transition is a response to increase of pressure and then the sign of $\Delta_r S$ using Clapeyron-Clausius relation.

Problem C.4 As discussed in Sect. 1.7.2, Fe^{2+} in ferropericlase, $(Fe,Mg)O$, undergoes gradual spin transition at the conditions in the Earth's lower mantle. Calculate the effect of change of the d-electron configuration on the Gibbs free energy of $(Fe_xMg_{1-x})O$ with x = 0.1 at 2000 K as a function of the extent of spin transition.

C.3. Estimation of Enthalpy of Mixing

The thermodynamic mixing properties of a solid solution are determined on the basis of calorimetric data (Fig. 9.4) and/or experimental phase equilibrium data. The latter topic is discussed in Sect. 10.12. In the absence of these data, and considering the urgency of the problem at hand, one is forced to explore methods of estimation of the mixing properties, of which the enthalpy of mixing is the most difficult problem to handle. Several empirical methods have been devised to estimate ΔH^{xs} from size mismatch data that were reviewed by Ganguly and Saxena (1987). More recently, quantum-chemical methods have been employed to estimate enthalpy of mixing of silicate minerals (Panero et al. 2006). However, powerful as the latter method is, it is highly specialized and is not within the reach of those who usually use thermodynamic properties to calculate phase relations of minerals. Thus, we discuss below some relatively simple but effective methods for the estimation of enthalpy of mixing of solid solutions.

C.3.1. Elastic Effect

The process of formation of a solid solution $[XA\phi + (1 - X)B\phi \rightarrow (A_xB_{1-x})\phi]$ can be formally thought to consist of the following steps: (a) change of the molar volumes of the end-member components to the molar volume V of the solid solution and (b) mixing of these components to form a chemically homogeneous mixture. These two steps may be referred to as the elastic and chemical components of the total enthalpy of mixing, as schematically shown below.

$$
\begin{array}{cc}
\text{Step (a)} & \text{Step (b)} \\
\end{array}
$$

$$
\left.
\begin{array}{l}
XV_{A\phi}^o \quad \Rightarrow XV_{A\phi} \\[2mm]
(1-X)V_{B\phi}^o \Rightarrow (1-X)V_{B\phi}
\end{array}
\right\} \longrightarrow \quad V_{(A_xB_{1-x})\phi}
$$

$$
\begin{array}{cc}
\Delta H_m \text{ (elastic)} & \quad \Delta H_m \text{ (chemical)}
\end{array}
$$

Expanding on the formulation of Ferreira et al. (1988), the discussion of which is beyond the scope of this section, Ganguly et al. (1993) showed that the elastic enthalpy of mixing is given by

$$
\Delta H_m(\text{elastic}) = \frac{1}{p}\left[(1 - X')\int_0^{X'} XZ(x)dX + X'\int_{X'}^1 (1 - X)Z(x)dX\right] \qquad (C.3.1)
$$

where X' is the mole fraction of $A\phi$,

$$
Z = \frac{\beta_T}{V}\left(\frac{dV_m}{dX}\right) \quad (\beta_T : \text{isothermal bulk modulus})
$$

and p is the number of substituting atoms in the nearest neighbor cell. This formulation accounts for the effect of multi-atom interactions that lead to a relaxation of the elastic energy. (The expression within the square brackets yields the elastic energy of mixing, ignoring the effects of multi-atom interactions, and is due to Ferreira et al. 1988).

The concept of nearest-neighbor cell is illustrated in Fig. C.2, using the unit cell of pyrope, $Mg_3Al_2Si_3O_{12}$ (Ganguly et al. 1993). The Mg atoms form triangular nearest neighbor cells, some of which are shown in the figure. The nearest neighbor

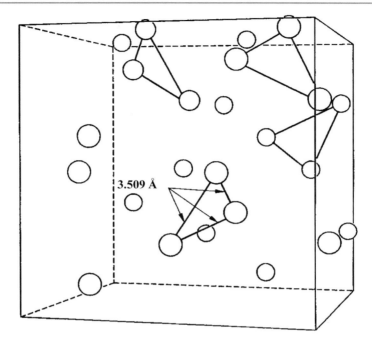

Fig. C.2 Unit cell of pyrope showing only Mg atoms. The nearest neighbor Mg-atoms form triangular clusters, and are 3.509 Å apart from each other. From Ganguly et al. (1993). With permission from Mineralogical Society of America

Mg atoms are 3.509 Å apart. Divalent cations like Fe^{2+}, Ca^{2+}, Mn^{2+} substitute for Mg^{2+} in garnet solid solution. Thus, the substituting ions in an aluminosilicate garnet solid solution form a three-atom (or three-ion) nearest neighbor cluster, so that p = 3.

Let us now calculate ΔH_m(elastic) for the pyrope-grossular solid solution. Ganguly et al. (1993) experimentally determined the $V_m - X$ relation of pyrope-grossular solid solution, and found that the data can be fitted well by a subregular model,

$$\Delta V_m^{xs} = W_{CaMg}^v X_{Mg} + W_{MgCa}^v X_{Ca},$$

with $W_{CaMg}^v = 0.36 \pm 0.23$ and $W_{MgCa}^v = 1.73 \pm 0.3$ cm^3/mol (12 oxygen per formula unit). Writing $V_m = X V_1^o + (1 - X) V_2^o + \Delta V^{xs}$, and expressing the last term according to a subregular model, we have

$$\frac{dV}{dX} = \left(V_1^o - V_2^o \right) + W_{12} + A_1 + A_2 \qquad \text{(C.3.2)}$$

where $A_1 = 2(W_{21} - 2W_{12})X$ and $A_2 = 3(W_{12} - W_{21})X^2$.

For simplicity, it was assumed that ratio β_T/V changes linearly with composition in the pyrope-grossular join. The assumption implies that the slope of the V-P curve at any P-T condition changes linearly with composition. Substituting Eq. (C.3.2) in the right hand side of Eq. (C.3.1) and using the available data for bulk moduli for pyrope and grossular (taken from the literature), ΔH_m(elastic) was evaluated by carrying out the required integration. The results, which are illustrated in Fig. 9.4 as "Elastic effect", are in good agreement with the calorimetric and optimized data. The interested reader is also referred to the work of Myhill (2018) for a more elaborate treatment of elastic effect on mixing properties of a solid solution.

C.3.2. Crystal-Field Effect

In the absence of crystal-field effect (see Sect. 1.7.2 for discussion of crystal field effect), the chemical effect usually does not usually seem to be significant compared to the elastic effect, as the data presented in Fig. 9.4 suggest (the calorimetric and optimized data show the net enthalpy of mixing due to both elastic and chemical effects). When there is a crystal effect, as in the pyrope-almandine or uvarovite-grossular solid solutions, the net enthalpy of mixing should incorporate the change of crystal field stabilization energy, ΔCFSE, of a transition metal cation. Thus, we write

$$\Delta H_m^{xs} \approx \Delta H_m(\text{elastic}) + \Delta CFSE(X) \qquad (C.3.3)$$

To evaluate the last term, let us consider the formation of a garnet solid solution from the pyrope (Prp) and almandine (Alm) end members, as follows

$$X(Fe_3Al_2Si_3O_{12}) + (1 - X)(Mg_3Al_2Si_3O_{12}) = (Fe_xMg_{1-x})_3Al_2Si_3O_{12}$$

where X is the mole fraction of the Alm component and equals Fe/(Fe+Mg) (note that in Alm, iron is in divalent state). In general, we can write

$$\Delta CFSE(X) = X[CFSE(X) - CFSE(X = 1)] \qquad (C.3.4)$$

where CFSE(X) is the crystal field stabilization energy per mole of Fe^{2+} at the specified composition (X). As the composition of the solid solution changes, the bond distances of the transition metal centered polyhedra also change leading to a change in the magnitude of crystal-field splitting (Δ) and hence of CFSE. The change in CFSE of Fe^{2+} in Alm-Prp and Alm-Spess joins, as determined by Geiger and Rossman (1994) from spectroscopic measurements, is illustrated in Fig. C.3 (Al, Py and Sp correspond to almandine, pyrope and spessartine, respectively, in Fig. C.3).

Let us now illustrate the calculation of CFSE(X) using the Alm-Prp solid solution as an example. The divalent cations in garnet occupy eight-coordinated dodecahedral sites. In a **regular** dodecahedral site, the d-orbitals split into two

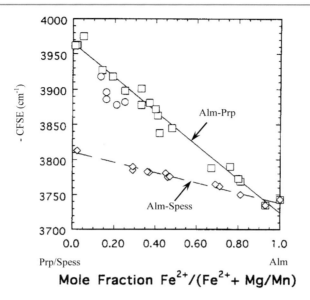

Fig. C.3 Change in the crystal field stabilization energy (CFSE) in the pyrope-almandine (Mg-Fe^{2+}) and spessartine-almandine (Mn^{2+}-Fe^{2+}) binary joins of garnet as a function of composition. Modified from Geiger and Rossman (1994)

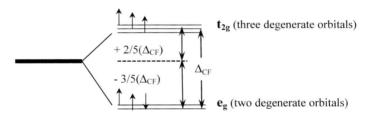

Fig. C.4 Splitting of five degenerate orbitals of Fe^{2+} in a regular dodecahedral crystal field and the distribution of electrons in a high-spin configurations, with spin pairing only in a lower e_g orbital

groups, with the upper (t_{2g}) and lower (e_g) groups containing three and two degenerate orbitals, respectively (Fig. C.4) (see Sect. 1.7.2 for further discussion). Since it has 6 d-electrons, in the high spin state Fe^{2+} has three electrons in the *two* lower e_g and three electrons in the *three* upper t_{2g} orbitals (if Δ_{CF} is the crystal field splitting, then the doubly degenerate e_g orbitals are $3/5(\Delta)$ below and the triply degenerate t_{2g} orbitals are $2/5(\Delta)$ above the initial pre-splitting energy level – see Sect. 1.7.2). Thus, in a regular dodecahedral site, the CFSE of Fe^{2+} in the high spin state is given by

$$\text{CFSE(X)} = 3\left(-\frac{3}{5}\Delta_{\text{CF}}(\text{X})\right) + 3\left(\frac{2}{5}\Delta_{\text{CF}}(\text{X})\right) = -\frac{3}{5}\Delta_{\text{CF}}(\text{X}) \qquad \text{(C.3.5)}$$

where $\Delta_{\text{CF}}(\text{X})$ is the crystal-field splitting at the composition X of the solid solution and the numbers outside the parentheses are the number of electrons in the e_g and t_{2g} orbitals (Δ_{CF} is usually given in the unit of cm^{-1} (wave number), which is converted to J/mol upon multiplication by 11.96). If the degeneracies of the t_{2g} and e_g orbitals are partly or completely removed due distortion of the polyhedra (Fig. 1.11), then the CFSE(X) can be easily estimated with appropriate modification of the above scheme that should be obvious.

In the absence of detailed experimental measurement of CFSE as a function of composition in a solid solution series, the change of Δ as a function of composition may be approximated according to the form $\Delta(x) \propto R^{-n}$, where R is the mean metal-ligand bond distance at the composition X. From the CFSE data in garnet and corundum, Ganguly and Saxena (1987) found n to vary between 1.12 and 1.74.

Geiger and Rossman (1994) calculated $\Delta\text{CFSE(X)}$ data in the Alm-Prp and Alm-Sps (Sps: Spessartine) solid solution series according to the method discussed above, and found that the $\Delta\text{CFSE(X)}$ can be described very well according to (symmetric) Regular Solution model such that $\Delta\text{CFSE(mix)} = W_{\text{CF}}\text{X}(1 - \text{X})$, with $W_{\text{CF}}(\text{Fe-Mg}) = -2.8$ kJ/cation-mole and $W_{\text{CF}}(\text{Fe-Mn}) = -0.8$ kJ/cation-mol, as illustrated in Fig. C.5a.

Qualitatively, one can predict if the crystal-field effect has a positive or negative contribution to the enthalpy of mixing by noting whether the average metal-ligand bond distance is larger or smaller in the solid solution compared to that in the end

Fig. C.5 a Contributions of the effects of elastic strain and crystal field stabilization energy (CFSE) to the enthalpy of mixing in the Alm-Prp join of garnet solid solution, in the framework of the regular solution model; **b** the combined elastic and crystal-field effects and comparison with the enthalpy of mixing in this binary join, as deduced by Hackler and Wood (1989) from experimental element partitioning data between garnet and olivine

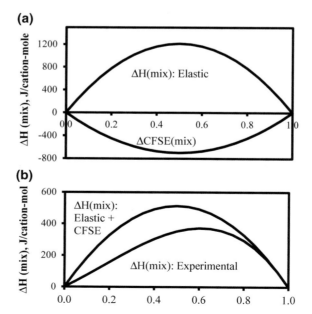

member of the transition metal compound. For example, in the Alm-Prp series, substitution of Mg for Fe^{2+} decreases the metal-oxygen bond distance. Consequently, one would expect the crystal field splitting in the Fe^{2+}-centered sites to increase, as observed, with increasing Prp/Alm ratio of the binary solid solution. Thus, according to Eq. C.3.5, CFSE(X) becomes more negative than CFSE for pure almandine, and consequently ΔCFSE(X) < 0 (Eq. C.3.4), thus contributing towards a negative deviation from ideality.

Note, however, that the CFSE data in the Alm-Sps join is not compatible with this expectation on the effect of change of metal-oxygen bond distance on CFSE. Because Mn^{2+} is a d^5 ion, it has zero CFSE (Eq. C.3.5) so that the change of CFSE in the Alm-Sps join is completely determined by that of Fe^{2+} (d^6) as a function of the change of metal-oxygen bond distance of the Fe^{2+}-centered dodecahedral sites. Since the 8-coordinated ionic radius of Mn^{2+} (0.96 Å) is slightly greater than that of Fe^{2+}(0.92 Å) (Shanon 1970), the Fe^{2+}-O bond distance in the dodecahedral site should be expected to slightly increase, and hence the CFSE of Fe^{2+} to slightly decrease with increasing Sps/Alm ratio of the binary solid solution. In contrast, the data illustrated in Fig. C.3 shows the opposite trend (slight increase of CFSE with Sps/Alm ratio). The reason for this unexpected trend of the experimental data remains unresolved (see discussion in Geiger and Rossman 1994).

C.3.3. Combination of Elastic and Crystal Field Effects

In this section we explore the combination of elastic and crystal field effects on the enthalpy of mixing between two end-member components, one of which represents a transition metal end member that has crystal field stabilization energy, CFSE. For this purpose, we choose the system Alm-Prp. The contribution to the ΔH(mix) due to elastic strain has been calculated from the physical properties of the end-members following the method discussed in the Sect. C.3.1, and is illustrated in Fig. C.5a along with ΔCFSE(mix).

Hackler and Wood (1989) experimentally determined the Fe-Mg partitioning between garnet and olivine and deduced from these data, in conjunction with the available mixing property data in olivine, the Fe-Mg interaction parameters in the binary garnet solid solution. They found that the mixing property of garnet has a small positive deviation from ideality that can be represented very well by a Sub-Regular solution model (see Sect. 9.2.2), with the following interaction parameters: $W_{MgFe}^G = 2117$ J/cation-mol and $W_{FeMg}^G = 695$ J/cation-mol. Using these data, one can calculate the molar excess Gibbs free energy of mixing, ΔG_m^{xs}, according to Eq. (9.2.9). Since ΔS_m^{xs} in this join is negligible, the ΔG_m^{xs} effectively equals its enthalpy of mixing. The experimentally constrained ΔH(mix) data versus composition, which is illustrated in Fig. C.5b, shows reasonably good agreement with ΔH(mix) obtained from combining the elastic and crystal field contributions Fig. C.5a).

The near ideality of the mixing behavior in the Fe-Mn binary join of garnet (Ganguly et al. 1996) is also well approximated by a combination of elastic and crystal field effects. Thus, a combination of these two effects should provide a good approximation to the mixing behavior of binary solid solutions, and provides a likely explanation for the small positive deviation from ideality or near ideal behavior of the mixing properties of ferromagnesian compounds; the positive deviation from ideality due to the elastic strain is almost compensated by the negative deviation arising from crystal field effect, as illustrated for the Alm-Prp join (Fig. C.5a).

C.3.3. Ab Initio Calculation of Thermodynamic Properties

There has been considerable effort in recent years in the Earth Sciences community to calculate thermodynamic properties of minerals from fundamental principles (e.g. Jacobs et al. 2013; Tsuchiya et al. 2004; Ottonello 2018). These calculations play important roles in the development of thermodynamic databases and phase equilibrium calculations, especially when adequate experimentally constrained data for important mineral phases are lacking, and also lead to predictions of yet unknown phase transformations that have considerable geophysical implications.

Much of the impetus over the past few decades in the computation of chemical and physical properties of materials have come from the development of **Density Functional Theory** (DFT) and commercial computational software packages incorporating this theory and statistical thermodynamics. The solution of the celebrated wave equation derived by Erwin Schrödinger (1887–1961) in 1926 (Nobel prize: 1933) yields, in principle, the electronic structure of a system from which all of its material properties can be calculated. However, solution of the wave equation for a real multi-electron system is a very difficult task. And the difficulty factor increases exponentially with the number of electrons and hits what has been called an "exponential wall" by Kohn (1999) for many or most systems of practical interest, defying the capacity of even the most advanced computer systems. A practical solution to this problem is afforded by the density functional theory that was developed by Walter Kohn (1923–2016; Nobel Prize 1998), Pierre Hohenburg and Lu Sham in two seminal papers in 1964 and 1965 (Hohenberg and Kohn 1964; Kohn and Sham 1965).

Instead of dealing with many electron wave function, which is a mathematical construct, DFT deals with electron density, which is a real property of a system. It first shows, following the lead from a work by Llewellyn Thomas and Enrico Fermi (known as Thomas-Fermi theorem), that energy of a molecule and all of its ground state properties can be determined from its spatially varying **electron density**, n($\mathbf{r}$), such as illustrated in Fig. C.6 for the mineral jadeite. It then develops a practical strategy for approximating the electron density of a real N-electron system by using "effective" single particle wave function for each electron, as if the electronic system is non-interacting, and taking care of the effect of the electronic interactions (both exchange and correlation) from single particle solutions of the wave equation for all N electrons.

The word functional means function of a function; the energy, E, of a system is a function of its electron density, which, in turn, is a function of the position in the system. Thus, energy is a functional of the electron density, E = f[n($\mathbf{r}$)], and the

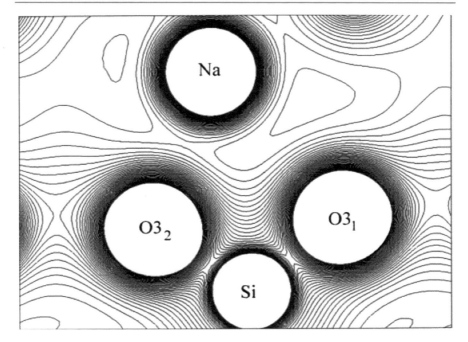

Fig. C.6 DFT based result of calculation of electron density in a plane of jadeite, $NaAlSi_2O_6$. The counters are displayed from 0 to 0.2 electron/Å^3 at 0.005 intervals. From Downs (2003). With permission from American Mineralogist

"correct" electron density, $n(\mathbf{r})$, of the system, within the framework of the approximations, is that at which its energy functional is at a minimum. For additional readings, the interested reader with reasonable background of quantum mechanics is referred to Kohn (1999: Nobel lecture), Oganov et al. (2002) and Zangwill (2015).

As an example of many applications of DFT to problems of interest in the Earth and Planetary Sciences, I show in Fig. C.7 the calculated heat capacities (C_p, C_v) of olivine, α-Mg_2SiO_4, by Ottonello et al. (2008).The calculated results are found to be in excellent agreement with the calorimetric data (Robie et al. 1982). Using DFT, Ottonello et al. (2008) also calculated the complete set of thermo-physical and -chemical properties of all of the Mg_2SiO_4 polymorphs (α-Mg_2SiO_4: Olivine; β-Mg_2SiO_4: Wadsleyite; γ-Mg_2SiO_4: Ringwoodite) that constitute the most abundant phases in the Earth's upper mantle and transition zone (see Box 7.1), as well as in the Martian mantle (Bertka and Fei 1997), and used these data to calculate the P-T stability fields of the polymorphs. Also using thermodynamic properties derived within the framework of DFT, Belmonte (2017) calculated the univariant reaction boundary marking the breakdown of Ringwoodite (Rng) to MgO (Periclase: Per) + $MgSiO_3$ (Perovskite: Prv) that primarily defines the 660 km discontinuity in the Earth's mantle. The experimental results for Rng to Per + Prv transition show

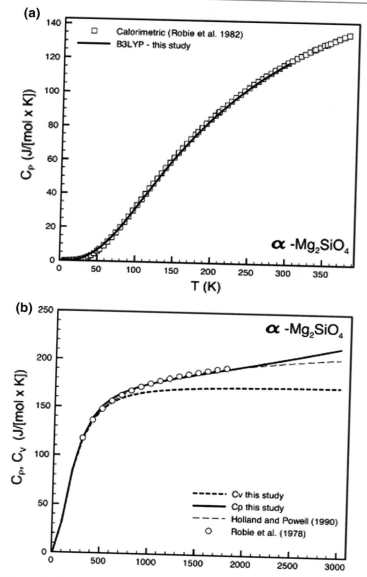

Fig. C.7 Comparison between results of DFT based calculations ("this study") of heat capacity of olivine (α-Mg$_2$SiO$_4$) and calorimetric data of Robie et al. (2008) (Ottonello et al. 2008). The C$_p$ data from Holland and Powell in (**b**) refer to that in their optimized thermochemical data base

considerable variation and the independent theoretical results allow one to isolate the more credible data set.

DFT based calculations of thermodynamic properties have also led to successful predictions of new metallic compounds and structures that are more stable at certain

P-T conditions than other known isochemical phases or phase assemblages. A mineralogical example that is of considerable geophysical interest is the DFT-based prediction of the likely transition of $MgSiO_3$ perovskite (Prv), which is the most dominant constituent of the Earth's lower mantle, to a "post-perovskite" phase (PPrv) at the P-T condition near the core-mantle boundary. This transition was discovered almost simultaneously by Murakami et al. (2004), Litaka et al. (2004) and Oganov and Ono (2004), and helped explain the seismic shear wave discontinuity and bulk sound velocity at the top of what is known as the D″ layer near the core-mantle boundary (see Box 7.1). The work of the last group was guided by DFT-based calculation of the enthalpy of transition as a function of pressure.

A PPrv structure of $MgSiO_3$ at very high pressure was anticipated from the behavior of other systems such as Fe_2O_3 that transforms from ilmenite to perovskite structure under pressure, just as $MgSiO_3$, but also shows a $CaIrO_3$-type *Cmcn* structure above 60 GPa (in both Prv and PPrv, Mg and Si are in six coordinated sites, but in PPrv the MgO_6 polyhedra form an edge sharing chain as in rutile). The DFT based calculations showed that at 0 K, the enthalpy change, Δ_rH, of the reaction $MgSiO_3$(Prv) = $MgSiO_3$(PPrv) decreased almost linearly as a function of pressure (Fig. C.8) and transitioned from a positive to negative value at 84–99 GPa. At T = 0, $\Delta_rG = \Delta_rH$ and hence the PPrv transition should be expected to take place within this pressure range at very low temperature.

To guide the extremely difficult high P-T experimental studies on Prv/PPrv transition, a rough estimate of the transition pressures at conditions comparable to the core-mantle boundary of the Earth may be made by calculating the Clapeyron slope, $dP/dT = \Delta_rS/\Delta_rV$ (Eq. 6.4.4) of the transition and anchoring it to equilibrium pressure at 0 K. Oganov and Ono (2004) used DFT to calculate the Δ_rS and

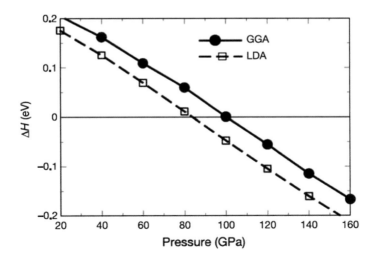

Fig. C.8 Calculated variation of the enthalpy change of perovskite to post-perovskite transition as a function of pressure at 0 K (Oganov and Ono, 2004). GGA and LDA stand for Generalized Gradient Approximation and Local Density Approximation, respectively, and relate to assumptions about the nature of electron density distribution in DFT based methods

combined that with $\Delta_r V$ (determined from the unit cell volume of a synthetic PPrv) to calculate a P-T slope of 9.56–9.85 bar/K. However, since there is no change of coordination number in this phase transition, a Clapeyron slope can also be calculated in a simple way by expressing $\Delta_r S$ in terms of $\Delta_r V$ according to the relation $\Delta_r S = (\alpha/\beta)\Delta_r V$ so that $dP/dT = \alpha/\beta \approx 10$ bar/K (see Sect. C.1.2) that is in very good agreement with the result derived by Oganov and Ono (2004).

Appendix D
Solutions of Selected Problems

Problem 1.1 Since the work is reversible, $W^+ = \int P_g dV_g$ (g: gas). For one mole of an ideal gas, $PV = RT$ so that $PdV = RdT - VdP$. Thus, $W^+ = R\int dT - \int V_g dP_g$.

For the path $A \rightarrow B$, $P = P_1$ ($dP = 0$), so that

$$W^+ (A \rightarrow B) = R(T_2 - T_1),$$

whereas for $B \rightarrow D$, $T = T_2$ ($dT = 0$) and hence $V = RT_2/P$. Thus,

$$W^+ (B \rightarrow D) = -\int_{P_1}^{P_2} VdP = -RT_2 \int_{P_1}^{P_2} \frac{dP}{P} = -RT_2 \ln \frac{P_2}{P_1}$$

Thus,

$$W^+ (A \rightarrow B \rightarrow D) = R(T_2 - T_1) - RT_2 \ln \frac{P_2}{P_1}$$

Similarly,

$$W^+ (A \rightarrow C \rightarrow D) = -RT_1 \ln \frac{P_2}{P_1} + R(T_2 - T_1)$$

Problem 1.2 The work performed by the force of friction is negative since the direction of displacement is exactly opposite to that of the frictional force.

From Eq. (1.4.1), $W = F_{friction} \Delta X \cos(180) = -F_{friction} \Delta X$.

Problem 2.4 Since the expansion is adiabatic, $\delta q = 0$. Also, since there is no external restraint, $\delta \omega^+ = 0$. Consequently, $\Delta U = Q - W^+ = 0$. To calculate ΔT, we proceed as follows.

© Springer Nature Switzerland AG 2020
J. Ganguly, *Thermodynamics in Earth and Planetary Sciences*,
Springer Textbooks in Earth Sciences, Geography and Environment,
https://doi.org/10.1007/978-3-030-20879-0

$$U = f(T, V); \text{ thus,}$$

$$dU = \left(\frac{\partial U}{\partial T}\right)_V dT + \left(\frac{\partial U}{\partial V}\right)_T dV = 0$$

Now since the internal energy of an ideal gas depends only on T, the second term on the right after the first equality is zero. However, $(\partial U/\partial T)_V \neq 0$ since U depends on T. Thus, $\Delta T = 0$.

To calculate ΔV, we use the ideal gas relation $PV = nRT$ so that at constant T ($R = 83.14$ bar-cm^3),

$$\Delta V = V_2 - V_1 = nRT\left(\frac{1}{P_2} - \frac{1}{P_1}\right) = 314.27 \text{ L}$$

Problem 3.3

(a) $dH = d(U + PV)$. Now since $\delta q = 0$ (adiabatic condition), $dU = \delta q - PdV = -PdV$. Thus, $dH_{\text{system}} = (VdP)_{\text{gas}}$. Also $dH = C_p dT$. Thus, using ideal gas relation $PV = RT$, $(C_p/T)dT = (R/P)dP$ so that $\int C_p d \ln T = R \int d \ln P$. Thus,

$$\left(C_p\right)_{\text{sys}} \ln \frac{T_f}{T_i} = R \ln \frac{P_f}{P_i}$$

where $\left(C_p\right)_{\text{sys}} \approx \left(C_p\right)_{\text{cyl}} = 85$ J/mol-K. Substituting the values of initial temperature, and initial and final pressures, the above relation yields $T_f = 215$ K.

(b) Using the first law, $dU = -\delta w^+$, where δw^+ is the P-V work done by the gas (Eq. 2.1.2) under adiabatic condition ($\delta q = 0$). The internal energy of an ideal gas depends only on its temperature. Thus, $dU_{\text{gas}} = (C_V)_{\text{gas}}dT = (C_P - R)_{\text{gas}}dT$ (see Eq. 3.7.11). Accordingly, $\Delta U_{\text{gas}} = (35 - 8.314)(T_f - T_i) = -2266$ J/mol. **Thus, W^+(the total work done by the gas) = 2266 J/mol**.

(c) $(\Delta S)_{\text{sys}} = \int_{T_i}^{T_f} \left(C_p\right)_{\text{sys}} d \ln T \approx (85) \ln \frac{215}{300} = -26.8$ J/mol.

Problem 4.3 $\ln v_i = f(P, T)$. Thus,

$$d \ln v_i = \left(\frac{\partial \ln v_i}{\partial P}\right)_T dP + \left(\frac{\partial \ln v_i}{\partial T}\right)_P dT$$

so that

$$\left(\frac{\partial \ln v_i}{\partial T}\right)_v = \left(\frac{\partial \ln v_i}{\partial P}\right)_T \left(\frac{\partial P}{\partial T}\right)_v + \left(\frac{\partial \ln v_i}{\partial T}\right)_T$$

Using now the property of implicit function (f(P,V,T) = 0) (Appendix B.4)

$$\left(\frac{\partial P}{\partial T}\right)_v = -\frac{\left(\frac{\partial P}{\partial V}\right)_T}{\left(\frac{\partial T}{\partial V}\right)_P} = \alpha k_T$$

(the last equality follows from Eqs. (3.7.1) and (3.7.2)).
Combination of the last two equations yields the desired relation.

Problem 6.8 From the combined statement of the first and second law of thermodynamics (Eq. 2.7.4), $dU = TdS - PdV$ for a system restricted to P-V work in a reversible process. Thus,

$$\left(\frac{\partial U}{\partial V}\right)_{T=0} = -P$$

Thus, at $T = 0$, the negative of the derivative of the U versus V curve of a phase defines the pressure of the phase at a specific value of V. For the phases I and II related by a first order phase transition, common tangency to a U-V plot determines a unique pressure experienced by both phases at absolute zero under equilibrium condition. Thus, this pressure must be the pressure of phase transition at $T = 0$. However, this transition pressure does not hold at any other temperature (the volumes of the two phases at the common tangent determines the ΔV of transition at $T = 0$.

The appropriate thermodynamic potential that should be used to determine the pressure of phase transition at an arbitrary temperature is H (Helmholtz free energy); it is related to T and V according to $dH = -PdV - SdT$ (Eq. 3.1.9) so that $(\partial F/\partial V)_T = -P$.

Problem 7.1 By definition, $\Delta P_{th} = \int (\alpha k_T)dT$. Using Eq. (3.8.2), the parenthetical term can be expressed as $\alpha k_T = \Gamma_{th}C_v/V$. Thus, assuming $\Gamma_{th}C_v$ and V to be constants, we obtain

$$\Delta P_{th} = \frac{\Gamma_{th}}{V} \int_{T_1}^{T_2} C_v dT = \frac{\Gamma_{th}}{V} \int_{T_1}^{T_2} dU \ ;$$

the last equality follows from Eq. (3.7.6) so that

$$\Delta P_{th} = \frac{\Gamma_{th}\Delta U}{V}$$

Problem 8.2 In an open system, G and V are related according to $dG = -SdT + VdP + \Sigma\mu_i dn_i$. Thus,

$$\left(\frac{\partial G}{\partial P}\right) = V$$

Differentiating with respect to n_i,

$$\frac{\partial}{\partial n_i}\left(\frac{\partial G}{\partial P}\right) = \left(\frac{\partial V}{\partial n_i}\right) = v_i$$

Since the order of differentiation us inconsequential, we can rewrite the above equation as

$$\frac{\partial}{\partial P}\left(\frac{\partial G}{\partial n_i}\right) = v_i$$

Thus,

$$\left(\frac{\partial \mu_i}{\partial P}\right) = v_i$$

It is now left to the reader to prove the other two equalities.

Problem 8.3 To derive the desired relation, we first consider a ternary solution and then generalize the result for an n-component system.

$$\left(\frac{\partial Y_m}{\partial X_i}\right)_{P,T,X_{j\neq i}} = y_i^o + \sum W_{ij}X_{j\neq i}$$

Thus, for a ternary solution, we have from (8.3.8)

$$
\begin{aligned}
y_1 = Y_m + &\left[y_1^o + W_{12}X_2 + W_{13}X_3\right] \\
- &\left[X_1\left(y_1^o + W_{12}X_2 + W_{13}X_3\right) + X_2\left(y_2^o + W_{21}X_1 + W_{23}X_3\right)\right. \\
&\left. + X_3\left(y_3^o + W_{31}X_1 + W_{32}X_2\right)\right]
\end{aligned}
$$

Noting that $W_{ij} = W_{ji}$, the terms within the last square brackets can be rearranged to yield $[\ldots] = \sum X_k y_i^o + 2\sum W_{kj}X_k X_j$. Substituting the given expression of Y_m, and rearranging/cancelling terms, we obtain the desired expression.

Problem 10.5 Assuming ideality of mixing ($\gamma_i = 1$),we have from the equilibrium $H_2O = H_2 + 1/2\,O_2$,

$$f(H_2O) = \frac{f(H_2)f(O_2)^{1/2}}{K}$$

$$P_T \approx P(H_2O) + P(H_2) = \frac{f(H_2O)}{\Phi(H_2O)} + \frac{f(H_2)}{\Phi(H_2)}$$

Combining these two relations

$$P_T \approx \frac{f(H_2)[f(O_2)]^{\frac{1}{2}}}{K[\Phi(H_2O)]} + \frac{f(H_2)}{\Phi(H_2)} = f(H_2)\left[\frac{[f(O_2)]^{\frac{1}{2}}}{K[\Phi(H_2O)]} + \frac{1}{\Phi(H_2)}\right]$$

Thus, using the relation $f(H_2) = P(H_2)\Phi(H_2)$, which holds for the condition of ideal mixing, so that $f(H_2) = P_T[X(H_2)]\Phi(H_2)$,

$$X(H_2) = \frac{1}{\psi + 1},$$

where $\psi = (f(O_2)^{1/2}/K)(\Phi_{H2}/\Phi_{H2O})$.

From Fig. 10.5, $f(O_2) \sim 10^{-27}$ and 10^{-22} at IW and QFM buffers, respectively. Choosing 1 bar, T, pure phase as the standard states of the volatile species, we have from JANAF Thermochemical Table, K (1 bar, 800 K) = $-\Delta_r G^\circ$ (1 bar, 800 K)/RT = $5.2(10^{-14})(\Delta_r G^\circ = -\Delta_f G(H_2O) = -203,496$ J/mol). Thus, using the fugacity coefficient data provided in the problem ($\Phi(H_2) = 1.60$ and $\Phi(H_2O) = 0.405$ at 2 kb, 800 K) and also assuming *ideal gas* behavior of the end member species, we get $X(H_2) = 0.001(0.005)$ and $0.293(0.621)$ at the $f(O_2)$ conditions defined by the OFM and IW buffers, where the parenthetical numbers apply to *ideal gas* model. Thus, changing the $P(O_2)$ from QFM to IW buffer would cause significant depression of dehydration temperature of a mineral (**Comment**: Such effect of $P(O_2)$ on dehydration temperature of hydrous minerals has been demonstrated in a number of experimental studies, beginning with the pioneering study of Eugster and Wones (1962) on the stability of ferruginous biotite, annite).

Problem 10.14 From Eq. (10.13.8), $(\partial I_{n_k})_{P,T,n_{j \neq k}} = \partial G - \mu_k \partial n_k - n_k \partial \mu_k$. The last term on the right is zero since μ_k is constant. Using the relations G = U – TS + PV and $dU = \delta q - PdV + \Sigma \mu_i dn_i$, we obtain $(\partial I_{n_k})_{P,T,n_{j \neq k}} = \delta q - TdS$. See Sect. 3.2 after this to complete the proof. Also derive the last equation using the preceding steps.

Problem 14.3 At constant T-V condition, the chemical potential of a species is given by (see Sect. 8.1: Box 8.1.2)

$$\mu_i = \left(\frac{\partial F}{\partial n_i}\right)_{T,V,n_j \neq n_i}$$

where F is the Helmholtz free energy and n_i is the number of moles of the component i. F can be expressed in terms of partition function as $F = -k_B T \ln Q$ (Box 14.2.1). For a volatile species, i, $Q_i = (q_i)^{n_i}/n_i!$ so that $\ln Q_i = n_i \ln q_i - \ln(n_i)! = n_i \ln q_i - [n_i \ln(n_i) - n_i]$, the terms within the square bracket resulting from the Sterling's approximation of $\ln n_i!$. Thus,

$$\frac{\partial \ln Q_i}{\partial n_i} = \ln q_i - \ln(n_i) - n_i \frac{\partial \ln(n_i)}{\partial n_i} + 1$$

The last two terms cancel out since $n_i/\partial n_i = 1/\partial \ln(n_i)$. Thus, $\mu_i = -k_B T(\ln q_i/n_i)$.

Problem 14.4 Using the given vibrational frequencies and degeneracies, we calculate the following values of the product terms of the frequencies/functions of the frequencies and from those, the reduced partition function ratio (RPFR), f, according to Eq. (14.4.11), for the exchange reaction $CH_4 + DBr = CH_3D + HBr$. The results of calculations at 300 K are given below. Similar calculations were carried out up to 800 K at 100 K interval.

	T, K	Πv_j	$\Pi(e^{-h v_j/2k_B T})$	$\Pi(1 - e^{-h v_j/k_B T})$	f
CH_3D	300	$4.207\ (10^{29})$	$1.816\ (10^{-20})$	0.993	38.936
CH_4	300	$2.236\ (10^{29})$	$3.312\ (10^{-19})$	0.989	
DBr	300	1880	1.103E−02	1.000	4.494
HBr	300	2650	1.741E−03	1.000	

Thus, K (300 K) = f(methane)/f(bromide) = 8.663. The calculations of f take care of the effect of symmetry numbers of the heavy and light molecules of methane and bromide ($\sigma = 3$ for CH_3D around the axis of symmetry along the CD bond; $\sigma = 12$ for CH_4, being equal to 3 along each CH bond in the tetrahedral molecule; $\sigma = 1$ for both DBr and HBr).
The $\ln K$ versus $1/T$ and versus $1/T^2$ are illustrated below.

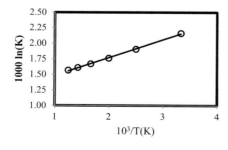
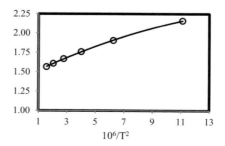

ln K versus $1/T$ and $1/T^2$ plots for hydrogen isotopic fractionation between methane and hydrogen bromide (Problem 14.4).

Problem B.1 (Appendix B) To test the validity of the statement that dZ is an exact differential, we find out if the differential satisfies the reciprocity relation, $\partial Z_x/\partial y = \partial Z_y/\partial x$.

$$\frac{\partial Z_x}{\partial y} = \frac{\partial(xy^2 - x^2)}{\partial y} = 2xy; \quad \frac{\partial Z_y}{\partial x} = \frac{\partial(x^2 y + y^2)}{\partial x} = 2xy$$

Hence dZ is indeed an exact differential. To find the function Z, we write

$$Z_x \equiv \frac{\partial Z}{\partial x} = xy^2 - x^2$$

Thus, at constant x, $\int \partial Z = \int (xy^2 - x^2)\partial x$ so that $Z = \frac{1}{2}(x^2 y^2) - x^3/3 + I$ (integration constant). The last term is independent of x, but must be a function of y so that $Z_y = x^2 y + y^2$. This condition is satisfied if $I = y^3/3$. Thus,

$$Z = \frac{(xy)^2}{2} - \frac{x^3}{3} + \frac{y^3}{3}$$

Now since dZ is an exact differential, the integral of dZ depends only on the initial and final conditions. Thus,

$$\int_{0,0}^{0,a} dZ = \frac{a^3}{3}$$

Comment: Even though the integral is independent of any path, the latter has been specified in the problem to see if you choose to follow a difficult track by carrying out a line integral of dZ, which of course will yield the correct answer, if done correctly, but would imply that you have not grasped the property of an exact differential.

References

Barrer RM, Bartholomew RF, Rees LVC (1963) Ion exchange in porous crystals. Part II. The relationship between self and exchange-diffusion coefficients. J Phys Chem Solids 24:309–317

Bertka CM, Fei Y (1997) Mineralogy of the Martian interior up to core-mantle boundary. J Geophys Res 102:5251–5264

Chakraborty S (1994) Relationships between thermodynamic mixing and diffusive transport in multicomponent solutions: Some constraints and potential applications. J. Phys. Chem. 98 (18):4923–4926

Chakraborty S (1995) Relationships between thermodynamic mixing and diffusive transport in multicomponent solutions: some constraints and potential applications. J Phys Chem 98 (18):4923–4926

Chakraborty S, Ganguly J (1991) Compositional zoning and cation diffusion in garnets. In: Ganguly J (ed) Diffusion, atomic ordering and mass transport. Advances in physical geochemistry, vol 8. Springer, New York, Berlin, Heidelberg, pp 120–175

Chakraborty S, Ganguly J (1994) A method to constrain thermodynamic mixing properties and diffusion coefficients in multicomponent solutions. Mater Sci Forum 155(156):279–283

Chermak JA, Rimstidt DJ (1989) Estimating the thermodynamic properties (ΔG_f^o and ΔH_f^o) of silicate minerals from the sum of polyhedral contributions. Am Mineral 74:1023–1031

Chermak JA, Rimstidt DJ (1990) Estimating the free energy of formation of silicate minerals at high temperatures from the sum of polyhedral contributions. Am Mineral 75:1376–1380

Darken LS (1948) Diffusion, mobility and their interrelation through free energy in binary metallic systems. Am Inst Min Metall Eng Trans 175:184–201

Downs RT (2003) Topology of the pyroxenes as a function of pressure, temperature, and composition as determined from the procrystal electron density. Am Mineral 88:556–566

Eugster HP, Wones DR (1962) Stability relations of the ferruginous biotite, annite. J Petrol 3:82–125

Fabrichnaya O, Saxena SK, Richet P, Westrum EF (2004) Thermodynamic data, models, and phase diagrams in multicomponent oxide systems. Springer

Ferreira LG, Mbaye AA, Zunger A (1988) Chemical and elastic effects on isostructural phase diagrams: the ε-G approach. Phys Rev B 37:10547–10570

Fyfe WS, Turner FJ, Verhoogen J (1958) Metamorphic reactions and metamorphic facies. Geol Soc Am 73:259

Ganguly J (2002) Diffusion kinetics in minerals: principles and applications to tectono-metamorphic processes. In: Gramaciolli CM (ed) Energy modelling in minerals. EMU notes mineral, vol 4. Eőtvős Univ Press, pp 271–309

Ganguly J, Cheng W, O'Neill HStC (1993) Syntheses, volume, and structural changes of garnets in pyrope-grossular join: implications for stability and mixing properties. Am Mineral 78:583–593

© Springer Nature Switzerland AG 2020
J. Ganguly, *Thermodynamics in Earth and Planetary Sciences*,
Springer Textbooks in Earth Sciences, Geography and Environment,
https://doi.org/10.1007/978-3-030-20879-0

Ganguly J, Cheng W, Tirone M (1996) Thermodynamics of aluminosilicate garnet solid solution: new experimental data, an optimized model, and thermometric applications. Contrib Mineral Petrol 126:137–151

Ganguly J, Saxena SK (1987) Mixtures and mineral reactions. Springer, Berlin, Heidelberg, New York, Paris, Tokyo, p 291

Geiger CA, Rossman (1994) Crystal field stabilization energies of almandine-pyrope and almandine-spessartine garnets determined by FTIR near infrared measurements. Phys Chem Min 21:516–525

Hackler RT, Wood BJ (1989) Experimental determination of iron and magnesium exchange between garnet and olivine and estimation of Fe-Mg mixing properties in garnet. Am Mineral 74:994–999

Haselton HT, Newton RC (1980) Thermodynamics of pyrope-grossular garnets and their stabilities at high temperatures and pressures. J Geophys Res 85:6973–6982

Hazen RM (1985) Comparative crystal chemistry and the polyhedral approach. In: Kieffer SW, Navrotsky A (eds) Microscopic to macroscopic. Reviews in mineralogy, vol 14. Mineral Soc America, pp 317–345

Helgeson HC, Delany JM, Nesbitt HW, Bird DK (1978) Summary and critique of the thermodynamic properties of rock-forming minerals. Am J Sci 278(A):1–229

Hohenburg PC, Kohn W (1964) Inhomogeneous electron gas. Phys Rev 136, B864–871

Holland TJB (1989) Dependence of entropy on volume of silicate and oxide minerals. Am Mineral 74:5–13

Holland TJB, Powell R (2011) An improved and extended internally consistent thermodynamic dataset for phases of petrological interest, involving a new equation of state for solids. J Met Geol 29:333–383

Jacobs MHG, Schmid-Fetzer R, van der Berg AP (2013) An alternative use of Kieffer's lattice dynamic model using vibrational density of states for constructing thermodynamic databases. Phys Chem Min 40:207–227

Kohn W (1999) Electronic structure of matter – wave functions and density functionals. Nobel Lecture

Kohn W, Sham LJ (1965) Self-consistent equations including exchange and correlation effects. Phys Rev 140:A1133–A1138

Lasaga AC (1998) Kinetic theory in earth sciences. Princeton Univ Press, New Jersey, p 811

Lesher CE, Walker D (1991) Thermal diffusion in petrology. In: Ganguly J (ed) Diffusion, atomic ordering and mass transport. Advances in physical geochemistry. Springer, pp 396–141

Lindsley DH, Grover JE, Davidson PM (1981) The thermodynamics of $Mg_2Si_2O_6 - CaMgSi_2O_6$ join: a review and a new model. In: Newton RC, Navrotsky A, Wood BJ (eds) Advances in Phys Geochem, vol1. Springer-Verlag, pp 149–175

Litaka T, Hirose K, Kawamura J, Murakami M (2004) The elasticity of $MgSiO_3$ post-perovskite phase in the Earth's lowermost mantle. Nature 430:442–445

Manning JR (1968) Diffusion kinetics for atoms in crystals. Van Nostrand, Princeton, New Jersey

Mueller RF (1962) Energetics of certain silicate solid solutions. Geochim Cosmochim Acta 26:581–598

Murakami M, Hirose K, Sata N, Ohishi Y, Kawamura K (2004) Phase transformation of $MgSiO_3$ perovskite in the deep mantle. Science 304:855–858

Myhill R (2018) The elastic solid solution model for minerals at high pressures and temperatures. Contrib Min Petrol 173(12) https://doi.org/10.1007/s00410-017-1436-z

Newton RC, Thompson AB, Krupka KM (1977) Heat capacity of synthetic $Mg_3Al_2Si_3O_{12}$ from 350–1000 K and the entropy of pyrope. Trans Am Geophys Union EOS 58:523

Oganov AR, Ono S (2004) Theoretical and experimental evidence for a post-perovskite phase of $MgSiO_3$ in Earth's D'' layer. Nature 430:445–448

Oganov AR, Brodholt JP, Price GD (2002) Ab initio theory of phase transitions and thermochemistry of minerals. In: Gramaccioli CM (ed) Energy modelling in minerals. European mineralogical union notes in mineralogy, vol 4. Eötvös University Press, Budapest, pp 83–160

Onsager L (1945) Theories and problems of liquid diffusion. N Y Acad Sci Ann 46:241–265

Ottonello G (2018) Ab initio reactivity of Earth's materials. Rivista del Nuovo Cimento 41:225–289

Ottonello G, Civaleri B, Ganguly J, Zuccolini MV, Noel Y (2008) Thermophysical properties of the α-β-γ polymorphs of Mg_2SiO_4. Phys Chem Min 36:87–106

Panero WR, Akber-Knutson S, Stixrude L (2006) Al_2O_3 incorporation in $MgSiO_3$ perovskite and ilmenite. Earth Planet Sci Lett 252:152–161

Pauling L (1960) The nature of the chemical bond. Cornell Univ Press, p 644

Robie RA, Hemingway BS, Takei H (1982) Heat capacities and entropies of Mg_2SiO_4, Mn_2SiO_4 and Co_2SiO_4 between 5 and 380 K. Am Mineral 67:470–482

Shanon RD (1970) Revised values of ionic radii. Acta Cryst B26:1046–1048

Spera FJ, Trial AF (1993) Verification of Onsager reciprocal relations in a molten silicate solution. Science 259:204–206

Tsuchiya T, Tsuchita J, Umemoto K, Wentzcovitch RM (2004) Phase transition in $MgSiO_3$ perovskite in the earth's lower mantle. Earth Planet Sci Lett 224:241–248

Van Hinsberg VJ, Vriend SP, Schumacher JC (2005) A new method to calculate end-member thermodynamic properties of minerals from their constituent polyhedra I: enthalpy, entropy and molar volume. J Metamorph Geol 23:165–179

Wood BJ (1981) Crystal field electronic effects on the thermodynamic properties of Fe^{2+} minerals. In: Newton RC, Navrotsky A, Wood BJ (eds) Advances in physical geochemistry, vol 1. Springer, pp 66–84.

Zangwill A (2015) A half century of density functional theory. Phys Today 68:34–39

Zhang X, Ganguly J, Ito M (2010) Ca-Mg diffusion in diopside: tracer and chemical inter-diffusion coefficients. Contrib Mineral Petrol 159:175–186

Author Index

A

Abers, G., 150
Achilli, A., 278
Adams, L. H., 195, 196
Ader, M., 275, 276
Adkins, J., 537
Affek, H., 537
Agee, C. B., 417, 418
Agren, J., 286
Agrinier, P., 275, 276
Aitta, A., 146
Akaogi, M., 97
Akber-Knutson, S., 574
Akella, J., 153
Allan, N. L., 412
Allegre, C. J., 389, 391
Anderson, D. L., 153, 176–178, 187, 195, 196, 387
Anderson, G. M., 364, 436
Anderson, O., 74, 154, 177, 178, 351
Anderson, O. L., 87, 88, 126, 177
Aranovich, L. Y., 304, 305
Artioli, G., 44, 45
Asaduzzaman, A., 476, 477, 536, 538, 539, 541
Asahara, Y., 410
Asimow, P. D., 192
Atake, T., 97
Atherton, M. P., 493
Atkinson, M. J., 436
Auroux, A., 509

B

Bailey, S. M., 446, 459–461
Baker, D. R., 498
Balan, E., 529, 536
Balassone, G, 301, 309
Balau, J. R., 487, 488
Ball, D. G., 269

Barbier, B., 44, 45
Barrer, R. M., 551
Bartholomew, R. F., 551
Basu, S., 310
Becke, F., 480
Becker, H-W, 411
Bekenstein, J. D., 36
Bell, P. M., 366, 368
Belmonte, D., 125, 581
Belonoshko, A. B., 120, 121, 163, 167
Berman, R. G., 85, 95, 296
Berner, R. A., 507
Bertka, C. M., 581
Bethe, H., 16
Bhattacharya, P. K., 160
Bhattacharyya, S., 160
Bhui, U. K., 268
Bickle, M. J., 189, 190, 377
Bigeleisen, J., 519, 520, 527, 533–535, 541
Birch, F., 122–125, 165, 197–198, 199, 200
Bird, D.K., 95, 571
Blanchard, M., 529
Blander, M., 300, 308, 312
Blundy, J. D., 412–416
Boehler, R., 70, 155, 186, 187, 188, 189
Bogotka, S. A., 533
Bonifacie, M., 275, 276
Borel, J. P., 497
Born, M., 64, 85
Bose, K., 92, 149, 150, 151, 168, 172
Bottinga, Y., 214
Bowen, N. L., 350–351, 352, 354, 363, 364
Bowers, T. S., 444
Brady, J., 500, 501, 502
Brenan, J. M., 488
Brett, R., 366, 368
Brewar, L., 272, 274
Brice, J. C., 414
Brock, E. E., 109

© Springer Nature Switzerland AG 2020
J. Ganguly, *Thermodynamics in Earth and Planetary Sciences*,
Springer Textbooks in Earth Sciences, Geography and Environment,
https://doi.org/10.1007/978-3-030-20879-0

Subject Index

Note: Locators in bold refers to definitions

A

Absolute
   temperature scale, 26–29
   zero (unattainibility of), 91, 92
Acidity ocean water, 436, 437
Activity, **229**, 244, 245
   determination of, **232**
   mean ion, 431, 432
Activity coefficient
   definition of, **203**
   and excess free energy, 238
   mean ion, 431, 442, 444
   in a reciprocal solution, 284–288
   retrieval from phase equilibria, 381–383
   of species in sea water, 463
Activity diagram, 444–452
Adiabat, 185–189, 211–214
Adiabatic
   decompression, 210–213
   flow, 55, 201–204
   process, 24, 26, 31, 69, 91
   wall/enclosure, 4–5, 23, 30, 49–50
Adsorption, 469–472
   chemical (chemisorption), 476
   Gibbs equation of, 472
Affinity (of a reaction), **321–322**
   and entropy production, 543–547
Aluminum avoidance principle, 40–42
Anharmonicity
   of lattice vibration, 11–12, 69, 84, 520, 521
   effects, 534
Arrow of time, 31
Assimilation, 393–395

Associated solutions, 306–309
Athermal solution, **301**, 309
Azeotropic minimum, 353–355, 364
Azeotropy, 353–354

B

Barometric formula, 271
Bernoulli equation, 206–207
Beta decay, 23
Binary solutions
   activity relations, 232–233, 238
   critical condition, 257–264
   mixing properties of, 236–239
   partial molar properties of, 222–226, 227–228
   stability conditions of, 250–254
Binodal condition, **259**, 260–264
Binodes, 253
Boltzmann relation, **31**, 37, 39, 284, 570
   distribution law, 513
   factor, 275, **514**
   relation, 32
Boltzmann's H-theorem, 45
Bulk modulus, **68**, 72, 73, 123
   divergence near critical point, 105–106
   pressure derivative of, 124, 165
   relation with sound velocity, 193–195

C

Cahn function, 265
Canonical ensemble, 514, 515
Carbonate compensation depth, 436
Carnot cycle, **26**, 27–29, 53

© Springer Nature Switzerland AG 2020
J. Ganguly, *Thermodynamics in Earth and Planetary Sciences*,
Springer Textbooks in Earth Sciences, Geography and Environment,
https://doi.org/10.1007/978-3-030-20879-0

Printed in the United States
by Baker & Taylor Publisher Services